APPLIED DISCRETE STRUCTURES
for
COMPUTER SCIENCE

APPLIED DISCRETE STRUCTURES for COMPUTER SCIENCE

Alan Doerr

Kenneth Levasseur

DEPARTMENT OF MATHEMATICS
UNIVERSITY OF LOWELL

SCIENCE RESEARCH ASSOCIATES, INC.
Chicago, Henley-on-Thames, Sydney, Toronto
A Subsidiary of IBM

Acquisition Editor	Alan W. Lowe
Project Editor	Mary C. Konstant
Compositor	Interactive Composition Corp.
Illustrator	Alex Teshin
Cover and Text Designer	Carol Harris

The SRA Computer Science Series
William A. Barrett and John D. Couch, *Compiler Construction: Theory and Practice*
Marilyn Bohl and Arline Walter, *Introduction to PL/1 Programming and PL/C*
Alan W. Doerr and Kenneth M. Levasseur, *Applied Discrete Structures for Computer Science*
Mark Elson, *Concepts of Programming Languages*
Mark Elson, *Data Structures*
Peter Freeman, *Software Systems Principles: A Survey*
Philip Gilbert, *Software Design and Development*
A. N. Habermann, *Introduction to Operating System Design*
Harry Katzan, Jr., *Computer Systems Organization and Programming*
Henry Ledgard and Michael Marcotty, *The Programming Language Landscape*
Stephen M. Pizer, *Numerical Computing and Mathematical Analysis*
Harold S. Stone, *Introduction to Computer Architecture, Second Edition*
Gregory F. Wetzel and William G. Bulgren, *The Algorithmic Process: An Introduction to Problem Solving*

Library of Congress Cataloging in Publication Data

Doerr, Alan W., 1938–
    Applied discrete structures for computer science.

    Bibliography: p.
    Includes index.
    1. Electronic data processing—Mathematics.
I. Levasseur, Kenneth M., 1950–   . II. Title.
QA76.9.M35D64 1985   511   84–23491
ISBN 0–574–21755–X

To our families

Donna, Christopher, Melissa, and Patrick Doerr

and

Karen, Joseph, Kathryn, and Matthew Levasseur

Contents

Preface

This book in discrete mathematics is intended to supply the typical freshman or sophomore in computer science and related disciplines with a first exposure to the mathematical topics essential to their study of computer science or digital logic. It also provides students who are preparing for an advanced-degree program with the background necessary for further study in theoretical computer science. It can be used in either a one- or a two-semester course. Written for the student, this text is the synthesis of many years of experience in teaching this and related courses to students at all undergraduate levels. It offers a unified treatment of the material outlined in all current national recommendations on discrete methods and applied algebra. A major feature of this text is its versatility. Sufficient topics have been included to accommodate students with varied backgrounds.

Chapter Coverage

Chapters 1 and 2 cover elementary concepts in sets and combinatorics. We have found that, although most, if not all, authors assume knowledge of these topics, the typical student needs considerable exposure to them. Applications relevant to computer science students are initiated in these first chapters.

Chapter 3, on logic, lays the framework for all subsequent material. The detail with which logic is developed stresses its overwhelming importance.

Chapter 4 expands on the first chapter on set theory. It utilizes the chapter on logic to further develop concepts in proofs and is an initiation to the topic of algebraic systems.

Students may find some of the topics in Chapters 1 through 4 somewhat theoretical in nature. So Chapter 5, "Introduction to Matrix Algebra," gives them a necessary "breather" from these theoretical concepts. Although some authors utilize matrix algebra in graph theory, none of them review this topic. We have found that many students are completely unfamiliar with matrix algebra.

Chapters 6 and 7 introduce the student to the concepts of relations and functions. By studying the numerous examples, students will become comfortable with these topics, which are crucial to their understanding of the remainder of this text.

Chapter 8 begins with a discussion of how recursion appears in algorithms, definitions, functions, proofs, etc. Major applications are recurrence relations and generating functions. Numerous examples of a variety of recurrence relations are presented in considerable detail in Sections 8.3 and 8.4.

Chapters 9 and 10 are an introduction to graphs and trees. They focus on the description of basic problems involving graphs and trees and their applications.

Chapter 11 is a formalization of concepts of algebraic structures that were introduced in previous chapters. This is done concretely through an introduction to the theory of groups. The chapter culminates in a description of how the idea of an algebraic structure has been adopted to object-oriented computer design.

Chapter 12 is a further development of matrix algebra. The first half of the chapter includes methods for solving systems of equations and how they can be used to compute matrix inverses. The second half is a development of the diagonalization process, including a brief introduction to vector spaces. Applications to recurrence relations and graph theory are given.

In Chapter 13, Boolean algebras are introduced naturally as an algebraic system, motivated by the similarities of logic and set theory. The focus is on examples and illustrations, while theory is explored. Logic design is a culminating application.

Chapter 14 covers the topics of monoids, languages, and finite-state machines and how they are interrelated.

In Chapter 15, we continue our discussion of groups with a further development of the theory and applications that include computations by homomorphic images and coding theory.

Chapter 16 is intended to introduce the student to basic concepts of ring and fields. The key ideas are developed by relying on the student's knowledge of high-school algebra. Polynomials, formal power series, and finite fields are discussed.

Features

Readability. Our students, who we feel are representative of typical undergraduate computer science students, have found all of the texts that we have used difficult to read. We believe that this occurs because typical students lack the background that most authors of discrete mathematics texts assume they have. The chapters on basic set theory, combinatorics, logic, and matrix algebra assist students who are weak in these areas. We have found that time devoted to these topics is well spent and pays off when more abstract topics are covered. Another factor that affects readability is the quantity and quality of examples that are relevant to the material being introduced. By providing numerous, clear examples, we hope that we have made this material more accessible to most students.

Applications. Whenever a major theoretical topic is covered, it is reinforced with at least one application to computer science so that students are able to apply key concepts immediately.

Pascal Notes. Discussions relating to Pascal and other programming languages appear throughout the text, but are clearly marked so that they can be avoided, if desired. In many cases, the Pascal Notes should be understandable to anyone who has had a course in a high-level programming language. We expect that the Pascal Notes will be of use to many students immediately. Some of our own students have commented that the Pascal Notes have affected the way that they write programs.

Coverage. This text is a synthesis of all national guidelines for the discrete methods/applied algebra sequence. Through our applications-oriented, hands-on approach, we feel that these guidelines can be followed without automatically losing a significant percentage of the students because they cannot follow explanations.

Exercises. With the exception of the two opening chapters, exercises immediately follow most sections. Problem sets for the first two chapters appear at the end of the chapter. The problem sets are divided into three sections. Section A consists of problems that all students should be able to do. They are often of a computational nature. Section B consists of a mixture of computational and theoretical problems that the average student should find difficult, but not impossible. Section C consists of challenging problems that suggest extensions of topics appearing in the text or introduce secondary topics.

This book ends with solutions and hints to selected exercises, a table of symbols, a bibliography, and an index.

Suggestions for Classroom Coverage

The material in this book is sufficient to fill two semesters for students who have a reasonable background in algebra. For a one-semester course, the instructor could choose from a variety of options to adapt the material to students' needs. For example:

Chapters 1, 2, 3, 4, 5, 6, 7, and 8:   An introductory one-semester
                                            course for the typical student

Chapters 1, 2, 3, 5, 6, 8, 9, and 10:   A non-algebraic introduction for the
                                            typical student

Chapters 2, 3, 6, 7, 8, 9, 10, and 13:   A non-algebraic approach for more
                                            advanced students

Chapters 4, 6, 7, 8.1, 11, 13, 14,   An introductory applied algebra
    15, and 16:   course for advanced students

Chapter-by-Chapter Comments

Chapters 1 and 2: These chapters have been written for the student who has no background in sets and elementary combinatorics. Upper-level students can be assigned these chapters for review.

Chapter 3: Nearly all of this chapter is essential, but care must be taken to avoid getting bogged down here. Section 3.5, on mathematical systems, may be covered lightly if proofs will not be emphasized. Section 3.8, on quantifiers, may also be covered lightly if the instructor does not habitually use them.

Chapter 4: Much of this chapter can be skipped if proofs will not be emphasized. There are two exceptions, however. The laws of set theory should be examined and compared to the laws of logic, and the concept of a partition should be discussed. The sections on minsets and duality can be omitted if Boolean algebras are not included in your course.

Chapter 5: This chapter is written for the student who has no background in matrix algebra. For many students, it can be used as a reading assignment.

Chapters 6 and 7: These chapters should be covered in their entirety.

Chapter 8: We feel that the first three sections of this chapter are essential for most students. Unless the focus of the course is algorithmic, Section 8.4 can be completely omitted. We believe that a brief introduction (approximately three lecture hours) to generating functions (Section 8.5) is essential.

Chapter 9: The first two sections of this chapter should be given as a reading assignment with minimal classroom discussion (one lecture hour).

Chapter 10: Classroom coverage of this chapter will depend on the students' previous exposure to trees in a course such as data structures.

Chapter 11: This chapter is essential to the appropriate coverage of Chapters 12 through 16, any of which can be covered independently.

Chapters 12 through 16: In the typical one-semester course, there may be time to cover one of these chapters in detail. Nearly all of these chapters can be covered in a two-semester course. In an applied algebra course, the main concentration would be on these chapters.

Acknowledgments

In preparing this text, the authors have taken advantage of suggestions from a variety of people. We are indebted to our editors, Alan Lowe and Mary Konstant, our colleagues in the mathematics and computer science departments, our students, and a number of anonymous reviewers. We also would like to acknowledge the helpful comments provided by the following reviewers: David C. Buchthal, University of Akron; John Morrison, Towson State University; and Francis L. Schneider, Furman University.

A.W.D.
K.M.L.

chapter

1

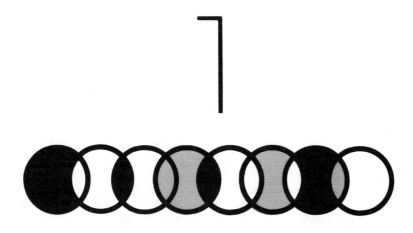

SET THEORY I

1 GOALS

In this chapter we will cover some of the basic set language and notation that will be used throughout the text. Venn diagrams will be introduced in order to give the reader a clear picture of set operations.

SET NOTATION AND DESCRIPTION

The term *set* is intuitively understood by most people to mean a collection of objects which are called *elements* (of the set). This concept is the starting point on which we will build more complex ideas, much as in geometry, where the concepts of point and line are left undefined.

Because a set is such a simple notion, you may be surprised to learn that it is one of the most difficult concepts for mathematicians to define to their own liking. For example, the description above is not a proper definition because it requires the definition of a collection. (How would you define "collection"?) Even deeper problems arise when you consider the possibility that a set could contain itself. Although these problems are of real concern to some mathematicians, they will not be of any concern to us.

Our first concern will be how to describe a set; that is, how do we most conveniently describe a set and the elements that are in it? If we are going to discuss a set for any length of time, we usually give it a name in the form of a capital letter (or occasionally some other symbol). In discussing set A, if x is an element of A, then we will write $x \in A$. On the other hand, if x is not an element of A, we write $x \notin A$. The most convenient way of describing the elements of a set will vary depending on the specific set.

Method 1: Enumeration. When the elements of a set are enumerated (or listed) it is traditional to enclose them in braces. For example the set of binary digits is $\{0, 1\}$ and the set of decimal digits is $\{0, 1, 2, 3, 4, 5, 6, 7, 8, 9\}$. The choice of a name for these sets would be arbitrary; but it would be "logical" to call them B and D, respectively. The choice of a set name is much like the choice of an identifier name in programming. Some large sets can be enumerated without actually listing all the elements. For example, the letters of the alphabet and the integers from 1 to 100 could be described as

$$A = \{a, b, c, \ldots, x, y, z\}, \text{ and}$$
$$G = \{1, 2, \ldots, 99, 100\}.$$

The three consecutive dots are called an ellipsis. We use them when it is clear what elements are included but not listed. An ellipsis is used in two other situations. To enumerate the positive integers, we would write $\{1, 2, 3, \ldots\}$, indicating that the list goes on infinitely. If we want to list a more general set, such as the integers between 1 and n, where n is some undetermined positive integer, we might write $\{1, \ldots, n\}$.

Method 2: Standard Symbols.   Frequently used sets are usually given symbols that are reserved for them alone. For example, since we will be referring to the positive integers throughout this book, we will use the symbol **P** instead of writing $\{1, 2, 3, \ldots\}$. A few of the other sets of numbers that we will use frequently are:

$\mathbf{N}$ = *natural numbers* = $\{0, 1, 2, 3, \ldots\}$.
$\mathbf{Z}$ = the *integers* = $\{\ldots, -3, -2, -1, 0, 1, 2, 3, \ldots\}$.
$\mathbf{Q}$ = the *rational numbers*.
$\mathbf{R}$ = the *real numbers*.
$\mathbf{C}$ = the *complex numbers*.

Method 3: Set-Builder Notation.   Another way of describing sets is to use *set-builder notation*. For example, we could define the rational numbers as

$$\mathbf{Q} = \{a/b : a, b \in \mathbf{Z}, b \neq 0\}.$$

Note that in the set-builder description for the rational numbers:

(1) a/b indicates that a *typical* element of the set is a "fraction."
(2) The colon is read "such that" or "where" and is used interchangeably with a vertical line, $|$.
(3) $a, b \in \mathbf{Z}$ is an abbreviated way of saying $a \in \mathbf{Z}$ *and* $b \in \mathbf{Z}$.
(4) All commas in mathematics are read as "and."

The important fact to keep in mind in set notation, or in any mathematical notation, is that it is meant to be a help, not a hindrance. We hope that notation will assist us in a more complete understanding of the collection of objects under consideration and will enable us to describe it in a concise manner. However, brevity of notation is not the aim of sets. If you prefer to write $a \in \mathbf{Z}$ and $b \in \mathbf{Z}$ instead of $a, b \in \mathbf{Z}$, you should do so. Also, there are frequently many different, and equally good, ways of describing sets. For example, $\{x \in \mathbf{R} : x^2 - 5x + 6 = 0\}$ and $\{x : x \in \mathbf{R}, x^2 - 5x + 6 = 0\}$ both describe the solution set $\{2, 3\}$.

A proper definition of the real numbers is beyond the scope of this text. It is sufficient to think of the real numbers as the set of points on a number line. The complex numbers can be defined using set-builder notation as $\mathbf{C} = \{a + bi : a, b \in \mathbf{R}\}$, where i is the square root of -1.

Definition: Finite Set. *A set is finite if it has a finite number of elements. Any set that is not finite is an infinite set.*

Definition: Cardinality. *Let A be a finite set. The number of different elements in A is called its cardinality and is denoted by #A.*

As we will see later, there are different infinite cardinalities. We can't make this distinction now, so we will restrict cardinality to finite sets until later.

SUBSETS

Definition:　Subset. *Let A and B be sets. We say that A is a subset of B (notation A ⊆ B) if and only if every element of A is an element of B.*

Example 1.1.

  (a) If $A = \{3, 5, 8\}$ and $B = \{5, 8, 3, 2, 6\}$, then $A \subseteq B$.
  (b) $\mathbf{N} \subseteq \mathbf{Z} \subseteq \mathbf{R} \subseteq \mathbf{C}$.
  (c) If $A = \{3, 5, 8\}$ and $B = \{5, 3, 8\}$, then $A \subseteq B$ and $B \subseteq A$.

Definition:　Equality. *Let A and B be sets. We say that A is equal to B (notation A = B) if and only if every element of A is an element of B and conversely every element of B is an element of A; that is, A ⊆ B and B ⊆ A.*

Example 1.2.

  (a) In Example 1.1c, $A = B$. Note that the ordering of the elements is unimportant.
  (b) The number of times that an element appears in an enumeration doesn't affect a set. For example, if $A = \{1, 5, 3, 5\}$ and $B = \{1, 5, 3\}$, then $A = B$. Warning to readers of other texts: Some books introduce the concept of a multiset, in which the number of occurrences of an element matters.

A few comments are in order about the expression *if and only if* as used in our definitions. This expression means "is equivalent to saying" or, more exactly, that the word (or concept) being defined can at any time be replaced by the defining expression. Conversely, the expression which defines the word (or concept) can be replaced by the word.

Occasionally there is need to discuss the set which contains no elements, namely the *empty set*, which is denoted by the Norwegian letter Ø. This set is also called the *null set*.

It is clear, we hope, from the definition of a subset that given any set A we have $A \subseteq A$ and $Ø \subseteq A$. Because of this, given any set A, both $Ø$ and A are called *improper subsets* of A. If $B \subseteq A$, $B \neq Ø$, and $B \neq A$, then B is called a *proper subset* of A.

BASIC SET OPERATIONS

Definition:　Intersection. *Let A and B be sets. The intersection of A and B (denoted by A ∩ B) is the set of all elements which are in both A and B. That is, $A \cap B = \{x : x \in A \text{ and } x \in B\}$.*

Example 1.3.

(a) Let $A = \{1, 3, 8\}$ and $B = \{-9, 22, 3\}$. Then $A \cap B = \{3\}$.
(b) Let $A = \{(x, y) : x + y = 7, \ x, y \in \mathbf{R}\}$
and $B = \{(x, y) : x - y = 3, \ x, y \in \mathbf{R}\}$. Then $A \cap B = \{(5, 2)\}$.
(c) $\mathbf{Z} \cap \mathbf{Q} = \mathbf{Z}$.
(d) Let $A = \{3, 5, 9\}$ and $B = \{-5, 8\}$. Then $A \cap B = \emptyset$.

Definition: Disjoint Sets. *Two sets are disjoint if they have no elements in common (as in Example 1.3d); i.e., $A \cap B = \emptyset$.*

Definition: Union. *Let A and B be sets. The union of A and B (denoted by $A \cup B$) is the set of all elements which are in A or in B or in both A and B. That is, $A \cup B = \{x : x \in A \text{ or } x \in B\}$.*

It is important to note in the set-builder notation for $A \cup B$, the word *or* is used in the *inclusive* sense; it includes the case where x is in both A and B.

Example 1.4.

(a) If $A = \{2, 5, 8\}$ and $B = \{7, 5, 22\}$, then $A \cup B = \{2, 5, 8, 7, 22\}$.
(b) $\mathbf{Z} \cup \mathbf{Q} = \mathbf{Q}$.
(c) $A \cup \emptyset = A$ for any set A.

Frequently, when doing mathematics, we need to establish a *universe*, or set of elements under discussion. For example, the set $A = \{x : 81x^4 - 16 = 0\}$ contains different elements depending on what kinds of numbers we allow ourselves to use in solving the equation $81x^4 - 16 = 0$. This set of numbers would be our universe. For example, if the universe is the integers, then A is $\emptyset$. If our universe is the rational numbers, then A is $\{2/3, -2/3\}$, and if the universe is the complex numbers, then A is $\{2/3, -2/3, 2i/3, -2i/3\}$.

Definition: Universe. *The universe, or universal set, is the set of all elements under discussion for possible membership in a set.*

We reserve the letter U for a universe in general discussions.

VENN DIAGRAMS

When working with sets, as in other branches of mathematics, it is often quite useful to be able to draw a picture or diagram of the situation under consideration. A diagram of a set is call a *Venn diagram*. The universal set U is represented by the interior of a rectangle and the sets by disks inside the rectangle.

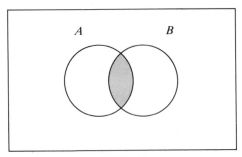

FIGURE 1.1
Venn diagram for intersection

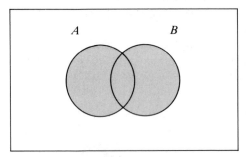

FIGURE 1.2
Venn diagram for union

Example 1.5.

(a) $A \cap B$ is illustrated in Figure 1.1 by shading the appropriate region.
(b) $A \cup B$ is illustrated in Figure 1.2.

In a Venn diagram, the region representing $A \cap B$ does not appear empty; however, in some instances it will represent the empty set. The same is true for any other region in a Venn diagram.

Definition: Complement. *Let A and B be sets. The complement of A relative to B (notation B − A) is the set of elements that are in B and not in A; i.e., $B - A = \{x : x \in B \text{ and } x \notin A\}$. If U is the universal set, then $U - A$ is denoted by A^c and is called simply the complement of A. $A^c = \{x \in U \mid x \notin A\}$.*

Example 1.6.

(a) Let $B = \{1, 2, 3, \ldots ,10\}$ and $A = \{2, 4, 6, 8, 10\}$. Then $B - A = \{1, 3, 5, 7, 9\}$ and $A - B = \emptyset$.
(b) If $U = \mathbf{R}$, then $\mathbf{Q}^c = $ the irrational numbers.

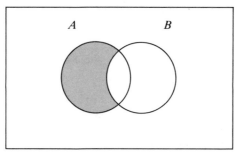

FIGURE 1.3
Venn diagram for $A - B$

(c) $U^c = \emptyset$ and $\emptyset^c = U$.
(d) The Venn diagram of $A - B$ is represented in Figure 1.3.
(e) The Venn diagram of A^c is represented in Figure 1.4.
(f) If $A \subseteq B$, then the Venn diagram of $B - A$ is in Figure 1.5.
(g) In the universe of integers, the set of even integers, $\{\ldots, -4, -2, 0, 2, 4, \ldots\}$, has the set of odd integers as its complement.

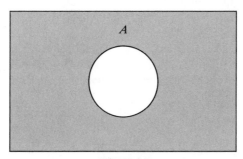

FIGURE 1.4
Venn diagram for A^c

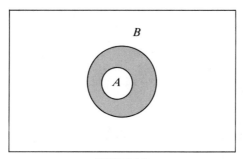

FIGURE 1.5
Venn diagram for $B - A$, where A is contained in B

Definition: Cartesian Product. *Let A and B be sets. The Cartesian product of A and B, denoted by A × B, is defined as A × B = {(a, b) : a ∈ A and b ∈ B}; that is, A × B is the set of all possible ordered pairs whose first component comes from A and whose second component comes from B.*

Example 1.7.

(a) Let $A = \{1, 2, 3\}$ and $B = \{4, 5\}$. Then $A \times B = \{(1, 4), (1, 5), (2, 4), (2, 5), (3, 4), (3, 5)\}$. Note that $\#(A \times B) = 6 = (\#A)(\#B)$.
(b) $A \times A = \{(1, 1), (1, 2), (1, 3), (2, 1), (2, 2), (2, 3), (3, 1), (3, 2), (3, 3)\}$. Note that $\#(A \times A) = 9 = (\#A)(\#A)$.

We can define the Cartesian product of three (or more) sets similarly. For example, $A \times B \times C = \{(a, b, c) : a \in A, b \in B, c \in C\}$. It is common to denote $A \times A$ by A^2, $A \times A \times A$ by A^3, etc., so that if $n \geq 2$, $A^n = \{(a_1, a_2, \ldots, a_n) : \text{each } a_i \in A\}$.

PASCAL NOTE

In Pascal, the type declaration S = SET OF 0..11 establishes a data type whose variables are subsets of the universe of integers from 0 to 11. In general, the universe must be a finite enumerated scalar type: boolean, character, integer subrange, or user-defined. If the variable declaration A, B, C : S is included in a program, then all of the standard set operations are available to compute values for these variables. The Pascal notation for these operations differs from the standard notation. For example, consider the sequence of statements A := [0, 2, 4, 6, 8, 10]; B := [0, 3, 6, 9]; C := A ∗ B; A := −A; B := A + B. This sequence assigns the values {0, 2, 4, 6, 8, 10} and {0, 3, 6, 9} to A and B, then assigns the value A ∩ B to C, the value of A^c to A and the value of A ∪ B to B. The values of A, B, and C at the end of these statements would be [1, 3, 5, 7, 9, 11], [0, 1, 3, 5, 6, 7, 9, 11], and [0, 6], respectively. Most Pascal systems will not allow you to read or write set-type variables. To write out a set value, you would have to test each of the integers from 0 to 11 for membership in a set. To test for whether integer x belongs to set A, the expression x IN A would be used.

EXERCISES FOR CHAPTER 1   *all*

A Exercises

1. Describe geometrically the following sets:
    (a) $\{x \in \mathbf{R} : |x| \leq 3\}$
    (b) $\{x \in \mathbf{Z} : |x| \leq 3\}$

(c) $\{x \in \mathbf{N} : |x| \le 3\}$

(d) $\{(x, y) : x, y \in \mathbf{R}, x^2 + y^2 = 25\}$

2. Enumerate the elements in the following sets:

(a) $\{x \in \mathbf{R} \mid x^2 - 3x + 2 = 0\}$

(b) $\{x \in \mathbf{R} \mid x^2 + 1 = 0\}$

(c) $\{x \in \mathbf{C} \mid x^2 + 1 = 0\}$

3. Describe the following sets using set-builder notation.

(a) $\{3, 5, 7, 9, \ldots , 77, 79\}$

(b) the rational numbers that are strictly between -1 and 1

(c) the even integers

4. Let $A = \{0, 2, 3\}$, $B = \{2, 3\}$, and $C = \{1, 5, 9\}$. Determine which of the following statements are true. Give reasons for your answers.

(a) $3 \in A$.

(b) $\{3\} \in A$.

(c) $\{3\} \subseteq A$.

(d) $B \subseteq A$.

(e) $A \subseteq B$.

(f) $\emptyset \subseteq C$.

(g) $\emptyset \in A$.

(h) $(B \cap A) \subseteq C$.

5. Let A, B, and C be as in Exercise 4 and let the universal set $U = \{0, 1, 2, \ldots , 9\}$. Determine:

(a) $A \cap B$

(b) $A \cup B$

(c) $B \cup A$

(d) $A \cup C$

(e) $A \cap C$

(f) $A - B$

(g) $C \times \{8\}$

(h) $B - A$

(i) A^c

(j) C^c

(k) $A \times B$

(l) B^2

(m) B^3

6. Let A, B, and C be as in Exercise 4, let $D = \{3, 2\}$, and let $E = \{2, 3, 2\}$. Determine which of the following are true. Give reasons for your decisions.

(a) $A = B$.

(b) $B = C$.

(c) $B = D$.

(d) $B = E$.

(e) $A \cap B = B \cap A$.

(f) $A \cup B = B \cup A$.

(g) $A - B = B - A$.

(h) $A \times B = B \times A$.

7. Let $U = \{0, 1, 2, 3, 4, 5, 6, 7, 8, 9\}$, $A = \{x \in U : x \text{ is a multiple of } 3\}$, and $B = \{x \in U : x^2 - 5 \ge 0\}$. Determine:

(a) $A \cup B$

(b) $A \cap B$

(c) B^c

8. Let $A = \{+, -\}$ and $B = \{00, 01, 10, 11\}$.

(a) List the elements of $A \times B$.

(b) How many elements do A^4 and $(A \times B)^3$ have?

B Exercises

9. Given that U = all students at a university, A = day students, B = mathematics majors, and C = graduate students, draw individual Venn diagrams to illustrate the following sets:

    (a) evening students
    (b) undergraduate mathematics majors

10. Let the sets A, B, C, and U be as in Exercise 9. Let $\#U = 16,000$, $\#A = 9000$, $\#B = 300$, and $\#C = 1000$. Also assume that the number of day students who are mathematics majors is 250, 50 of which are graduate students, and that the total number of day graduate students is 700. Determine the number of students who are:

(a) evening students
(b) non-mathematics majors
(c) undergraduates (day or evening)
(d) day graduate non-mathematics majors
(e) evening graduate students
(f) evening graduate mathematics majors
(g) evening undergraduate non-mathematics majors

11. Let A and B be sets. When are $A \times B$ and $B \times A$ equal?

C Exercises

12. One reason that we left the definition of a set vague is Russell's Paradox. Many mathematics books contain an account of this paradox. Two references are Stoll and Quine. Find one such reference and read it.

13. Use a Pascal FOR loop to write out the integers that belong to a set variable of type SET OF 0 . . 11.

chapter

2

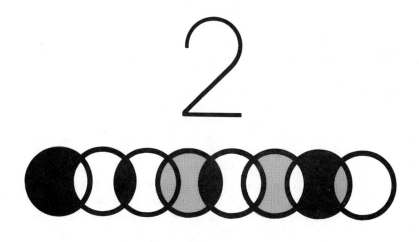

COMBINATORICS

2 GOALS

Throughout this book we will be counting things. In this chapter we will outline some of the tools that will help us count.

Counting occurs not only in highly sophisticated applications of mathematics to engineering and computer science but also in many basic applications. Like many other powerful and useful tools in mathematics, the concept is simple; we only have to recognize when and how it can be applied. Consider the following examples:

Example 2.1. A snack bar serves 5 different sandwiches and 3 different beverages. How many different lunches can a person order?

One way of determining the number of possible lunches is by listing or enumerating all the possibilities. One systematic way of doing this is by means of a tree (Figure 2.1).

Every path that begins at the position labeled START and goes to the right can be interpreted as a choice of 1 of the 5 sandwiches followed by a choice of 1 of the 3 beverages. Note that considerable work is required to arrive at the number 15 this way, but we also get more than just a number. The result is a complete list of all possible lunches. If we need to answer a question that starts with "How many . . . ," enumeration would be done only as a last resort. In a later chapter we will examine more enumeration techniques.

An alternate method of solution for this example is to make the simple observation that there are 5 different choices for sandwiches and 3 different choices for beverages, so there are $5 \cdot 3 = 15$ different lunches that can be ordered.

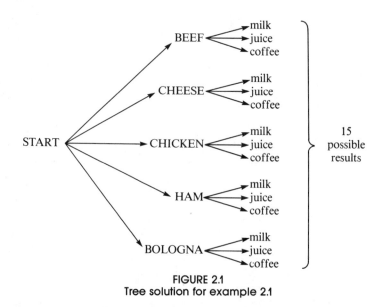

FIGURE 2.1
Tree solution for example 2.1

Example 2.2. A person is to complete a true-false questionnaire consisting of 10 questions. How many different ways are there to answer the questionnaire? Since each question can be answered either of 2 ways (true or false) and there are a total of 10 questions, there are $2 \cdot 2 \cdot 2 \cdot 2 \cdot 2 \cdot 2 \cdot 2 \cdot 2 \cdot 2 \cdot 2 = 2^{10}$ different ways of answering the questionnaire. The reader is encouraged to visualize the tree diagram of this example.

We formulize the procedures developed in the previous examples with the following rule and its extension.

RULE OF PRODUCTS

Rule of Products: *If two operations must be performed, and if the first operation can always be performed p_1 different ways and the second operation can always be performed p_2 different ways, then there are $p_1 p_2$ different ways that the two operations can be performed.*

Note: It is important that p_2 does not depend on the option that is chosen in the first operation. Another way of saying this is that p_2 is independent of the first operation. If p_2 is dependent on the first operation, then the rule of products does not apply.

Extended Rule of Products: *The rule of products can be extended to include sequences of more than two operations. If n operations must be performed and the number of options for each operation are $p_1, p_2 \ldots, $ and p_n, respectively, then the n operations can be performed*

$$p_1 \times p_2 \times \cdots \times p_n \text{ ways.}$$

Example 2.3. A questionnaire contains 4 questions that have 2 possible answers and 3 questions with 5 possible answers. There are $2 \cdot 2 \cdot 2 \cdot 2 \cdot 5 \cdot 5 \cdot 5 = 2^4 5^3 = 2000$ different ways to answer the questionnaire.

Many counting problems involve selecting a subset of a set and placing the subset in order. We may wish to know how many different ways we can arrange, or order, 3 different objects, for example, the letters a, b, and c. Certainly abc, bac, and cba are 3 such arrangements. In fact, there are $3 \cdot 2 \cdot 1 = 6$ different arrangements, or orders, of the elements in the set $A = \{a,b,c\}$.

Permutations

Definition: Permutation. *If A is a set, a permutation of A is an ordering of the elements of A.*

Example 2.4. The alphabetical ordering of the players of a baseball team is one permutation of the set of players. Other orderings of the player's names might be done by batting average, age, or height. The information that determines the ordering is called the _key_. We would expect that each key would give a different permutation of the names. If there are 25 players on the team, there are $25 \cdot 24 \cdot 23 \cdot \ldots \cdot 3 \cdot 2 \cdot 1$ different permutations of the players.

Example 2.5. The selection of the president, secretary, and treasurer for a club is a permutation of three members of the club. The set that is being permuted is the set of officers. In this case, the officers are not put into positions 1, 2, and 3; their positions are determined by the offices they hold.

From the rule of products we get the following rule for counting permutations: Before we develop a formula for permutations of k elements taken from a set of n elements, we look at some useful notation.

Definition: Factorials. _If n is a positive integer, then n factorial is the product of the first n positive integers and is denoted n!. Additionally, we define zero factorial to be 1._

$$0! = 1.$$
$$1! = 1.$$
$$2! = 1 \cdot 2 = 2.$$
$$3! = 1 \cdot 2 \cdot 3 = 6.$$

Note that 4! is 4 times 3!, or 24, and 5! is 5 times 4!, or 120. In addition, note that as n grows in size, $n!$ grows extremely quickly. For example, $11! = 39{,}916{,}800$. If the answer to a problem happens to be 25!, for example, you would never be expected to write that number out completely. However, if a problem had as its answer $25!/23!$ it can be reduced to $25 \cdot 24$, or 600.

From the rule of products we get the following rule for counting permutations: _The number of possible permutations of k elements taken from a set of n elements is denoted by_

$$P(n \,;\, k) = n(n - 1) \cdots (n - k + 1) = \frac{n!}{(n - k)!}.$$

This formula is proven in chapter 3.

Example 2.6. There are $P(25; 3) = 25 \cdot 24 \cdot 23$ ways to choose 3 officers for a club of 25 members.

Example 2.7. There are $P(25; 25) = \dfrac{25!}{0!}$ ways of putting 25 names in order.

Example 2.8. Suppose that you walk into a theater that has a single row of 6 seats. If 2 of the seats are to be occupied, how many seating arrangements are possible?

The first person who walked into the theater had 6 seats to choose from and the second person had 5 choices. The number 30 is our answer. Note that $30 = P(6; 2)$.

COMBINATIONS

Example 2. 9. Now consider the situation where the theater is so dark that the 2 individuals are indistinguishable, as in Figure 2.2. In this case, how many seating arrangements are possible? The answer is 15. This is due to the fact that 2 different seating arrangements in the lighted theater account for a single arrangement in the darkened theatre.

What made this problem different from previous ones is that we only wanted to know how many ways 2 of the 6 seats could be selected. The order in which they were selected or who sat in them didn't matter. In terms of sets, the problem was to choose a subset of the seats that contained 2 elements. Since the selection of a subset of a certain size from a set is a standard operation, we make the following definition.

Definition: Combination. *A combination of k elements from a set A is a subset of A containing k elements.*

Combination Rule: *The number of different combinations of k elements from a set of cardinality n is denoted by the binomial coefficient $\binom{n}{k}$ where*

$$\binom{n}{k} = \frac{n!}{(n-k)!\, k!} = \frac{n(n-1)\cdots(n-k+1)}{k(k-1)\cdots 2\cdot 1}.$$

Example 2. 10. A committee usually starts as an unstructured set of persons selected from a larger membership. Thus, before any officers are elected, a committee can be thought of as a combination. If a club of 25 members has a 5-member social committee, there are $\binom{25}{5} = 53{,}130$ different possible social committees. If any structure or restriction is placed on the way the social committee is to be selected, the number of possible committees

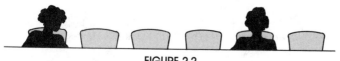

FIGURE 2.2
Illustration for example 2.9

will change. For example, if the club has a rule that the treasurer must be on the social committee, then the number of possibilities is reduced to $\binom{24}{4} = 10{,}626$. If we further require that a chairperson other than the treasurer be selected for the social committee, we have $\binom{24}{4} \cdot 4 = 52{,}504$ different possible social committees.

The choice of the 4 non-treasurers accounts for the $\binom{24}{4}$ and the choice of a chairperson accounts for the 4.

Example 2. 11.   There is $\binom{n}{0} = 1$ way of choosing a combination of zero elements from a set of n.

Remark: The term *binomial coefficient* is used for numbers of the form $\binom{n}{k}$ because they are the numbers that appear as coefficients when an expression such as $(x + y)^n$ is expanded.

In more complex counting problems you must decompose the objects that are being counted into two or more subsets. Then, provided none of the sets overlap, you can add up the cardinalities of the subsets to solve the problem.

Law of Addition: *If the elements of a finite set A can be separated into n subsets $B_1, B_2, \ldots, B_n$, each of which is disjoint from one another, then the cardinality of A is the sum of the cardinalities of the B_i;*

$$\#A = \#B_1 + \cdots + \#B_n.$$

Note: Another way of expressing the condition on the B_i's above is that for any two different integers, j and k, between 1 and n,

$$B_j \cap B_k = \phi \text{ and } B_1 \cup B_2 \cup \cdots \cup B_n = A.$$

Example 2. 12.   The number of students in a class could be determined by adding the numbers of students who are freshmen, sophmores, juniors, and seniors and those who belong to none of these categories. However, you probably couldn't add the students by major since some students may have double majors.

POWER SETS

Definition: Power Set. *If A is any set, the power set of A is the set of all subsets of A, including the empty set and A itself. It is denoted $\mathcal{P}(A)$.*

Example 2.13.   If $A = \{1, 2\}$, $\mathcal{P}(A) = \{\emptyset, \{1\}, \{2\}, \{1, 2\}\}$.

We end this chapter by deriving the formula for the number of elements in the power set of a finite set.

A word of warning: To understand the last line of this discussion, you may have to learn or review the binomial theorem: *if n is a positive integer and x and y are real numbers, then* $(x + y)^n =$

$$\binom{n}{0}y^n + \binom{n}{1} \times y^{n-1} + \binom{n}{2}x^2y^{n-2} + \cdots + \binom{n}{n-1}x^{n-1}y + \binom{n}{n}x^n.$$

Believe the following theorem on the basis of our example and the exercises. Right now, the proof is not as important as the formula.

Theorem 2. 1. *If A is a finite set, then* $\mathcal{P}(A) = 2^{\#A}$.

Let's consider a concrete example first. If $A = \{1, 2, 3, 4, 5\}$, the elements of $\mathcal{P}(A)$ can be separated into those with zero elements, one element, . . . , and five elements. The numbers of subsets of this kind are the binomial coefficients $\binom{5}{0}$, $\binom{5}{1}$, . . . , and $\binom{5}{5}$.

$$\binom{5}{0} + \binom{5}{1} + \cdots + \binom{5}{5} = 1 + 5 + 10 + 10 + 5 + 1 = 32 = 2^5.$$

Note that $5 = \#A$. No matter what $\#A$ is, we can do the same thing. If $\#A = n$, then we can break the power set of A into $n + 1$ subsets and obtain

$$\binom{n}{0} + \binom{n}{1} + \cdots + \binom{n}{n} = (1 + 1)^n = 2^n.$$

EXERCISES FOR CHAPTER 2

1–22

A Exercises

1. Automobile license plates in Massachusetts usually consist of 3 digits followed by 3 letters. The first digit is never zero. How many different plates of this type could be made?

2. The judiciary committee at a college is made up of 3 faculty members and 4 students. If 10 faculty members and 25 students have been nominated for the committee, how many judiciary committees are possible?

3. (a) Suppose that a single character is stored in a computer using eight bits. Then each character is represented by a different sequence of eight 0's and 1's called a bit pattern. How many different bit patterns are there? (That is, how many different characters could be represented?)
    (b) How many bit patterns have exactly three 1's?
    (c) How many different bit patterns have an even number of 1's?

4. (a) How many ways can 3 persons be seated in a car with 4 seats?
         Assume that someone must drive.
    (b) What is your answer if only 2 of the 3 persons have a driver's
         license?

5. How many ways can you arrange the letters in the following words?
    (a) COMBINE
    (b) SUBSET
    (c) NOON

6. Derive the formula $\#\mathcal{P}(A) = 2^{\#A}$ using the rule of products.

7. How many subsets of $\{1. \ . \ .10\}$ contain at least 7 elements?

8. How many *proper* subsets of $\{1, 2, 3, 4, 5\}$ contain the numbers 1 and
    5? How many of them also *do not* contain the number 2?

9. As a freshman, suppose that you had to take 2 of 4 lab science courses,
    1 of 2 literature courses, 2 of 3 math courses, and 1 of 7 physical
    education courses. Disregarding possible time conflicts, how many dif-
    ferent schedules do you have to choose from?

10. Suppose that you are about to flip a coin and then roll a die. Let
    $A = \{HEAD, TAIL\}$ and $B = \{1, 2, 3, 4, 5, 6\}$.
    (a) What is $\#(A \times B)$?
    (b) How could you interpret the set $A \times B$?

11. Consider three persons, A, B, and C, who are to be seated in a row of
    3 chairs. Suppose that A and B are identical twins. How many seating
    arrangements of these persons can there be:
    (a) If you are a total stranger?
    (b) If you are A and B's mother?
    (This problem is designed to show you that there are different correct
    answers to some problems.)

12. Convince yourself that:
    (a) $\dbinom{n}{k} = \dbinom{n}{n-k}$.
    (b) $\dbinom{n}{1} = n$.

13. Use the binomial theorem to calculate $(9998)^3$. Note that
    $(9998)^3 = (10000 - 2)^3$.

14. If a raffle has 3 different prizes and there are 1000 raffle tickets sold, how
    many different ways can the prizes be distributed?

15. The congressional committees on mathematics and computer science are

made up of 5 congressmen each, and a congressional rule is that the two committees must be disjoint. If there are 385 members of congress, how many ways could the committees be selected?

16. (a) How many ways can you arrange the letters in the word CONGRESS?
    (b) In how many of these arrangements are the two S's not together?

17. How many ways can a student do a 10-question true-false exam if he or she can choose not to answer any number of questions?

18. Suppose you have a choice of fish, lamb, or beef for a main course, a choice of peas or carrots for a vegetable, and a choice of pie, cake, or ice cream for dessert. If you must order 1 item from each category, how many different dinners are possible?

19. Let $T = \{1, 2, 3, 4, 5\}$. How many subsets of T have less than 4 elements?

20. A questionnaire contains 6 questions each having yes-no answers. For each yes response, there is a follow-up question with 4 possible responses. In how many ways can the questionnaire be answered?

21. Suppose you have a choice of vanilla, chocolate, or strawberry for ice cream, a choice of peanuts or walnuts for chopped nuts, and a choice of hot fudge or marshmallow for topping. If you must order one item from each category, how many different sundaes are possible?

22. (a) A person flips a fair (evenly balanced) coin 4 times in succession. How many different results are possible? Consider the outcomes of H, H, T, H and H, T, H, H different. Illustrate the answer by constructing a tree diagram labeling each branch by H (heads) or T (tails).
    (b) Since the coin is fair, the probability of obtaining a heads (or tails) in any given flip of the coin is $1/2$. What is the probability of obtaining H, H, T, H? H, T, T, H? What is the probability of obtaining 2 heads? Three tails?

B Exercises

23. There are 10 points $P_1, \ldots, P_{10}$ on a plane, no 3 on the same line.
    (a) How many lines are determined by the points?
    (b) How many triangles are determined by the points?

24. How many ways can you separate a set with n elements into 2 nonempty subsets if the order of the subsets is immaterial? What if the order of the subsets is important?

25. How many ways can n persons be grouped into pairs, when n is even?

C Exercises

26. (a) Write a Pascal program to determine the largest value of n for which
         $n!$ is less than or equal to MAXINT, the largest integer of type
         INTEGER.
    (b) Write a Pascal function to compute $\binom{n}{k}$ that is better than this:

```
FUNCTIONAL BINOMIAL (N,K:NATURAL):INTEGER;
VAR
   I,NUM,DI,D2:INTEGER;
BEGIN
   NUM: = 1; D1: = 1; D2 = 1;
   FOR I: = 1 TO N DO NUM: = NUM*I;
   FOR I: = 1 TO N-K DO D1: = D1*I;
   FOR I: = 1 TO K DO D2: = D2*I;
   BINOMIAL: = NUM DIV (D1*D2)
END;
```

        What's wrong with this one?

chapter

3

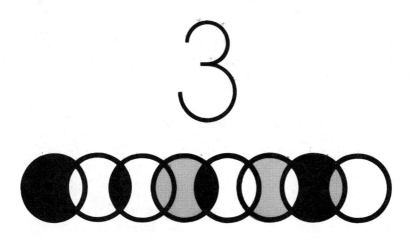

LOGIC

3 GOALS

In this chapter, we will introduce some of the basic concepts of mathematical logic. In order to fully understand some of the later concepts in this book, you must be able to recognize valid logical arguments. Although these arguments will usually be applied to mathematics, they employ the same techniques that are used by a lawyer in a courtroom or a physician examining a patient. An added reason for the importance of this chapter is that the circuits that make up digital computers are designed using the same algebra of propositions that we will be discussing.

3.1 Propositions and Logical Operators

PROPOSITIONS

Definition: Proposition. *A proposition is a sentence to which one and only one of the terms true or false can be meaningfully applied.*

Example 3.1.1. "Four is even," "$4 \in \{1, 3, 5\}$," and "$43 \geq 21$" are propositions.

In traditional logic, a declarative statement with a definite truth value is considered a proposition. Although our ultimate aim is to discuss mathematical logic, we won't separate ourselves completely from the traditional setting. This is natural because the basic assumptions, or postulates, of mathematical logic are modeled after the logic we use in everyday speech. Since compound sentences are frequently used in everyday speech, we expect that logical propositions contain connectives like the word *and*. The statement "Black holes exist and Mars supports life" is a proposition and, hence, must have a definite truth value. Whatever that truth value is, it should be the same as the truth value of "Mars supports life and black holes exist."

LOGICAL OPERATORS

There are several ways in which we commonly combine simple statements into compound ones. The words *and, or, not, if . . . then,* and *if and only if* can be added to one or more propositions to create a new proposition. To avoid any confusion, we will precisely define each one's meaning and introduce its standard symbol. With the exception of negation (*not*), all of the operators act on pairs of propositions. Since each proposition has two possible truth values, there are four ways that truth can be assigned to two propositions. In defining the effect that a logical operator has on two propositions, the result must be specified for all four cases. The most convenient way of doing this is with a truth table, which we will illustrate by defining the word *and*.

Definition: Conjunction (And). *If p and q are propositions, their conjunction, p and q (denoted p $\wedge$ q), is defined by the truth table in Table 3.1.1.*

Notes:

(a) To read this truth table, you must realize that any one line represents a case: one possible set of values for p and q.

(b) The numbers 0 and 1 are used to denote false and true, respectively. This is consistent with the way that many programming languages treat logical, or Boolean, variables since a single bit, 0 or 1, can represent a truth value. In Pascal, the ordinal function on the set of Boolean values is defined by ORD(FALSE) = 0 and ORD(TRUE) = 1.

(c) For each case, the symbol under p represents the truth value of p. The same is true for q. The symbol under $p \wedge q$ represents the truth value of $p \wedge q$ for that case. For example, the second row of the truth table represents the case in which p is false, q is true, and the resulting truth value for $p \wedge q$ is false. As in everyday speech, $p \wedge q$ is true only when both propositions are true.

(d) Just as the letters x, y, and z are frequently used in algebra to represent numeric variables, p, q, and r seem to be the most commonly used symbols for logical variables. When we say that p is a logical variable, we mean that any proposition can take the place of p.

(e) One final comment: The order in which we list the cases in a truth table is standardized in this book. If the truth table involves two simple propositions, the numbers under the simple propositions can be interpreted as the two-digit binary integers in increasing order, 00, 01, 10, and 11, for 0, 1, 2, and 3.

Definition: Disjunction (Or). *If p and q are propositions, their disjunction is p or q, denoted p $\vee$ q, and is defined by the truth table in Table 3.1.2.*

<table>
<tr><td colspan="3" align="center">**TABLE 3.1.1**
Truth Table for And</td><td colspan="3" align="center">**TABLE 3.1.2**
Truth Table for Or</td></tr>
<tr><td>p</td><td>q</td><td>$p \wedge q$</td><td>p</td><td>q</td><td>$p \vee q$</td></tr>
<tr><td>0</td><td>0</td><td>0</td><td>0</td><td>0</td><td>0</td></tr>
<tr><td>0</td><td>1</td><td>0</td><td>0</td><td>1</td><td>1</td></tr>
<tr><td>1</td><td>0</td><td>0</td><td>1</td><td>0</td><td>1</td></tr>
<tr><td>1</td><td>1</td><td>1</td><td>1</td><td>1</td><td>1</td></tr>
</table>

TABLE 3.1.3
Truth Table for Not

p	$\sim p$
0	1
1	0

Note: The only case in which disjunction is false is when *both* propositions are false. This interpretation of the word *or* is called the *nonexclusive or*. The *exclusive or* will be discussed when we consider logical design in Chapter 13.

Definition: Negation (Not). *If p is a proposition, its negation, not p, is denoted ~p and is defined by the truth table in Table 3.1.3.*

Note: Negation is the only operator that acts on a single proposition; hence only two cases are needed.

The Conditional Operator (If . . . Then).   Consider the following propositions from everyday speech:

(a) I'm going to quit if I don't get a raise.
(b) If I pass the final, then I'll graduate.
(c) I'll be going to the movies provided that my car starts.

All three propositions are conditional, they can all be restated to fit into the form *if* CONDITION, *then* CONCLUSION. For example, statement a can be rewritten as "If I don't get a raise, then I'm going to quit."
    A conditional statement is meant to be interpreted as a guarantee; if the condition is true, then the conclusion is expected to be true. It says no more and no less.

Definition: Conditional Operator. *The conditional statement if p then q, denoted p → q, is defined by the truth table in Table 3.1.4.*

TABLE 3.1.4
Truth Table for If . . . then

p	q	$p \rightarrow q$
0	0	1
0	1	1
1	0	0
1	1	1

TABLE 3.1.5
Truth Table for If and Only If

p	q	$p \leftrightarrow q$
0	0	1
0	1	0
1	0	0
1	1	1

Example 3.1.2.   Assume your instructor gave you the following promise: "If you receive a grade of 95 or better in the final examination, then you will receive an A in this course." Your instructor has made a promise to you (placed a *condition* with you). If you fulfill his or her condition, you expect the *conclusion* (the receipt of an A) to be forthcoming. Your graded final has been returned to you. Has your instructor told the truth (kept the promise) or is your instructor guilty of a falsehood?

Case I: Your final exam score was less than 95 (the condition is false) and you did not receive an A (the conlusion is false). The instructor told the truth.

Case II: Your final exam score was less than 95, yet you received an A for the course. The instructor told the truth. (Perhaps your overall course average was excellent.)

Case III: Your final exam score was greater than 95, but you did not receive an A. The instructor lied.

Case IV: Your final exam score was greater than 95, and you received an A. The instructor told the truth.

Note: The only case in which a conditional proposition is false is when the condition is true and the conclusion is false.

Definition: Biconditional Operator (If and Only If). *If p and q are propositions, the biconditional statement p if and only if q, denoted p $\leftrightarrow$ q, is defined by the truth table in Table 3.1.5.*

$p \leftrightarrow q$ is true when p and q have the same truth values.

3.2   Truth Tables and Propositions Generated by a Set

Consider the compound proposition $c : (p \wedge q) \vee (\sim q \wedge r)$, where p, q, and r are propositions. This is an example of a proposition generated by p, q, and r. We will define this terminology later in the section. Since each of the three simple propositions has two possible truth values, it follows that there are eight different combinations of truth values that determine a value for c. These values can be obtained from a truth table for c. To construct the truth

TABLE 3.2.1
Truth Table for c

p	q	r	$(p \wedge q)$	$\sim q$	$(\sim q \wedge r)$	.c
0	0	0	0	1	0	0
0	0	1	0	1	1	1
0	1	0	0	0	0	0
0	1	1	0	0	0	0
1	0	0	0	1	0	0
1	0	1	0	1	1	1
1	1	0	1	0	0	1
1	1	1	1	0	0	1

table, we build c from p, q, and r and from the logical operators. The result is Table 3.2.1. Strictly speaking, the first three columns and the last column make up the truth table for c. The other columns are work space needed to build up to c.

Note that the first three columns of the truth table are an enumeration of the eight three-digit binary integers. This standardizes the order in which the cases are listed. In general, if c is generated by n simple propositions, then the truth table for c will have 2^n rows with the first n columns being an enumeration of the n digit binary integers. At a glance we can see that for exactly four of the eight cases, c will be true. For example, if p and r are true and q is false (the sixth case), then c is true.

Let S be any set of propositions. We will give two definitions of a proposition generated by S. The first is a bit imprecise, but should be clear. The second definition is called a *recursive definition*. If you find it confusing, use the first definition and reread the second occasionally.

Definition: Proposition Generated by S.

(1) A proposition generated by S is any valid combination of propositions in S with conjunction, disjunction, and negation.
(2) (a) If $p \in S$, then p is a proposition generated by S.
  (b) If x and y are propositions generated by S, then so are (x), $\sim x$, $(x) \vee (y)$, and $(x) \wedge (y)$.

Note: We have not included the conditional and biconditional in the definition because they can both be obtained from conjunction, disjunction, and negation, as we will see later.

If S is a finite set, then we may use slightly different terminology. For example, if $S = \{p, q, r\}$, we might say that a proposition is generated by p, q, and r instead of $\{p, q, r\}$. One other variation that we will use is that the parentheses that are added when taking a disjunction or conjunction will be dropped if the lack of them causes no confusion. For example, the conjunction of p and $q \lor r$ will be written $p \land (q \lor r)$ instead of $(p) \land (q \lor r)$. Note that dropping both sets would be confusing; in fact it would be incorrect. It is customary to use the following hierarchy for interpreting propositions, with parentheses overriding this order:

First: Negation
Second: Conjunction
Third: Disjunction

Within any level of the hierarchy, work from left to right. Using these rules, $p \land q \lor r$ is taken to mean $(p \land q) \lor r$. These precedence rules are universal and are exactly those used in many computer languages such as Pascal.

Example 3.2.1:   A few shortened expressions and their fully parenthesized versions:

(a) $p \land q \land r$ is $((p) \land (q)) \land r$.
(b) $\sim p \lor \sim r$ is $(\sim p) \lor (\sim r)$.
(c) $\sim \sim p$ is $\sim (\sim p)$.

A proposition generated by a set S need not include each element of S in its expression. For example, $\sim q \land r$ is a proposition generated by p, q, and r. In this context, its truth table is the first three columns and the sixth column of the truth table for $(p \land q) \lor (\sim q \land r)$.

EXERCISES FOR SECTIONS 3.1 AND 3.2

A Exercises

1.   Construct the truth tables of:
   (a) $\sim (p \land q)$ (d) $(p \land q) \lor (q \land r) \lor (r \land p)$
   (b) $p \land (\sim q)$ (e) $(\sim p) \lor (\sim q)$
   (c) $(p \land q) \land r$ (f) $p \lor q \lor r \lor s$

2. Rewrite the following with as few extraneous parentheses as possible.
   (a) $(\sim ((p) \land (r))) \lor (s)$
   (b) $((p) \lor (q)) \land ((r) \lor (q))$

3. In what order are the operations in the expression $p \lor \sim q \lor r \land \sim p$ performed?

4.   The following four statements all carry the same meaning:
   If p then q.
   p is a sufficient condition for q.

q is a necessary condition for p.

p only if q.

Rewrite the following statements in terms of the other three.

(a) If an integer is a multiple of 4, then it is even.

(b) The fact that a polygon is a square is a sufficient condition that it is a rectangle.

(c) If $x = 5$ then $x^2 = 25$.

(d) If $x^2 - 5x + 6 = 0$, then $x = 2$ or $x = 3$.

(e) $x^2 = y^2$ is a necessary condition for $x = y$.

(f) If every continuous function on S is bounded, then S is compact.

5. Write the converse of Statements 4.a through 4.f using:

(a) "if . . . then . . ." language

(b) "necessary/sufficient" language

6. Rewrite the following using "if . . . then . . ." language.

(a) A necessary and sufficient condition for a triangle to be equilateral is that it be equiangular.

(b) A necessary and sufficient condition for a finite monoid to be a group is that its identity element is its only idempotent.

3.3 Equivalence and Implication

TAUTOLOGIES

Consider two propositions generated by p and q: $\sim(p \wedge q)$ and $(\sim p) \vee (\sim q)$. At first glance, they are different propositions. In form, they are different; but they have the same meaning. One way to see this is to substitute actual propositions for p and q, such as:

$$p: \text{ I've been to Toronto; and}$$

$$q: \text{ I've been to Chicago.}$$

Then $\sim(p \wedge q)$ translates to "I haven't been to both Toronto and Chicago," while $(\sim p) \vee (\sim q)$ is "I haven't been to Toronto or I haven't been to Chicago." Determine the truth values of these propositions. Naturally, they will be true for some people and false for others. What is important is that no matter what truth values they have, $\sim(p \wedge q)$ and $(\sim p) \vee (\sim q)$, will have the same truth value. The easiest way to see this is by examining the truth tables of these propositions (Table 3.3.1).

In all four cases, $\sim(p \wedge q)$ and $(\sim p) \vee (\sim q)$ have the same truth value. Then when the biconditional operator is applied to them, the result is a value of true in all cases.

TABLE 3.3.1
Truth Tables for $\sim(p \wedge q)$ and $(\sim p) \vee (\sim q)$

p	q	$p \wedge q$	$\sim(p \wedge q)$	$\sim p$	$\sim q$	$(\sim p) \vee (\sim q)$
0	0	0	1	1	1	1
0	1	0	1	1	0	1
1	0	0	1	0	1	1
1	1	1	0	0	0	0

Definition: Tautology. *An expression involving logical variables that is true in all cases is called a tautology.*

Example 3.3.1. All of the following are tautologies:

(a) $\sim(p \wedge q) \leftrightarrow (\sim p) \vee (\sim q)$.
(b) $p \vee \sim p$.
(c) $(p \wedge q) \rightarrow p$.
(d) $q \rightarrow (p \vee q)$.
(e) $(p \vee q) \leftrightarrow (q \vee p)$.

Their truth tables consist of a column of ones (trues).

Definition: Contradiction. *An expression involving logical variables that is false for all cases of its truth table is called a contradiction.*

Example 3.3.2. $p \wedge \sim p$ and $(p \vee q) \wedge (\sim p) \wedge (\sim q)$ are contradictions.

Definition: Equivalence. *Let S be a set of propositions and let r and s be propositions generated by S. r and s are equivalent if r $\leftrightarrow$ s is a tautology. The equivalence of r and s is denoted r $\Leftrightarrow$ s.*

Example 3.3.3. The following are all equivalences:

(a) $(p \wedge q) \vee (\sim p \wedge q) \Leftrightarrow q$.
(b) $p \rightarrow q \Leftrightarrow \sim q \rightarrow \sim p$.
(c) $p \vee q \Leftrightarrow q \vee p$.

All tautologies are equivalent to one another. We will use the number 1 to symbolize a tautology.

Example 3.3.4. $p \vee \sim p \Leftrightarrow 1$.

All contradictions are equivalent to one another. We will use the number 0 to symbolize a contradiction.

Example 3.3.5. $p \wedge \sim p \Leftrightarrow 0$.

Equivalence is to logic what equality is to algebra. Just as there are many ways of writing an algebraic expression, the same logical meaning can be expressed in many different ways.

IMPLICATION

Example 3.3.6. Consider the two propositions:

x: The money is behind Door A; and

y: The money is behind Door A or Door B.

Imagine that you were told that there is a large sum of money behind one of two doors marked A and B, and that one of the two propositions x and y is true and the other is false. Which door would you choose? All that you need to realize is that if x is true, then y will also be true. Since we know that this can't be the case, y must be the true proposition and the money is behind Door B.

This is an example of a situation in which the truth of one proposition leads to the truth of another. Certainly, y can be true when x is false; but x can't be true when y is false. In this case, we say that x implies y.

Look at the truth table of $p \rightarrow q$ in Table 3.3.1. If p implies q, then the third case can be ruled out, since it is the case that makes a conditional proposition false.

Definition: Implication. *Let S be a set of propositions and let r and s be propositions generated by S. We say that r implies s if $r \rightarrow s$ is a tautology. We write $r \Rightarrow s$ to indicate this implication.*

Example 3.3.7. A commonly used implication is that p implies $p \vee q$. Its truth is verified by the truth table in Table 3.3.2.

If we let p represent "The money is behind Door A" and q represent "The money is behind Door B," $p \Rightarrow (p \vee q)$ is a formalized version of the

TABLE 3.3.2
Truth Table for $p \rightarrow (p \vee q)$

p	q	$p \vee q$	$p \rightarrow (p \vee q)$
0	0	0	1
0	1	1	1
1	0	1	1
1	1	1	1

reasoning used in Example 3.3.6. A common name for this implication is *disjunctive addition*. In the next section we will consider some of the most commonly used implications and equivalences.

When we defined what we mean by a proposition generated by a set in Section 3.2, we didn't include the conditional and biconditional operators. This was because of the two equivalences $p \rightarrow q \Leftrightarrow (\sim p) \vee q$ and $p \leftrightarrow q \Leftrightarrow (p \wedge q) \vee (\sim p \wedge \sim q)$. Therefore, any proposition that includes the conditional or biconditional operators can be written in an equivalent way using only conjunction, disjunction, and negation. We could even dispense with disjunction since $p \vee q$ is equivalent to a proposition that uses only conjunction and negation.

PASCAL NOTE

The logical operators that can be used to create Boolean expressions in Pascal are AND, OR, and NOT. If you want to use any other logical operator, you must define the operator as a function. For example, the biconditional operator would be

```
FUNCTION BICOND(P, Q: BOOLEAN) :BOOLEAN;
BEGIN
        IF P = Q THEN BICOND := TRUE
                 ELSE BICOND := FALSE
END;
```

Since conditional (IF . . THEN) statements in Pascal are not propositions, there is no correspondence between them and the conditional operator. For example, consider

```
IF X < 0 THEN X := -X.
```

The "THEN clause," $X := -X$, is executed if and only if $X < 0$ is true. Therefore, conditional statements are more similar to the biconditional than the conditional.

EXERCISES FOR SECTION 3.3

A Exercises

1. Given the following propositions generated by p, q, and r, which are equivalent to one another?

(a) $(p \wedge r) \vee q$ (e) $(p \vee q) \wedge (r \vee q)$

(b) $p \vee (r \vee q)$ (f) $r \rightarrow p$

(c) $r \wedge p$ (g) $r \vee \sim p$

(d) $\sim r \vee p$ (h) $p \rightarrow r$

2. Of the propositions in Exercise 1, which ones imply $p \lor q$?

3. Of the propositions in Exercise 1, which are implied by $p \lor q$?

4. How large is the largest set of propositions generated by p and q with the property that no two elements are equivalent?

5. Find a proposition that is equivalent to $p \lor q$ and uses only conjunction and negation.

6. (a) Construct the truth table for $x : (p \land (\sim q)) \lor (r \land p)$
    (b) Give an example other than x itself of a proposition generated by p, q, and r that is equivalent to x.
    (c) Give an example of a proposition other than x itself that implies x.
    (d) Give an example of a proposition other than x that is implied by x.

7. Suppose that x and y are propositions generated by p, q, r, and s. Explain what it means for x and y to be equivalent. How do you determine if x and y are equivalent?

C Exercise

8. Write a Pascal function that simulates the conditional operator.

3.4 The Laws of Logic

In this section, we will list the most basic equivalences and implications of logic. Most of the equivalences listed in Table 3.4.1 should be obvious to the reader. Remember, 0 stands for contradiction, 1 for tautology. They are similar to the algebraic laws, such as $a + (b + c) = (a + b) + c$, the associative law of addition. In fact, associativity of both conjunction and disjunction are among the laws of logic. Notice that with one exception, the laws are paired in such a way that exchanging the symbols $\land$, $\lor$, 1, and 0 for $\lor$, $\land$, 0, and 1, respectively, in any law gives you a second law. For example, $p \lor 0 \Leftrightarrow p$ results in $p \land 1 \Leftrightarrow p$. This is called a *duality principle*. For now, think of it as a way of remembering two laws for the price of one. We will leave it to the reader to verify a few of these laws with truth tables. However, the reader should note that the duality principle can only be applied to equivalences. If one attempts to apply duality to the true statement $p \land q \Rightarrow p$, we obtain $p \lor q \Rightarrow p$, which is false.

Example 3.4.1. The identity law:

p	1	$p \land 1$	$(p \land 1) \leftrightarrow p$
0	1	0	1
1	1	1	1

therefore, $(p \land 1) \Leftrightarrow p$.

TABLE 3.4.1
Basic Logical Laws

Commutative Laws

$p \vee q \Leftrightarrow q \vee p$ $\qquad\qquad\qquad$ $p \wedge q \Leftrightarrow q \wedge p$

Associative Laws

$(p \vee q) \vee r \Leftrightarrow p \vee (q \vee r)$ $\qquad$ $(p \wedge q) \wedge r \Leftrightarrow p \wedge (q \wedge r)$

Distributive Laws

$p \wedge (q \vee r) \Leftrightarrow (p \wedge q) \vee (p \wedge r)$

$\qquad\qquad\qquad$ $p \vee (q \wedge r) \Leftrightarrow (p \vee q) \wedge (p \vee r)$

Identity Laws

$p \vee 0 \Leftrightarrow p$ $\qquad\qquad\qquad$ $p \wedge 1 \Leftrightarrow p$

Negation Laws

$p \wedge \sim p \Leftrightarrow 0$ $\qquad\qquad\qquad$ $p \vee \sim p \Leftrightarrow 1$

Idempotent Laws

$p \vee p \Leftrightarrow p$ $\qquad\qquad\qquad$ $p \wedge p \Leftrightarrow p$

Null Laws

$p \wedge \emptyset \Leftrightarrow \emptyset$ $\quad$ 0 = False $\qquad\qquad$ $p \vee 1 \Leftrightarrow 1$

Absorbtion Laws

$p \wedge (p \vee q) \Leftrightarrow p$ $\qquad\qquad\qquad$ $p \vee (p \wedge q) \Leftrightarrow p$

DeMorgan's Laws

$\sim(p \vee q) \Leftrightarrow (\sim p) \wedge (\sim q)$ $\qquad$ $\sim(p \wedge q) \Leftrightarrow (\sim p) \vee (\sim q)$

Involution Law

$\sim(\sim p) \Leftrightarrow p$

Some of the logical laws in Table 3.4.2 might be less obvious to you. For any that you are not comfortable with, substitute actual propositions for the logical variables. For example, if p is "John owns a pet store" and q is "John likes pets," the detachment law should make sense.

PASCAL NOTE

The two looping statements WHILE and REPEAT . . UNTIL are controlled by Boolean expressions. The choice of which of the two to use in a given situation is purely a matter of convenience. To convert a program segment with one of these loops to use the other is usually quite simple. The Boolean

TABLE 3.4.2
Common Implications and Equivalence

Detachment

$$(p \to q) \land p \Rightarrow q$$

Contrapositive

$$(p \to q) \land \sim q \Rightarrow \sim p$$

Disjunctive Addition

$$p \Rightarrow (p \lor q)$$

Conjunctive Simplification

$$(p \land q) \Rightarrow p \quad \text{and} \quad (p \land q) \Rightarrow q$$

Disjunctive Simplification

$$(p \lor q) \land \sim p \Rightarrow q \quad \text{and} \quad (p \lor q) \land \sim q \Rightarrow p$$

Chain Rule

$$(p \to q) \land (q \to r) \Rightarrow (p \to r)$$

Conditional Equivalences

$$(p \to q) \Leftrightarrow (\sim q \to \sim p) \Leftrightarrow (\sim p \lor q)$$

Biconditional Equivalences

$$(p \leftrightarrow q) \Leftrightarrow ((p \to q) \land (q \to p)) \Leftrightarrow ((p \land q) \lor (\sim p \land \sim q))$$

condition governing the original loop is negated and used as the condition for the new loop. Sometimes there are other minor changes to be made, particularly if input is involved in the loop, but the negation of the Boolean condition is usually the key change.

In the following segment, we search for the location within an array of integers for a value N. If the value is found, the location is I, an index of the array and the value of FOUND is true. If no such value is found, FOUND stays false.

```
FOUND := FALSE;
I := 1;
REPEAT
     IF A[I] = N THEN FOUND := TRUE
                 ELSE I := I+1
UNTIL FOUND OR (I > MAX)
        (* MAX = NUMBER OF ELEMENTS IN A *)
```

To convert this segment to use a WHILE loop, the exit condition (FOUND OR (I > MAX)) is negated. Its negation is much easier to read if we apply DeMorgan's Law. The result is

```
WHILE (NOT FOUND) AND (I <= MAX) DO
    IF A[I]  =  N THEN FOUND := TRUE
            ELSE I := I+1;
```

EXERCISES FOR SECTION 3.4

B Exercises

1. Describe how duality can be applied to implications if we introduce the symbol ⇐, read "is implied by."

C Exercises

2. (a) What laws are used to convert the following segments to use a REPEAT . . UNTIL loop? Write out the converted segment in (a) so that you don't try to read at EOF (End of File).

```
WHILE NOT EOF
    DO      BEGIN
                READ(CH);
                COUNT := COUNT +11
            END
```

(b)
```
S:=0; k :=1; n :=100;

WHILE k:= n DO
    BEGIN
        S := S + k;
        k := k + 1
    END
```

3.5 Mathematical Systems   (6m iT)

In this section, we present an overview of what a mathematical system is and how logic plays an important role in one. The axiomatic method that we will use here will not be duplicated with as much formality anywhere else in the book, but we hope that an emphasis on how mathematical facts are developed and organized will help to unify the concepts that we present. The system of propositions and logical operators that we have developed will serve as a model for our discussion. Roughly, a mathematical system can be defined as follows.

Definition: Mathematical System. *A mathematical system consists of:*

(1) A set or universe, U.

(2) Definitions—sentences that explain the meaning of concepts that relate to the universe. Any term used in describing the universe itself is said to be undefined. All definitions are given in terms of these undefined concepts of objects.

(3) Axioms—assertions about the properties of the universe and rules for creating and justifying more assertions. These rules always include the system of logic that we have developed to this point.

(4) Theorems—the additional assertions mentioned above.

Example 3.5.1.   In Euclidean geometry the universe consists of points and lines (two undefined terms). Among the definitions is a definition of parallel lines and among the axioms is the axiom that two distinct parallel lines never meet.

Example 3.5.2.   *Propositional calculus* is a formal name for the logical system that we've been discussing. The universe consists of propositions. The axioms are the truth tables for the logical operators and the key definitions are those of equivalence and implication. We use propositions to describe any other mathematical system; therefore, this is the minimum amount of structure that a mathematical system can have.

Definition: Theorem. *A true proposition derived from axioms of mathematical system is called a theorem.*

Virtually all theorems can be expressed in terms of a finite number of propositions, $p_1, p_2 \ldots, p_n$, called the *premises*, and a proposition, C, called the *conclusion*. These theorems take the form

$$p_1 \wedge p_2 \wedge \cdots \wedge p_n \Rightarrow C,$$

or more informally,

$$p_1, p_2, \ldots, \text{ and } p_n \text{ imply } C.$$

For a theorem of this type, we say that the premises imply the conclusion.

When a theorem is stated, it is assumed that the axioms of the system are true. In addition, any previously proven theorem can be considered an extension of the axioms and can be used in demonstrating that the new theorem is true. When the proof is complete, the new theorem can be used to prove subsequent theorems. A mathematical system can be visualized as an inverted pyramid with the axioms at the base and the theorems expanding out in various directions (Figure 3.5.1).

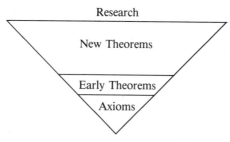

FIGURE 3.5.1
The body of knowledge in a mathematical system

PROOFS

Definition: Proof. *A proof of a theorem is a finite sequence of logically valid steps that demonstrate that the premises of a theorem imply the conclusion.*

Exactly what constitutes a proof is not always clear. For example, a research mathematician might require only a few steps to prove a theorem to a colleague, but might take an hour to give an effective proof to a class of students. Therefore, what constitutes a proof often depends on the audience. But the audience is not the only factor. One of the most famous theorems in graph theory, The Four Color Theorem, was finally proven in 1976, after a century of effort by many mathematicians. Part of the proof consisted of having a computer check many different graphs for a certain property. Without the aid of the computer, this checking would have taken years. In the eyes of some mathematicians, this proof was considered questionable. Shorter proofs have been developed since 1976 and there is no controversy associated with The Four Color Theorem at this time. (It is stated in Chapter 9.)

PROOFS IN PROPOSITIONAL CALCULUS

Theoretically, you can prove anything in propositional calculus with truth tables. In fact, the laws of logic stated in Section 5.4 are all theorems. Propositional calculus is one of the few mathematical systems for which any valid sentence can be determined true or false by mechanical means. A program to write truth tables is not too difficult to write; however, what can be done theoretically is not always practical. For example,

$$a, a \rightarrow b, b \rightarrow c, \ldots, x \rightarrow y, y \rightarrow z \Rightarrow z$$

is a theorem in propositional calculus. However, suppose that you wrote such a program and you had it write the truth table for

$$(a \wedge (a \rightarrow b) \wedge \cdots \wedge (y \rightarrow z)) \rightarrow z.$$

The truth table will have 2^{26} cases. At one thousand cases per second, it would take approximately 64,000 seconds (about 18 hours) to verify the theorem. Now if you decided to check a similar theorem,

$$p_1, p_1 \rightarrow p_2, \ldots, p_{99} \rightarrow p_{100} \Rightarrow p_{100},$$

you would really have time trouble. The fastest computer imaginable would surely break down before checking the 2^{100} cases that need to be checked. For most of the remainder of this section, we will discuss an alternate method for proving theorems in propositional calculus. It is the same method that we will use in a less formal way for proofs in other systems. Formal axiomatic methods would be too unwieldy to actually use in later sections. However, none of the theorems in later chapters would be stated if they couldn't be proven by the axiomatic method.

We will introduce two types of proof here, direct and indirect.

DIRECT PROOFS

A *direct proof* is a proof in which the truth of the premises of a theorem are shown to directly imply the truth of the theorem's conclusion.

Example 3.5.3 Theorem: $p \rightarrow r$, $q \rightarrow s$, $p \lor q \Rightarrow s \lor r$. The direct proof of this theorem is:

Step	Proposition	Justification
(1)	$p \lor q$	Premise
(2)	$\sim p \rightarrow q$	(1), conditional rule
(3)	$q \rightarrow s$	Premise
(4)	$\sim p \rightarrow s$	(2), (3), chain rule
(5)	$\sim s \rightarrow p$	(4), conditional rule
(6)	$p \rightarrow r$	Premise
(7)	$\sim s \rightarrow r$	(5), (6), chain rule
(8)	$s \lor r$	(7), conditional rule #

Note that # marks the end of a proof.

Rules for Formal Proofs. *Example 3.5.3 illustrates the usual method of formal proof in a formal mathematical system. The rules governing these proofs are:*

(1) A proof must end in a finite number of steps.
(2) Each step must be either a premise or a proposition that is implied from previous steps using any valid equivalence or implication.
(3) For a direct proof, the last step must be the conclusion of the theorem. For an indirect proof (see below), the last step must be a contradiction.

Justification Column.   The column labeled "justification" is analogous to the comments that appear in most good computer programs. They simply make the proof more readable.

Example 3.5.4.   Here are two direct proofs of $\sim p \vee q$, $s \vee p$, $\sim q \Rightarrow s$:

(1) $\sim p \vee q$ Premise
(2) $\sim q$ Premise
(3) $\sim p$ Disjunctive simplification, (1), (2)
(4) $s \vee p$ Premise
(5) s ·   Disjunctive simplification, (3), (4). #

You are invited to justify the steps in this second proof:

(1) $\sim p \vee q$
(2) $\sim q \rightarrow \sim p$
(3) $s \vee p$
(4) $p \vee s$
(5) $\sim p \rightarrow s$
(6) $\sim q \rightarrow s$
(7) $\sim q$
(8) s. #

CONDITIONAL CONCLUSIONS

The conclusion of a theorem is often a conditional proposition. The condition of the conclusion can be included as a premise in the proof of the theorem. The object of the proof is then to prove the consequence of the conclusion. This rule is justified by the logical law

$$p \rightarrow (h \rightarrow c) \Leftrightarrow (p \wedge h) \rightarrow c.$$

Example 3.5.5.   The following proof of $p \rightarrow (q \rightarrow s)$, $\sim r \vee p$, $q \Rightarrow r \rightarrow s$ includes r as a fourth premise. The truth of s concludes the proof.

(1) $\sim r \vee p$ Premise
(2) r Added premise
(3) p (1), (2), disjunction simplification
(4) $p \rightarrow (q \rightarrow s)$ Premise
(5) $q \rightarrow s$ (3), (4), detachment
(6) q Premise
(7) s (5), (6), detachment. #

INDIRECT PROOFS

Consider a theorem $P \Rightarrow C$, where P represents $P_1 \wedge P_2 \wedge \cdots P_n$, the premises. The method of *indirect proof* is based on the equivalence $P \rightarrow C \Leftrightarrow \sim(P \wedge \sim C)$.

In words, this logical law states that if $P \Rightarrow C$, then $P \wedge \sim C$ is always false; i.e., $P \wedge \sim C$ is a contradiction. This means that a valid method of proof is to negate the conclusion of a theorem and add this negation to the premises. If a contradiction can be implied from this set of propositions, the proof is complete. For the proofs in this section, a contradiction will often take the form $t \wedge \sim t$. For proofs involving numbers, a contradiction might be $1 = 0$ or $0 < 0$. Indirect proofs involving sets might conclude with $x \in \phi$ or $(x \in A \text{ and } x \in A')$. Indirect proofs are often more convenient than direct proofs.

Example 3.5.6   Here is an example of an indirect proof of the theorem in Example 3.5.3:

(1)	$\sim(s \vee r)$	Negated conclusion
(2)	$\sim s \wedge \sim r$	DeMorgan's Law, (1)
(3)	$\sim s$	Conjunctive Simplification, (2)
(4)	$q \rightarrow s$	Premise
(5)	$\sim q$	Contrapositive, (3), (4)
(6)	$\sim r$	Conjunctive Simplification, (2)
(7)	$p \rightarrow r$	Premise
(8)	$\sim p$	Contrapositive, (6), (7)
(9)	$(\sim p) \wedge (\sim q)$	Conjunctive, (5), (8)
(10)	$\sim(p \vee q)$	DeMorgan's Law, (9)
(11)	$p \vee q$	Premise
(12)	0	(10), (11) #

PROOF STYLE

The rules allow you to list the premises of a theorem immediately; however, a proof is much easier to follow if the premises are only listed when they are needed.

Example 3.5.7.   Here is an indirect proof of $a \rightarrow b$, $\sim(b \vee c) \Rightarrow \sim a$:

(1)	a	Negation of the conclusion
(2)	$a \rightarrow b$	Premise
(3)	b	(1), (2), detachment
(4)	$b \vee c$	(3), disjunctive addition
(5)	$\sim(b \vee c)$	Premise
(6)	0	(4), (5) #

As we mentioned at the outset of this section, we are only presenting an overview of what a mathematical system is. For greater detail on axiomatic theories, see Stoll (1961). An excellent description of how propositional calculus plays a part in artificial intelligence is contained in Hofstadter (1980). If you enjoy the challenge of constructing proofs in propositional calculus, you should enjoy the game WFF'N PROOF (1962), by L. E. Allen.

EXERCISES FOR SECTION 3.5

A Exercises

1. Prove with truth tables:

   (a) $p \lor q, ~q \Rightarrow p$.
   (b) $p \to q, ~q \Rightarrow ~p$.

2. Give direct and indirect proofs of:

   (a) $a \to b, c \to b, d \to (a \lor c), d \Rightarrow b$.
   (b) $(p \to q) \land (r \to s), (q \to t) \land (s \to u), ~(t \land u), p \to r \Rightarrow ~p$.
   (c) $p \to (q \to r), ~s \lor p, q \Rightarrow s \to r$.
   (d) $p \to q, q \to r, ~(p \land r), p \lor r \Rightarrow r$.
   (e) $((~q), (p \to q), (p \lor t)) \Rightarrow t$.

3. Are the following arguments valid? If valid, construct a formal proof; if not valid, explain why.

   (a) If wages increase, then there will be inflation. The cost of living will not increase if there is no inflation. Wages will increase. Therefore, the cost of living will increase.
   (b) If the races are fixed or the casinos are crooked, then the tourist trade will decline. If the tourist trade decreases, then the police will be happy. The police force is never happy. Therefore, the races are not fixed.

4. Describe how $p_1, p_1 \to p_2, p_2 \to p_3, \ldots, p_{99} \to p_{100} \Rightarrow p_{100}$ could be proven in 199 steps.

5. Rewrite all the statements in Exercises 3a and 3b using "necessary/sufficient" language.

6. Determine the validity of the following argument:
   For students to do well in a discrete mathematics course, it is necessary that they study hard. Students who do well in courses do not skip classes. Students who study hard do well in courses. Therefore students who do well in a discrete mathematics course do not skip class.

3.6   Propositions over a Universe

Example 3.6.1. Consider the sentence "He was a member of the Boston Red Sox." There is no way that we can assign a truth value to this sentence unless "he" is specified. For that reason, we would not consider it a proposition. However, "he" can be considered a variable that holds a place for any name. We might want to restrict the value of "he" to all names in the major-league baseball record books. If that is the case, we say that the sentence is a proposition over the set of major-league baseball players, past and present.

Definition: Proposition over a Universe. *Let U be a non-empty set. A proposition over U is a sentence that contains a variable that can take on any value in U and which has a definite truth value as a result of any such substitution.*

Example 3.6.2.

(a) A few propositions over the integers are $4x^2 - 3x = 0, 0 \leq n \leq 5, k$ is a multiple of 3.
(b) A few propositions over the rational numbers are $4x^2 - 3x = 0$, $y^2 = 2, (s - 1)(s + 1) = s^2 - 1$.
(c) A few propositions over the subsets of **P** are $(A = \phi) \vee (A = \mathbf{P})$, $3 \in A, A \cap \{1, 2, 3\} \neq \phi$.

All of the laws of logic that we listed in Section 3.4 are valid for propositions over a universe. For example, if p and q are propositions over the integers, we can be certain that $p \wedge q \Rightarrow p$, because $(p \wedge q) \rightarrow p$ is a tautology and is true no matter what values the variables in p and q are given. If we specify p and q to be $p(n) : n \leq 4$ and $q(n) : n \leq 8$, we can also say that p implies $p \wedge q$. This is not a usual implication, but for the propositions under discussion, it is true. One way of describing this situation in general is with truth sets.

TRUTH SET

Definition: Truth Set. *If p(n) is a proposition over U, the truth set of p(n) is $T_{p(n)} = \{a \in U \mid p(a) \text{ is true}\}$.*

Example 3.6.3. The truth set of the proposition $\{1, 2\} \cap A = \phi$ taken as a proposition over the power set of $\{1, 2, 3, 4\}$ is $\{\phi, \{3\}, \{4\}, \{3, 4\}\}$.

Example 3.6.4 In the universe **Z** (the integers), the truth set of $4x^2 - 3x = 0$ is $\{0\}$. If the universe is expanded to the rational numbers, the truth set becomes $\{0, 3/4\}$. The term *solution set* is often used for the truth set of an equation such as the one above.

Definition: Tautology and Contradiction. *A proposition over U is a tautology if its truth set is U. It is a contradiction if its truth set is empty.*

Example 3.6.5. $(s - 1)(s + 1) = s^2 - 1$ is a tautology over the rationals. $x^2 - 2 = 0$ is a contradiction over the rationals.

The truth sets of compound propositions can be expressed in terms of the truth sets of simple propositions. For example, if $a \in T_{p \wedge q}$, then a makes $p \wedge q$ true. Therefore, a makes both p and q true, which means that $a \in T_p \cap T_q$. This explains why the truth set of the conjunction of two propositions equals the intersection of the truth sets of the two propositions. The following list summarizes the connection between compound and simple truth sets:

$$T_{p \wedge q} = T_p \cap T_q$$
$$T_{p \vee q} = T_p \cup T_q$$
$$T_{\sim p} = T_p^c$$
$$T_{p \leftrightarrow q} = (T_p \cap T_q) \cup (T_p^c \cap T_q^c)$$
$$T_{p \to q} = T_p^c \cup T_q$$

Equivalence. Two propositions are equivalent if $p \leftrightarrow q$ is a tautology. In terms of truth sets, this means that p and q are equivalent if $T_p = T_q$.

Example 3.6.6.

(a) $n + 4 = 9$ and $n = 5$ are equivalent propositions over the integers.
(b) $A \cap \{4\} \neq \phi$ and $4 \in A$ are equivalent propositions over the power set of the natural numbers.

Implication. If p and q are propositions over U, p implies q if $p \to q$ is a tautology. Since the truth set of $p \to q = T_p^c \cup T_q$, the Venn diagram for $T_{p \to q}$ in Figure 3.6.1 shows that $p \Rightarrow q$ when $T_p \subseteq T_q$.

Example 3.6.7.

(a) Over the natural numbers: $n \leq 4 \Rightarrow n \leq 8$ since $\{0, 1, 2, 3, 4\} \subseteq \{0, 1, 2, 3, 4, 5, 6, 7, 8\}$.
(b) Over the power set of the integers: $\#(A^c) = 1$ implies $A \cap \{0, 1\} \neq \phi$.
(c) $A \subseteq$ even integers $\Rightarrow A \cap$ odd integers $= \phi$.

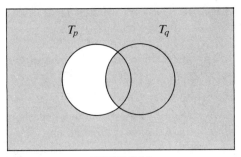

FIGURE 3.6.1
Venn diagram for $T_{p \to q}$

PASCAL NOTE

Propositions in Pascal are called *Boolean expressions*. A few examples are N = 0, K IN S and NOT (P OR Q). For these expressions, N would have to be numeric (REAL or INTEGER), S must be of set type, K must be of S's base type, and P and Q must be BOOLEAN. Boolean expressions are usually used as conditions in IF . . THEN . . ELSE statements and are used to control WHILE and REPEAT . . UNTIL loops. Boolean valued functions make it possible to construct very complex expressions yet refer to them with a single identifier. The following definition of PERFECTSQUARE is a Pascal version of the proposition over **P**: "n is a perfect square."

```
(POSITIVE = 0..MAXINT)
FUNCTION PERFECTSQUARE (N: POSITIVE) : BOOLEAN;
BEGIN
IF SQR(TRUNC(SQRT(N))=N THEN  PERFECTSQUARE:=TRUE
                        ELSE PERFECTSQUARE:=FALSE
END
```

EXERCISES FOR SECTION 3.6

A Exercises

1. If $U = \mathcal{P}\,\{1, 2, 3, 4\}$, what are the truth sets of the following propositions?
   (a) $A \cap \{2, 4\} = \phi$.
   (b) $3 \in A$ and $1 \notin A$.
   (c) $A \cup \{1\} = A$.
   (d) A is a proper subset of $\{2, 3, 4\}$.
   (e) $\#A^c = \#A$.

2. Over the universe of positive integers:
   $p(n) : n$ is prime and $n < 32$.
   $q(n) : n$ is a power of 3.
   $r(n) : n$ is a divisor of 27.
   (a) What are the truth sets of these propositions?
   (b) Which of the three propositions implies one of the others?

3. If $U = \{0, 1, 2\}$, how many propositions over U could you list without listing two that are equivalent?

4. Given the propositions over the natural numbers $p : n < 4$, $q : 2n > 17$, and $r : n$ is a divisor of 18, what are the truth sets of:
   (a) q
   (b) $p \wedge q$
   (c) r
   (d) $q \rightarrow r$

5. Suppose that s is a proposition over $\{1, \ldots, 8\}$. If $T_s = \{1, 2, 5, 7\}$, give two examples of propositions that are equivalent to s.

6. (a) Determine the truth sets of the following propositions over the positive integers:
   $p(n) : n$ is a perfect square and $n < 100$.
   $q(n) : n = \#\mathscr{P}(A)$ for some set A.
   (b) Determine $T_{p \wedge q}$ for p and q above.

7. Let the universe be $\mathbf{Z}$, the set of integers. Which of the following propositions are equivalent over $\mathbf{Z}$?
   (a) $0 < n^2 \le 4$.
   (b) $0 < n^3 \le 8$.
   (c) $0 < n \le 2$.

3.7 Mathematical Induction (omit)

In this section, we will examine mathematical induction, a technique for proving propositions over the positive integers. Mathematical (or finite) induction reduces the proof that all of the positive integers belong to a truth set to a finite number of steps.

Example 3.7.1. Consider the following proposition over the positive integers, which we will label $p(n)$: The sum of the positive integers from 1 to n is $n(n + 1)/2$. This is a well-known formula that is quite simple to verify for a given value of n. For example, $p(5)$ is: The sum of the positive integers from 1 to 5 is $5(5 + 1)/2$. Indeed, $1 + 2 + 3 + 4 + 5 = 15 = 5(6)/2$. Unfortunately, this doesn't serve as a proof that $p(n)$ is a tautology. All that we've established is that 5 is in the truth set of $p(n)$. Since the positive integers are infinite, we certainly can't use this approach to obtain a proof of the formula.

An Analogy: Mathematical induction is often useful in overcoming a problem such as this one. A proof by mathematical induction is similar to knocking over a row of closely spaced dominos that are standing on end. To knock over the five dominos in Figure 3.7.1, all you need to do is push Domino 1 to the right. To be assured that they all will be knocked over, some work must be done ahead of time. The dominos must be positioned so that if any domino is pushed to the right, it will push the next domino in the line.

Now imagine the propositions $p(1)$, $p(2)$, $p(3)$, . . . to be an infinite line of dominos. Let's see if these propositions are in the same formation as the dominos were. First, we will focus on one specific point of the line: $p(99)$ and $p(100)$. We are not going to prove that either of these propositions is true, just that the truth of $p(99)$ implies the truth of $p(100)$. In terms of our analogy, if $p(99)$ is knocked over, it will knock over $p(100)$.

In proving $p(99) \Rightarrow p(100)$, we will use $p(99)$ as our premise. We must prove: The sum of the positive integers from 1 to 100 is $100(100 + 1)/2$. We start by observing that the sum of the positive integers from 1 to 100 is $(1 + 2 + \cdots + 99) + 100$. That is, the sum of the positive integers from 1 to 100 equals the sum of the sum of the first ninety-nine and 100. We can now apply our premise, $p(99)$, to the sum $1 + 2 + \cdots + 99$. After rearranging our numbers, we obtain the desired expression for $1 + 2 + \cdots + 100$:

$$1 + 2 + \cdots + 99 + 100 = 99(99 + 1)/2 + 100$$
$$= 99(100)/2 + 2(100)/2$$
$$= 100(101)/2$$
$$= 100(100 + 1)/2.$$

What we've just done is analogous to checking two dominos in a line and finding that they are properly positioned. Since we are dealing with an infinite line, we must check all pairs at once. This is accomplished by proving that $p(n) \Rightarrow p(n + 1)$ for all $n \geq 1$:

FIGURE 3.7.1
Illustration of example 3.7.1

$$1 + 2 + \cdots + n + (n + 1) = (1 + 2 + \cdots + n) + (n + 1)$$
$$= n(n + 1)/2 + (n + 1) \; (p(n) \text{ used here})$$
$$= n(n + 1)/2 + 2(n + 1)/2$$
$$= (n + 1)(n + 2)/2$$
$$= ((n + 1)(n + 1) + 1)/2.$$

They are all lined up! Now look at $p(1)$: The sum of the positive integers from 1 to 1 is $1(1 + 1)/2$. Clearly, $p(1)$ is true. This sets off a chain reaction. Since $p(1) \Rightarrow p(2)$, $p(2)$ is true. Since $p(2) \Rightarrow p(3)$, $p(3)$ is true; and so on. #

THE PRINCIPLE OF MATHEMATICAL INDUCTION

The Principle of Mathematical Induction.   *Let p(n) be a proposition over the positive integers, then p(n) is a tautology if*

(a) p(1) is true, and
(b) n ≥ 1 and p(n) ⇒ p(n + 1).

Note: The truth of $p(1)$ is called the *basis for the induction proof.* The premise that $p(n)$ is true in Statement b is called the *induction hypothesis.* The proof that $p(n)$ implies $p(n + 1)$ is called the *induction step of the proof.* Despite our analogy, the basis is usually done first in an induction proof. The order doesn't really matter.

Example 3.7.2.   Consider the implication over the positive integers $p(n)$:

$$q_0 \to q_1, \; q_1 \to q_2, \; \ldots, \; q_{n-1} \to q_n, \; q_0 \Rightarrow q_n.$$

A proof that $p(n)$ is a tautology follows.

Basis: $p(1)$ is $q_0 \to q_1$, $q_0 \Rightarrow q_1$. This is the logical rule of detachment which we know is true. If you haven't done so yet, write out the truth table of $((q_0 \to q_1) \land q_0) \to q_1$ to verify this step.

Induction: Assume that $n \geq 1$ and $p(n)$ is true. We want to prove that $p(n + 1)$ must be true. That is:

$$q_0 \to q_1, \; \ldots, \; q_{n-1} \to q_n, \; q_n \to q_{n+1}, \; q_0 \Rightarrow q_{n+1}$$

Here is a direct proof of $p(n + 1)$:

Steps	Proposition(s)	Justification
$(1) - (n + 1)$	$q_0 \rightarrow q_1, \ldots, q_{n-1} \rightarrow q_n, q_0$	Premises
$(n + 2)$	q_n	$(1) - (n + 1), p(n)$
$(n + 3)$	$q_n \rightarrow q_{n+1}$	Premise
$(n + 4)$	q_{n+1}	$(n + 2), (n + 3),$ detachment #

Example 3.7.3. For all $n \geq 1$, $n^3 + 2n$ is a multiple of 3. An inductive proof follows:

Basis: $1^3 + 2(1) = 3$ is a multiple of 3.

Induction: Assume that $n \geq 1$ and $n^3 + 2n$ is a multiple of 3. Consider $(n + 1)^3 + 2(n + 1)$. Is it a multiple of 3?

$$(n + 1)^3 + 2(n + 1) = (n^3 + 3n^2 + 3n + 1) + (2n + 2)$$
$$= (n^3 + 2n) + 3n^2 + 3n + 3$$
$$= (n^3 + 2n) + 3(n^2 + n + 1).$$

Yes, $(n + 1)^3 + 2(n + 1)$ is the sum of two multiples of 3; therefore, it is also a multiple of 3. #

VARIATIONS

Now we will discuss some of the variations of the principle of mathematical induction. The first simply allows for universes that are similar to **P**, like $\{-2, -1, 0, 1, \ldots\}$ or $\{5, 6, 7, 8, \ldots\}$.

Principle of Mathematical Induction (Generalized). *If $p(n)$ is a proposition over $\{k_0, k_0 + 1, k_0 + 2, \ldots\}$, where k_0 is any integer, then $p(n)$ is a tautology if*

   (1) $p(k_0)$ is true, and
   (2) $k \geq k_0$ and $p(k) \Rightarrow p(k + 1)$.

Example 3.7.4.   In Chapter 2, we stated that the number of different permutations of k elements taken from an n element set, $P(n, k)$, can be computed with the formula $n!/(n - k)!$. We can prove this statement by induction   on n.   For $n \geq 0$,   let $q(n)$ be   the   proposition "$P(n, k) = n!/(n - k)!$ for $k = 0, \ldots, n$."

   Basis: $q(0)$ states that $P(0, 0) =$ the number of ways that no elements can be selected from $\emptyset$ and arranged in order $= 0!/0! = 1$, which is true. A general law in combinatorics is that there is exactly one way of doing nothing.

Induction: Assume that $q(n)$ is true for some natural number n. It is left for us to prove that this assumption implies that $q(n + 1)$ is true. Suppose that we have a set of cardinality $n + 1$ and want to select and arrange k of its elements. There are two cases to consider, the first of which is easy. If $k = 0$, then there is one way of selecting zero elements from the set; hence $P(n + 1, 0) = 1 = (n + 1)!/(n + 1 - 0)!$ and the formula works in this case.

The more challenging case is when k is positive. Here we count the value of $P(n + 1, k)$ by counting the number of ways that the first element in the arrangement can be filled and then counting the number of ways that the remaining $k - 1$ elements can be filled in using the induction hypothesis. There are $n + 1$ possible choices for the first element. Since that leaves n elements to fill in the remaining $k - 1$ positions, there are $P(n, k - 1)$ ways of completing the arrangement. By the rule of products,

$$P(n + 1, k) = (n + 1)P(n, k - 1)$$
$$= \frac{(n + 1)n!}{(n - (k - 1))!}$$
$$= \frac{(n + 1)!}{((n + 1) - k)!} \cdot \text{\#}$$

A second variation allows for the expansion of the induction hypothesis. The course-of-values principle includes the previous generalization.

The Course-of-Values Principle. *If $p(n)$ is a proposition over $\{k_0, k_0 + 1, k_0 + 2, \ldots\}$, then $p(n)$ is a tautology if*

(1) $p(k_0)$ is true, and
(2) $k \geq k_0, p(k_0), p(k_0 + 1), \ldots, p(k) \Rightarrow p(k + 1)$.

Example 3.7.5. A prime number is defined as a positive integer that has exactly two positive divisors, 1 and itself. There are an infinite number of primes. The list of primes starts with 2, 3, 5, 7, 11, ... The proposition over 2, 3, 4, . . . that we will prove here is $p(n) : n$ can be written as the product of one or more primes. In most texts, the assertion that $p(n)$ is a tautology would appear as:

Theorem. *Every positive integer greater than or equal to 2 has a prime decomposition.*

If you were to encounter this theorem outside the context of a discussion of mathematical induction, it might not be obvious that the proof can be done by induction. Recognizing when an induction proof is appropriate is mostly a matter of experience. Now on to the proof!

Basis: Since 2 is a prime, it is already decomposed into primes (one of them).

Induction: Suppose that for some $k \geq 2$ all of the integers $2, 3, \ldots, k$ have a prime decomposition. (Note the course-of-value hypothesis.) Consider $k + 1$. Either $k + 1$ is prime or it isn't. If $k + 1$ is prime, it is already decomposed into primes. If not, then $k + 1$ has a divisor, d, other than 1 and $k + 1$. Hence, $k + 1 = cd$ where both c and d are between 2 and k. By the induction hypothesis, c and d have prime decompositions, $c_1 c_2 \cdots c_n$ and $d_1 d_2 \cdots d_m$, respectively. Therefore, $k + 1$ has the prime decomposition $c_1 c_2 \cdots c_n d_1 d_2 \cdots d_m$. #

HISTORICAL NOTE

Mathematical induction originated in the late nineteenth century. Two mathematicians who were prominent in its development were R. Dedekind and G. Peano. Dedekind developed a set of axioms that describe the positive integers. Peano refined these axioms and gave a logical interpretation to them. The axioms are usually called the Peano Postulates.

Peano's Postulates.   The system of positive integers consists of a nonempty set, **P**; a least element of **P**, denoted 1; and a "successor function," s, with the properties

(1) If $k \in \mathbf{P}$, then there is an element of **P** called the *successor of k*, denoted $s(k)$.
(2) No two elements of **P** have the same successor.
(3) No element of **P** has 1 as its successor.
(4) If $S \subseteq \mathbf{P}$, $1 \in S$, and $k \in S \Rightarrow s(k) \in S$, then $S = \mathbf{P}$.

Notes:

(a) You might recognize $s(k)$ as simply being $k + 1$. The successor function is exactly the same as the standard Pascal function SUCC as applied to INTEGER type, with the exception that SUCC(MAXINT) is undefined.
(b) Axiom 4, mentioned above, is the one that makes mathematical induction possible. In an induction proof, we simply apply that axiom to the truth set of a proposition.

EXERCISES FOR SECTION 3.7

A Exercises

1. Prove that the sum of the first n odd integers equals n^2.

2. Prove   that   if $n \geq 1$,   then $1(1!) + 2(2!) + \cdots + n(n!) = (n + 1)! - 1$.

3. Prove that if $n \geq 2$, then the generalized DeMorgan's Law is true:
$\sim(p_1 \wedge p_2 \wedge \cdots \wedge p_n) \Leftrightarrow (\sim p_1) \vee (\sim p_2) \vee \cdots \vee (\sim p_n)$.

4. The number of strings of n zeros and ones that contain an even number of ones is 2^{n-1}. Prove this fact by induction for $n \geq 1$.

5. Prove that for $n \geq 1$: $\sum_{k=1}^{n} k^2 = n(n + 1)(2n + 1)/6$.

6. Suppose that there are n people in a room, $n \geq 1$, and that they all shake hands with one another. Prove that $(n - 1)n/2$ handshakes will have occurred.

7. Prove that for $n \geq 1$: $1 + 2 + 4 + \cdots + 2^n = 2^{n+1} - 1$.

8. Use mathematical induction to show that $\dfrac{1}{1 \cdot 2} + \dfrac{1}{2 \cdot 3} + \dfrac{1}{3 \cdot 4} +$
$\cdots + \dfrac{1}{n(n + 1)} = \dfrac{n}{n + 1}$ for all $n \geq 1$.

9. Let $p(n)$ be "$8^n - 3^n$ is a multiple of 5." Prove that $p(n)$ is a tautology over **N**.

B Exercise

10. A recursive definition is similar to an inductive proof. It consists of a basis, usually the simple part of the definition, and the recursion, which defines complex objects in terms of simpler ones. For example, if x is a real number and n is a positive integer, we can define x^n as follows:
Basis: $x^1 = x$.
Recursion: If $n > 1$, $x^n = x^{n-1}x$.
Therefore, $x^3 = (x^2)x = (x^1 x)x = ((x)x)x$.

Proofs involving objects that are defined recursively are often inductive. Prove that if n, m, ε **P**, $x^{n+m} = x^n x^m$. Hint: $p(m)$ is the proposition that you should prove is a tautology. More on recursion is presented in Chapter 8.

C Exercise

11. Let S be a finite set and let P_n be defined recursively by $P_1 = S$ and $P_n = S \times P_{n-1}$, $n > 1$.
    (a) List the elements of P_3 for the case $S = \{a, b\}$.
    (b) Determine the formula for $\#P_n$, given that $\#S = k$ and prove your formula by induction.

3.8 Quantifiers

As we saw in Section 3.6, if $p(n)$ is a proposition over a universe U, its truth set $T_{p(n)}$ is equal to a subset of U. In many cases, such as when $p(n)$ is an equation, we are most concerned with whether $T_{p(n)}$ is empty or not. In other cases, we might be interested in whether $T_{p(n)} = U$; i.e., $p(n)$ is a tautology. Since the conditions $T_{p(n)} \neq \emptyset$ and $T_{p(n)} = U$ are so often an issue, we have a special system of notation for them.

THE EXISTENTIAL QUANTIFIER

If $p(n)$ is a proposition over U with $T_{p(n)} \neq \emptyset$, we commonly say "There exists an n in U such that $p(n)$ (is true)." We abbreviate this sentence with the symbols $(\exists\ n)_U(p(n))$. $\exists$ is called the *existential quantifier*.

Example 3.8.1.

(a) $(\exists\ k)_Z(k^2 - k - 12 = 0)$ is another way of saying that there is an integer that solves the equation $k^2 - k - 12 = 0$. The fact that two such integers exist doesn't affect the truth of this proposition in any way.

(b) $(\exists\ k)_Z(3k = 102)$ simply states that 102 is a multiple of 3, which is true. On the other hand, $(\exists\ k)_Z(3k = 100)$ states that 100 is a multiple of 3, which is false.

(c) $(\exists\ x)_R(x^2 + 1 = 0)$ is false since the solution set of the equation $x^2 + 1 = 0$ is empty. It is common to write $(\nexists\ x)_R(x^2 + 1 = 0)$ in this case.

There are a wide variety of ways that you can write a proposition with an existential quantifier. Table 3.8.1 contains a list of different formats that could be used for both the existential and universal quantifiers.

THE UNIVERSAL QUANTIFIER

If $p(n)$ is a proposition over U with $T_{p(n)} = U$, we commonly say "For all n in U, $p(n)$ (is true)." We abbreviate this proposition with the symbols $(\forall n)_U(p(n))$. $\forall$ is termed the *universal quantifier*.

TABLE 3.8.1
Formats for Existential and Universal Quantifiers

Universal Quantifier	Existential Quantifier
$(\forall n)_U(p(n))$	$(\exists n)_U(p(n))$
$(\forall n \in U)(p(n))$	$(\exists n \in U)(p(n))$
$\forall n \in U, p(n)$	$\exists n \in U$ such that $p(n)$
$p(n), \forall n \in U$	$p(n)$ is true for some $n \in U$
$p(n)$ is true for all $n \in U$	

Example 3.8.2.

(a) We can say that the square of every real number is non-negative symbolically with $(\forall x)_R(x^2 \geq 0)$.
(b) $(\forall n)_Z(n + 0 = 0 + n = n)$ says that the sum of zero and any integer n is n. This fact is called the *identity property of zero for addition*.

THE NEGATION OF QUANTIFIED PROPOSITIONS

When you negate a quantified proposition, the existential and universal quantifiers complement one another.

Example 3.8.3. Over the universe of animals, define $F(x) : x$ is a fish and $W(x)$: lives in the water. We know that the proposition $W(x) \rightarrow F(x)$ is not always true. In other words, $(\forall x)(W(x) \rightarrow F(x))$ is false. Another way of stating this fact is that there exists an animal that lives in the water and is not a fish; i.e., $\sim(\forall x)(W(x) \rightarrow F(x)) \Leftrightarrow (\exists x)(\sim(W(x) \rightarrow F(x))) \Leftrightarrow (\exists x)(W(x) \wedge \sim F(x))$.

Note that the negation of a universally quantified proposition is an existentially quantified proposition. In addition, when you negate an existentially quantified proposition, you obtain a universally quantified proposition. Symbolically,

$$\sim((\forall n)_U(p(n))) \Leftrightarrow (\exists n)_U(\sim p(n)), \text{ and}$$
$$\sim((\exists n)_U(p(n))) \Leftrightarrow (\forall n)_U(\sim p(n)).$$

Example 3.8.4.

(a) The ancient Greeks first discovered that $\sqrt{2}$ is an irrational number; that is, $\sqrt{2}$ is not a rational number. $\sim((\exists r)_Q(r^2 = 2))$ and $(\forall r)_Q(r^2 \neq 2)$ both state this fact symbolically.
(b) $\sim((\forall n)_p(n^2 - n + 41 \text{ is prime}))$ is equivalent to $(\exists n)_p(n^2 - n + 41$ is composite). They are both either true or false.

MULTIPLE QUANTIFIERS

If a proposition has more than one variable, then you can quantify it more than once. For example, $p(x, y) : x^2 - y^2 = (x + y)(x - y)$ is a tautology over the set of all pairs of real numbers because it is true for each pair (x, y) in $\mathbf{R} \times \mathbf{R}$. Another way to look at this proposition is as a proposition with two variables. The assertion that $p(x, y)$ is a tautology could be quantified as $(\forall x)_R((\forall y)_R(p(x, y)))$ or $(\forall y)_R((\forall x)_R(p(x, y)))$.

In general, multiple universal quantifiers can be arranged in any order without logically changing the meaning of the resulting proposition. The same is true for multiple existential quantifiers. For example, $p(x, y): x + y = 4$ and $x - y = 2$ is a proposition over $\mathbf{R}^2$. $(\exists x)_{\mathbf{R}}((\exists y)_{\mathbf{R}}(x + y = 4$ and $x - y = 2))$ and $(\exists y)_{\mathbf{R}}((\exists x)_{\mathbf{R}}(x + y = 4$ and $x - y = 2))$ are equivalent. A proposition with multiple existential quantifiers such as this one says that there are simultaneous values for the quantified variables that make the proposition true. A similar example is $q(x, y): 2x - y = 2$ and $4x - 2y = 5$, which is always false; that is,

$$\sim((\exists x)_{\mathbf{R}}((\exists y)_{\mathbf{R}}(q(x, y)))) \Leftrightarrow \sim((\exists y)_{\mathbf{R}}((\exists x)_{\mathbf{R}}(q(x, y))))$$
$$\Leftrightarrow (\forall y)_{\mathbf{R}}(\sim((\exists x)_{\mathbf{R}}(q(x, y))))$$
$$\Leftrightarrow ((\forall y)_{\mathbf{R}}((\forall x)_{\mathbf{R}}(\sim q(x, y))))$$
$$\Leftrightarrow ((\forall x)_{\mathbf{R}}((\forall y)_{\mathbf{R}}(\sim q(x, y)))).$$

When existential and universal quantifiers are mixed, the order cannot be exchanged without possibly changing the meaning of the proposition. For example, let $\mathbf{R}^+$ be the positive real numbers. $x: (\forall a)_{\mathbf{R}^+} ((\exists b)_{\mathbf{R}^+} (ab = 1))$ and $y: (\exists b)_{\mathbf{R}^+} ((\forall a)_{\mathbf{R}^+} (ab = 1))$ have different meanings; x is true, while y is false.

TIPS ON READING MULTIPLY QUANTIFIED PROPOSITIONS

It is understandable that you would find propositions such as x difficult to read. The trick to deciphering these expressions is to "peel" one quantifier off the proposition just as you would peel off the layers of an onion (but quantifiers shouldn't make you cry). Since the outermost quantifier in x is universal, x says that $z(a): (\exists b)_{\mathbf{R}^+} (ab = 1)$ is true for each value that a can take on. Now take the time to select a value for a, like 6. For the value that we selected, we get $z(6): (\exists b)_{\mathbf{R}^+} (6b = 1)$, which is obviously true since $6b = 1$ has a solution in the positive real numbers. We will get that same truth value, true, no matter which positive real number we choose for a; therefore, $z(a)$ is a tautology over $\mathbf{R}^+$ and we are justified in saying that x is true. The key to understanding propositions like x on your own is to experiment with actual values for the outermost variables as we did above.

Now consider y. To see that y is false, we peel off the outer quantifier. Since it is an existential quantifier, all that y says is that some positive real number makes $w(b): (\forall a)_{\mathbf{R}^+} (ab = 1)$ true. Choose a few values of b to see if you can find one that makes $w(b)$ true. For example, if we pick $b = 2$, we get $(\forall a)_{\mathbf{R}^+} (2a = 1)$, which is false, since $2a$ is almost always different from 1. You should be able to convince yourself that no value of b will make $w(b)$ true; therefore, y is false.

Another way of convincing yourself that y is false is to convince yourself that $\sim y$ is true:

$$\sim y \Leftrightarrow (\forall b)_{\mathbf{R}^+} \sim ((\forall a)_{\mathbf{R}^+} (ab = 1))$$
$$\Leftrightarrow (\forall b)_{\mathbf{R}^+} ((\exists a)_{\mathbf{R}^+} (ab \neq 1)).$$

For each value of b, a value for a that makes $ab \neq 1$ is $1 + 1/b$; therefore, $\sim y$ is true.

EXERCISES FOR SECTION 3.8

A Exercises

1. Over the universe of books, define the propositions $B(x)$: x has a blue cover, $M(x)$: x is a mathematics book, $U(x)$: x is published in the United States, and $R(x, y)$: The bibliography of x includes y. Translate into words:

(a) $(\exists x)(\sim B(x))$.

(c) $(\exists x)(M(x) \wedge \sim B(x))$.

(b) $(\forall x)(M(x) \wedge U(x) \rightarrow B(x))$.

(d) $(\exists y)((\forall x)(M(x) \rightarrow R(x, y)))$.

Express using quantifiers:

(e) Every book with a blue cover is a mathematics book.

(f) There are mathematics books that are published outside of the United States.

(g) Not all books have bibliographies.

2. Use quantifiers to say that $\sqrt{3}$ is not a rational number.

3. Over the universe of real numbers, use quantifiers to say that the equation $a + x = b$ has a solution for all values of a and b. Hint: You will need three quantifiers.

4. What do the following propositions say, where U is the power set of $\{1, 2, \ldots, 9\}$? Which of the propositions are true?

(a) $(\forall A)_U(\#A \neq (A^c))$.

(b) $(\exists A)_U(\exists B)_U(\#A = 5, \#B = 5 \text{ and } A \cap B = \emptyset)$.

(c) $(\forall A)_U(\forall B)_U(A - B = B^c - A^c)$.

5. Let $M(x)$ be "x is a mammal," let $A(x)$ be "x is an animal," and let $W(x)$ be "x is warm-blooded."

(a) Translate into a formula: Every mammal is warm-blooded.

(b) Translate into English: $(\exists x)(A(x) \wedge (\sim M(x)))$.

6. Translate into your own words and indicate whether it is true or false that $(\exists u)_Z(4u^2 - 9 = 0)$.

7. Use universal quantifiers to state that the sum of any two rational numbers is rational.

8. Let $C(x)$ be "x is cold-blooded," let $F(x)$ be "x is a fish," and let $S(x)$ be "x lives in the sea."

(a) Translate into a formula: Every fish is cold-blooded.

(b) Translate into English: $(\exists x)(S(x) \wedge \sim F(x))$ and $(\forall x)(F(x) \rightarrow S(x))$.

chapter

4

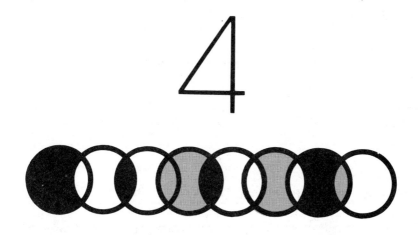

MORE ON SETS

4

GOALS

In this chapter we shall look more closely at some basic facts about sets. One question we could ask ourselves is: Can we manipulate sets similarly to the way we manipulated expressions in basic algebra, or to the way we manipulated propositions in logic? In basic algebra we are aware that $a \cdot (b + c) = a \cdot b + a \cdot c$ for all real numbers a, b, and c. In logic we verified an analogue of this statement, namely, $(p \wedge (q \vee r)) \Leftrightarrow ((p \wedge q) \vee (p \wedge r))$, where p, q, and r were arbitrary propositions. If A, B, and C are arbitrary sets, is $A \cap (B \cup C) = (A \cap B) \cup (A \cap C)$? How do we convince ourselves of its truth or falsity? Let us consider some approaches to this problem, look at their pros and cons, and determine their validity. Many of the ideas expressed are true, in general, in mathematics. Partitions of sets and minsets will be introduced.

4.1 Methods of Proof for Sets

There are a variety of ways that we could attempt to prove that the distributive law for intersection over union is true; i.e., that for any three sets A, B, and C, $A \cap (B \cup C) = (A \cap B) \cup (A \cap C)$. We start with a common "non-proof" and then work towards more acceptable methods.

EXAMPLES AND COUNTEREXAMPLES

We could, for example, let $A = \{1, 2\}$, $B = \{5, 8, 10\}$, and $C = \{3, 2, 5\}$ and determine whether the distributive law is true. Obviously, in doing this we will have only determined that the distributive law is true for this *one* example. It does *not* prove the distributive law for all possible sets A, B, and C and hence is an *invalid* method of proof. However, trying a few examples has considerable merit insofar as it makes us more comfortable with the statement in question, and indeed if the statement is *not* true for the example, we have disproved the statement.

Definition: Counterexample. *An example which disproves a statement is called a counterexample.*

Example 4.1.1. From basic algebra we learned that multiplication is distributive over addition. Is addition distributive over multiplication; i.e., is $a + (b \cdot c) = (a + b) \cdot (a + c)$? If we choose the values $a = 3$, $b = 4$, and $c = 1$, we find that $3 + (4 \cdot 1) \neq (3 + 4)(3 + 1)$. Therefore, this set of values serves as a counterexample to a distributive law of addition over multiplication.

PROOF USING VENN DIAGRAMS

In this method, we illustrate both sides of the statement via a Venn diagram and determine whether both Venn diagrams give us the same "picture." For example, the left side of the distributive law is developed in Figure 4.1.1 and the right side in Figure 4.1.2. Note that the final results give you the same shaded area.

The advantage of this method is that it is relatively quick and mechanical. The disadvantage is that it is workable only if there are a small number of sets under consideration. Also, the method *illustrates* rather than proves facts. Many mathematicians do not consider it a valid method of proof.

PROOF USING SET-MEMBERSHIP TABLES

Let A be a subset of a universal set U and let $u \in U$. To use this method we note that exactly one of the following is true: $u \in A$ or $u \notin A$. Denote the situation where $u \in A$ by 1 and that where $u \notin A$ by 0. Working with two sets, A and B, and if $u \in U$, there are four possible outcomes of "where u can

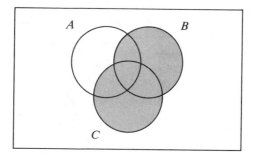

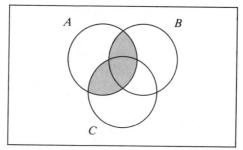

FIGURE 4.1.1
Left side of distributive law developed

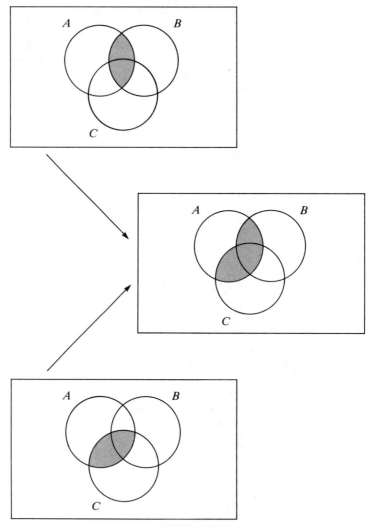

FIGURE 4.1.2
Right side of distributive law developed

be." What are they? The set-membership table for $A \cup B$ is :

A	B	$A \cup B$
0	0	0
0	1	1
1	0	1
1	1	1

This table illustrates that $u \in A \cup B$ if and only if $u \in A$ or $u \in B$.

In order to prove the distributive law via a set-membership table, write out the table for each side of the set statement to be proved and note that if S and T are two columns in a table, then the set statement S is equal to the set statement T if and only if corresponding entries in each column are the same.

To prove $A \cap (B \cup C) = (A \cap B) \cup (A \cap C)$, first note that the statement involves three sets, A, B, and C. So there are $2^3 = 8$ possibilities for the membership of an element in the sets.

A	B	C	$B \cup C$	$A \cap B$	$A \cap C$	$A \cap (B \cup C)$	$(A \cap B) \cup (A \cap C)$
0	0	0	0	0	0	0	0
0	0	1	1	0	0	0	0
0	1	0	1	0	0	0	0
0	1	1	1	0	0	0	0
1	0	0	0	0	0	0	0
1	0	1	1	0	1	1	1
1	1	0	1	1	0	1	1
1	1	1	1	1	1	1	1

Since each entry in Column 7 is the same as the corresponding entry in Column 8, we have shown that $A \cap (B \cup C) = (A \cap B) \cup (A \cap C)$ for any sets A, B, and C. The main advantage of this method is that it is mechanical. The main disadvantage is that it is reasonable to use only for a relatively small number of sets. If we are trying to prove a statement involving five sets, there are $2^5 = 32$ rows! Also, this method tends to distract a person from learning and understanding the core concepts of the subject matter, namely, definitions and theorems.

PROOF USING DEFINITIONS

This method involves using definitions and basic concepts to prove the given statement. This procedure forces one to learn, relearn, and understand basic definitions and concepts. It helps individuals to focus their attention on the main ideas of each topic and therefore is the most useful method of proof. One does not learn a topic by memorizing or occasionally glancing at core topics, but by using them in a variety of contexts. The word *proof* panics most people; however, everyone can become comfortable with proofs. Do not expect to prove every statement immediately. In fact, it is not our purpose to prove every theorem or fact encountered, only those which illustrate methods and/or basic concepts. Throughout the text we will focus in on main techniques of proofs. Let's illustrate by proving the distributive law.

Proof Technique 1. State or restate the theorem so you understand what is given (the hypothesis) and what you are trying to prove (the conclusion).

Theorem 4.1.1. *If A, B, and C are sets, then A $\cap$ (B $\cup$ C) = (A $\cap$ B) $\cup$ (A $\cap$ C).*

Assume: A, B, and C are sets.

Prove: $A \cap (B \cup C) = (A \cap B) \cup (A \cap C)$.

Commentary: What am I trying to prove? What types of objects am I working with: sets? real numbers? propositions? The answer is sets: sets of elements that can be anything you care to imagine. The universe from which we draw our elements plays no part in the proof of this theorem.

We need to show that the two sets are equal. Let's call them the left-hand set ($L.H.S.$) and the right-hand set ($R.H.S.$) To prove that $L.H.S. = R.H.S.$, we must prove two things: (a) $L.H.S. \subseteq R.H.S.$ and (b) $R.H.S. \subseteq L.H.S.$

To prove Part a (and, similarly, Part b), we must show that each element of $L.H.S.$ is an element of $R.H.S.$ Once we have diagnosed the problem we are ready to begin.

Proof of Theorem 4.1.1: We must prove:

(a) $A \cap (B \cup C) \subseteq (A \cap B) \cup (A \cap C)$.

Let $x \in A \cap (B \cup C)$ to show $x \in (A \cap B) \cup (A \cap C)$
$x \in A \cap (B \cup C)$.

Definition of $\cap$, $\cup$	$\Rightarrow x \in A$ and ($x \in B$ or $x \in C$).
Distributive law of logic	$\Rightarrow (x \in A$ and $x \in B)$ or $(x \in A$ and $x \in C)$.
Definition of $\cap$	$\Rightarrow x \in (A \cap B)$ or $x \in (A \cap C)$.
Definition of $\cup$	$\Rightarrow x \in (A \cap B) \cup (A \cap C)$.

and (b) $(A \cap B) \cup (A \cap C) \subseteq A \cap (B \cup C)$.

Let $x \in (A \cap B) \cup (A \cap C)$ to show $x \in A \cap (B \cup C)$
$x \in (A \cap B) \cup (A \cap C)$.

Why?	$\Rightarrow$	$(x \in A \cap B)$ or $(x \in A \cap C)$.
Why?	$\Rightarrow$	$(x \in A$ and $x \in B)$ or $(x \in A$ and $x \in C)$.
Why?	$\Rightarrow$	$x \in A$ and $(x \in B$ or $x \in C)$.
Why?	$\Rightarrow$	$x \in A$ and $(x \in B \cup C)$.
Why?	$\Rightarrow$	$x \in A \cap (B \cup C)$. #

Proof Technique 2.

(1) To prove that $A \subseteq B$, we must show that if $x \in A$, then $x \in B$.
(2) To prove that $A = B$, we must show:
    (a) $A \subseteq B$, and
    (b) $B \subseteq A$.

To further illustrate the Proof-by-Definition technique, let's prove the following:

Theorem 4.1.2. *Let A, B, and C be sets, then* $A \times (B \cap C) = (A \times B) \cap (A \times C)$.

Commentary: We again ask ourselves: What are we trying to prove? What types of objects are we dealing with? We realize that we wish to prove two facts: (a) *L.H.S.* $\subseteq$ *R.H.S.*, and (b) *R.H.S.* $\subseteq$ *L.H.S.*

To prove Part a (and, similarly, Part b), we'll begin the same way. Let ____ $\in$ *L.H.S.* to show ____ $\in$ *R.H.S.* What should ____ be? What does a typical object in *L.H.S.* look like?

Proof of Theorem 4.1.2: We must prove:

(a) $A \times (B \cap C) \subseteq (A \times B) \cap (A \times C)$.
Let $(x, y) \in A \times (B \cap C)$ to prove $(x, y) \in (A \times B) \cap (A \times C)$.

$(x, y) \in A \times (B \cap C)$.

Why?	$\Rightarrow$	$x \in A$ and $y \in (B \cap C)$.
Why?	$\Rightarrow$	$x \in A$ and $(y \in B$ and $y \in C)$.
Why?	$\Rightarrow$	$(x \in A$ and $y \in B)$ and $(x \in A$ and $y \in C)$.
Why?	$\Rightarrow$	$(x, y) \in (A \times B)$ and $(x, y) \in (A \times C)$.
Why?	$\Rightarrow$	$(x, y) \in (A \times B) \cap (A \times C)$.

and

(b) $(A \times B) \cap (A \times C) \subseteq A \times (B \cap C)$.
Let $(x, y) \in (A \times B) \cap (A \times C)$ to prove $(x, y) \in A \cap (B \times C)$.

$(x, y) \in (A \times B) \cap (A \times C)$.

Why?	$\Rightarrow$	$(x, y) \in (A \times B)$ and $(x, y) \in (A \times C)$.
Why?	$\Rightarrow$	$(x \in A$ and $y \in B)$ and $(x \in A$ and $y \in C)$.
Why?	$\Rightarrow$	$x \in A$ and $(y \in B$ and $y \in C)$.
Why?	$\Rightarrow$	$x \in A$ and $y \in (B \cap C)$.
Why?	$\Rightarrow$	$(x, y) \in A \times (B \cap C)$. #

EXERCISES FOR SECTION 4.1

A Exercises

1. Supply all answers to the question "Why?" in the proofs of Theorems 4.1.1 and 4.1.2.

2. Prove the following:
   (a) Let A, B, and C be sets. If $A \subseteq B$ and $B \subseteq C$, then $A \subseteq C$.
   (b) Let A and B be sets. Then $A - B = A \cap B^c$.
   (c) Let A, B, and C be sets. $(A \subseteq B$ and $A \subseteq C)$ if and only if $A \subseteq B \cap C$.
   (d) Let A and B be sets. $A \subseteq B$ if and only if $B^c \subseteq A^c$.
   (e) Let A, B, and C be sets. If $A \subseteq B$, then $A \times C \subseteq B \times C$.

3. Disprove the following:
   - (a) $A - B = B - A$.
   - (b) $A \times B = B \times A$.
   - (c) $A \cap B = A \cap C$ implies $B = C$.

4. Let A and B be sets. Write the following in "if . . . then . . ." language:
   - (a) $x \in B$ is a sufficient condition for $x \in A \cup B$.
   - (b) $A \cap B \cap C = \emptyset$ is a necessary condition for $A \cap B = \emptyset$.
   - (c) $A \cup B = B$ is a necessary and sufficient condition for $A \subseteq B$.

   B Exercises

5. Prove by induction that if $A, B_1, B_2, \ldots, B_n$ are sets, $n \geq 2$, then $A \cap (B_1 \cup B_2 \cup \cdots \cup B_n) = (A \cap B_1) \cup (A \cap B_2) \cup \cdots (A \cap B_n)$.

4.2 Laws of Set Theory

The following basic set laws can be derived using either the Basic Definition or the Set-Membership approach and can be illustrated by Venn diagrams.

Commutative Laws

(1) $A \cup B = B \cup A$.	(1') $A \cap B = B \cap A$.

Associative Laws

(2) $A \cup (B \cup C)$ $= (A \cup B) \cup C$.	(2') $A \cap (B \cap C)$ $= (A \cap B) \cap C$.

Distributive Laws

(3) $A \cap (B \cup C)$ $= (A \cap B) \cup (A \cap C)$.	(3') $A \cup (B \cap C)$ $= (A \cup B) \cap (A \cup C)$.

Identity Laws

(4) $A \cup \emptyset = \emptyset \cup A = A$.	(4') $A \cap U = U \cap A = A$.

Complement Laws

(5) $A \cup A^c = U$.	(5') $A \cap A^c = \emptyset$.

Indempotent Laws

(6) $A \cup A = A$.	(6') $A \cap A = A$.

Null Laws

(7) $A \cup U = U$.	(7') $A \cap \emptyset = \emptyset$.

Absorption Laws

(8) $A \cup (A \cap B) = A$.	(8') $A \cap (A \cup B) = A$.

DeMorgan's Laws

(9) $(A \cup B)^c = A^c \cap B^c$.	(9') $(A \cap B)^c = A^c \cup B^c$.

Involution Law

(10) $(A^c)^c = A$.

It is quite clear that most of these laws resemble or, in fact, are analogues of laws in basic algebra and the algebra of propositions.

PROOFS USING PREVIOUSLY PROVEN THEOREMS

Once a few basic laws or theorems have been established, we frequently use them to prove additional theorems. This method of proof is sometimes more efficient than that of Proof by Definition. To illustrate, let us prove the following:

Theorem 4.2.1. *Let A and B be sets. Then* $(A \cap B) \cup (A \cap B^c) = A$.

Proof: $(A \cap B) \cup (A \cap B^c) = A \cap (B \cup B^c)$.   Why?
$\qquad\qquad\qquad\qquad\quad = A \cap U$.   Why?
$\qquad\qquad\qquad\qquad\quad = A$.   Why? #

PROOF USING THE INDIRECT METHOD

The procedure one most frequently uses to prove a theorem in mathematics is the Direct Method, as illustrated in Theorems 4.1.1 and 4.1.2. Occasionally there are situations where this method is not applicable. Consider the following:

Theorem 4.2.2. *Let A, B, C be sets. If* $A \subseteq B$ *and* $B \cap C = \emptyset$, *then* $A \cap C = \emptyset$.

Commentary: The usual and first approach would be to assume $A \subseteq B$ and $B \cap C = \emptyset$ is true and to attempt to prove $A \cap C = \emptyset$ is true. You should try this so you can experience difficulties in applying the Direct Method to this problem.

The Indirect Method is as follows: If we assume the theorem false and determine that this cannot occur—that is, if we obtain a contradiction—then the theorem must be true. This approach is on sound logical footing since it is exactly the same method of indirect proof that we discussed in Section 3.5.

Proof of Theorem 4.2.2: Assume $(A \subseteq B$ and $B \cap C = \emptyset)$ and $A \cap C \neq \emptyset$. To prove that this cannot occur, let $x \in A \cap C$. Why is this possible?

$x \in A \cap C.$

Why?	$\Rightarrow$	$x \in A$ and $x \in C.$
Why?	$\Rightarrow$	$x \in B$ and $x \in C.$
Why?	$\Rightarrow$	$x \in B \cap C.$

But this cannot occur. Why? Hence the theorem is true. #

EXERCISES FOR SECTION 4.2

In the exercises that follow it is most important that you outline the logical procedures or methods you use.

A Exercises

1. Supply all answers to the question "Why?" in Theorems 4.2.1 and 4.2.2.
2. Use previously proven theorems to prove:
    (a) $A \cup (B - A) = A \cup B.$
    (b) $A - B = B^c - A^c.$
    (c) $(A \subseteq B), A \cap C \neq \emptyset \Rightarrow B \cap C \neq \emptyset.$
    (d) $A \cap (B - C) = (A \cap B) - (A \cap C).$
    (e) $A - (B \cup C) = (A - B) \cap (A - C).$

3. *Hierarchy of Set Operations.* The rules that determine the order of evaluation in a set expression that involves more than one operation are similar to the rules for logic. In the absence of parentheses, complementations are done first, intersections second, and unions third. Parentheses are used to override this order. If the same operation appears two or more consecutive times, evaluate from left to right. In what order are the following expressions performed?
    (a) $A \cup B^c \cap C.$
    (b) $A \cap B \cup C \cap B.$
    (c) $A \cup B \cup C^c.$

C Exercise

4. There are several ways that can be used to format the proofs in this chapter. One that should be familiar to you from Chapter 3 is illustrated with the following proof.
    Alternate proof of Part a in Theorem 4.1.2:

(1) $x \in A \cap (B \cup C)$ Premise

(2) $(x \in A) \wedge (x \in B \cup C)$ (1), definition of intersection

(3) $(x \in A) \wedge$
$((x \in B) \vee (x \in C))$ (2), definition of union

(4) $((x \in A) \wedge (x \in B)) \vee$
$((x \in A) \wedge (x \in C))$ (3), distributing $\wedge$ over $\vee$

(5) $(x \in A \cap B) \vee$
$(x \in A \cap C)$ (4), definition of intersection

(6) $x \in (A \cap B) \cup (A \cap C)$ (5), definition of union

Prove Part b of Theorem 4.1.2 and Theorem 4.2.2 with this format.

4.3 Partitions of Sets

When confronted by a complicated task the human mind will frequently "divide and conquer"; that is, subdivide the task into subproblems and solve each part separately. A task or set can be subdivided, or partitioned, in many different ways, depending on what the observer would like to accomplish.

Example 4.3.1. Consider the set of students in a classroom. How can the instructor partition this set? The instructor could partition it into subsets, or blocks, where each block contains students: (a) in the same row, or (b) who have the same color eyes, or (c) in specific weight categories, or (d) in specific height categories, etc.

One could also think of the concept of partitioning a set as a "packaging problem." How can one "package" a carton of, say, twenty-four cans? We could use: four six-packs, three eight-packs, two twelve-packs, etc. In all cases: (a) the sum of all cans in all packs must be twenty-four, and (b) a can must be in one and only one pack.

Definition: Partition. *Let A be a set. A partition of A is any set of nonempty subsets (or blocks) $A_1, A_2, \ldots$ of A such that:*

(a) $A_1 \cup A_2 \cup \cdots = A$, *and*
(b) *the subsets A_i are mutually disjoint; that is, $A_i \cap A_j = \emptyset$ for $i \neq j$.*

Example 4.3.2. Let $A = \{a, b, c\}$. Then $\{\{a\}, \{b, c\}\}$ is a partition of A. We see that we have partitioned A into two blocks, namely $\{a\}$ and $\{b, c\}$. Note how our definition of a partition allows for the possibility that A can be broken up into an infinite number of blocks. Of course, if A is finite, as in Example 4.3.2, the number of blocks can be no larger than $\#A$.

Example 4.3.3. Two examples of partitions of $\mathbf{Z}$ are $\{\{n\}\,|\,n \in \mathbf{Z}\}$ and $\{\{n\,|\,n \in \mathbf{Z},\ n < 0\},\ \{0\},\ \{n\,|\,n \in \mathbf{Z},\ n > 0\}\}$. The set of subsets $\{\{n \in \mathbf{Z}\ |\,n \geq 0\},\ \{n \in \mathbf{Z}\,|\,n \leq 0\}\}$ is not a partition because the two subsets have a non-empty intersection. A second example of a non-partition is $\{\{n \in \mathbf{Z} : |n| = k\} : k = -1, 0, 1, 2, \ldots\}$. One of the blocks, $\{n \in \mathbf{Z} : |n| = -1\} = \emptyset$.

EXERCISES FOR SECTION 4.3

A Exercises

1. Find all partitions of the set $A = \{a, b, c\}$.

2. Which of the following collection of subsets of the plane, $\mathbf{R} \times \mathbf{R}$, are partitions?
    (a) $\{\{(x, y)\,|\,x + y = c\}\,|\,c \in \mathbf{R}\}$
    (b) the set of all circles in $\mathbf{R} \times \mathbf{R}$
    (c) the set of all circles centered at $(0, 0)$, together with $\{(0, 0)\}$
    (d) $\{\{(x, y)\}\,|\,(x, y) \text{ is included in } \mathbf{R} \times \mathbf{R}\}$

3. A student, on an exam paper, defined the term *partition* the following way: "Let A be a set. A partition of A is any set of non-empty subsets A_1, $A_2, \ldots$ of A such that each element of A is in one and only one of the subsets A_i." Is this definition correct? Why?

4. Do the subsets A_1, A_2, A_3, and A_4 in Figure 4.3.1 constitute a partition of A?

B Exercises

5. The definition of $\mathbf{Q} = \{a/b\,|\,a, b \in \mathbf{Z},\ b \neq 0\}$ given in Chapter 1 is, at best, awkward since, if we use the definition to list elements in $\mathbf{Q}$, we will have duplications, i.e., $-1/-2$, $1/2$, $4/8$, etc. Try to write a more precise definition of $\mathbf{Q}$ so there is no duplication of elements.

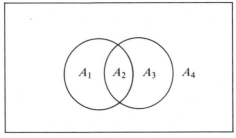

FIGURE 4.3.1
Exercise

6. Let $\{A_1, A_2, \ldots, A_n\}$ be a partition of set A and let B be any non-empty subset of A. Prove that $\{A_i \cap B \mid A_i \cap B \neq \emptyset\}$ is a partition of $A \cap B$.

4.4 Minsets

Let B_1 and B_2 be subsets of a set A. We note that the Venn diagram of Figure 4.4.1 is naturally partitioned into the subsets A_1, A_2, A_3, and A_4. Further we observe that A_1, A_2, A_3, and A_4 can be described with B_1 and B_2 as follows:

$$A_1 = B_1 \cap B_2^c.$$
$$A_2 = B_1 \cap B_2.$$
$$A_3 = B_1^c \cap B_2.$$
$$A_4 = B_1^c \cap B_2^c.$$

Each of the A_i's is called a *minset* or *minterm* generated by B_1 and B_2. We note that each minset (minterm) is formed by taking the intersection of two sets where each may be either B_1 or its complement B_1^c and B_2 or its complement B_2^c. Note also, given two sets B_1 and B_2, there are 2^2 minsets (or minterms).

The reader should note that if we apply all possible combinations of the operations intersection, union, and complementation to the sets B_1 and B_2 of Figure 4.4.1, the smallest sets generated will be exactly the minsets, the minimum sets. Hence the derivation of the term *minset*.

Next consider the Venn diagram containing three sets, B_1, B_2 and B_3. What are the minsets generated by B_1, B_2, and B_3? How many are there? Following the procedures outlined above, we note that

$$B_1 \cap B_2 \cap B_3^c$$
$$B_1 \cap B_2^c \cap B_3$$
$$B_1 \cap B_2^c \cap B_3^c$$

are three of the $2^3 = 8$ minsets. See Exercise 1 of this section.

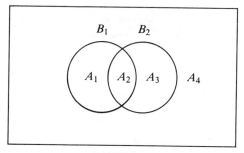

FIGURE 4.4.1
Illustration of minsets

Definition: Minset. *Let $\{B_1, B_2, \ldots, B_n\}$ be a set of subsets of a set A. A set of the form $D_1 \cap D_2 \cap \cdots \cap D_n$ where each D_i may be either B_i or B_i^c is called a minset, or minterm, generated by $B_1, B_2, \ldots, B_n$.*

Example 4.4.1. For another view, consider the following: Let $A = \{1, 2, 3, 4, 5, 6\}$ with subsets $B_1 = \{1, 3, 5\}$ and $B_2 = \{1, 2, 3\}$. How can we, using set operations applied to B_1 and B_2, produce all of A efficiently *without* duplication? We note that:

$$B_1 \cap B_2 = \{1, 3\},$$
$$B_1^c \qquad = \{2, 4, 6\}, \text{ and}$$
$$B_2^c \qquad = \{4, 5, 6\}.$$

We have produced the elements of A but we have the elements 4 and 6 repeated in two sets. In place of B_1^c and B_2^c, let us try $B_1^c \cap B_2$ and $B_1 \cap B_2^c$, respectively:

$$B_1^c \cap B_2 = \{2\}, \text{ and}$$
$$B_1 \cap B_2^c = \{5\}.$$

We have now produced the elements 1, 2, 3, 5 using $B_1 \cap B_2$, $B_1^c \cap B_2$, and $B_1 \cap B_2^c$, yet we have not listed the elements 4 and 6. $B_1 \cup B_2, B_1 \cup B_2^c$, etc., will produce duplications of listed elements and will not produce 6. We note that $B_1^c \cap B_2^c = \{4, 6\}$, exactly the elements we need. Each element of A appears exactly once in one of the four minsets $B_1 \cap B_2, B_1^c \cap B_2, B \cap B_2^c$, and $B_1^c \cap B_2^c$. Hence we have a partition of A.

Theorem 4.4.1. *Let A be a set and let $B_1, \ldots, B_n$ be subsets of A. The set of nonempty minsets (minterms) generated by $B_1, \ldots, B_n$ is a partition of A.*

The proof of this theorem is left to the reader with the hint that it can be done by induction, using Exercise 6 from Section 4.3 as a tool.

The most significant fact about minsets is that any subset of A that can be obtained from $B_1, \ldots, B_n$ using the standard set operations can be obtained in a standard form by taking the union of selected minsets.

Definition: Normal Form. *A set is said to be in minset normal (or canonical) form when it is expressed as the union of distinct non-empty minsets or it is Ø.*

EXERCISES FOR SECTION 4.4.

A Exercises

1. Consider the subsets $A = \{1, 7, 8\}$, $B = \{1, 6, 9, 10\}$, and $C = \{1, 9, 10\}$, where $U = \{1, 2, \ldots, 10\}$.

(a) List the non-empty minsets generated by A, B, and C.

(b) How many elements of the power set of U can be generated by A, B, and C? Compare this number with $\#\mathcal{P}(U)$. Give an example of one subset that cannot be generated by A, B, and C.

2. Partition the set of strings of 0's and 1's of length two or less, using the minsets generated by $B_1 = \{s : s \text{ has length 2}\}$, and $B_2 = \{s : s \text{ starts with a 0}\}$.

3. (a) Partition $\{1, 2, \ldots , 9\}$ into the minsets generated by $B_1 = \{5, 6, 7\}$, $B_2 = \{2, 4, 5, 9\}$, and $B_3 = \{3, 4, 5, 6, 8, 9\}$.

(b) How many different subsets of $\{1, 2, \ldots ,9\}$ can you create using B_1, B_2, B_3 with the standard set operations?

(c) Do there exist subsets C_1, C_2, C_3 that will generate every subset of $\{1, 2, \ldots ,9\}$?

4. Let B_1, B_2, and B_3 be subsets of a universal set U.

(a) Find all minsets generated by B_1, B_2, B_3.

(b) Illustrate via a Venn diagram all minsets obtained in Part a.

(c) Express the following sets in minset normal form: B_1^c, $B_1 \cap B_2$, $B_2^c \cap B_3^c$.

5. (a) Partition $A = \{0, 1, 2, 3, 4, 5\}$ with the minsets generated by $B_1 = \{0, 2, 4\}$ and $B_2 = \{1, 5\}$.

(b) How many different subsets of A can you generate from B_1 and B_2?

B Exercises

6. If $\{B_1, B_2, \ldots , B_n\}$ is a partition of A, how many minsets are generated by B_1, B_2, $\ldots$, B_n?

7. Prove Theorem 4.4.1.

C Exercise

8. Let S be a finite set of n elements. Let B_i, $i = 1, 2, 3, \ldots , k$ be k non-empty subsets of S. There are 2^{2^k} minset normal forms generated by the k subsets. The number of subsets of S is 2^n. Since we can make $2^{2^k} > 2^n$ by choosing $k \geq \log_2 n$, it is clear that two distinct minset normal-form expressions do not always equal distinct subsets of S. Even for $k < \log_2 n$, it may happen that two distinct minset normal-form expressions equal the same subset of S. Determine necessary and sufficient conditions for distinct normal-form expressions to equal distinct subsets of S.

4.5 The Duality Principle

In Section 4.2 we observed that each of the set laws labeled 1 through 9 had an analogue 1' through 9'. We notice that each of the laws in Column 2 can be obtained from the corresponding law in Column 1 by replacing U by ∩, ∩ by U, Ø by U, U by Ø, and leaving the complement as it is stated.

Definition: Duality Principle for Sets. *Let S be any identity involving sets and the operations c, ∩, and U. If S* is obtained from S by making the substitutions ∩ → U, U → ∩, Ø → U, U → Ø, then the Statement S* is also true and it is called the dual of the Statement S.*

Example 4.5.1. The dual of $(A \cap B) \cup (A \cap B^c) = A$ is $(A \cup B) \cap (A \cup B^c) = A$.

One should not underestimate the importance of this concept. It gives us a whole second set of identities, theorems, and concepts. For example, we can write the dual of minset and minset normal form to obtain what is called *maxset* and *maxset normal form.*

EXERCISES FOR SECTION 4.5

   A Exercises

1. State the dual of:
    (a) $A \cup (B \cap A) = A$.
    (b) $A \cup ((B^c \cup A) \cap B)^c = U$.
    (c) $(A \cup B^c)^c \cap B = A^c \cup B$.

2. Consider Table 3.4.1 and then write the principle of duality for logic.

3. Write the dual of:
    (a) $(p \vee \sim((\sim q \vee p) \wedge q)) \Leftrightarrow 1$.
    (b) $(\sim(p \wedge (\sim q)) \vee q) \Leftrightarrow ((\sim p) \vee q)$.

   B Exercises

4. Use the principle of duality and the definition of minset to write the definition of maxset. (Hint, just replace ∩ by U.)

5. Let $A = \{1, 2, 3, 4, 5, 6\}$ and let $B_1 = \{1, 3, 5\}$ and $B_2 = \{1, 2, 3\}$. Find the maxsets generated by B_1 and B_2. Note the set of maxsets does not constitute a partition of A. Can you explain why?
    (a) Write out the definition of maxset normal form.
    (b) Repeat Problem 4 of Section 4.4 for maxsets.

6. Is the dual of Exercise 6 of Section 4.3 true? Why?

chapter

5

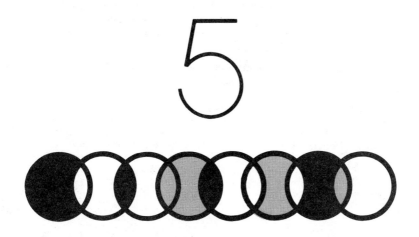

INTRODUCTION TO
MATRIX ALGEBRA

5 GOALS

The purpose of this chapter is to introduce you to matrix algebra, which has many applications. You are already familiar with several algebras: elementary algebra, the algebra of logic, the algebra of sets. We hope that as you studied the algebra of logic and the algebra of sets, you compared them with elementary algebra and noted that the basic laws of each are similar. We will see that matrix algebra is also similar. As in previous discussions, we begin by defining the objects in question and the basic operations.

5.1 Basic Definitions

Definition: Matrix. *A matrix is a rectangular array of elements of the form*

$$A = \begin{bmatrix} A_{11} & A_{12} & A_{13} & \cdots & A_{1n} \\ A_{21} & A_{22} & A_{23} & \cdots & A_{2n} \\ A_{31} & A_{32} & A_{33} & \cdots & A_{3n} \\ \cdot & \cdot & \cdot & \cdots & \cdot \\ A_{m1} & A_{m2} & A_{m3} & \cdots & A_{mn} \end{bmatrix}.$$

A convenient way of describing a matrix in general is to designate each entry via its position in the array. That is, the Entry A_{34} is the entry in the third row and fourth column of the matrix A. Since it is rather cumbersome to write out the large rectangular array above each time we wish to discuss the generalized form of a matrix, it is common practice to replace the above by $A = [A_{ij}]$. We will assume that each entry A_{ij} ($1 \leq i \leq m$, $1 \leq j \leq n$) is a real number. However, entries can come from any set, for example, the set of complex numbers.

Definition: Order. *The matrix A above has m rows and n columns. It therefore, is called an m × n (read "m by n") matrix, and it said to be of order m × n.*

We will use $M_{m \times n}$ (**R**) to stand for the set of all $m \times n$ matrices whose entries are real numbers.

Example 5.1.1.

$$A = \begin{bmatrix} 2 & 3 \\ 0 & -5 \end{bmatrix}, \qquad B = \begin{bmatrix} 0 \\ 1/2 \\ 15 \end{bmatrix}, \qquad \text{and} \qquad D = \begin{bmatrix} 1 & 2 & 5 \\ 6 & -2 & 3 \\ 4 & 2 & 8 \end{bmatrix}$$

are 2 × 2, 3 × 1, and 3 × 3 matrices respectively.

Since we now understand what a matrix looks like, we are in a position to investigate the operations of matrix algebra for which users have found the most applications.

Example 5.1.2. First we ask ourselves: Is the matrix $A = \begin{bmatrix} 1 & 2 \\ 3 & 4 \end{bmatrix}$ equal to the matrix $B = \begin{bmatrix} 1 & 2 \\ 3 & 5 \end{bmatrix}$? Next, is the matrix $A = \begin{bmatrix} 1 & 2 & 3 \\ 4 & 5 & 6 \end{bmatrix}$ equal to the matrix $B = \begin{bmatrix} 1 & 2 \\ 4 & 5 \end{bmatrix}$? Why not? We formalize in the following definition.

Definition: Equality. *The matrix A is said to be equal to the matrix B (written A = B) if and only if:*

(1) A and B have the same order, and
(2) corresponding entries are equal: that is, $A_{ij} = B_{ij}$ for all i and j.

5.2 Addition and Scalar Multiplication

Example 5.2.1. Concerning addition, it seems natural that if

$$A = \begin{bmatrix} 1 & 0 \\ 2 & -1 \end{bmatrix} \quad \text{and} \quad B = \begin{bmatrix} 3 & 4 \\ -5 & 2 \end{bmatrix}, \quad \text{then}$$

$$A + B = \begin{bmatrix} 1+3 & 0+4 \\ 2+(-5) & (-1)+2 \end{bmatrix} = \begin{bmatrix} 4 & 4 \\ -3 & 1 \end{bmatrix}. \text{ If however,}$$

$$A = \begin{bmatrix} 1 & 2 & 3 \\ 0 & 1 & 2 \end{bmatrix} \quad \text{and} \quad B = \begin{bmatrix} 3 & 0 \\ 2 & 8 \end{bmatrix}, \text{ can we find } A + B?$$

Definition: Addition. *Let A and B be m × n matrices. Then A + B is an m × n matrix where $(A + B)_{ij} = A_{ij} + B_{ij}$ (read "the ith jth entry of the matrix A + B is obtained by adding the ith jth entry of A to the ith jth entry of B").*

It is clear from Example 5.2.1 and the definition of addition that $A + B$ is defined if and only if A and B are of the same order.

Another frequently used operation is that of multiplying a matrix by a number, commonly called a *scalar*. Unless specified otherwise, we will assume that all scalars are real numbers.

Example 5.2.2. If $c = 3$ and if $A = \begin{bmatrix} 1 & -2 \\ 3 & 5 \end{bmatrix}$ and we wish to find cA, it seems natural to multiply each entry of A by 3 so that $3A = 3\begin{bmatrix} 1 & -2 \\ 3 & 5 \end{bmatrix} = \begin{bmatrix} 3 & -6 \\ 9 & 15 \end{bmatrix}$.

Definition: Scalar Multiplication. *Let A be an m × n matrix and c a scalar. Then cA is the m × n matrix obtained by multiplying c times each entry of A; that is $(cA)_{ij} = cA_{ij}$.*

5.3 Multiplication of Matrices

A definition which is more awkward to motivate (and we will not attempt to do so here) is the product of two matrices. The reader will see in further illustrations and concepts that if we define the product of matrices the following way, we will obtain a very useful algebraic system which is quite similar to elementary algebra.

Definition: Multiplication. *Let A be an $m \times n$ matrix and let B be an $n \times p$ matrix. Then the product of A and B, denoted by AB, is an $m \times p$ matrix whose ith jth entry*

$$\begin{aligned}(AB)_{ij} &= A_{i1}B_{1j} + A_{i2}B_{2j} + \cdots + A_{in}B_{nj} \\ &= \sum_{k=1}^{n} A_{ik}B_{kj}\end{aligned}$$

for $1 \leq i \leq m$, $1 \leq j \leq p$.

The mechanics of computing one entry in the product of two matrices is illustrated in Figure 5.3.1. The computation of a product can take a considerable amount of time in comparison to the time required to add two matrices.

Col. 2 ↓

Row 1 →

$$\begin{pmatrix} 1 & -1 & 0 \\ -1 & 2 & -1 \\ 0 & -1 & 1 \end{pmatrix} \begin{pmatrix} -6 & 2 & 4 \\ 3 & 3 & 6 \\ 1 & 4 & 5 \end{pmatrix}$$

Row 1, Col. 2 of Product ↓

$$= \begin{pmatrix} * & (1)(2) + (-1)(3) + (0)(4) & * \\ * & * & * \\ * & * & * \end{pmatrix}$$

$$= \begin{pmatrix} * & -1 & * \\ * & * & * \\ * & * & * \end{pmatrix}$$

FIGURE 5.3.1
Computation of one ninth of the product of two three-by-three matrices

Suppose that A and B are $n \times n$ matrices; then $(AB)_{ij}$ is determined using n multiplications and $n - 1$ additions. The full product takes n^3 multiplications and $n^3 - n^2$ additions. This compares with n^2 additions for the sum of two $n \times n$ matrices. The product of two 10 by 10 matrices will require 1000 multiplications and 900 additions, clearly a job that you would assign to a computer. The sum of two matrices requires a more modest 100 additions.

This analysis is based on the assumption that matrix multiplication will be done using the formula that is given in the definition. There are more advanced methods that, in theory, reduce multiplication times. Strassen's algorithm (see Aho, Hopcroft, and Ullman, 1974) computes the product of two n by n matrices using no more than a multiple of $n^{2.8}$ operations. That is, there exists a number C (which is large) such that no more than $C\,n^{2.8}$ operations are needed to complete a multiplication. Since C would be so large, this algorithm is useful only if n is large.

Example 5.3.1.

Let $A =$ a 3×2 matrix $\begin{bmatrix} 1 & 0 \\ 3 & 2 \\ -5 & 1 \end{bmatrix}$ and $B =$ a 2×1 matrix $\begin{bmatrix} 6 \\ 1 \end{bmatrix}$.

Then $AB = \begin{bmatrix} 1 & 0 \\ 3 & 2 \\ -5 & 1 \end{bmatrix} \begin{bmatrix} 6 \\ 1 \end{bmatrix}$

$= $ a 3×1 matrix $\begin{bmatrix} (1)(6) + (0)(1) \\ (3)(6) + (2)(1) \\ (-5)(6) + (1)(1) \end{bmatrix} = \begin{bmatrix} 6 \\ 20 \\ -29 \end{bmatrix}$.

Remarks:

(1) The product AB is defined only if A is a $m \times \underline{n}$ matrix and B is an $\underline{n} \times p$ matrix; that is, the two "inner" numbers must be the same. Furthermore, the order of the product matrix AB is the "outer" numbers, in this case $m \times p$.

(2) It is wise to first obtain the order of the product matrix. For example, if A is a 3×2 matrix and B is a 2×2 matrix, then AB is a 3×2 matrix of the form

$$AB = \begin{bmatrix} C_{11} & C_{12} \\ C_{21} & C_{22} \\ C_{31} & C_{32} \end{bmatrix}.$$

Then to obtain, for example, C_{31}, we multiply corresponding entries in the third row of A times the first column of B and add the results.

78

Example 5.3.2.

Let $A = \begin{bmatrix} 1 & 0 \\ 0 & 3 \end{bmatrix}$ and $B = \begin{bmatrix} 3 & 0 \\ 2 & 1 \end{bmatrix}$.

Then $AB = \begin{bmatrix} (1)(3) + (0)(2) & (1)(0) + (0)(1) \\ (0)(3) + (3)(2) & (0)(0) + (3)(1) \end{bmatrix}$

$= \begin{bmatrix} 3 & 0 \\ 6 & 3 \end{bmatrix}$.

Note: $BA = \begin{bmatrix} 3 & 0 \\ 2 & 3 \end{bmatrix} \neq AB$.

Remarks:

(1) An $n \times n$ matrix is called a *square matrix*.
(2) If A is a square matrix, AA is defined and is denoted by A^2, and $AAA = A^3$. Similarly, $AAA \ldots A = A^n$, where A is multiplied by itself n times.
(3) The $m \times n$ matrices each of whose entries is 0 is denoted by $\mathbf{0}_{m \times n}$ or simply $\mathbf{0}$, when no confusion arises.

EXERCISES FOR SECTION 5.3

A Exercises

1. Let $A = \begin{bmatrix} 1 & -1 \\ 2 & 3 \end{bmatrix}$, $B = \begin{bmatrix} 0 & 1 \\ 3 & -5 \end{bmatrix}$, and $C = \begin{bmatrix} 0 & 1 & -1 \\ 3 & -2 & 2 \end{bmatrix}$.

Determine:
(a) AB and BA
(b) $A + B$ and $B + A$
(c) If $c = 3$, show that $c(A + B) = cA + cB$.
(d) Show that $(AB)C = A(BC)$.
(e) A^2C
(f) $B + 0$
(g) $A\mathbf{0}_{2 \times 2}$ and $\mathbf{0}_{2 \times 2}A$, where $\mathbf{0}_{2 \times 2}$ is the 2 × 2 zero matrix
(h) $0A$, where 0 is the real number (scalar) zero
(i) Let $c = 2$ and $d = 3$. Show that $(c + d)A = cA + dA$.

2. Let $A = \begin{bmatrix} 2 & 0 \\ 0 & 3 \end{bmatrix}$. Find a matrix B such that $AB = I$ and $BA = I$, where $I = \begin{bmatrix} 1 & 0 \\ 0 & 1 \end{bmatrix}$.

3. Find $A\,I$ and $B\,I$ where I is as in Exercise 2,

$$A = \begin{bmatrix} 1 & 8 \\ 9 & 5 \end{bmatrix}, \quad \text{and } B = \begin{bmatrix} -2 & 3 \\ 5 & -7 \end{bmatrix}.$$

What do you notice?

4. Find A^3 if $A = \begin{bmatrix} 1 & 0 & 0 \\ 0 & 2 & 0 \\ 0 & 0 & 3 \end{bmatrix}$. What is A^{15} equal to?

5. (a) Determine I^2, I^3, if $I = \begin{bmatrix} 1 & 0 & 0 \\ 0 & 1 & 0 \\ 0 & 0 & 1 \end{bmatrix}$.

   (b) What is I^n equal to for any $n \geq 1$?
   (c) Prove your answer to Part b by induction.

6. Let $A = \begin{bmatrix} 1 & 0 & 2 \\ 2 & -1 & 5 \\ 3 & 2 & 1 \end{bmatrix}$, $B = \begin{bmatrix} 0 & 2 & 3 \\ 1 & 1 & 2 \\ -1 & 3 & -2 \end{bmatrix}$,

and $C = \begin{bmatrix} 2 & 1 & 2 & 3 \\ 4 & 0 & 1 & 1 \\ 3 & -1 & 4 & 1 \end{bmatrix}$. Compute, if possible:

   (a) $A - B$
   (b) AB
   (c) $AC - BC$ (f) $C \begin{bmatrix} 2 \\ 1 \\ 0 \\ -1 \end{bmatrix}$
   (d) ABC
   (e) $CA - CB$

7. (a) If $A = \begin{bmatrix} 2 & 1 \\ 1 & -1 \end{bmatrix}$, $X = \begin{bmatrix} x_1 \\ x_2 \end{bmatrix}$, and $B = \begin{bmatrix} 3 \\ 1 \end{bmatrix}$, show that

   $A X = B$ is a way of expressing the system

   $$2x_1 + 1x_2 = 3$$
   $$1x_1 - 1x_2 = 1 \text{ using matrices.}$$

   (b) Express the following systems of equations using matrices:

   (i) $2x_1 - x_2 = 4$
   $$x_1 + x_2 = 0$$

   (ii) $x_1 + x_2 + 2x_3 = 1$
   $$x_1 + 2x_2 - x_3 = -1$$
   $$x_1 + 3x_2 + x_3 = 5$$

   (iii) $x_1 + x_2 = 3$
   $$x_2 = 5$$
   $$x_1 + 3x_3 = 6$$

5.4 Special Types of Matrices

We have already investigated one special type of matrix, namely the zero matrix, and found that it behaves in matrix algebra in an analogous fashion to the real number 0; that is, as the additive identity. We will now investigate the properties of a few other special matrices.

Definition: Diagonal Matrix. *A square matrix D is called a diagonal matrix if $D_{ij} = 0$ whenever $i \neq j$.*

Example 5.4.1.

$$A = \begin{bmatrix} 1 & 0 & 0 \\ 0 & 2 & 0 \\ 0 & 0 & 5 \end{bmatrix}, \quad B = \begin{bmatrix} 3 & 0 & 0 \\ 0 & 0 & 0 \\ 0 & 0 & -5 \end{bmatrix}, \text{ and } I = \begin{bmatrix} 1 & 0 & 0 \\ 0 & 1 & 0 \\ 0 & 0 & 1 \end{bmatrix} \text{ are}$$

all diagonal matrices.

In Example 5.4.1, the 3×3 diagonal matrix I whose diagonal entries are all 1's has the singular property that for any other 3×3 matrix A we have $A I = I A = A$. For example:

Example 5.4.2.

$$\text{If } A = \begin{bmatrix} 1 & 2 & 5 \\ 6 & 7 & -2 \\ 3 & -3 & 0 \end{bmatrix}, \text{ then}$$

$$AI = \begin{bmatrix} 1 & 2 & 5 \\ 6 & 7 & -2 \\ 3 & -3 & 0 \end{bmatrix} \text{ and}$$

$$IA = \begin{bmatrix} 1 & 2 & 5 \\ 6 & 7 & -2 \\ 3 & -3 & 0 \end{bmatrix}.$$

In other words, the matrix I behaves in matrix algebra like the real number 1; that is, as a multiplicative identity. In matrix algebra the matrix I is called simply the *identity matrix*. Convince yourself that if A is any $n \times n$ matrix $A I = I A = A$.

Definition: Identity Matrix. *The $n \times n$ diagonal matrix whose diagonal components are all 1's is called the identity matrix and is denoted by I or I_n.*

In the set of real numbers we realize that, given a non-zero real number x, there exists a real number y such that $x y = y x = 1$. We know that real numbers commute under multiplication so that the two equations can be summarized as $x\,y = 1$. Further we know that $y = x^{-1} = 1/x$. Do we have an analogous situation in $M_{n \times n}(\mathbf{R})$? Can we define the multiplicative inverse of an $n \times n$ matrix A? It seems natural to imitate the definition of multiplicative inverse in the real numbers.

Definition: Matrix Inverse. *Let A be a $n \times n$ matrix. If there exists an $n \times n$ matrix B such that $A\,B = B\,A = I$, then B is the multiplicative inverse of A (called simply the inverse of A) and is denoted by A^{-1} (read "A inverse").*

When we are doing computations involving matrices, it would be helpful to know that when we find A^{-1}, the answer we obtain is the only inverse of the given matrix.

Remark: Those unfamiliar with the laws of matrices should go over the proof of Theorem 5.4.1 after they have familiarized themselves with Section 5.5.

Theorem 5.4.1. *The inverse of an $n \times n$ matrix A, when it exists, is unique.*

Proof: Let A be an $n \times n$ matrix. Assume to the contrary, that A has two (different) inverses, say B and C. Then

$$
\begin{aligned}
B &= BI & &\text{Identity property of } I \\
&= B(AC) & &\text{Assumption that } C \text{ is an inverse of } A \\
&= (BA)C & &\text{Associativity of matrix multiplication} \\
&= IC & &\text{Assumption that } B \text{ is an inverse of } A \\
&= C & &\text{Identity property of } I \ \#
\end{aligned}
$$

Example 5.4.3. Let $A = \begin{bmatrix} 2 & 0 \\ 0 & 3 \end{bmatrix}$. What is A^{-1}? Without too much difficulty, by trial and error, we determine that $A^{-1} = \begin{bmatrix} 1/2 & 0 \\ 0 & 1/3 \end{bmatrix}$.

If $A = \begin{bmatrix} 1 & 2 \\ -3 & 5 \end{bmatrix}$, what is A^{-1}? Here the answer is considerably more difficult. In order to understand more completely the notion of the inverse of a matrix, it would be beneficial to have a formula which will enable us to compute the inverse of at least a 2×2 matrix. To do this, we need to recall the definition of the determinant of a 2×2 matrix. Appendix A gives a more complete description of the determinant of a 2×2 and higher-order matrices.

Definition: Determinant (2×2 Matrix). *Let $A = \begin{bmatrix} a & b \\ c & d \end{bmatrix}$. The determinant of the matrix A, written det (A) or $|A|$, is a real number and is equal to det $(A) = ad - bc$.*

Example 5.4.4.

$$\text{If } A = \begin{bmatrix} 1 & 2 \\ -3 & 5 \end{bmatrix}, \text{ then } det(A) = (1)\,(5) - (2)\,(-3)$$

$$= 11.$$

$$\text{If } A = \begin{bmatrix} 1 & 2 \\ 2 & 4 \end{bmatrix}, \text{ then } det(A) = 0.$$

Theorem 5.4.2. Let $A = \begin{bmatrix} a & b \\ c & d \end{bmatrix}$. If $det(A) \neq 0$, then $A^{-1} = 1/det(A) \begin{bmatrix} d & -b \\ -c & a \end{bmatrix}$.

Proof: See Exercise 4.

Example 5.4.5.

$$\text{If } A = \begin{bmatrix} 1 & 2 \\ -3 & 5 \end{bmatrix}, \text{ then}$$

$$A^{-1} = 1/11 \begin{bmatrix} 5 & -2 \\ 3 & 1 \end{bmatrix} = \begin{bmatrix} 5/11 & -2/11 \\ 3/11 & 1/11 \end{bmatrix}.$$

The reader should verify that $A A^{-1} = I$ and $A^{-1}A = I$.

Example 5.4.6. If $A = \begin{bmatrix} 1 & 2 \\ 2 & 4 \end{bmatrix}$, then $det(A) = 0$, so that $1/det(A)$ becomes $1/0$, which does not exist, so that A^{-1} does not exist.

Remarks:

(1) It is clear from Example 5.4.6 and Theorem 5.4.1 that if A is a 2×2 matrix and if $det(A) = 0$, then A^{-1} does not exist.
(2) A formula for the inverse of $n \times n$ matrices $n \geq 3$ can be derived which also involves $1/det(A)$. Hence, in general, if the determinant of a matrix is zero, the matrix does not have an inverse.
(3) In Chapter 12 we will develop a technique to compute the inverse of a higher-order matrix, if it exists.
(4) Matrix inversion comes first in the hierarchy of matrix operations; therefore, AB^{-1} is $A(B^{-1})$.

EXERCISES FOR SECTION 5.4

A Exercises

1. For the given matrices A find A^{-1} if it exists and verify that $AA^{-1} = A^{-1}A = I$. If A^{-1} does not exist explain why.

(a) $A = \begin{bmatrix} 1 & 3 \\ 2 & 1 \end{bmatrix}$.

(b) $A = \begin{bmatrix} 1 & -3 \\ 0 & 1 \end{bmatrix}$.

(c) $A = \begin{bmatrix} 6 & -3 \\ 8 & -4 \end{bmatrix}$.

(d) $A = \begin{bmatrix} 1 & 0 \\ 0 & 1 \end{bmatrix}$.

(e) Use the definition of the inverse of a matrix to find A^{-1}:

$$A = \begin{bmatrix} 3 & 0 & 0 \\ 0 & 1/2 & 0 \\ 0 & 0 & -5 \end{bmatrix}.$$

2. Let $A = \begin{bmatrix} 2 & 3 \\ 1 & 4 \end{bmatrix}$ and $B = \begin{bmatrix} 3 & -3 \\ 2 & 1 \end{bmatrix}$. Verify that $(AB)^{-1} = B^{-1}A^{-1}$.

3. Let A and B be $n \times n$ invertible matrices. Prove that $(AB)^{-1} = B^{-1}A^{-1}$. (Hint: Use Theorem 5.4.1.) Why is the right side of the above statement written "backwards"? Is this necessary?

4. Let $A = \begin{bmatrix} a & b \\ c & d \end{bmatrix}$. Derive the formula for A^{-1}.

5. (a) Let A and B be as in Problem 5.4.2 and show that $det(AB) = (detA)(detB)$.

   (b) It can be shown that the statement in Part a is true for all $n \times n$ matrices. Let A be any invertible $n \times n$ matrix. Prove that $det(A^{-1}) = (det\ A)^{-1}$. Note: $det(I) = 1$, see Appendix A.

   (c) Verify that the equation in Part b is so for the matrix in Exercise 1a of this section.

6. Prove by induction that $\begin{bmatrix} a & 0 \\ 0 & b \end{bmatrix}^n = \begin{bmatrix} a^n & 0 \\ 0 & b^n \end{bmatrix}$ for $n \geq 1$.

7. Prove: If the determinant of a matrix A is zero, then A does not have an inverse. (Hint: Use the indirect method of proof and Exercise 5.)

C Exercise

8. (a) Let A, B, and D be $n \times n$ matrices. Assume that B is invertible. If $A = BDB^{-1}$, prove by induction that $A^m = BD^mB^{-1}$ is true for $m \geq 1$.

   (b) Given that $A = \begin{bmatrix} -8 & 15 \\ -6 & 11 \end{bmatrix} = B\begin{bmatrix} 1 & 0 \\ 0 & 2 \end{bmatrix}B^{-1}$, where $B = \begin{bmatrix} 5 & 3 \\ 3 & 2 \end{bmatrix}$, what is A^{10}?

5.5   Laws of Matrix Algebra

The following is a summary of the basic laws of matrix operations. Assume that the indicated operations are defined; that is, that the orders of the matrices A, B, and C are such that the operations make sense.

(1) $A + B = B + A$.
(2) $A + (B + C) = (A + B) + C$.
(3) $c(A + B) = cA + cB$, where $c \in \mathbf{R}$.
(4) $(c_1 + c_2)A = c_1A + c_2A$, where $c_1, c_2 \in \mathbf{R}$.
(5) $c_1(c_2A) = (c_1c_2)A$, where $c_1, c_2 \in \mathbf{R}$.
(6) $0A = \mathbf{0}$, where $\mathbf{0}$ is the zero matrix.
(7) $0A = \mathbf{0}$, where 0 is on the left is the number 0.
(8) $A + \mathbf{0} = A$.
(9) $A + (-1)A = \mathbf{0}$.
(10) $A(B + C) = AB + AC$.
(11) $(B + C)A = BA + CA$.
(12) $A(BC) = (AB)C$.

Example 5.5.1.   If we wished to write out each of the above laws carefully, we would specify the orders of the matrices. For example, Law 10 should read:

(10) Let A, B, and C be $m \times n$, $n \times p$, and $n \times p$ matrices (Why?) respectively. Then $A(B + C) = AB + AC$.

Remarks:

(1) We notice the absence of the "law" $AB = BA$. Why?
(2) Is it really necessary to have *both* a right (No. 11) and a left (No. 10) distributive law? Why?
(3) What does Law 8 define? What does Law 9 define?

EXERCISES FOR SECTION 5.5

A Exercises

1.   Rewrite the above laws specifying as in Example 5.5.1 the orders of the matrices.

2.   Verify each of the above laws via examples.

3.   Let $A = \begin{bmatrix} 1 & 2 \\ 0 & -1 \end{bmatrix}$, $B = \begin{bmatrix} 3 & 7 & 6 \\ 2 & -1 & 5 \end{bmatrix}$, and $C = \begin{bmatrix} 0 & -2 & 4 \\ 7 & 1 & 1 \end{bmatrix}$. Find:

(a) $AB + AC$
(b) $A(B + C)$
(c) A^{-1}

(d) Solve the equation $AX = D$, where A is as above, $X = \begin{bmatrix} x_1 \\ x_2 \end{bmatrix}$, and $D = \begin{bmatrix} 3 \\ -2 \end{bmatrix}$.

4. Let $A = \begin{bmatrix} 7 & 4 \\ 2 & 1 \end{bmatrix}$ and $B = \begin{bmatrix} 3 & 5 \\ 2 & 4 \end{bmatrix}$. Find:

(a) $3A + 2B$
(b) AB
(c) BA
(d) B^{-1}
(e) $det((AB)^{10})$

5.6 Matrix Oddities

We have seen that matrix algebra is similar in many ways to elementary algebra. Indeed, if we want to solve the matrix equation $AX = B$ for the variable X, we imitate the procedure used in elementary algebra for solving the equation $ax = b$. Notice how exactly the same properties are used in the following detailed solutions of both equations.

Solution of $ax = b$		Solution of $AX = B$
$ax = b$		$AX = B$
$a^{-1}(ax) = a^{-1}b$ if $a \neq 0$		$A^{-1}(AX) = A^{-1}B$ if A^{-1} exists
$(a^{-1}a)x = a^{-1}b$	associative law	$(A^{-1}A)X = A^{-1}B$
$1\,x = a^{-1}b$	definition of inverse	$IX = A^{-1}B$
$x = a^{-1}b$	definition of identity	$X = A^{-1}B$

Certainly the solution process for $AX = B$ is the same as that of $ax = b$.

The solution of $xa = b$ is $x = ba^{-1} = a^{-1}b$. In fact, we usually write the solution of both equations as $x = b/a$. In matrix algebra, the solution of $XA = B$ is $X = BA^{-1}$, which is not necessarily equal to $A^{-1}B$. So in matrix algebra, since the commutative law (under multiplication) is not true, we have to be more careful in the methods we use to solve equations.

It is clear from the above that if we wrote the solution of $AX = B$ as $X = B/A$, we would not know how to interpret the answer B/A. Does it mean $A^{-1}B$ or BA^{-1}? Because of this, A^{-1} is *never* written as $1/A$.

Some of the main dissimilarities between matrix algebra and elementary algebra are that in matrix algebra:

(1) AB may be different from BA.
(2) There exist matrices A and B such that $A\,B = 0$, and yet $A \neq 0$ and $B \neq 0$.
(3) There exist matrices A where $A \neq 0$, and yet $A^2 = 0$.
(4) There exist matrices A where $A^2 = A$ with $A \neq 0$ and $A \neq I$.
(5) There exist matrices A where $A^2 = I$, where $A \neq I$ and $A \neq -I$.

EXERCISES FOR SECTION 5.6

A Exercises

1. Discuss each of the above "oddities" with respect to elementary algebra.

2. Determine 2×2 matrices which show each of the above "oddities" are true.

B Exercises

3. Prove the following implications, if possible:
    (a) $A^2 = A$ and $det(A) \neq 0 \Rightarrow A = I$.
    (b) $A^2 = I$ and $det(A) \neq 0 \Rightarrow A = I$ or $A = -I$.

4. Let M_n be the set of real $n \times n$ matrices. Let $P \subseteq M_n$ be the subset of matrices defined by $A \in P$ iff $A^2 = A$. Let $Q \subseteq P$ be defined by $A \in Q$ if and only if $det\ A \neq 0$.
    (a) Determine the cardinality of Q.
    (b) Consider the special case $n = 2$ and prove that a sufficient condition for $A \in P \subseteq M_2$ is that A has a zero determinant (i.e., A is singular) and $tr(A) = 1$ where $tr(A) = a_{11} + a_{22}$ is the sum of the main diagonal elements of A.
    (c) Is the condition of Part b a necessary condition?

C Exercises

5. Write each of the following systems in the form $AX = B$, and then solve the systems using matrices.
    (a) $2x_1 + x_2 = 3$.
         $x_1 - x_2 = 1$.
    (b) $2x_1 - x_2 = 4$.
         $x_1 - x_2 = 0$.
    (c) $2x_1 + x_2 = 1$.
         $x_1 - x_2 = 1$.
    (d) $2x_1 + x_2 = 1$.
         $x_1 - x_2 = -1$.
    (e) $3x_1 + 2x_2 = 1$.
         $6x_1 + 4x_2 = -1$.

6. Recall that $p(x) = x^2 - 5x + 6$ is called a polynomial, or, more specifically, a polynomial over $\mathbf{R}$, where the coefficients are elements of $\mathbf{R}$ and $x \in \mathbf{R}$. Also, think of the method of solving, and solutions of, $x^2 - 5x + 6 = 0$. We would like to define the analogous situation for 2×2 matrices. First define where A is a 2×2 matrix $p(A) = A^2 - 5A + 6I$. Discuss the method of solving and the solutions of $A^2 - 5A + 6I = 0$.

chapter

6

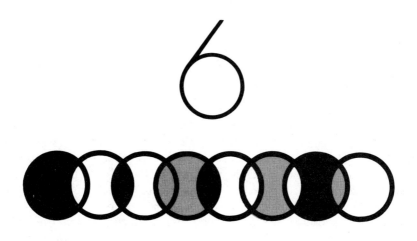

RELATIONS

6 GOALS

One only completely understands a set of objects if the structure of that set is made clear by the interrelationships between its elements. For example, the individuals in a crowd can be compared by height, by nationality, or through several other criteria. In mathematics, such comparisons are called *relations*. The goal of this chapter is to develop the language, tools, and concepts of relations.

6.1 Basic Definitions

In Chapter 1 we introduced the concept of the Cartesian product of sets. Let's assume that a person owns three shirts and two pairs of slacks. More precisely, let A = {blue shirt, tan shirt, mint green shirt} and B = {grey slacks, tan slacks}. Then certainly $A \times B$ is the set of all possible combinations (six) of shirts and slacks that the individual can wear. However, the individual may wish to restrict himself or herself to combinations which are color coordinated, or "related." This may not be all possible pairs in $A \times B$ but will certainly be a subset of $A \times B$. For example, one such subset may be {(blue shirt, grey slacks), (blue shirt, tan slacks), (mint green shirt, tan slacks)}.

Definition: Relation. *Let A and B be sets. A relation from A into B is any subset of A $\times$ B.*

Example 6.1.1. Let A = {1, 2, 3} and B = {4, 5}. Then {(1, 4), (2, 4), (3, ·5)} is a relation from A into B. Of course, there are many others we could describe; 64, to be exact.

Example 6.1.2. Let A = {2, 3, 5, 6} and define a relation r from A into A by $(a, b) \in r$ if and only if a divides evenly into b. So r = {(2, 2), (3, 3), (5, 5), (6, 6), (2, 6), (3, 6)}.

Definition: Relation on a Set. *A relation from a set A into itself is called a relation on A.*

Sometimes it is helpful to illustrate a relation. Consider Example 6.1.1. A picture of r can be drawn as in Figure 6.1.1. The arrows indicate that 1 is related to 4 under r. Also, 2 is related to 4 and 3 is related to 5 under r, while the upper arrow denotes that r is a relation from the whole set A into the set B.

A typical element in a relation r is an ordered pair (x, y). In some cases, r can be described by actually listing the pairs which are in r, as in the previous examples. This may not be convenient if r is relatively large. Other

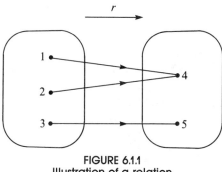

FIGURE 6.1.1
Illustration of a relation

notations are used depending on personal preference or past practice. Consider the following relations on the real numbers:

$$r = \{(x, y) \mid y \text{ is the square of } x\}, \text{ and}$$
$$s = \{(x, y) \mid x \leq y\}.$$

The notation $(4, 16) \in r$ or $(3, 7.2) \in s$ makes sense in both cases. However, r would be more naturally expressed as $r(x) = x^2$ or $r(x) = y$, where $y = x^2$. But this notation when used for s is at best awkward. The notation $x \leq y$ is clear and self-explanatory; it is a better notation to use than $(x, y) \in s$.

Many of the relations we will work with "resemble" the relation $\leq$, so $x \, s \, y$ is a commonly used way to express the fact that x is related to y through the relation s.

RELATION NOTATION

Let s be a relation from a set A into a set B. Then the fact that $(x, y) \in s$ is frequently written $x \, s \, y$.

Let $A = \{2, 3, 5, 8\}$, $B = \{4, 6, 16\}$, and $C = \{1, 4, 5, 7\}$; let r be the relation "divides," denoted by $\mid$, from A into B; and let s be the relation $\leq$ from B into C. So $r = \{(2, 4), (2, 6), (2, 16), (3, 6), (8, 16)\}$ and $s = \{(4, 4), (4, 5), (4, 7), (6, 7)\}$.

Notice from Figure 6.1.2 that we can, for certain elements of A, go through elements in B to results in C. That is:

$$2 \mid 4 \text{ and } 4 \leq 4.$$
$$2 \mid 4 \text{ and } 4 \leq 5.$$
$$2 \mid 4 \text{ and } 4 \leq 7.$$
$$2 \mid 6 \text{ and } 6 \leq 7.$$
$$3 \mid 6 \text{ and } 6 \leq 7.$$

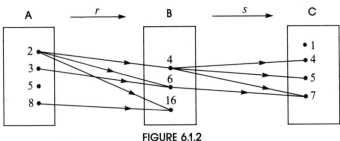

FIGURE 6.1.2
Illustration of relation "divides"

Based on this observation, we can define a new relation, call it rs, from A into C. In order for (a, c) to be in rs, it must be possible to travel along a path in Figure 6.1.2 from a to c. In other words, $(a, c) \in rs$ if and only if $(\exists b)_B$ (arb and bsc). The name rs was chosen solely because it reminds us that this new relation was formed by the two previous relations r and s. The complete listing of all elements in rs is $\{(2, 4), (2, 5), (2, 7), (3, 7)\}$. We summarize in a definition.

Definition: Composition. *Let r be a relation from a set A into a set B, and let s be a relation from B into a set C. The composition of r and s, written rs, is the set of pairs of the form $(a, c) \in A \times C$, where $(a, c) \in rs$ if and only if there exists $b \in B$ such that $(a, b) \in r$ and $(b, c) \in s$.*

Remark: A word of warning to those readers familiar with composition of functions. (For those who are not, disregard this remark. It will be repeated at an appropriate place in Chapter 7.) As indicated above, the traditional way of describing a composition of two relations is rs where r is the first relation and s the second. However, function composition is traditionally expressed "backwards"; that is, as sr (or $s \circ r$), where r is the first function and s is the second.

EXERCISES FOR SECTION 6.1

A Exercises

1. For each of the following relations r defined on **P**, determine which of the given ordered pairs belong to r.
   (a) $x \, r \, y$ iff $x \mid y$; $(2, 3), (2, 4), (2, 8), (2, 17)$
   (b) $x \, r \, y$ iff $x \leq y$; $(2, 3), (3, 2), (2, 4), (5, 8)$
   (c) $x \, r \, y$ iff $y = x^2$; $(1, 1), (2, 3), (2, 4), (2, 6)$

2. The following relations are on $\{1, 3, 5\}$. Let r be the relation $x \, r \, y$ iff $y = x + 2$ and s the relation $x \, s \, y$ iff $x \leq y$.

(a) Find *rs*.

(b) Find *sr*.

(c) Illustrate *rs* and *sr* via a diagram.

(d) Is the relation (set) *rs* equal to the relation *sr*? Why?

3. Let $A = \{1, 2, \ldots, 5\}$ and define r on A by xry iff $x+1 = y$. We define $r^2 = rr$, $r^3 = r^2r$, etc. Find:

(a) r

(b) r^2

(c) r^3

4. Given s and t, relations on $\mathbf{Z}$, $s = \{(1, n) : n \in \mathbf{Z}\}$ and $t = \{(n, 1) : n \in \mathbf{Z}\}$, what are st and ts?

B Exercises

5. Let ρ be the relation on the power set $\mathscr{P}(S)$ of a finite set S of cardinality n. Define ρ by $(A, B) \in \rho$ iff $A \cap B = \emptyset$.

(a) Consider the specific case $n = 3$ and determine the cardinality of ρ.

(b) What is the cardinality of ρ for an arbitrary n? Express your answer in terms of n. (Hint: A particularly concise closed form expression can be obtained, if one considers binomial expressions.)

6. Let r_1, r_2, and r_3 be relations on any set A. Prove: If $r_1 \subseteq r_2$, then $r_1 r_3 \subseteq r_2 r_3$.

6.2 Graphs of Relations

In this section we will give a brief explanation of procedures for graphing a relation. A graph is nothing more than an illustration that gives us, at a glance, a clearer idea of the situation under consideration. A road map indicates where we have been and how to proceed to reach our destination. A flow chart helps us to zero in on the procedures to be followed to code a problem. The graph of the function $y = 2x + 3$ in algebra helps us to understand how the function behaves. Indeed, it tells us that the graph of this function is a straight line. The pictures of relations in the previous section gave us an added insight into what a relation is. They indicated that there are several different ways of graphing relations. We will investigate two additional methods.

Example 6.2.1. Let $A = \{0, 1, 2, 3\}$, and let r be the relation $\{(0, 0), (0, 3), (1, 2), (2, 1), (3, 2), (2, 0)\}$. The elements of A are called the *vertices of the graph*. Place the vertices, enclosed in a circle or denoted by a *point*. Connect vertex a to vertex b with an *arrow*, called *an edge of the graph*, going

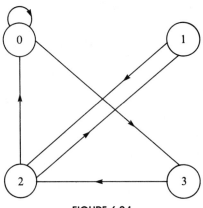

FIGURE 6.2.1
Example 6.2.1, directed graph

from vertex *a* to vertex *b* if and only if *a r b*. This type of graph of a relation *r* is called a *directed graph or digraph*. The result is Figure 6.2.1.

The actual location of the vertices is immaterial. The main idea is to place the vertices in such a way that the graph is easy to read, a help to you. Obviously, after a rough-draft graph of a relation, we may decide to relocate and/or order the vertices so that the final result will be neater. Figure 6.2.1 could be presented as in Figure 6.2.2.

A vertex of a graph is also called a *node,* a *point,* or a *junction.* An edge of a graph is also referred to as an *arc,* a *line,* or a *branch.* Do not be concerned if two graphs of a given relation look different. It is a nontrivial problem to determine if two graphs are graphs of the same relation.

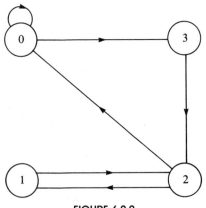

FIGURE 6.2.2
Another way of expressing 6.2.1

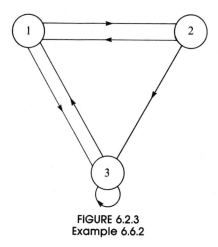

FIGURE 6.2.3
Example 6.6.2

Example 6.2.2. Consider the relation s whose digraph is Figure 6.2.3. What information does this give us? Certainly we know that s is a relation of a set A, where $A = \{1, 2, 3\}$ and $s = \{(1, 2), (2, 1), (1, 3), (3, 1), (2, 3) (3, 3)\}$.

Example 6.2.3. Let $B = \{a, b\}$, and let $A = \mathcal{P}(B) = \{\varnothing, \{a\}, \{b\}, \{a, b\}\}$. Then $\subseteq$ is a relation on A whose digraph is Figure 6.2.4.

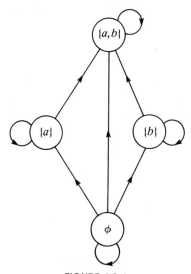

FIGURE 6.2.4
Example 6.2.3

This graph is helpful insofar as it reminds us that each set is a subset of itself (How?) and shows us at a glance the relationship between the various subsets in $\mathscr{P}(B)$. Some relations, such as this one, can also be conveniently depicted by what is called a *Hasse*, or *ordering*, *diagram*. To read a Hasse diagram for a relation on a set A, remember:

(1) Each vertex of A must be related to itself, so the arrows from a vertex to itself are not necessary.

(2) If vertex b appears above vertex a and if vertex a is connected to vertex b by an edge, then $a\ r\ b$, so direction arrows are not necessary.

(3) If vertex c is above vertex a and if c is connected to a by a sequence of edges, then $a\ r\ c$.

(4) The vertices (or nodes) are denoted by *points* rather than by "circles."

The Hasse diagram of the directed graph depicted in Figure 6.2.4 is Figure 6.2.5.

Example 6.2.4. Consider the relation s whose Hasse diagram is Figure 6.2.6. How do we read this diagram? What is A? What is s? What does the digraph of s look like? Certainly $A = \{a, b, c, d, e\}$ and $a\ s\ c,\ c\ s\ d,\ a\ s\ d,\ a\ s\ e$, etc., so

$$s = \{(a, a), (b, b), (c, c), (d, d), (e, e), (a, c),$$
$$(a, d), (a, e), (a, b), (c, d), (c, e), (d, e), (b, e)\}.$$

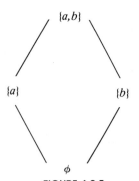

FIGURE 6.2.5
Hasse diagram of the directed
graph depicted in Figure 6.2.4

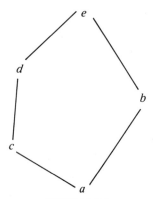

FIGURE 6.2.6
Example 6.2.4

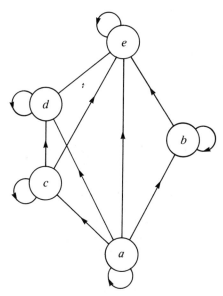

FIGURE 6.2.7
Diagraph of Figure 6.2.6

The digraph for s is Figure 6.2.7. It is certainly more complicated to read than the Hasse diagram.

EXERCISES FOR SECTION 6.2

A Exercises

1. Let $A = \{1, 2, 3, 4\}$, and let r be the relation $\leq$ on A. Draw the digraph and the Hasse diagram of r.

2. Let $B = \{2, 3, 4, 6, 12, 36, 48\}$, and let s be the relation $|$, "divides," on B. Draw the Hasse diagram of s.

3. Draw the Hasse diagram of the relation $\subseteq$ on $\mathscr{P}(A)$, where $A = \{a, b, c\}$.

4. (a) Let A be the set of strings of 0's and 1's of length 3 or less. Define the relation of d on A by $x\, d\, y$ if x is contained within y. For example, $(01)\, d(101)$. Draw the Hasse diagram for this relation.
   (b) Do the same for the relation p defined by $x\, p\, y$ if x is a prefix of y. For example, $(10)\, p\, (101)$, but $(01)\, p\, (101)$ is false.

5. Draw the digraph for the relation ρ in Exercise 5 of Section 6.1, where $S = \{a, b\}$. Explain why a Hasse diagram could not be used to depict ρ.

6.3 Properties of Relations

Consider the set $B = \{1, 2, 3, 4, 6, 12, 36, 48\}$ and the relations "divides" and $\leq$ on B. We notice that these two relations on B have several properties in common. In fact:

(1) Every element in B divides itself and is less than or equal to itself. This is called the *reflexive property*.

(2) If we search for "two" elements from B where the first divides the second and the second divides the first, then we are forced to choose the same first and second number. The reader can verify that a similar result is true for the relation $\leq$ on B. This is called the *antisymmetric property*.

(3) Next if we choose three numbers from B such that the first divides (or is $\leq$) the second and the second divides (or is $\leq$) the third, then this forces the first number to divide (or be $\leq$) the third. This is called the *transitive property*.

Sets on which relations are defined which satisfy the above properties are of special interest to us. More detailed definitions follow.

Definitions: Reflexive, Antisymmetric, and Transitive. *Let A be a set and r a relation on A, then:*

(1) r is called reflexive if and only if a r a for all a $\in$ A.

(2) r is called antisymmetric if and only if whenever a r b and a $\neq$ b then b r a is false; or equivalently whenever a r b and b r a then a = b. (The reader is encouraged to think about both conditions since they are frequently used.)

(3) r is called transitive if and only if whenever a r b and b r c then a r c.

A word of warning about antisymmetry: Students frequently find it difficult to understand this definition. Keep in mind that this term is defined through an "If ... then ... " statement. The question that you must ask is: Is it true that whenever there are elements a and b from A where $a \, r \, b$ and $a \neq b$, it follows that b is not related to a? If so, then the relation r is antisymmetric. Another way to determine whether a relation is antisymmetric is to examine its graph. The relation is not antisymmetric if there exists a pair of vertices that are connected by edges in both directions. Note that the negative of antisymmetric is **not** symmetric. We will define the symmetric property later.

Definition: Partial Ordering. *A relation on a set A which is reflexive, antisymmetric, and transitive is called a partial ordering on A. A set on which there is a partial ordering relation defined is called a partially ordered set or poset.*

Example 6.3.1. Let A be a set. Then $\mathcal{P}(A)$ together with the relation $\subseteq$ is a poset. To prove this we show that the three properties hold:

(1) Let $B \in \mathcal{P}(A)$. We must show that $B \subseteq B$. This is true by definition of subset. Hence, the relation is reflexive.
(2) Let $B_1, B_2 \in \mathcal{P}(A)$ and assume (Why?) that $B_1 \subseteq B_2$ and $B_1 \neq B_1$. Could it be that $B_2 \subseteq B_1$? No. Why? Hence, the relation is anti-symmetric.
(3) Let $B_1, B_2, B_3 \in \mathcal{P}(A)$ and assume that $B_1 \subseteq B_2$ and $B_2 \subseteq B_3$. Does it follow that $B_1 \subseteq B_3$? Yes. Hence, the relation is transitive.

Example 6.3.2. Consider the relation s defined by the Hasse diagram in Figure 6.2.6. A relation defined by a Hasse diagram is always a partial ordering. Let's convince ourselves of this.

(1) First, s is reflexive. If this is not clear draw the digraph of the above relation.
(2) Next, s is antisymmetric. From the diagram, can we find two different elements, say c_1 and c_2, such that $c_2 \ s \ c_1$ and $c_1 \ s \ c_2$? No. If this argument is not clear, try the following: From the diagram find "two" elements such that the first is related through s to the second and the second is related to the first. The only elements that work are pairs of identical elements.
(3) Finally, s is transitive. How do we show this? We must show that for any three elements chosen such that the first is related to the second and the second is related to the third, then the first must be related to the third. Consider, for example, the three elements c, d, and e. From the diagram, $c \ s \ d$ and $d \ s \ e$, so the hypothesis is satisfied. We must show that $c \ s \ e$. This is true from the diagram. How? If this is not clear, the digraph of s may be helpful. Does this one example show transitivity?

Another property that is frequently referred to is that of symmetry.

Definition: Symmetry. *Let r be a relation on a set A. r is called symmetric if and only if whenever $a \ r \ b$, it follows that $b \ r \ a$.*

Consider the relation $=$ defined on **R**. Certainly $a = b$ implies that $b = a$ so $=$ is a symmetric relation on **R**.

Surprisingly, $=$ is also an antisymmetric relation on **R**. This is due to the fact that the condition of the antisymmetry property, $a = b$ and $a \neq b$, is a contradiction. Remember, a conditional proposition is always true when the condition is false. So a relation can be both symmetric and antisymmetric on a set! Again recall that these terms are not negatives of each other.

Definition: Equivalence Relation. *A relation r on a set A is called an equivalence relation if and only if it is reflexive, symmetric, and transitive.*

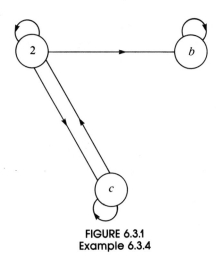

FIGURE 6.3.1
Example 6.3.4

The classic example of an equivalence relation is the relation $=$ on **R**. In fact, the term *equivalence relation* is used because those relations which satisfy the definition behave quite like the $=$ relation.

Example 6.3.3. Let **Z*** be the set of non-zero integers. One of the most basic equivalence relations in mathematics is the relation q on $\mathbf{Z} \times \mathbf{Z}^*$ defined by $(a, b)\, q\, (c, d)$ if and only if $ad = bc$. We will leave it to the reader to verify that q is indeed an equivalence relation. Two ordered pairs, (a, b) and (c, d), are related if the fractions a/b and c/d are numerically equal.

Example 6.3.4. Consider the relation s described by the digraph in Figure 6.3.1. This relation is reflexive (Why?), not symmetric (Why?), and not transitive (Why?). Is s an equivalence relation? A partial ordering?
  A classic example of a partial ordering relation is $\leq$ on **R**. Indeed, when graphing partial ordering relations, it is natural to "plot" the elements from the given poset starting with the "least" element to the "greatest" and to use terms like "least," "greatest," etc. Because of this the reader should be forewarned that many texts use the $\leq$ notation when describing an arbitrary partial ordering. This can be quite confusing for the novice, so we continue to use the general notation r, s, etc., when speaking of relations.

EXERCISES FOR SECTION 6.3

  A Exercises

  1. (a) Let $B = \{a, b\}$ and $A = \mathscr{P}(B)$. Draw the Hasse diagram for $\subseteq$ on A.

      (b) Let $A = \{1, 2, 3, 6\}$. Show that $|$, "divides," is a partial ordering on A. Draw the Hasse diagram.

(c) Compare the graphs of Parts *a* and *b*.

2. (a) Repeat Exercise 6.3.1a for $B = \{a, b, c\}$.
   (b) Repeat Exercise 6.3.1b for $A = \{1, 2, 3, 5, 6, 10, 15, 30\}$.

3. (a) Consider the relations defined by the digraphs in Figure 6.3.2. Determine whether the given relations are reflexive, symmetric, antisymmetric, or transitive. Try to develop procedures for determining the validity of these properties from the graphs.
   (b) Which of the above are graphs of equivalence relations or of partial ordering relations?
   (c) List the set of ordered pairs in each of the above relations.

4. Determine which of the following are equivalence relations and/or partial ordering relations for the given sets.
   (a) $A = \{$lines in the plane$\}$; $x \, r \, y$ if and only if x is parallel to y.
   (b) $A = \mathbf{R}$; $x \, r \, y$ if and only if $|x - y| \leq 7$

5. Let $A = \{0, 1, 2, 3\}$ and let $r = \{(0, 0), (1, 1), (2, 2), (3, 3), (1, 2), (2, 1), (3, 2), (2, 3), (3, 1), (1, 3)\}$.
   (a) Show that r is an equivalence relation on A.
   (b) Let $a \in A$ and define $c\,(a) = \{b \in A \,|\, a \, r \, b\}$. $c\,(a)$ is called the *equivalence class* of the element a under r. Find $c\,(a)$ for each element $a \in A$.
   (c) Show that $\{c\,(a) \,|\, a \in A\}$ forms a partition of A for this set A.
   (d) Let r be an equivalence relation on an arbitrary set A. Prove that the set of all equivalence classes constitutes a partition of A.

6. For the set of cities on a map, consider the relation $x \, r \, y$ if and only if city x is connected by a road to city y. A city is considered to be connected to itself and two cities are connected even though there are cities on the road between them. Is this an equivalence relation or a partial ordering? Explain.

7. Consider the following relation on $\{1, 2, 3, 4, 5, 6\}$ $r = \{(i, j): |i-j| = 2\}$.
   (a) Is r reflexive?
   (b) Is r symmetric?
   (c) Is r transitive?
   (d) Draw a graph of r.

8. Consider the following relations on $\mathbf{Z}_8 = \{0, 1, \ldots, 7\}$. Which are equivalence relations? For the equivalence relations, list the equivalence classes.
   (a) $a \, r \, b$ iff the English spellings of a and b begin with the same letter.
   (b) $a \, s \, b$ iff $a-b$ is a positive integer.

9. Define t on $A = \{1, 2, \ldots, 9\}$ by $x t y$ iff $x + y = 10$. Is t an equivalence relation on A? If yes, list its equivalence classes. If no, why not?

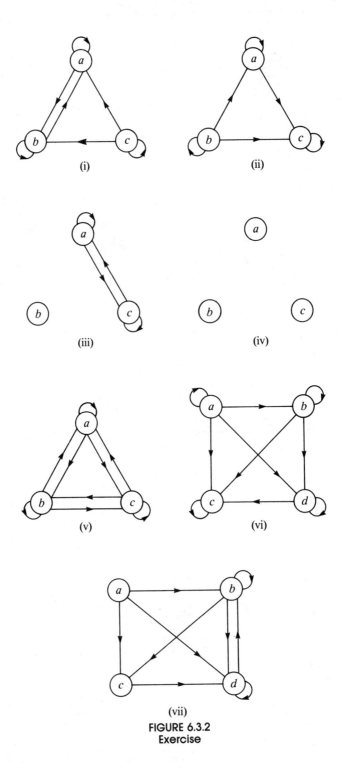

(i)

(ii)

(iii)

(iv)

(v)

(vi)

(vii)

FIGURE 6.3.2
Exercise

B Exercises

10. Let $A = \{q_i \mid i = 1, 2, 3, \ldots, 8\}$, where each q_i is a proposition. Let the relation on A be $\Rightarrow$.

   (a) Verify that $q \rightarrow q$ is a tautology, thereby showing that $\Rightarrow$ is reflexive on A.
   (b) Prove that $\Rightarrow$ is antisymmetric on A. Note we do not use $=$ when speaking about propositions, but, rather, $\Leftrightarrow$.
   (c) Prove that $\Rightarrow$ is transitive on A.
   (d) Given that q_i is the proposition $n \le i$ on $\mathbf{N}$, draw the Hasse diagram.

C Exercises

11. For those familiar with linked lists, it may be clear that an alternate way of representing a digraph is via a linked list. Represent the digraph in Figure 6.3.3 by a linked list.

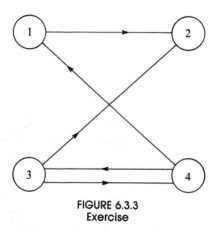

FIGURE 6.3.3
Exercise

12. Let $S = \{a, b, c, d, e, f, g\}$ be a poset $[S, \le]$ with the Hasse diagram shown in Figure 6.3.4. Another relation $r \subseteq S \times S$ is defined as follows: $(x, y) \in r$ if and only if there exists $z \in S$ such that $z \le x$ and $z \le y$ in the poset $[S, \le]$.

   (a) Prove that r is reflexive.
   (b) Prove that r is symmetric.
   (c) A *compatible* with respect to relation r is any subset Q of set S such that $x \in Q$ and $y \in Q \Rightarrow (x, y) \in r$. A compatible Q is a *maximal compatible* if Q is not a proper subset of another compatible. Give all maximal compatibles with respect to relation r defined above.

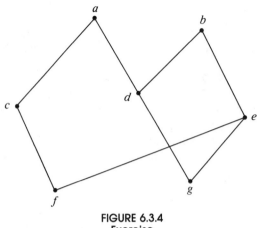

FIGURE 6.3.4
Exercise

(d) Discuss a characterization of the set of maximal compatibles for relation r when $[S, \leq]$ is a general finite poset. What conditions, if any, on a general finite poset $[S, \leq]$ will make r an equivalence relation?

6.4 Matrices of Relations

We have discussed two of the many possible ways of representing a relation, namely as a digraph or as a set of ordered pairs. In the exercises of Section 6.3 we mentioned a third, that is, as a linked list. In this section we will discuss the representation of relations by matrices and some of its applications.

Definition: Adjacency Matrix. *Let A and B be finite sets or orders m and n, respectively, and let r be a relation from A into B. Then r can be represented by a matrix R called the adjacency matrix (or the Boolean matrix or the relation matrix) of r.*

To construct the matrix R, proceed as follows:

Let $A = \{a_1, a_2, \ldots, a_m\}$ and $B = \{b_1, b_2, \ldots, b_n\}$. Then R is an $m \times n$ matrix whose ith jth entry

$$R_{ij} = \begin{cases} 1 \text{ if } a_i \ r \ b_j; \\ 0 \text{ otherwise.} \end{cases}$$

Example 6.4.1. Let $A = \{2, 5, 6\}$ and let r be the relation $\{(2, 2) \ (2, 5), (5, 6), (6, 6)$ on A. Since r is a relation from A into the same set A (the B of the definition), we have $a_1 = 2$, $a_2 = 5$, and $a_3 = 6$ and $b_1 = 2$, $b_2 = 5$, and $b_3 = 6$. Next, since:

$$2r2, \text{ we have } R_{11} = 1;$$
$$2r5, \text{ we have } R_{12} = 1;$$
$$5r6, \text{ we have } R_{23} = 1; \text{ and}$$
$$6r6, \text{ we have } R_{33} = 1.$$

All other entries of R are 0. So

$$R = \begin{bmatrix} 1 & 1 & 0 \\ 0 & 0 & 1 \\ 0 & 0 & 1 \end{bmatrix}.$$

From the definition of r we note that

$$r^2 = \{(2, 2)\ (2, 5)\ (2, 6)\ (5, 6)\ (6, 6)\}.$$

The adjacency matrix of r^2 is

$$R = \begin{bmatrix} 1 & 1 & 1 \\ 0 & 0 & 1 \\ 0 & 0 & 1 \end{bmatrix}.$$

We do not write R^2 only for notational purposes. In fact, R^2 can be obtained from the matrix product RR; however, we must use a slightly different form of arithmetic.

Definition: Boolean Arithmetic. *Boolean arithmetic is the arithmetic defined on $\{0, 1\}$ using Boolean addition and Boolean multiplication, defined as:*

$$0 + 0 = 0 \qquad 0 + 1 = 1 + 0 = 1 \qquad 1 + 1 = 1$$
$$0 \cdot 0 = 0 \qquad 0 \cdot 1 = 1 \cdot 0 = 0 \qquad 1 \cdot 1 = 1.$$

From Chapter 3 it is clear that this is the "arithmetic of logic," where + replaces "or" and • replaces "and."

Example 6.4.2.

$$\text{If } R = \begin{bmatrix} 0 & 1 & 0 & 0 \\ 1 & 0 & 1 & 0 \\ 0 & 1 & 0 & 1 \\ 0 & 0 & 1 & 0 \end{bmatrix} \quad \text{and} \quad S = \begin{bmatrix} 0 & 1 & 1 & 1 \\ 0 & 0 & 1 & 1 \\ 0 & 0 & 0 & 1 \\ 0 & 0 & 0 & 0 \end{bmatrix},$$

$$\text{then } RS = \begin{bmatrix} 0 & 0 & 1 & 1 \\ 0 & 1 & 1 & 1 \\ 0 & 0 & 1 & 1 \\ 0 & 0 & 0 & 1 \end{bmatrix} \quad \text{and} \quad SR = \begin{bmatrix} 1 & 1 & 1 & 1 \\ 0 & 1 & 1 & 1 \\ 0 & 0 & 1 & 0 \\ 0 & 0 & 0 & 0 \end{bmatrix}, \text{ using}$$

Boolean arithmetic.

Theorem 6.4.1.   *Let A_1, A_2 and A_3 be finite sets where r_1 is a relation from A_1 into A_2 and r_2 is a relation from A_2 into A_3. If R_1 and R_2 are the adjacency matrices of r_1 and r_2 respectively, then the product R_1R_2 is the adjacency matrix of the composition r_1r_2.*

Remark: A convenient help in constructing the adjacency matrix of a relation from a set A into a set B is to write the elements from A in a column preceding the first column of the adjacency matrix and the elements of B in a row above the first row. Initially, the matrix in Example 6.4.1 would be

$$R = \begin{array}{c} \\ 2 \\ 5 \\ 6 \end{array} \overset{\displaystyle 2 \quad 5 \quad 6}{\left[\right]},$$

and R_{ij} is 1 if and only if $(a_i, b_j) \in r$. So that, since the pair $(2, 5) \in r$, the entry of R corresponding to the row labeled 2 and the column labeled 5 in the matrix is a 1.

EXERCISES FOR SECTION 6.4

A Exercises

1. Let $A_1 = \{1, 2, 3, 4\}$, $A_2 = \{4, 5, 6\}$, and $A_3 = \{6, 7, 8\}$. Let r_1 be the relation from A_1 into A_2 defined by $r_1 = \{(x, y) \mid x + y = 2\}$, and let r_2 be the relation from A_2 into A_3 defined by $r_2 = \{(x, y) \mid y - x = 1\}$.
    (a) Determine the adjacency matrices of r_1 and r_2.
    (b) Use the definition of composition to find r_1r_2.
    (c) Verify the result in part by finding the product of the adjacency matrices of r_1 and r_2.

2. (a) Determine the adjacency matrix of each relation given via the digraphs in Exercise 3 of Section 6.3.
    (b) Using the matrices found in Part a above, find r^2 of each relation in Exercise 3 of Section 6.3.
    (c) Find the digraph of r^2 directly from the given digraph and compare your results with those of Part b.

3. Suppose that the matrices in Example 6.4.2 are relations on $\{1, 2, 3, 4\}$. What relations do R and S describe?

4. Let $D =$ days of the week (M–F),
    $\qquad W = \{1, 2, 3\}$ employees of a programming tutorial center, and
    $\qquad V =$ languages offered
    $\qquad = \{A(da), B(asic), C(obol), F(ortran), L(isp), P(ascal)\}$.

    We define s (schedule) from D into W by dsw if w is scheduled to work on day d. We also define r from W into V by $w\, r\, 1$ if w can tutor students in language 1. If

$$
\begin{array}{c}
\quad\quad 1\ \ 2\ \ 3 \\
S = \begin{array}{c} M \\ T \\ W \\ Th \\ F \end{array}
\begin{bmatrix}
1 & 0 & 1 \\
0 & 1 & 1 \\
1 & 0 & 1 \\
0 & 1 & 0 \\
1 & 1 & 0
\end{bmatrix}
\end{array}
\qquad
\text{and } R = \begin{array}{c} 1 \\ 2 \\ 3 \end{array}
\begin{array}{cccccc}
A & B & C & F & L & P \\
\begin{bmatrix}
0 & 1 & 1 & 0 & 0 & 1 \\
1 & 1 & 0 & 1 & 0 & 1 \\
0 & 1 & 0 & 0 & 1 & 1
\end{bmatrix}
\end{array},
$$

determine SR and give an interpretation of this relation.

5. How many different reflexive, symmetric relations are there on a set with three elements? (Hint: Consider the possible matrices.)

6. Let $A = \{a, b, c, d\}$. Let r be the relation on A with adjacency matrix

$$
\begin{array}{c}
\quad\ a\ \ b\ \ c\ \ d \\
\begin{array}{c} a \\ b \\ c \\ d \end{array}
\begin{bmatrix}
1 & 0 & 0 & 0 \\
0 & 1 & 0 & 0 \\
1 & 1 & 1 & 0 \\
0 & 1 & 0 & 1
\end{bmatrix}
\end{array}
$$

(a) Explain why r is a partial order on A.
(b) Draw its Hasse diagram.

7. Define relations P and Q on $\{1, 2, 3, 4\}$ by $P = \{(a, b) : |a-b| = 1\}$ and $Q = \{(a, b) : a - b$ is even $\}$.

(a) Represent P and Q as both graphs and matrices.
(b) Determine PQ, P^2, and Q^2 and represent them clearly in any way.

B Exercise

8. (a) Prove that if r is a transitive relation on a set A, then $r^2 \subseteq r$.
(b) Find an example of a transitive relation for which $r^2 \neq r$.

6.5 Closure Operations on Relations

In Section 6.1, we studied relations and one important operation on relations, namely composition. This operation enabled us to generate new relations from previously known relations. In Section 6.3, we discussed some key properties of relations. We now wish to consider the situation of constructing a new relation r^+ from a previously known relation r where, first, r^+ contains r and, second, r^+ satisfies the transitive property.

Consider a telephone network in which the main office a is connected to, and can communicate to, individuals b and c. Both b and c can communicate to another person, d; however, the main office cannot communicate with d. Assume communication is only one way, as indicated. This situation can be described by the relation $r = \{(a, b), (a, c), (b, d), (c, d)\}$. We would like to change the system so that the main office a can communicate with person *d* and still maintain the previous system. We, of course, want the most economical system.

This can be rephrased as follows: Find the smallest relation r^+ which contains r as a subset and which is transitive; $r^+ = \{(a, b), (a, c), (b, d),$ $(c, d), (a, d)\}$.

Definition: Transitive Closure. *Let A be a set and r be a relation on A. The transitive closure of r, denoted by r^+, is the smallest relation which contains r as a subset and which is transitive.*

Example 6.5.1. Let $A = \{1, 2, 3, 4\}$, and let $S = \{(1, 2), (2, 3), (3, 4)\}$ be a relation on A. This relation is called the *successor relation on A* since each element is related to its successor. How do we compute S^+?

By inspection we note that $(1, 3)$ must be in S^+. Let's analyze why. This is so since $(1, 2) \in S$ and $(2, 3) \in S$, and the transitive property forces $(1, 3)$ to be in S^+. In general, it follows that if $(a, b) \in S$ and $(b, c) \in S$, then $(a, c) \in S^+$. This condition is exactly the membership requirement for the pair (a, c) to be in the composition $S S = S^2$. So every element in S^2 must be an element in S^+. So far, S^+ contains at least $S \cup S^2$. In particular, for this example, since $S = \{(1, 2), (2, 3), (3, 4)\}$ and $S^2 = \{(1, 3), (2, 4)\}$, we have $S \cup S^2 = \{(1, 2), (2, 3), (3, 4), (1, 3), (2, 4)\}$.

Is the relation $S \cup S^2$ transitive? Again, by inspection, $(1, 4)$ is not an element of $S \cup S^2$, but it must be an element of S^+ since $(1, 3)$ and $(3, 4)$ are required to be in S^+. From above, $(1, 3) \in S^2$ and $(3, 4) \in S$, and the composite $S^2 S = S^3$ produces $(1, 4)$.

This shows that $S^3 \subseteq S^+$. This process must be continued until the resulting relation is transitive. If A is finite, as is true in this example, the transitive closure will be obtained in a finite number of steps. Here, $S^+ = S \cup S^2 \cup S^3$.

Theorem 6.5.1. *If S is a relation on a set A and if $\#A = n$, then the transitive closure $S^+ = S \cup S^2 \cup S^3 \cup \ldots \cup S^n$.*

Let's now consider the matrix analogue of the transitive closure.

Example 6.5.2. Consider the relation $r = \{(1, 4), (2, 1), (2, 2), (2, 3), (3, 2), (4, 3), (4, 5), (5, 1)\}$ on the set $A = \{1, 2, 3, 4, 5\}$. The matrix R of r is

$$\begin{bmatrix} 0 & 0 & 0 & 1 & 0 \\ 1 & 1 & 1 & 0 & 0 \\ 0 & 1 & 0 & 0 & 0 \\ 0 & 0 & 1 & 0 & 1 \\ 1 & 0 & 0 & 0 & 0 \end{bmatrix}.$$

Recall that $r^2, r^3, \ldots$ can be determined through computing the matrices R^2, $R^3, \ldots$. Here,

$$R^2 = \begin{bmatrix} 0 & 0 & 1 & 0 & 1 \\ 1 & 1 & 1 & 1 & 0 \\ 1 & 1 & 1 & 0 & 0 \\ 1 & 1 & 0 & 0 & 0 \\ 0 & 0 & 0 & 1 & 0 \end{bmatrix}, \qquad R^3 = \begin{bmatrix} 1 & 1 & 0 & 0 & 0 \\ 1 & 1 & 1 & 1 & 1 \\ 1 & 1 & 1 & 1 & 0 \\ 1 & 1 & 1 & 1 & 0 \\ 0 & 0 & 1 & 0 & 1 \end{bmatrix},$$

$$R^4 = \begin{bmatrix} 1 & 1 & 1 & 1 & 0 \\ 1 & 1 & 1 & 1 & 1 \\ 1 & 1 & 1 & 1 & 1 \\ 1 & 1 & 1 & 1 & 1 \\ 1 & 1 & 0 & 0 & 0 \end{bmatrix}, \quad \text{and} \quad R^5 = \begin{bmatrix} 1 & 1 & 1 & 1 & 1 \\ 1 & 1 & 1 & 1 & 1 \\ 1 & 1 & 1 & 1 & 1 \\ 1 & 1 & 1 & 1 & 1 \\ 1 & 1 & 1 & 1 & 0 \end{bmatrix}.$$

How do we relate $\bigcup\limits_{i=i}^{5} r^i$ to the powers of R?

Theorem 6.5.2. *Let R^+ be the matrix of r^+, the transitive closure of r, which is a relation on a set of n elements. Then $R^+ = R + R^2 + \cdots + R^n$, where addition is done using Boolean arithmetic.*

Using this theorem, we find R^+ is the 5×5 matrix consisting of all 1's, thus, r^+ is all of $A \times A$.

WARSHALL'S ALGORITHM

Let r be a relation on the set $\{1, 2, \ldots, n\}$ with relation matrix R. The matrix of the transitive closure R^+, can be computed by the equation $R^+ = R + R^2 + \cdots + R^n$. By using ordinary polynomial evaluation methods, you can compute R^+ with $n - 1$ matrix multiplications: $R^+ = R(I + R(I + \cdots R(I + R))) \cdots))$. For example, if $n = 3$, $R^+ = R(I + R(I + R))$.

Warshall's algorithm makes use of the fact that if T is a relation matrix, $T + T = T$ due to the fact that $1 + 1 = 1$ in Boolean arithmetic. Let $S_k = R + R^2 + \cdots + R^k$. Then

$$R = S_1$$
$$S_1(I + S_1) = R(I + R) = R + R^2 = S^2$$
$$S_2(I + S_2) = (R + R^2)(I + R + R^2)$$
$$= (R + R^2) + (R^2 + R^3) + (R^3 + R^4)$$
$$= R + R^2 + R^3 + R^4$$
$$= S_4$$
$$S_4(I + S_4) = S_8.$$

Note how each matrix multiplication doubles the number of terms that have been added to the sum that you currently have computed. In algorithmic form, we can compute R^+ as follows.

Algorithm 6.5.1: Washall's Algorithm—Version 1. *Let R be a known relation matrix and let* R^+ *be its transive closure matrix, which is to be computed.*

    1.0. $T := R$

    2.0. Repeat

           2.1 $S := T$

           2.2 $T := S(I + S)$(*using Boolean arithmetic*)

           Until $T = S$

    3.0. Terminate with $R^+ = T$.

Notes:

(a) Often the higher-powered terms in S_n do not contribute anything to R^+. When the condition $T = S$ becomes true in Step 2, this is an indication that no higher-powered terms are needed.

(b) To compute R^+ using Version 1 of Warshall's Algorithm, you need to perform no more than $\lceil \log_2 n \rceil$ matrix multiplications, where $\lceil x \rceil$ is the least integer that is greater than or equal to x. For example, if r is a relation on 25 elements, no more than $\lceil \log_2 25 \rceil = 5$ matrix multi-plications are needed.

    A second, improved version of Warshall's Algorithm reduces computation time to the time that it takes to perform one matrix multi-plication.

Algorithm 6.5.2: Warshall's Algorithm—Version 2. Let R be a known relation matrix and let R^+ be its transive closure matrix, which is to be computed.

    1.0 $T := R$

    2.0 FOR $k := 1$ to n DO

           FOR $i := 1$ to n DO

                 FOR $j := 1$ to n DO

                      $T[i, j] := T[i, j] + (T[i, k] \cdot T[k, j])$

    3.0 Terminate with $R^+ = T$.

EXERCISES FOR SECTION 6.5

    A Exercises

1. Let A and S be as in Example 6.5.1. Compute S^+ as in Example 6.5.2. Verify your results by checking against the relation S^+ obtained in Example 6.5.1.

2. Let A and r be as in Example 6.5.2. Compute the relation r^+ as in Example 6.5.1. Verify your results.

3. (a) Write the digraphs of the relations S, S^2, S^3 and S^+ of Example 6.5.1.
   (b) Verify that in terms of the graph of S, a $S^+ b$ if and only if b is reachable from a using a path of any finite length.

4. Let r be the relation represented by the digraph in Figure 6.5.1.
   (a) Find r^+.
   (b) Determine the digraph of r^+ directly from the digraph of r.
   (c) Verify your result in Part b by computing the digraph from your result in Part a.

5. (a) Define reflexive closure and symmetric closure by imitating the definition of transitive closure.
   (b) Use your definitions to compute the reflexive and symmetric closures of Examples 6.5.1 and 6.5.2.
   (c) What are the transitive reflexive closures of these examples?
   (d) Convince yourself that the reflexive closure of the relation $<$ on the set of positive integers $\mathbf{P}$ is $\leq$.

6. What common relations on $\mathbf{Z}$ are the transitive closures of the following relations?
   (a) aSb if and only if $a + 1 = b$.
   (b) aRb if and only if $|a - b| = 2$.

B Exercise

7. (a) Let A be any set and r a relation on A. Prove $(r^+)^+ = r^+$.
   (b) Is the transitive closure of a symmetric relation symmetric and reflexive? Explain.

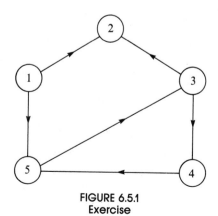

FIGURE 6.5.1
Exercise

chapter

7

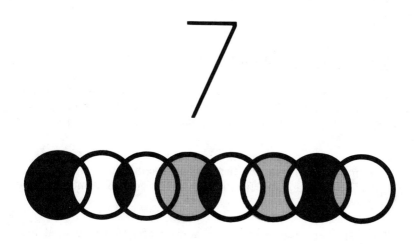

FUNCTIONS

7

GOALS

In this chapter we will consider some basic concepts of the relations which are called functions. A large variety of mathematical ideas and applications can be more completely understood when expressed through the function concept.

7.1 Definition of a Function and Notation

Definition: Function. *A function from a set A into a set B is a relation from A into B such that each element of A is related to exactly one element of the set B. The set A is called the domain of the function and the set B is called the codomain.*

This definition of a function is the standard one that appears in most texts. The reader should note that a function f is a relation from A into B with two important restrictions:

1. Each element in the set A, the domain of f, must be used by the procedure or rule defining f, and
2. The phrase "is related to exactly one element of the set B" means that if $(a, b) \in f$ and $(a, c) \in f$, then $b = c$.

Example 7.1.1. Let $A = \{-2, -1, 0, 1, 2\}$ and $B = \{0, 1, 2, 3, 4\}$, and define s as $s = \{(-2, 4), (-1, 1), (0, 0), (1, 1), (2, 4)\}$, which is a function from A into B.

Example 7.1.2. Let $\mathbf{R}$ be the real numbers. Then $L = \{(x, 3x) \mid x \in \mathbf{R}\}$ is a function from $\mathbf{R}$ into $\mathbf{R}$, or, more simply, L is a function *on* $\mathbf{R}$.

We will use a different system of notation for functions than the one we used for relations. If f is a function from the set A into the set B, we will write $f : A \rightarrow B$.

The reader is probably more familiar with the procedure for describing functions that is used in basic algebra or calculus courses, for example, $y = x^2$, $f(x) = 2x + 1$ and $g(x) = 1/x$. Here the domain was assumed to be those elements of $\mathbf{R}$ whose substitutions for x make sense and the codomain was assumed to be $\mathbf{R}$. We, however, will list the domain and codomain in addition to describing what the function does in order to define a function. The terms *mapping, map,* and *transformation* are also used for functions.

One way to imagine a function and what it does is to think of it as a machine. The machine could be mechanical, electronic, hydraulic, or abstract. Imagine that the machine only accepts certain objects as raw materials or input. The possible raw materials make up the domain. Given some input, the machine produces a finished product which depends on the input. The possible finished products that we imagine could come out of this process make up the codomain.

Example 7.1.3. $f: \mathbf{R} \to \mathbf{R}$ defined by $f(x) = x^2$ is an alternate (and preferred) description of $f = \{(x, x^2) \mid x \in \mathbf{R}\}$.

Definition: Image of an Element. *Let $f: A \to B$, (read "let f be a function from the set A into the set B"). If $a \in A$, then $f(a)$ is used to denote that element of B that a is related to. $f(a)$ is called the image of a, or, more precisely, the image of a under f. We write $f(a) = b$ to indicate that the image of a is b.*

In Example 7.1.3, the image of 2 under f is 4; that is, $f(2) = 4$. In example 7.1.1, the image of -1 under s is 1; that is, $s(-1) = 1$.

Definition: Range of a Function. *The range of a function is the set of images of its domain, denoted $f(domain)$. If $f: X \to Y$, then $f(X) = \{f(a) \mid a \in X\} = \{b \in Y \mid \exists a \in X \text{ such that } f(a) = b\}$.*

Note that the range of a function is a subset of its codomain. $f(X)$ is also read as "the image of the set X under the function f," or simply "the image of f."

In Example 7.1.1, $s(A) = \{0, 1, 4\}$, 2 and 3 are not images of any element of A. In addition, note that both 1 and 4 are related to more than one element of the domain $s(1) = s(-1) = 1$ and $s(2) = s(-2) = 4$. Reread the definition of a function if you feel that this violates the definition of a function.

In Example 7.1.2, the range of L is equal to its codomain, $\mathbf{R}$. If b is any real number, we can demonstrate that it belongs to $L(\mathbf{R})$ by finding a real number x for which $L(x) = b$. By the definition of L, $L(x) = 3x$, which leads us to the equation $3x = b$. This equation always has a solution $b/3$; thus $L(\mathbf{R}) = \mathbf{R}$.

The formula that we used to describe image of a real number under L, $L(x) = 3x$ is preferred over the set notation for L due to its brevity. Anytime that a function can be described with a rule or formula, we will use this form of a description. In the first example, the image of each element of A is its square. To describe that fact, we write $S(a) = a^2 (a \in A)$, or $S: A \to B$ defined by $S(a) = a^2$.

There are many ways that a function can be described. The complexity of the function often dictates its representation.

Example 7.1.4. Suppose a survey of 1000 persons is done asking how many hours of television each watches per day. Consider the function $W: \{0, 1, \ldots, 24\} \to \{0, \ldots, 1000\}$ defined by $W(a) = $ number of persons who gave a response of a hours. This function will probably have no formula such as the ones for s and L above. A bar graph might be the best way to represent W.

Example 7.1.5. Consider the function $m\colon \mathbf{P} \to \mathbf{Q}$ defined by the set $m = \{(1, 1), (2, 1/2), (3, 9), (4, 1/4), (5, 25), \ldots \}$. No single simple formula could describe m, but if we assume that the pattern given continues, we can write

$$m(x) = \begin{cases} x^2 & \text{if } x \text{ is odd;} \\ 1/x & \text{if } x \text{ is even.} \end{cases}$$

FUNCTIONS OF TWO VARIABLES

If the domain of a function is the Cartesian product of two sets, then our notation and terminology is changed slightly. For example, consider the function $C\colon \mathbf{N} \times \mathbf{N} \to \mathbf{N}$ defined by $C((n_1, n_2)) = n_1^2 + n_2^2 - n_1 n_2 + 10$. For this function, we would drop one set of parentheses and write $C(4, 2) = 22$, not $C((4, 2)) = 22$. We call C a function of two variables. From one point of view, this function is no different from any others that we have seen. The elements of the domain happen to be slightly more complicated. On the other hand, we can look at the individual components of the ordered pairs as being separate. If we interpret C as giving us the cost of producing quantities of two products, we can imagine varying n_1, while n_2 is fixed or vice versa.

PASCAL NOTE

The heading of a Pascal function lists the name, domain, and codomain of the function, in that order. For example, FUNCTION GAUSS (X:REAL): REAL is a heading for a function called GAUSS with a real domain and codomain.

FUNCTION MAX (FIRST, SECOND: INTEGER): INTEGER describes a function with domain $\mathbf{Z} \times \mathbf{Z}$ and codomain $\mathbf{Z}$.

The analogy between mathematical functions and Pascal functions breaks down when a functional value depends on a global variable. An example of this situation is if Y is declared a global integer variable in the main program in which GAUSS is used and if the body of GAUSS is

```
BEGIN
    GAUSS:= 2*X+Y
END
```

Then GAUSS's domain is actually $\mathbf{R} \times \mathbf{R}$ even though the heading does not indicate it. In most cases, the use of a global variable in a function definition is not considered to be good programming style. Thus, the heading should describe the domain.

We close this section with two examples of relations which are not functions.

Example 7.1.6. Let $A = B = \{1, 2, 3\}$ and let $f = \{(1, 2), (2, 3)\}$. Here f is not a function since f does not act on, or "use," all elements of A.

Example 7.1.7. Let $A = B = \{1, 2, 3\}$ and let $g = \{(1, 2), (2, 3), (2, 1), (3, 2)\}$. We note that g acts on all of A. However, g is still not a function since $(2, 3) \in g$ and $(2, 1) \in g$; yet $3 \neq 1$.

EXERCISES FOR SECTION 7.1

A Exercises

1. Let $A = \{1, 2, 3, 4\}$ and $B = \{a, b, c, d\}$. Determine which of the following are functions. Explain.
   (a) $f \subseteq A \times B$, where $f = \{(1, a), (2, b), (3, c), (4, d)\}$.
   (b) $g \subseteq A \times B$, where $g = \{(1, a), (2, a), (3, b), (4, d)\}$.
   (c) $h \subseteq A \times B$, where $h = \{(1, a), (2, b), (3, c)\}$.
   (d) $k \subseteq A \times B$, where $k = \{(1, a), (2, b), (2, c), (3, a), (4, a)\}$.
   (e) $L \subseteq A \times A$, where $L = \{(1, 1), (2, 1), (3, 1), (4, 1)\}$.

2. Let A be a set and let S be any subset of A. Let $C_s : A \to \{0, 1\}$ be defined by

$$C_s(x) = \begin{cases} 1 & \text{if } x \in S; \\ 0 & \text{if } x \notin S. \end{cases}$$

   The function C_s is called the *characteristic* function of S.
   (a) If $a = \{a, b, c\}$ and $S = \{a, b\}$, find C_s.
   (b) If $A = \{a, b, c, d, e\}$ and $S = \{a, c, e\}$, find C_s.
   (c) If $A = \{a, b, c\}$, find C_ϕ, C_A.

3. Find the ranges of each of the relations which are functions in Exercise 1.

4. Find the ranges of the following functions on $\mathbf{Z}$:
   (a) $g = \{(x, 4x + 1)\,|\,x \in \mathbf{Z}\}$.
   (b) $h(x) =$ largest integer that is greater than or equal to $\sqrt{|x|}$.
   (c) $P(x) = x + 10$.

B Exercise

5. If $\#A$ and $\#B$ are both finite, how many different functions are there from A into B?

C Exercise

6. Let f be a function with domain A and codomain B. The relation $K \subseteq A \times A$ is defined on the domain of f by $(x, y) \in K$ iff $f(x) = f(y)$. The relation K is called the *kernel* of the function f.

(a) Prove that K is an equivalence relation.
(b) For the specific case of $A = \mathbf{Z}$, where $\mathbf{Z}$ is the set of integers, let $f:\mathbf{Z} \rightarrow \mathbf{Z}$ be defined by $f(s) = x^2$. Describe the equivalence classes of the kernel for this specific function.

7.2 Injective, Surjective, and Bijective Functions

Consider the following functions:

Example 7.2.1.

Let $A = \{1, 2, 3, 4\}$ and $B = \{a, b, c, d\}$, and define $f:A \rightarrow B$ by
$f(1) = a$
$f(2) = b$
$f(3) = c$
$f(4) = d$

Example 7.2.2.

Let $A = \{1, 2, 3, 4\}$ and $B = \{a, b, c, d\}$, and define
$g:A \rightarrow B$ by
$g(1) = a$
$g(2) = b$
$g(3) = a$
$g(4) = b$
$g(4) = b$

The function in the first example gives us more information about the set B than the second function. Since A clearly has four elements, f tells us that the set B contains at least four elements since each element of the set A is mapped onto one and only one element of the set B. The properties that f has and g does not have are the most basic properties that we look for in a function. The following definitions summarize the basic vocabulary for function properties.

Definitions: We say that a function $f:A \rightarrow B$ is:

(1) *injective, or one to one, if $a \neq b$ implies $f(a) \neq f(b)$ (or, equivalently, if $f(a) = f(b)$ then $a = b$) for all a and b in A.*
(2) *surjective, or onto, if the range of f is equal to B (or, equivalently, if for every $b \in B$, $b = f(a)$ for some $a \in A$).*
(3) *bijective, or a one-to-one correspondence, if f is both injective and surjective.*

Functions that are injective, surjective, or bijective are called *injections, surjections,* and *bijections,* respectively.

Example 7.2.3. The map f of Example 7.2.1 is bijective.

Example 7.2.4. Let $A = \{1, 2, 3\}$ and $B = \{a, b, c, d\}$, and define $f : A \to B$ by $f(1) = b, f(2) = c$, and $f(3) = a$. Then f is injective but not surjective.

Example 7.2.5. The characteristic function, C_S in Exercise 2 of Section 7.1, is surjective if S is a proper subset of A, but never injective if $\#A > 2$.

Example 7.2.6. Let $A =$ students in this classroom, let $B =$ chairs in this classroom, and let S be the function which maps each student into the chair he or she is sitting in. When is this function one to one? Onto?

Functions can also be used for counting the elements in large finite sets or in infinite sets. Let's say we wished to count the occupants in an auditorium containing 1500 seats. If each seat is occupied, the answer is obvious, namely 1500 people. What we have done is to set up a one-to-one correspondence, or bijection, from seats to people, as in Example 7.2.6. We formalize in a definition.

Definition: Cardinality. *Two sets are said to have the same cardinality if there exists a bijection between them.*

If a set is finite or has the same cardinality as the set of natural numbers, it is called a *countable set.* If a set is finite, we say that its cardinality is the number of elements that it has.

Readers who have studied analysis realize that the set of rational numbers is a countable set, while the set of real numbers is not a countable set.

Example 7.2.6. The alphabet $\{A, B, C, \ldots, Z\}$ has cardinality 26 through the following bijection:

$$
\begin{array}{cccc}
A & B & C & \ldots \ Z \\
\updownarrow & \updownarrow & \updownarrow & \updownarrow \\
1 & 2 & 3 & \ldots \ 26
\end{array}
$$

Example 7.2.6. The set $2\mathbf{P}$ of even positive integers has the same cardinality as the set $\mathbf{P}$ of positive integers. To prove this, we must find a map from $\mathbf{P}$ to $2\mathbf{P}$ and then prove the map is a bijection.

Define: $f : \mathbf{P} \to 2\mathbf{P}$ by $f(x) = 2x$.

(1) f is one to one.

Proof: Let $a, b \in \mathbf{P}$ and assume that $f(a) = f(b)$. We must prove that $a = b$.## $f(a) = f(b) \Rightarrow 2a = 2b \Rightarrow a = b$.

(2) f is onto.

Proof: Let $b \in 2\mathbf{P}$. We want to show that there exists an element $a \in P$
such that $f(a) = b$. If $b \in 2\mathbf{P}$, then $b = 2p$ for some $p \in \mathbf{P}$ by the definition
of $2\mathbf{P}$. So we have $f(p) = 2p = b$. Hence, each element of $2\mathbf{P}$ comes from
some element of $\mathbf{P}$.

We close this section with an example called the Pigeonhole Principle,
which has numerous applications even though it is an obvious, common-sense
statement. Never underestimate the importance of simple ideas. The Pigeonhole Principle states that if there are more pigeons than pigeonholes, then two
or more pigeons must share the same pigeonhole. A more rigorous mathematical statement of the principle follows.

Pigeonhole Principle: Let f be a function from a set X onto a set Y. If
$n \geq 1$ and if $(\#X) > n(\#Y)$, then there exists an element of Y that is the
image of at least $n + 1$ elements of X.

Example 7.2.7.   Assume that a room contains four students with the first
names John, James, and Mary. Prove that two students have the same first
name. We can visualize a mapping from the set of students to the set of first
names; each student has a first name. The pigeonhole principle applies with
$n = 1$, and we can conclude that two of the students have the same first name.

EXERCISES FOR SECTION 7.2

A Exercises

1. Determine which of the functions in Exercise 1 of Section 7.1 are one
    to one. Which are onto?

2. (a) Determine all bijections from the $\{1, 2, 3\}$ into $\{a, b, c\}$.
    (b) Determine all bijections from $\{1, 2, 3\}$ into $\{a, b, c, d\}$.

3. Which of the following are one to one, onto, or both?
    (a) $f_1 : \mathbf{R} \rightarrow \mathbf{R}$ defined by $f_1(x) = x^3 - x$.
    (b) $f_2 : \mathbf{Z} \rightarrow \mathbf{Z}$ defined by $f_2(x) = -x + 2$.
    (c) $f_3 : \mathbf{N} \times \mathbf{N} \rightarrow \mathbf{N}$ defined by $f_3(j, k) = 2^j 3^k$.
    (d) $f_4 : \mathbf{P} \rightarrow \mathbf{P}$ defined by $f_4(n) = \Box n/2 \Box$, where $\Box x \Box$ means the
        smallest integer not less than x.
    (e) $f_5 : \mathbf{N} \rightarrow \mathbf{N}$ defined by $f(n) = n^2 + n$.
    (f) $f_6 : \mathbf{N} \rightarrow \mathbf{N} \times \mathbf{N}$ defined by $f_6(n) = (n, |10 - n|)$.

4. Let $f : P(N) \rightarrow P(N)$ be defined by $f(A) = A \cup \{1\}$.
    (a) Verify that f is a function.
    (b) Determine whether f is a surjection.

5. Use the Pigeonhole Principle to prove that an injection cannot exist

between a finite set A and a finite set B if the cardinality of A is greater than the cardinality of B.

6. The important properties of relations are not generally of interest for functions. Most functions are not reflexive, symmetric, anti-symmetric, or transitive. Can you give examples of functions that do have these properties?

7. Let $A = B = \{1, 2, 3, 4, 5\}$. Define functions $f : A \rightarrow B$ (if possible) such that:

    (a) f is one to one and onto.
    (b) f is neither one to one nor onto.
    (c) f is one to one but not onto.
    (d) f is onto but not one to one.

8. (a) Let $A = B =$ the natural numbers. Define functions from A into B with the same properties as in exercise 7 (if possible).

    (b) Let A and B be finite sets where $\#A = \#B$. Is it possible to define a function $f : A \rightarrow B$ which is one to one but not onto? Is it possible to find a function f which is onto but not one to one?

9. Consider $f : \mathbf{N} \rightarrow \mathbf{Z}_{10}$ defined by $f(a) =$ the remainder after dividing 10 into a.

    (a) What is $f(23)$?
    (b) Is f an injection (one to one)?
    (c) Is f a surjection (onto)?
    (d) Describe the set of elements of $\mathbf{N}$ whose image is zero.

10. Which of the following are injections, surjections, or bijections on $\mathbf{R}$, the set of real numbers?

    (a) $f(x) = -2x$.
    (b) $g(x) = x^2 - 1$.
    (c) $h(x) = x$ if $x < 0$ and x^2 if $x \geq 0$.

11. In your own words, explain the statement "The sets of integers and even integers have the same cardinality."

12. (a) Prove that $\mathbf{P}$ is countable.
    (b) Prove that $\mathbf{Z}$ is countable.
    (c) Prove that the set of finite strings of 0's and 1's is countable.

7.3 Composition, Identity, and Inverse

Now that we have a good understanding of what a function is, our next step is to consider an important operation on functions. Our purpose is not to develop the algebra of functions as completely as we did for the algebras of logic, matrices, and sets, but the reader should be aware of the similarities between the algebra of functions and that of matrices. We first define equality of functions.

Definition: Equality. *Let f, g : A → B; that is, let f and g both be functions from the set A into the set B. The function f is equal to the function g, notation f = g, if and only if f(x) = g(x) for all x ∈ A.*

COMPOSITION

One of the most important operations on functions is that of composition.

Definition: Composition. *Let f : A → B and let g : B → C. Then the composition of f with g, written g∘f, is a function from A into C defined by (g∘f)(x) = g(f(x)), which is read "f followed by g of x is equal to g of f of x."*

The reader should note that it is traditional to write the composition of functions from right to left. Thus, in the above definition, the first function performed in computing g∘f, which is also written gf, is f. On the other hand, for relations, the composition rs is read from left to right, so that the first relation is r.

Example 7.3.1.

(a) Let $f: \{1, 2, 3\} \to \{a, b\}$ be defined by $f(1) = a$, $f(2) = a$, and $f(3) = b$. Let $g : \{a, b\} \to \{5, 6, 7\}$ be defined by $g(a) = 5$ and $g(b) = 7$. Then $g∘f: \{1, 2, 3\} \to \{5, 6, 7\}$ is defined by $(g∘f)(1) = 5$, $(g∘f)(2) = 5$ and $(g∘f)(3) = 7$. For example, $(g∘f)(1) = g(f(1)) = g(a) = 5$. Note that $f∘g$ is not defined. Why?

(b) Let $f : \mathbf{R} \to \mathbf{R}$ be defined by $f(x) = x^3$ and let $g : \mathbf{R} \to \mathbf{R}$ be defined by $g(x) = 3x + 1$. Then, since $(g∘f)(x) = g(f(x)) = g(x^3) = 3x^3 + 1$, we have $g∘f: \mathbf{R} \to \mathbf{R}$ is defined by $(g∘f)(x) = 3x^3 + 1$. Here $f∘g$ is also defined and $f∘g : \mathbf{R} \to \mathbf{R}$ is defined by $(f∘g)(x) = (3x + 1)^3$. Moreover, since $3x^3 + 1 \neq (3x + 1)^3$ for at least one real number, $g∘f \neq f∘g$ so that the commutative law is not true for functions under the operation of composition. However, the associate law is true for functions under the operation of composition.

Theorem 7.3.1. *Let $f : A \to B$, $g : B \to C$, and $h : C \to D$. Then $h(gf) = (hg)f$.*

Proof Technique: In order to prove that two functions are equal, we must use the definition of equality of functions. Assuming that the functions have the same domain, they are equal if, *for each domain element*, the images of that element under the two functions are equal.

Proof: We wish to prove that $(h(gf))(x) = ((hg)f)(x)$ for all $x \in A$.

$$(h(gf))(x) = h((gf)(x)) \quad \text{definition of composition of } h \text{ with } gf.$$
$$= h(g(f(x))) \quad \text{definition of composition of } g \text{ with } f.$$

Similarly,

$$((hg)f)(x) = (hg)(f(x)) \quad \text{definition of composition of } hg \text{ with } f.$$
$$= h(g(f(x))) \quad \text{definition of composition of } h \text{ with } g.$$

which is the same as the above.

If f is a function from a set A onto itself, we can find $f \circ f$, $f \circ f \circ f$, etc., which we write as f^2, f^3, ... This idea can be expressed more elegantly as follows: If $f : A \to A$, the repeated composition of f with itself is defined recursively as:

(1) $f^1(a) = f(a)$, $a \in A$, and
(2) $f^{n+1}(a) = f(f^n(a))$ for $n \geq 1$ and $a \in A$.

Two useful theorems concerning composition are given below. The proofs are left for the exercises.

Theorem 7.3.2. *If $f : A \to B$ and $g : B \to C$ are injections, then $gf : A \to C$ is an injection.*

Theorem 7.3.3. *If $f : A \to B$ and $g : B \to C$ are surjections, then $gf : A \to C$ is a surjection.*

We would now like to define the concepts of identity and inverse for functions under the operation of composition. The motivation and descriptions of the definitions of these terms come from the definitions of the terms in the set of real numbers and for matrices. For real numbers, the numbers 0 and 1 play the unique role that $x + 0 = 0 + x = x$ and $x \cdot 1 = 1 \cdot x = x$ for any real number x. 0 and 1 are the identity elements for the reals under the operations of addition and multiplication respectively. Similarly, the $n \times n$ zero matrix 0 and the $n \times n$ identity (multiplicative) matrix I are such that for any $n \times n$ matrix A, $A + 0 = 0 + A = A$ and $AI = IA = I$. Hence, an elegant way of defining the identity function under the operation of composition would be to imitate the above well-known facts.

IDENTITY FUNCTION

Definition: Identity. *The identity function, under composition, is a function from A onto A, denoted by i (or, more specifically, i_A) such that $f \circ i = i \circ f = f$ for all functions f from A to A.*

An alternate description is that the identity function on A is the function $i(a) = a$ for all $a \in A$. This can be proven from the above definition.

Example 7.3.2.   If $A = \{1, 2, 3\}$, then the identity function $i : A \to A$ is defined by $i(1) = 1$, $i(2) = 2$, and $i(3) = 3$.

Example 7.3.3.   The identity function on $\mathbf{R}$ is $i : \mathbf{R} \to \mathbf{R}$ defined by $i(x) = x$.

INVERSE FUNCTION

We will introduce the inverse of a function with a special case: the inverse of a function on a set. After you've taken the time to understand this concept, you can read about the inverse of a function from one set into another. The reader is encouraged to reread the definition of the inverse of a matrix in Section 5.4 to see that the following definition of the inverse function is a direct analogue of that definition.

Definition: Inverse Function. *Let $f : A \to A$. If there exists a function $g : A \to A$ such that $g \circ f = f \circ g = i_A$, then g is called the inverse of the function f and is denoted by f^{-1}, read "f inverse."*

An alternate description of the inverse of a function, which can be proven from the definition, is as follows:
Let $f : A \to A$ be such that $f(a) = b$. Then, when it exists, f^{-1} is a function from A to A such that $f^{-1}(b) = a$. Note that f^{-1} "undoes" what f does.

Example 7.3.4.   Let $A = \{1, 2, 3\}$ and let f be the function defined on A such that $f(1) = 2$, $f(2) = 3$, and $f(3) = 1$. Then $f^{-1} : A \to A$ is defined by $f^{-1}(1) = 3$, $f^{-1}(2) = 1$, and $f^{-1}(3) = 2$. Show that $f^{-1} \circ f = f \circ f^{-1} = i_A$.

Example 7.3.5.   If $g : \mathbf{R} \to \mathbf{R}$ is defined by $g(x) = x^3$, then g^{-1} is the function which undoes what g does. Since g cubes real numbers, g^{-1} must be the "reverse" process, namely, the cube root. Therefore, $g^{-1} : \mathbf{R} \to \mathbf{R}$ is defined by $g^{-1}(x) = \sqrt[3]{x}$. We should show that $g^{-1} \circ g = i$ and $g \circ g^{-1} = i$. We will do the first, and the reader is encouraged to do the second.

$$
\begin{aligned}
(g^{-1} \circ g)(x) &= g^{-1}(g(s)) && \text{Definition of composition} \\
&= g^{-1}(x^3) && \text{Definition of } g \\
&= \sqrt[3]{x^3} && \text{Definition of } g^{-1} \\
&= x && \text{Definition of cube root} \\
&= i(x) && \text{Definition of the identity function}
\end{aligned}
$$

so that $g^{-1} \circ g = i$. Why?

The definition of the inverse of a function alludes to the fact that not all functions have inverses. How do we determine when the inverse of a function exists?

Theorem 7.3.4. *Let* $f : A \rightarrow A$. f^{-1} *exists if and only if* f *is one to one and onto.*

Proof: ($\Rightarrow$) Assume that f^{-1} is a function. In this half of the proof, we wish to show that f is one to one and onto. To prove that f is one to one, let (a, b) and (c, b) be in f and show that $a = c$. If $(a, b), (c, b) \in f$, then (b, a) and (b, c) are in f^{-1}. Since f^{-1} is a function, a and c must be equal, so f is one to one. Next, to prove that f is onto, observe that for f^{-1} to be a function, it must use all of its domain, namely A. Let b be any element of A. Then b has an image under f^{-1}, $f^{-1}(b)$. Another way of writing this is $(b, f^{-1}(b)) \in f^{-1}$. By the definition of the inverse, this is equivalent to $(f^{-1}(b), b) \in f$. Hence, b is in the range of f. Since b was chosen arbitrarily, this shows that the range of f must be all of A.   -
  ($\Leftarrow$)   Assume f is one to one and onto to prove f^{-1} exists. We leave this half of the proof to the reader.   #

Definition: Permutation. *A bijection of a set A into itself is called a permutation of A.*

Next, we will consider the situation where $f : A \rightarrow B$ and B is not necessarily equal to A. How do we define the inverse in this case?

Definition: Inverse of a Function (General Case). *Let* $f : A \rightarrow B$. *If there exists a function* $g : B \rightarrow A$ *such the* $g \circ f = i_A$ *and* $f \circ g = i_B$, *then* g *is called the inverse of* f *and is denoted by* f^{-1}, *read "f inverse."*

Note the slightly more complicated condition for the inverse in this case because the domains of $f \circ g$ and $g \circ f$ are different if A and B are different.

Theorem 7.3.5. *Let* $f : A \rightarrow B$. f^{-1} *exists if and only if* f *is one to one and onto.*

Example 7.3.6.   Let $A = \{1, 2, 3\}$ and $B = \{a, b, c\}$. Define $f : A \rightarrow B$ by $f(1) = a$, $f(2) = b$, and $f(3) = c$. Then $g : B \rightarrow A$ defined by $g(a) = 1$, $g(b) = 2$, and $g(c) = 3$ is the inverse of f:

$$\left. \begin{array}{l} gf(1) = 1 \\ gf(2) = 2 \\ gf(3) = 3 \end{array} \right\} \Rightarrow gf = i_A \qquad \left. \begin{array}{l} fg(a) = a \\ fg(b) = b \\ fg(c) = c \end{array} \right\} \Rightarrow fg = i_B$$

Example 7.3.7.   The Pascal functions Ord and Chr are inverses of one another. The domain of Ord (and the codomain of Chr) is the character set on your computer. The codomain of Ord (and domain of Chr) depends on the size of your character set, usually $0 \ldots 63$ or $0 \ldots 127$.

EXERCISES FOR SECTION 7.3

A Exercises

1. Let $A = \{1, 2, 3\}$, $B = \{a, b, c, d\}$, and $C = \{7\}$. Define $f : A \rightarrow B$ by $f(1) = c, f(2) = b$, $f(3) = a$, and define $g : B \rightarrow C$ by $g(x) = 7$ for all $x \in B$.
   (a) Find $g \circ f$.
   (b) Does it make sense to discuss $f \circ g$? If not, why not?
   (c) Does f^{-1} exist? Why?
   (d) Does g^{-1} exist? Why?

2. Let $A = \{1, 2, 3\}$. Define $f : A \rightarrow A$ by $f(1) = 2$, $f(2) = 1$, and $f(3) = 3$. Find f^2, f^3, f^4, f^{-1}.

3. Let $A = \{1, 2, 3\}$.
   (a) List all permutations of A.
   (b) Find the inverse of each of the permutations of Part a.
   (c) Find the square of each of the permutations of Part a.
   (d) Show that the composition of any two permutations of A is a permutation of A.
   (e) Prove: Let A be any set where the $\#A = n$. Then the number of permutations of A is n!.

4. Define s, u, and d, all functions on the integers, by $s(n) = n^2$, $u(n) = n + 1$, and $d(n) = n - 1$. Determine:
   (a) *usd*
   (b) *sud*
   (c) *dsu*

5. Compare the definitions of identity function and inverse function with the definitions of the identity matrix and the inverse of a matrix in Chapter 5.

6. *Inverse images.* If f is any function from A into B, we can describe the inverse image of $f : B \rightarrow \mathcal{P}(A)$, which is also commonly denoted f^{-1}. If $b \in B$, $f^{-1}(b) = \{a \in A \mid f(a) = b\}$. If f does have an inverse, the inverse image of $b = \{f^{-1}(b)\}$.
   (a) Let $g : \mathbf{R} \rightarrow \mathbf{R}$ be defined by $g(x) = x^2$. What are $g^{-1}(4)$, $g^{-1}(0)$ and $g^{-1}(-1)$?
   (b) If $r(x) = \lfloor x \rfloor$, $x \in \mathbf{R}$, what is $r^{-1}(1)$?

7. Let f, g, and h all be functions from $\mathbf{Z}$ into $\mathbf{Z}$ defined by $f(n) = n + 5$, $g(n) = n - 2$, and $h(n) - n^2$. Define:
   (a) fg
   (b) f^3
   (c) fh

8. Define the following functions on the integers by $f(k) = k + 1$, $g(k) = 2k$, and $h(k) = \lceil k/2 \rceil$.
   (a) Which of these functions are one to one?
   (b) Which of these functions are onto?
   (c) Express in simplest terms the compositions fg, gf, gh, hg, and h^2.

B Exercises

9. State and prove a theorem on inverse functions analogous to Theorem 5.4.1.

10. Let f and g be functions whose inverses exist. Prove that $(g \circ f)^{-1} = f^{-1} \circ g^{-1}$. (Hint: See Exercise 3 of Section 5.4.)

11. Prove Theorems 7.3.2 and 7.3.3. Give an example of each theorem.

12. Prove the second half of Theorem 7.3.4.

chapter

8

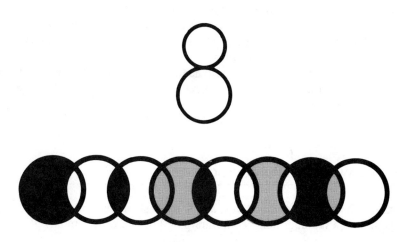

RECURSION AND
RECURRENCE RELATIONS

8

GOALS

An essential tool that anyone interested in computer science must master is how to think recursively. The ability to understand definitions, concepts, algorithms, etc., that are presented recursively and the ability to put thoughts into a recursive framework is essential in computer science. One of our goals in this chapter is to help the reader become more comfortable with recursion in its commonly encountered forms.

A second goal is to discuss recurrence relations. We will concentrate on methods of solving recurrence relations, including an introduction to generating functions.

8.1 The Many Faces of Recursion

Consider the following definitions, all of which should be somewhat familiar to you. When reading them, concentrate on how they are similar.

Example 8.1.1. A very common alternate notation for the binomial coefficient $\binom{n}{k}$ is $C(n, k)$. We will use the latter notation in this chapter. Here is a recursive definition of binomial coefficients:

$C(n, k)$ where $n \geq 0$, $k \geq 0$, and $n \geq k$ is given by
$C(n, n) = 1$;
$C(n, 0) = 1$; and
$C(n, k) = C(n - 1, k) + C(n - 1, k - 1)$ if $n > k > 0$.

POLYNOMIALS AND THEIR EVALUATION

Definition: Polynomial Expression (Non-Recursive). *An nth degree polynomial $(n \geq 0)$ is an expression of the form $a_n x^n + a_n x^{n-1} + \cdots + a_1 x + a_0$, where $a_n, a_{n-1}, \ldots, a_0$ are elements of some designated set of numbers called the set of coefficients and $a_n \neq 0$.*

Zeroth degree polynomials are called *constant polynomials* and are simply elements of the set of coefficients.

This definition is often introduced in algebra courses to describe expressions such as $f(n) = 4n^3 + 2n^2 - 8n + 9$, a third-degree, or cubic, polynomial; however, it has drawbacks when the variable is given a value and the expression must be evaluated. For example, suppose that $n = 7$. Your first impulse is likely to do this:

$$f(7) = 4(7^3) + 2(7^2) - 8(7) + 9 = 4(343) + 2(49) - 8(7) + 9 = 1423.$$

A count of the number of operations performed shows that five multiplications and three additions/subtractions were used. The following definition of a polynomial expression suggests another method of evaluation.

Definition: Polynomial Expression (Recursive). *Let S be a set of coefficients.*

(a) *A zeroth degree polynomial expression is an element of S.*
(b) *For $n \geq 1$, an nth degree polynomial expression is an expression of the form $p(x)x + a$ where $p(x)$ is an $(n - 1)$th degree polynomial expression and $a \in S$.*

We can easily verify that $f(n)$ is a third-degree polynomial expression:

$$\begin{aligned} f(n) &= (4n^2 + 2n - 8) \, n + 9 \\ &= ((4n + 2)n - 8) \, n + 9 \\ &= (((4)n) + 2) \, n - 8) \, n + 9. \end{aligned}$$

4 is a zeroth degree polynomial. Therefore $(4)n + 2$ is a first-degree polynomial; therefore, $((4)n + 2) \, n - 8$ is a second-degree polynomial; therefore, $f(n)$ is a third-degree polynomial. The final expression for $f(n)$ is called its *telescoping form*. If we use it to calculate $f(7)$, we need only three multiplications and three additions/subtractions. This is called *Horner's method* for evaluating a polynomial expression.

Example 8.1.2.

(a) The telescoping form of $p(x) = 5x^4 + 12x^3 - 6x^2 + x + 6$ is $(((((5) \, x + 12) \, x - 6) \, x + 1) \, x + 6$. Using Horner's method, $p(c)$ requires four multiplications and four additions/subtractions for any real number c.
(b) $g(x) = -x^5 + 3x^4 + 3x^3 + 2x^2 + x$ has the telescoping form $((((((-1) \, x + 3) \, x + 3) \, x + 2) \, x + 1) \, x + 0$.

If the coefficients and the degree of polynomial are represented with a Pascal record type such as

```
POLY = RECORD
        COEFF:ARRAY [0 .. MAX] OF REAL;
        DEGREE:  0 .. MAX
       END
```

then an iterative version of Horner's method is convenient. A Pascal function that uses this method is:

```
FUNCTION HORNER (P:  POLY;  C:  REAL):  REAL;
VAR
     I:  INTEGER;  H:  REAL;
BEGIN
     WITH P DO
          BEGIN
             H := COEFF [DEGREE];
             FOR I := DEGREE -1 DOWNTO 0 DO
                  H := H*C + COEFF [I];
             HORNER := H
          END
END;
```

Another representation of a polynomial would be as a linked list. The Pascal types

```
LINK = ↑TERM;
TERM = RECORD
            COEFF: REAL;
            NEXT: LINK
       END;
POLY = LINK
```

could be used to represent $p(x) = a_n x^n + \cdots + a_1 x + a_0$, as in Figure 8.1.1.

A recursive Pascal function to calculate $p(c)$ is:

```
FUNCTION HORNER (P:  POLY;  C:  REAL):  REAL;
BEGIN
     WITH P↑ DO
          IF NEXT = NIL THEN HORNER := COEFF
          ELSE HORNER := HORNER (NEXT, C) *C + COEFF
END
```

Example 8.1.3. A recursive algorithm for a binary search of a sorted file of records: Let $F:r(1), \ldots , r(n)$ represent a file of n records sorted by a numeric key in descending order. The jth record is $r(j)$, and we denote its

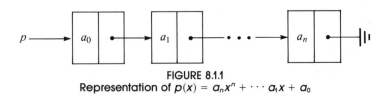

FIGURE 8.1.1
Representation of $p(x) = a_n x^n + \cdots a_1 x + a_0$

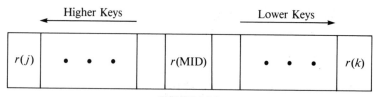

FIGURE 8.1.2
Illustration of BSRCH

key value by $r(j)$. key. For example, each record might contain data on the buildings in a city and the key value might be the height of the building. Then $r(1)$ would be the record for the tallest building. The algorithm BSRCH(j, k) can be applied to search for a record in F with key value C. This would be accomplished by the execution of BSRCH($1, n$). When the algorithm is completed, the variable FOUND will have a value of true if a record with the desired key value was found, and the value of $j*$ will be the index of a record with $r(j*)$. key $= C$. If FOUND stays false, no such record exists in the file. The general idea behind the algorithm is illustrated in Figure 8.1.2.

```
BSRCH( j ,  k ) :
1.0 FOUND := FALSE
2.0 IF  j ≤ k  THEN
       2.1 MID := ⌊( j + k )/2⌋
       2.2 IF  r(MID).key = C
              THEN 2.2.1 j* := MID
              2.2.2 FOUND := TRUE
              ELSE IF  r(MID).key < C
              THEN EXECUTE BSRCH( j ,  MID − 1 )
              ELSE EXECUTE BSRCH( MID + 1 ,  k )
```

PASCAL NOTE

If BSRCH is coded as a Pascal procedure, the variables $j*$, FOUND, and F should be variable parameters. This reduces the amount of memory needed and lets the values of $j*$ and FOUND in the final execution of BSRCH filter back to the original calling point in your program.

For the next two examples, consider a sequence of numbers to be a list of numbers consisting of a zeroth number, first number, second number, ... If a sequence is given the name N, the kth number of N is often written N_k.

Example 8.1.4. The Fibonacci sequence is the sequence F defined by

$$F_0 = 1, F_1 = 1 \quad \text{and} \quad F_k = F_{k-2} + F_{k-1}$$

for $k \geq 2$.

Example 8.1.5. Define the sequence of numbers B by

$$B_0 = 100 \qquad \text{and} \qquad B_k = (1.08)\, B_{k-1}$$

for $k \geq 1$.

RECURSION

All of the previous examples were presented recursively. That is, every "object" is described in one of two forms. One form is by a simple definition, which is usually called the basis for the recursion. The second form is by a recursive description in which objects are described in terms of themselves, with the following qualification. What is essential for a proper use of recursion is that the objects can be expressed in terms of simpler objects, where "simpler" means closer to the basis of the recursion. To avoid what might be considered a circular definition, the basis must be reached after a finite number of applications of the recursion.

Fibonacci Revisited. For the Fibonacci sequence, the basis is the specification of the first two numbers, F_0 and F_1. Consider the computation of F_3. The recursion reduces F_3 to $F_1 + F_2$. Note how we've moved backwards in the sequence from the third, number to the first, and second. Clearly, this is closer to the basis. The complete determination of F_3 is

$$
\begin{aligned}
F_3 &= F_1 + F_2 \\
&= F_1 + (F_0 + F_1) \quad \text{(all defined by the basis)} \\
&= 1 + (1 + 1) = 3.
\end{aligned}
$$

ITERATION

On the other hand, we could find a certain term in the Fibonacci sequence, say F_5, by starting with the basis terms and working forward as follows:

$$
\begin{aligned}
F_2 &= F_0 + F_1 = 1 + 1 = 2 \\
F_3 &= F_1 + F_2 = 1 + 2 = 3 \\
F_4 &= F_2 + F_3 = 2 + 3 = 5 \\
F_5 &= F_3 + F_4 = 3 + 5 = 8.
\end{aligned}
$$

This is called an *iterative computation* of the Fibonacci sequence. Here we start with the basis and work our way forward to a less simple number, such as F_5. Try to compute F_5 using the recursive definition for F as we did for F_3. It will take much more time than it would have taken to do the computations above. Iterative computations usually tend to be faster than computations that apply recursion. Therefore, one useful skill is being able to convert a recursive formula into a nonrecursive formula, such as one that requires only iteration or a faster method, if possible.

An iterative formula for $C(n, k)$ is also much more efficient than an application of the recursive definition. The recursive definition is not without its merits, however. First, the recursive equation is often useful in manipulating algebraic expressions involving binomial coefficients. Second, it gives us an insight into the combinatoric interpretation of $C(n, k)$. In choosing k elements from $\{1, 2, \ldots, n\}$, there are $C(n - 1, k)$ ways of choosing all k from $\{1, 2, \ldots, n - 1\}$, and there are $C(n - 1, k - 1)$ ways of choosing the k elements if n is to be selected and the remaining $k - 1$ elements come from $\{1, 2, \ldots, n - 1\}$. Note how we used the Law of Addition from Chapter 2 in our reasoning.

Binary Search Revisited. In the binary search-algorithm, the place where recursion is used is easy to pick out. When a record is examined and the key is not the one you want, the search is cut down to a subfile of no more than half the number of records that you were searching in before. Obviously, this is a simpler search. The basis is hidden in the algorithm. The two cases that complete the search can be thought of as the basis. Either (a) you find a record that you want, or (b) the subfile that you have been left to search in is empty ($j > k$).

BSRCH can be translated without much difficulty into any language that allows recursive calls to its subprograms. The advantage to such a program is that its coding would be much shorter than a nonrecursive program that does a binary search. However, in most cases the recursive version will be slower and require more memory at execution time.

INDUCTION AND RECURSION

The definition of the positive integers in terms of Peano's Postulates (Section 3.7) is a recursive definition. The basis element is the number 1 and the recursion is that if n is a positive integer, then so is its successor. In this case, n is the simple object and the recursion is of a forward type. Of course, the validity of an induction proof is based on our acceptance of this definition. Therefore, the appearance of induction proofs when recursion is used is no coincidence.

Example 8.1.6. Let B be the sequence in Example 8.1.5. Then $B_k = 100(1.08)^k$, $k \geq 0$. A proof by induction follows: If $k = 0$, then $B_0 = 100(1.08)^0 = 100$, as defined. Now assume that for some $k \geq 0$, the formular for B_k is true.

$$
\begin{aligned}
B_{k+1} &= (1.08)B_k \text{ by the recursive definition} \\
&= 1.08\,[(100)\,(1.08)^k] \text{ by the induction hypothesis} \\
&= 100\,(1.08)^{k+1}.
\end{aligned}
$$

The formula that we have just proven for B is called a *closed form expression*. It involves no recursion or summation signs.

Definition: Closed Form Expression. *Let $E = E(x_1, \ldots, x_n)$ by an alge-braic expression involving variables $x_1, \ldots, x_n$ which are allowed to take on values from some predetermined set. E is a closed form expression if there exists a number B such that the evaluation of E with any allowed values of $x_1, \ldots, x_n$ will take no more than B operations (alternatively, B time units).*

Example 8.1.7. A closed form expression for $\sum_{k=1}^{n} k = 1 + 2 + \cdots + n$ is $n(n+1)/2$.

EXERCISES FOR SECTION 8.1

A Exercises

1. By the recursive definition of binomial coefficients, $C(5, 2) = C(4, 2) + C(4, 1)$. Continue expanding $C(5, 2)$ to express it in terms of quantities defined by the basis. Check your result by applying the factorial definition of $C(n, k)$.

2. Define the sequence L by $L_o = 5$ and for $k \geq 1$, $L_k = 2L_{k-1} - 7$. Determine L_4. Prove by induction that $L_k = 7 - 2^{k+1}$.

3. Let $p(x) = x^5 + 3x^4 - 15x^3 + x - 10$.
    (a) Write $p(x)$ in telescoping form.
    (b) Calculate by hand $p(3)$.
    (c) Use a calculator to find $p(3)$ with its original form.
    (d) Compare your speed in Parts b and c.

B Exercises

4. Suppose that a file of nine records, $r(1), r(2), \ldots, r(9)$, is sorted by key in descending order so that $r(3).\text{key} = 12$ and $r(4).\text{key} = 10$. List the executions of BSRCH that would be needed to complete BSRCH(1, 9) for:
    (a) $C = 12$
    (b) $C = 11$
    Assume that distinct records have distinct keys.

5. What is wrong with the following definition of $f: \mathbf{R} \rightarrow \mathbf{R}? f(0) = 1$ and $f(x) = f(x/2)/2$.

8.2 Sequences, or Discrete Functions

Definition: Sequence of Integers. *A sequence of integers is a function from the natural numbers into the integers. That is, if S is a sequence of integers, $S : \mathbf{N} \rightarrow \mathbf{Z}$. The image of any natural number k can be written inter-changeably as $S(k)$ or S_k and is called the kth term of S. k itself is called the index or argument.*

Example 8.2.1.

(a) The sequence A defined by $A(k) = k^2 - k$, $k \geq 0$, is a sequence of integers.

(b) The sequence B defined recursively by $B(0) = 2$ and $B(k) = B(k-1) + 3$ for $k \geq 1$ is a sequence of integers. The terms of B can be computed either by applying the recursion formula or by interation. For example:

$$B(3) = B(2) + 3$$
$$= (B(1) + 3) + 3$$
$$= ((B(0) + 3) + 3) + 3$$
$$= ((2 + 3) + 3) + 3$$
$$= 11; \text{ or}$$
$$B(1) = B(0) + 3 = 2 + 3 = 5$$
$$B(2) = B(1) + 3 = 5 + 8 = 8$$
$$B(3) = B(2) + 3 = 8 + 3 = 11.$$

(c) Let C_r be the number of strings of zeros and ones with length r having no consecutive zeros. These terms define a sequence C of integers.

Remarks:

(1) A sequence is often called a *discrete function*.
(2) The domain of a sequence is occasionally different from the natural numbers. The most common variation is $\{k \mid k \geq k_0\}$ for some fixed integer k_0. For example, $\{1, 2, 3, \ldots\}$ or $\{-3, -2, -1, 0, 1, 2, \ldots\}$ could be the domain of a sequence. In real-life applications, the domain might be finite. For two integers, ℓ and h, with $\ell < h$, the domain can be $\{k \mid \ell \leq k \leq h\}$. In Pascal or Ada this set of integers can be specified as the subrange $\ell \mathinner{\ldotp\ldotp} h$.
(3) The codomain of a sequence can be any set. For example, a sequence of complex numbers would have the set of complex numbers as its codomain.
(4) Although it is important to keep in mind that a sequence is a function, another useful way of visualizing a sequence is as a list. For example, the sequence A could be written as $(0, 0, 2, 6, 12, 20, \ldots)$. Finite sequences can appear much the same way when they are the input to or output from a computer. The index of a sequence can be thought of as a time variable. Imagine the terms of a sequence flashing on a screen every second. The s_k would be what you see in the kth second. It is convenient to use terminology like this in describing sequences. For example, the terms that precede the kth term of A would be $A(0)$, $A(1), \ldots, A(k-1)$. They might be called the *earlier terms*.

A FUNDAMENTAL PROBLEM

Given the definition of any sequence, a fundamental problem that we will concern ourselves with is to devise a method for determining any specific term in a minimum amount of time. Generally, time can be equated with the number of operations needed. In counting operations, the application of a recursive formula would be considered an operation.

Example 8.2.2.

(a) The terms of A in Example 8.2.1 are very easy to compute because of the closed form expression. No matter what term you decide to compute, only two operations need to be performed.

(b) How to compute the terms of B is not so clear. Suppose that you wanted to know $B(100)$. One approach would be to apply the definition recursively $B(100) = B(99) + 3 = (B(98) + 3)) + 3 = \cdots$. The recursion equation for B would be applied 100 times and 100 additions would then follow. To compute $B(k)$ by this method, $2k$ operations are needed. An iterative computation of $B(k)$ is an improvement: $B(1) = 2 + 3 = 5$, $B(2) = 5 + 3 = 8, \ldots$. Only k additions are needed. This still isn't a good situation. As k gets large, we take more and more time to compute $B(k)$. The formula $B(k) = B(k - 1) + 3$ *is called a recurrence relation on B*. The process of finding a closed form expression for $B(k)$, one that requires no more than some fixed number of operations, is called *solving the recurrence relation*.

(c) The determination of C_k is a standard kind of problem in combinatorics. One solution is by way of a recurrence relation. In fact, many problems in combinatorics are most easily solved by first searching for a recurrence relation and then solving it. The following observation will suggest the recurrence relation that we need to determine C_k: If $k \geq 2$, then every string of zeros and ones with length k and no two consecutive 0's is either $1s_{k-1}$ or $01s_{k-2}$, where s_{k-1} and s_{k-2} are strings with no two consecutive zeros of length $k - 1$ and $k - 2$ respectively. From this observation we can see that $C_k = C_{k-2} + C_{k-1}$ for $k \geq 2$. The terms $C_0 = 1$ and $C_1 = 2$ are easy to determine by enumeration. Now, by iteration, any C_k can be easily determined. For example, $C_6 = 21$ can be computed with 5 additions. A closed form expression for C_k would be an improvement. Note that the recurrence relation for C_k is identical to the one for the Fibonacci sequence (Example 8.1.4). Only the basis is different.

EXERCISES FOR SECTION 8.2

A Exercises

1. Prove by induction that $B(k) = 3k + 2, k \geq 0$, is a closed form expression for the sequence B in Example 8.2.1.

2. (a) Consider sequence Q defined by $Q(k) = 2k + 9, k \geq 1$. Complete the table below and determine a recurrence relation that describes Q.

k	$Q(k)$	$Q(k) - Q(k-1)$
2		
3		
4		
5		
6		

   (b) Let $A(k) = k^2 - k, k \geq 0$. Complete the table below and determine a recurrence relation for A.

k	$A(k) - A(k-1)$	$A(k) - 2A(k-1) + A(k-2)$
2		
3		
4		
5		

3. Given k lines ($k \geq 1$) on a plane such that no two lines are parallel and no three lines meet at the same point, let $P(k)$ be the number of regions into which the lines divide the plane (including the infinite ones (see Figure 8.2.1).

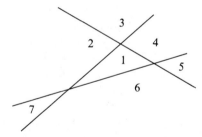

FIGURE 8.2.1
Exercise

Describe geometrically how the recurrence relation $P(k) = P(k-1) + k$ can be obtained. Given that $P(0) = 1$, determine $P(5)$.

4. A sample of a radioactive substance is expected to decay by 0.15 percent each hour. If w_t, $t \geq 0$ is the weight of the sample t hours into an experiment, write a recurrence relation for w.

B Exercise

5. Let $M(n)$ be the number of multiplications needed to evaluate an nth degree polynomial. Use the recursive definition of a polynomial expression to define M recursively.

8.3   Recurrence Relations

In this section we will begin our study of recurrence relations and their solutions. Our primary focus will be on the class of finite order linear recurrence relations with constant coefficients (shortened to finite order linear relations). First, we will examine closed form expressions from which these relations arise. Second, we will present an algorithm for solving them. In later sections we will consider some other common relations (8.4) and introduce two additional tools for studying recurrence relations: generating functions (8.5) and matrix methods (Chapter 12).

Definition: Recurrence Relation. *Let S be a sequence of numbers. A recurrence relation on S is a formula that relates all but a finite number of terms of S to previous terms of S. That is, there is a k_0 in the domain of S such that if $k > k_0$, then S(k) is expressed in terms of some (and possibly all) of the terms that preceed S(k). If the domain of S is {0, 1, 2, . . . }, the terms S(0), S(1), . . . , $S(k_0)$ are not defined by the recurrence formula. Their values are the initial conditions (or boundary conditions, or basis) that complete the definition of S.*

Example 8.3.1.

(a) The Fibonacci sequence is defined by the recurrence relation $F_k = F_{k-2} + F_{k-1}$, $k \geq 2$ and the initial conditions $F_0 = 1$ and $F_1 = 1$. The recurrence relation is called a second-order relation because F_k depends on the two previous terms of F. Recall that the sequence C in Section 8.2 can be defined with the same recurrence relation, but with different initial conditions.

(b) The relation $T(k) = 2(T(k-1))^2 - kT(k-3)$ is a third-order recurrence relation. If values of $T(0)$, $T(1)$, and $T(2)$ are specified, then T is completely defined.

(c) The recurrence relation $S(n) = S(\lfloor n/2 \rfloor) + 5$, $n \geq 0$, is of infinite order. To determine $S(n)$, you must go back $n - \lfloor n/2 \rfloor$ terms. Since $n - \lfloor n/2 \rfloor$ keeps getting larger and larger as n gets large, no finite order can be given to S.

SOLVING A RECURRENCE RELATION

Sequences are often most easily defined with a recurrence relation; however, the calculation of terms by directly applying a recurrence relation can be time consuming. The process of determining a closed form expression for the terms of a sequence from its recurrence relation is called *solving the relation*.

There is no single technique or algorithm that can be used to solve all recurrence relations. In fact, some recurrence relations cannot be solved. The relation that defines T above is one such example. Most of the recurrence relations that you are likely to encounter in the future as classified as finite order linear recurrence relations with constant coefficients. This class is the one that we will spend most of our time with in this chapter.

Definition: nth Order Linear Relation. *Let S be a sequence of numbers with domain $k \geq 0$. An nth order linear recurrence relation on S with constant coefficients is a recurrence relation that can be written in the form*

$$S(k) + C_1 S(k - 1) + \cdots + C_n S(k - n) = f(k) \qquad (k \geq n)$$

where $C_1, C_2, \ldots, C_n$ are numbers and f is a numeric function that is defined for $k \geq n$.

Important Note: We will shorten the name of this class of relations to *nth order linear relations*. Therefore, in further discussions, $S(k) + 2kS(k - 1) = 0$ would not be considered a first-order linear relation.

Example 8.3.2.

(a) The Fibonacci sequence is defined by the second-order linear relation
$F_i - F_{i-1} - F_{i-2} = 0$.
(b) The relation $P(j) + 2P(j - 3) = j^2$ is a third-order linear relation.
(c) The relation $A(k) = 2(A(k - 1) + k)$ can be rewritten as $A(k) - 2A(k - 1) = 2k$. Therefore, it is a first-order linear relation.

RECURRENCE RELATIONS OBTAINED FROM "SOLUTIONS"

Before giving an algorithm for solving finite order linear relations, we will examine a few recurrence relations that arise from certain closed form expressions. The closed form expressions are selected so that we will obtain finite order linear relations from them. This approach may seem a bit contrived, but if you were to write down a few simple algebraic expressions, chances are that most of them would be similar to the ones we are about to examine.

Example 8.3.3.

(a) Consider D, defined by $D(k) = 5 \cdot 2^k$, $k \geq 0$. If $k \geq 1$,

$$D(k) = 5 \cdot 2^k = 2 \cdot 5 \cdot 2^{k-1} = 2D(k - 1).$$

The relation $D(k) - 2D(k - 1) = 0$ and the initial condition $D(0) = 5$ define D.

(b) If $C(k) = 3^{k-1} + 2^{k+1} + k$, $k \geq 0$, quite a bit more algebraic manipulation is required to get our result:

$$C(k) = 3^{k-1} + 2^{k+1} + k$$
$$3C(k-1) = 3^{k-1} + 3 \cdot 2^k + 3(k-1) \qquad \text{(Subtract)}$$
$$C(k) - 3C(k-1) = -2^k - 2k + 3 \qquad \begin{array}{l}\text{(3^{k-1} term eliminated, a}\\ \text{first-order relation)}\end{array}$$

$$2C(k-1) - 3C(k-2) = -2^k - 4k + 10 \qquad \text{(Subtract)}$$
$$C(k) - 5C(k-1) + 6C(k-2) = 2k - 7 \qquad \begin{array}{l}\text{(A second-order linear}\\ \text{relation)}\end{array}$$

The recurrence relation that we have just obtained, defined for $k \geq 2$, together with the initial conditions $C(0) = 7/3$ and $C(1) = 5$, define C. We could do more algebra to obtain a third-order linear relation in this case.

Table 8.3.1 summarizes our results together with a few other examples that we will let the reader derive. Based on these results, we might conjecture that any closed form expression for a sequence that combines (a) exponential expressions such as 2^k, and (b) polynomial expressions will be solutions of finite order linear relations. Not only is this true, but the converse is true: a finite order linear relation defines a closed form expression that is similar to the ones that were just examined. The only additional information that is needed is a set of initial conditions.

Definition: Homogeneous Recurrence Relation. *An nth order linear relation is a homogeneous recurrence relation if $f(k) = 0$ for all k. For each recurrence relation $S(k) + C_1 S(k-1) + \cdots + C_n S(k-n) = f(k)$, the associated homogeneous relation is $S(k) + C_1 S(k-1) + \cdots + C_n S(k-n) = 0$.*

Example 8.3.4. $D(k) - 2D(k-1) = 0$ is a first-order homgeneous relation. Since it can also be written as $D(k) = 2D(k-1)$, it should be no surprise that it arose from an expression that involves powers of 2 (see

TABLE 8.3.1
Recurrence Relations Obtained from Certain Sequences

Closed Form Expression	Recurrence Relation
$D(k) = 5 \cdot 2^k$	$D(k) - 2D(k-1) = 0$
$C(k) = 3^{k-1} + 2^{k+1} + k$	$C(k) - 5C(k-1) + 6C(k-2) = 2k - 7$
$Q(k) = 2k + 9$	$Q(k) - Q(k-1) = 2$
$A(k) = k^2 - k$	$A(k) - 2A(k-1) + A(k-2) = 2$
$B(k) = 2k^2 + 1$	$B(k) - 2B(k-1) + B(k-2) = 4$
$G(k) = 2 \cdot 4^k - 5(-3)^k$	$G(k) - G(k-1) + 12G(k-2) = 0$
$J(k) = (3 + k)2^k$	$J(k) - 4J(k-1) + 4J(k-2) = 0$

Example 8.3.3a). More generally, you would expect that the solution of $L(k) = aL(k - 1)$ would involve a^k. Actually, the solution is $L(k) = L(0)a^k$, where the value of $L(0)$ is the initial condition.

Example 8.3.5. Consider the second-order homogeneous relation $S(k) - 7S(k - 1) + 12\ S(k - 2) = 0$ together with the initial conditions $S(0) = 4$ and $S(1) = 4$. From our discussion above, we can predict that the solution to this relation involves terms of the form ba^k, where b and a are non-zero constants that must be determined. If the solution were to equal this quantity exactly, then $S(k) = ba^k$, $S(k - 1) = ba^{k-1}$, and $S(k - 2) = ba^{k-2}$. If we substitute the exponential expressions into the original equation, we get an equation

$$ba^k - 7ba^{k-1} + 12ba^{k-2} = 0. \tag{8.3a}$$

Each term on the left-hand side has a factor of ba^{k-2}, which is non-zero. Dividing through by this common factor yields

$$a^2 - 7a + 12 = (a - 3)(a - 4) = 0. \tag{8.3b}$$

The only possible values of a are 3 and 4. Equation (8.3b) is called the *characteristic equation* of the recurrence relation. The fact is that our original recurrence relation is true for any sequence of the form $S(k) = b_1 3^k + b_2 4^k$, where b_1 and b_2 are real numbers. This set of sequences is called the *general solution of the recurrence relation*. If we didn't have initial conditions for S, we would stop here. The initial conditions make it possible for us to obtain definite values for b_1 and b_2.

$$\begin{cases} S(0) = 4 \\ S(1) = 4 \end{cases} \Rightarrow \begin{cases} b_1 3^0 + b_2 4^0 = 4 \\ b_1 3^1 + b_1 4^1 = 4 \end{cases} \Rightarrow \begin{cases} b_1 + b_2 = 4 \\ 3b_1 + 4b_2 = 4 \end{cases}$$

The solution of this set of simultaneous equations is $b_1 = 12$ and $b_2 = -8$. The solution is then $S(k) = 12\ 3^k - 8\ 4^k$.

Definition: Characteristic Equation. *The characteristic equation of the homogeneous nth order linear relation $S(k) + C_1 S(k - 1) + \cdots + C_n S(k - n) = 0$ is the nth degree polynomial equation*

$$a^n + \sum_{j=1}^{n} C_j a^{n-j} = a^n + C_1 a^{n-1} + \cdots + C_{n-1}a + C_n = 0.$$

The left-hand side of this equation is called the characteristic polynomial.

Example 8.3.6.

(a) The characteristic equation of $F(k) - F(k - 1) - F(k - 2) = 0$ is $a^2 - a - 1 = 0$.

(b) The characteristic equation of $Q(k) + 2Q(k - 1) - 3Q(k - 2) - 6 Q(k - 4) = 0$ is $a^4 + 2a^3 - 3a^2 - 6 = 0$. Note that the absence of a $Q(k - 3)$ term means that there is no a^{4-3} term in the characteristic equation.

Algorithm 8.3.1: Algorithm for Solving Homogeneous nth Order Linear Relations.

(a) Write out the characteristic equation of the relation $S(k) + C_1 S(k - 1) + \cdots + C_n S(k - n) = 0$, which is $a^n + C_1 a^{n-1} + \cdots + C_{n-1} a + C_n = 0$.

(b) Find all roots of the characteristic equation, called characteristic roots.

(c) If there are n distinct characteristic roots, $a_1, a_2, \ldots, a_n$, then the general solution of the recurrence relation is $S(k) = b_1 a_1^k + b_2 a_2^k + \cdots + b_n a_n^k$. If there are fewer than n characteristic roots, then at least one root is a multiple root. If a_j is a double root, then $b_j \, a_j^k$ is replaced with $(b_{j_0} + b_{j_1} k^1) a_j^k$. In general, if a_j is a root of multiplicity p, then $b_j a_j^k$ is replaced with $(b_{j0} + b_{j1}k + \cdots + b_{j(p-1)}k^{p-1})a_j^k$.

(d) If n initial conditions are given, obtain n linear equations in n unknowns (the b_j's from Step c) by substitution. If possible, solve these equations.

Although this algorithm is valid for all values of n, there are limits to the size of n for which the algorithm is feasible. Using just a pencil and paper, we can always solve second-order equations. The quadratic formula for the roots of $ax^2 + bx + c = 0$ is

$$\frac{-b \pm \sqrt{b^2 - 4ac}}{2a}.$$

The solutions of $a^2 + c_1 a + c_2 = 0$ are then

$$\tfrac{1}{2}(-c_1 + \sqrt{c_1^2 - 4c_2}) \quad \text{and} \quad \tfrac{1}{2}(-c_1 - \sqrt{c_1^2 - 4c_2}).$$

Although cubic and quartic formulas exist, they are too lengthy to introduce here. For this reason, the only higher-order relations ($n \geq 3$) that you could be expected to solve by hand are ones for which there is an easy factorization of the characteristic polynomial.

Example 8.3.7. Suppose that T is defined by $T(k) = 7T(k - 1) - 10T(k - 2)$, $T(0) = 4$ and $T(1) = 17$. We can solve this recurrence relation with Algorithm 8.3.1:

(a) The characteristic equation is $a^2 - 7a + 10 = 0$. Note that we had written the recurrence relation in "nonstandard" form. To avoid errors in this easy step, you might consider a rearrangement of the equation to, in this case, $T(k) - 7T(k - 1) + 10T(k - 2) = 0$.

(b) The characteristic roots are $\tfrac{1}{2}(7 + \sqrt{49 - 40}) = 5$ and $\tfrac{1}{2}(7 - \sqrt{49 - 40}) = 2$. These roots can be just as easily obtained by factoring the characteristic polynomial into $(a - 5)(a - 2)$.

(c) The general solution of the recurrence relation is $T(k) = b_1 2^k + b_2 5^k$.

(d) $\begin{Bmatrix} T(0) = 4 \\ T(1) = 17 \end{Bmatrix} \Rightarrow \begin{Bmatrix} b_1 2^0 + b_2 5^0 = 4 \\ b_1 2^1 + b_2 5^1 = 17 \end{Bmatrix} \Rightarrow \begin{Bmatrix} b_1 + b_2 = 4 \\ 2b_1 + 5b_2 = 17 \end{Bmatrix}$.

The simultaneous equations have the solution $b_1 = 1$, $b_2 = 3$. Therefore, $T(k) = 2^k + 3 \cdot 5^k$.

Here is one rule that might come in handy: *If the coefficients of the characteristic polynomial are all integers, with the constant term equal to m, then the only possible rational characteristic roots are divisors of m (both positive and negative).*

With the aid of a computer (or possibly only a calculator), we can increase n. Approximations of the characteristic roots can be obtained by any of several well-known methods, some of which are part of standard software packages. There is no general rule that specifies the values of n for which numerical approximations will be feasible. The accuracy that you get will depend on the relation that you try to solve. (See Exercise 17 of this section.)

Example 8.3.8. Solve $S(k) - 7S(k - 2) + 6S(k - 3) = 0$, where $S(0) = 8$, $S(1) = 6$, and $S(2) = 22$.

(a) The characteristic equation is $a^3 - 7a + 6 = 0$.
(b) The only rational roots that we can attempt are ± 1, ± 2, ± 3, and ± 6. By checking these, we obtain the three roots 1, 2, and -3.
(c) The general solution is $S(k) = b_1(1)^k + b_2 2^k + b_3(-3)^k$. The first term can simply be written b_1.

(d) $\begin{Bmatrix} S(0) = 8 \\ S(1) = 6 \\ S(2) = 22 \end{Bmatrix} \Rightarrow \begin{Bmatrix} b_1 + b_2 + b_3 = 8 \\ b_1 + 2b_2 - 3b_3 = 6 \\ b_1 + 4b_2 + 9b_3 = 22 \end{Bmatrix}$.

You can solve this system by elimination to obtain $b_1 = 5$, $b_2 = 2$, and $b_3 = 1$. Therefore, $S(k) = 5 + 2 \cdot 2^k + 1(-3)^k = 5 + 2^{k+1} + (-3)^k$.

Example 8.3.9. Solve $D(k) - 8D(k - 1) + 16D(k - 2) = 0$ where $D(2) = 16$ and $D(3) = 80$.

(a) Characteristic equation: $a^2 - 8a + 16 = 0$.
(b) $a^2 - 8a + 16 = (a - 4)^2$. Therefore, there is a double characteristic root, 4.
(c) General solution: $D(k) = (b_{10} + b_{11}k) 4^k$.

(d) $\begin{Bmatrix} D(2) = 16 \\ D(3) = 80 \end{Bmatrix} \Rightarrow \begin{Bmatrix} (b_{10} + b_{11}2) 16 = 16 \\ (b_{10} + b_{11}3) 64 = 80 \end{Bmatrix} \Rightarrow$

$\begin{Bmatrix} 16b_{10} + 32b_{11} = 16 \\ 64b_{10} + 192b_{11} = 80 \end{Bmatrix} \Rightarrow \begin{Bmatrix} b_{10} = 1/2 \\ b_{11} = 1/4 \end{Bmatrix}$.

Therefore $D(k) = (1/2 + (1/4)k) 4^k = (2 + k) 4^{k-1}$.

SOLUTION OF NONHOMOGENEOUS FINITE ORDER LINEAR RELATIONS

Our algorithm for nonhomogeneous relations will not be as complete as for the homogeneous case. This is due to the fact that different right-hand sides ($f(k)$'s) call for different procedures in obtaining a particular solution in Steps b and c.

Algorithm 8.3.2: Algorithm for Solving Nonhomogeneous Finite Order Linear Relations. To solve the recurrence relation $S(k) + C_1 S(k - 1) + \cdots + C_n S(k - n) = f(k)$:

(a) Write the associated homogeneous relation (i.e., $f(k) = 0$) and find its general solution (Steps a through c of Algorithm 8.3.1). Call this the *homogeneous solution.*

(b) Start to obtain what is called a *particular solution of the recurrence relation* by taking an educated guess at the form of a particular solution. For a large class of right-hand sides, this is not really a guess, since the particular solution is often the same type of function as $f(k)$ (see Table 8.3.2).

(c) Substitute your guess from Step b into the recurrence relation. If you made a good guess, you should be able to determine the unknown coefficients of your guess. If you made a wrong guess, it should be apparent from the result of this substitution, so go back to Step b.

(d) The general solution of the recurrence relation is the sum of the homogeneous and particular solutions. If no initial conditions are given, then you are finished. If n initial conditions are given, obtain n linear equations in n unknowns and solve the system, if possible, to get a complete solution.

TABLE 8.3.2
Particular Solutions for Given Right-hand Sides

Right-hand side	Form of particular solution
a constant, q	a constant, d
a linear function $q_0 + q_1 k$	a linear function $d_0 + d_1 k$
an mth degree polynomial $q_0 + q_1 k + \cdots + q_m k^m$	an mth degree polynomial $d_0 + d_1 k + \cdots + d_m k^m$
an exponential function $q\, a^k$	an exponential function $d\, a^k$

Example 8.3.10. $S(k) + 5S(k - 1) = 9$, $S(0) = 6$.

(a) The associated homogeneous relation, $S(k) + 5S(k - 1) = 0$ has characteristic equation $a + 5 = 0$; therefore, $a = -5$. The homogeneous solution is $S^{(h)}(k) = b_1(-5)^k$.

(b) Since the right-hand side is a constant, we guess that the particular solution will be a constant, d.

(c) If we substitute $S^{(p)}(k) = d$ into the recurrence relation, we get $d + 5d = 9$, or $6d = 9$; therefore, $S^{(p)}(k) = 1.5$.

(d) The general solution of the recurrence relation is $S(k) = S^{(h)}(k) + S^{(p)}(k) = b_1(-5)^k + 1.5$. The initial condition will give us one equation to solve in order to determine b_1:

$$S(0) = 6 \Rightarrow b_1(-5)^0 + 1.5 = 6 \Rightarrow b_1 + 1.5 = 6.$$

Therefore, $b_1 = 4.5$ and $S(k) = 4.5(-5)^k + 1.5$.

Example 8.3.11. Consider $T(k) - 7T(k - 1) + 10T(k - 2) = 6 + 8k$ with $T(0) = 1$ and $T(1) = 2$.

(a) From Example 8.3.7, we know that $T^{(h)}(k) = b_1 2^k + b_2 5^k$.

(b) Since the right-hand side is a linear polynomial, $T^{(p)}(k)$ is linear; i.e., of the form $d_0 + d_1 k$.

(c) Substitution yields:

$$(d_0 + d_1 k) - 7(d_0 + d_1(k - 1)) + 10(d_0 + d_1(k - 2))$$
$$= 6 + 8k$$
$$(4d_0 - 13d_1) + (4d_1) k = 6 + 8k.$$

Two polynomials are equal only if their coefficients are equal. Therefore,

$$\begin{cases} 4d_0 - 13d_1 = 6 \\ 4d_1 = 8 \end{cases} \Rightarrow \begin{cases} d_0 = 8 \\ d_1 = 2 \end{cases}.$$
$$T^{(p)}(k) = 8 + 2k.$$

(d) Use the general solution $T(k) = b_1 2^k + b_2 5^k + 8 + 2k$ and the initial conditions to get a final solution:

$$\begin{cases} T(0) = 1 \\ T(1) = 2 \end{cases} \Rightarrow \begin{cases} b_1 + b_2 + 8 = 1 \\ 2b_1 + 5b_2 + 10 = 2 \end{cases}$$
$$\Rightarrow \begin{cases} b_1 + b_2 = -7 \\ 2b_1 + 5b_2 = -8 \end{cases}$$
$$\Rightarrow \begin{cases} b_1 = -9 \\ b_2 = 2 \end{cases}.$$

Therefore, $T(k) = -9 \cdot 2^k + 2 \cdot 5^k + 8 + 2k$.

Example 8.3.12. Suppose that you open a savings account that pays an annual interest rate of 8%. In addition, suppose that you decide to deposit one dollar when you open the account and you intend to double your deposit each year. Let $B(k)$ be your balance after k years. B can be described by the relation $B(k) = 1.08 \, B(k - 1) + 2^k$. With $B(0) = 1$. If, instead of doubling the deposit each year, you deposited a constant amount, q, the 2^k term would be replaced with q. A sequence of regular deposits such as this is called an *annuity*.

Returning to the original situation, we can obtain a closed form expression for $B(k)$:

(a) $B^{(h)}(k) = b_1(1.08)^k$.

(b) $B^{(p)}(k)$ should be of the form $d2^k$.

(c) $d \, 2^k = 1.08(d \, 2^{k-1}) + 2^k$
$\Rightarrow (2d)2^{k-1} = (1.08d)2^{k-1} + 2 \cdot 2^{k-1}$
$\Rightarrow 2d = 1.08d + 2$
$\Rightarrow 0.92d = 2 \Rightarrow d = 2.174$ (to the nearest thousandth).
Therefore, $B^{(p)}(k) = (2.174)2^k$.

(d) $B(0) = 1 \Rightarrow b_1 + 2.174 = 1$
$b_1 = -1.174$
$B(k) = (-1.174)(1.08)^k + (2.174)2^k$.

Example 8.3.13. Find the general solution of $S(k) - 3 \, S(k - 1) - 4 \, S(k - 2) = 4^k$.

(a) The characteristic roots of the associated homogeneous relation are -1 and 4. Therefore, $S^{(h)}(k) = b_1(-1)^k + b_2 4^k$.

(b) A function of the form $d4^k$ will not be a particular solution of the nonhomogeneous relation since it solves the associated homogeneous relation (set $b_1 = 0$ in $S^{(h)}(k)$). When the right-hand side involves an exponential function with a base that equals a characteristic root, you should multiply your guess at a particular solution by k. Our guess at $S^{(p)}(k)$ would then be $dk4^k$. See below for a more complete description of this procedure.

(c) Substitute $dk4^k$ for $S(k)$:

$$dk4^k - 3d(k - 1)4^{k-1} - 4d(k - 2)4^{k-2} = 4^k$$
$$16dk4^{k-2} - 12d(k - 1)4^{k-2} - 4d(k - 2)4^{k-2} = 4^k$$

Each term on the left-hand side has a factor of 4^{k-2}:

$$16dk - 12d(k - 1) - 4d(k - 2) = 4^2$$
$$20d = 16$$
$$d = 0.8.$$

Therefore, $S^{(p)}(k) = 0.8k4^k$.

(d) The general solution of the recurrence relation is

$$S(k) = b_1(-1)^k + b_2 4^k + 0.8k4^k.$$

BASE OF RIGHT-HAND SIDE EQUAL TO CHARACTERISTIC ROOT

If the right-hand side of a nonhomogeneous relation involves an exponential with base a and a is a characteristic root of multiplicity p, then multiply your guess at a particular solution as prescribed in Table 8.3.2 by k^p, where k is the index of the sequence.

Example 8.3.14.

(a) If $S(k) - 9S(k - 1) + 20S(k - 2) = 2\ 5^k$, the characteristic roots are 4 and 5. $S^{(p)}(k)$ will be of the form $dk5^k$.

(b) If $S(n) - 6S(n - 1) + 9S(n - 2) = 3^{n+1}$, the only characteristic root is 3, but it is a double root (multiplicity 2). Therefore, the form of the particular solution is $dn^2 3^n$.

(c) If $Q(j) - Q(j - 1) - 12Q(j - 2) = (-3)^j + 6\ 4^j$, the characteristic roots are -3 and 4. The form of $S^{(p)}(j)$ will be $d_1 j(-3)^j + d_2 j\ 4^j$.

(d) If $S(k) - 9S(k - 1) + 8S(k - 2) = 9k + 1 = (9k + 1)\ 1^k$, the characteristic roots are 1 and 8. If the right-hand side is a polynomial, as it is in this case, then the exponential factor 1^k can be introduced. The particular solution will take the form $k(d_0 + d_1 k)$.

We conclude this section with a comment on the situation in which the characteristic equation gives rise to complex roots. If we restrict the coefficients of our finite order linear relations to real numbers, or even to integers, we can still encounter characteristic equations whose roots are complex. Here, we will simply take the time to point out that our algorithms are still valid with complex characteristic roots, but the customary method for expressing the solutions of these relations is different. Since an understanding of these representations requires some background in complex numbers, we will simply suggest that an interested reader can refer to a more advanced treatment of recurrence relations (see also difference equations).

EXERCISES FOR SECTION 8.3

A Exercises

Solve the following sets of recurrence relations and initial conditions.

1. $S(k) - 10S(k - 1) + 9S(k - 2) = 0$ $S(0) = 3,$
   $S(1) = 11$

2. $S(k) - 9S(k - 1) + 18S(k - 2) = 0$ $S(0) = 1, S(1) = 4$

3. $S(k) - 0.25S(k - 1) = 0$ $S(0) = 6$

4. $S(k) - 20S(k - 1) + 100S(k - 2) = 0$ $S(0) = 2,$
                                                                $S(1) = 30$

5. $S(k) - 2S(k - 1) + S(k - 2) = 2$ $S(0) = 25,$
                                                                $S(1) = 16$

6. $S(k) - S(k - 1) - 6S(k - 2) = -30$ $S(0) = 20,$
                                                                $S(1) = -5$

7. $S(k) - 5S(k - 1) = 5^k$ $S(0) = 3$

8. $S(k) - 5S(k - 1) + 6S(k - 1) = 2$ $S(0) = 1,$
                                                                $S(1) = -1$

9. $S(k) - 4S(k - 1) + 4S(k - 2) = 3k + 2^k$ $S(0) = 1, S(1) = 1$

10. $S(k) = rS(k - 1) + a$ $S(0) = 0,$
                                                                $r, a > 0, r \neq 1$

11. $S(k) - 4S(k - 1) - 11S(k - 2)$
    $+ 30S(k - 3) = 0$ $S(0) = 0,$
                                                                $S(1) = -35,$
                                                                $S(2) = -85$

12. Find a closed form expression for $P(k)$ in Exercise 3 of Section 8.2.

13. (a) Find a closed form expression for the terms of the Fibonacci se-
         quence (see Example 8.1.4).
    (b) The sequence C was defined by $C_r =$ the number of strings of zeros
         and ones with length r having no consecutive zeros (Example
         8.2.1c). Its recurrence relation is the same as that of the Fibonacci
         sequence. Determine a closed form expression for $C_r, r \geq 1$.

14. If

$$S(n) = \sum_{j=1}^{n} g(j),$$

S can be described with the recurrence relation $S(n) = S(n - 1)$
$+ g(n)$. For each of the following sequences that are defined using a
summation, find a closed form expression:

(a)

$$S(n) = \sum_{j=1}^{n} j, n \geq 1$$

(b)

$$Q(n) = \sum_{j=1}^{n} j^2, n \geq 1$$

(c)

$$P(n) = \sum_{j=0}^{n} (1/2)^j, \ n \geq 0$$

(d)

$$T(n) = \sum_{j=1}^{n} j^3, \ n \geq 1.$$

B Exercises

15. Let $D(n)$ be the number of ways that the set $\{1, 2, \ldots, n\}$, $n \geq 1$, can be partitioned into two non-empty subsets.
    (a) Find a recurrence relation for D. (Hint: It will be a first-order linear relation.)
    (b) Solve the recurrence relation.

16. If you were to deposit a certain amount of money at the end of each year for a number of years, this sequence of payment would be called an annuity (see Example 8.3.12.).
    (a) Find a closed form expression for the balance or value of an annuity that consists of payments of q dollars at a rate of interest of i (9.5% interest would correspond to $i = 0.095$). Note that for a normal annuity, the first payment is made after one year.
    (b) With an interest rate of 12.5%, how much would you need to deposit into an annuity to have a value of one million dollars after 18 years?
    (c) The payment of a loan is a form of annuity in which the initial value is some negative amount (the amount of the loan) and the annuity ends when the value is raised to zero. How much could you borrow if you can afford to pay $5000 per year for 25 years at 14% interest?

C Exercises

17. Suppose that C is a small positive number. Consider the recurrence relation $B(k) - 2B(k - 1) + (1 - C^2)B(k - 2) = C^2$, with initial conditions $B(0) = 1$ and $B(1) = 1$. If C is small enough, we might consider approximating the relation by replacing $1 - C^2$ with 1 and C^2 with 0. Solve the original relation and its approximation. Let B_a be the solution of the approximation. Compare closed form expressions for $B(k)$ and $B_a(k)$. Their forms are very different because the characteristic roots of the original relation were close together and the approximation resulted in one double characteristic root. If characteristic roots of a relation are relatively far apart, this problem will not occur. For example, the general solutions of $S(k) + 1.001S(k - 1) - 2.004002S(k - 2) = 0.0001$ and $S_a(k) + S_a(k - 1) - 2S_a(k - 2) = 0$ are similar.

8.4 Some Common Recurrence Relations

In this section we intend to examine a variety of recurrence relations that are not finite order linear with constant coefficients. For each part of this section, we will consider a concrete example, present a solution, and, if possible, examine a more general form of the original relation.

Example 8.4.1. Consider the homogeneous first-order linear relation $S(n) - nS(n - 1) = 0, n \geq 1$, with initial condition $S(0) = 1$. Upon close examination of this relation, we see that the nth term is n times the $(n - 1)$st term, which is a characteristic of n factorial. $S(n) = n!$ is a solution of this relation, for if $n \geq 1$,

$$
\begin{aligned}
S(n) = n! &= (n)(n - 1) \cdots (2)(1) \\
&= (n)(n - 1)! \\
&= nS(n - 1).
\end{aligned}
$$

In addition, since $0! = 1$, the initial condition is satisfied. It should be pointed out that from a computational point of view, our "solution" really isn't much of an improvement since the exact calculation of $n!$ takes n multiplications, which is the operation count for determining $S(n)$ by an iterative method.

If we examine a similar relation, $G(k) = 2^k G(k - 1)$, $k \geq 0$ with $G(0) = 1$, a table of values for G suggests a possible solution:

k	0	1	2	3	4	5
$G(k)$	1	2	2^3	2^6	2^{10}	2^{15}

The exponent of 2 in $G(n)$ is growing according to the relation $E(k) = E(k - 1) + k$, with $E(0) = 0$. Thus $E(k) = \frac{1}{2}k(k + 1)$ and $G(k) = 2^{\frac{1}{2}k(k+1)}$. Note that $G(k)$ could also be written as $(2^0)(2^1) \cdots (2^k)$ for $k \geq 0$, but this is not a closed form expression.

In general, the relation $P(n) = f(n) P(n - 1)$ $n \geq 0$ with $P(0) = f(0)$, where f is a function that is defined for all $n \geq 0$, has the "solution"

$$
P(n) = f(0) f(1) \cdots f(n) = \prod_{k=0}^{n} f(k), \quad n \geq 0.
$$

This product form of $P(n)$ is not a closed form expression because as n grows, the number of multiplications grow. Thus, it is really not a true solution. Often, as for $G(n)$ above, a closed form expression can be obtained from the product form.

Example 8.4.2. (Analysis of a Binary Search Algorithm). Suppose that you intend to use a binary search algorithm (see Example 8.1.3) on files of zero or more sorted records and that the records are stored in an array, so that you have easy access to each record. A natural question to ask, particularly in an analysis of algorithms course, is "How much time will it take to complete the search?" When a question like this is asked, the time that we refer to is often the so-called worst-case time. That is, if we were to search through n records, What is the longest amount of time that we will need to complete the search? In order to make an analysis such as this independent of the computer to be used, time is measured by counting the number of steps that are executed. Each step (or sequence of steps) is assigned an absolute time, or weight; therefore, our answer will not be in seconds, but in absolute time units. If the steps in two different algorithms are assigned weights that are consistent, then analyses of the algorithms can be used to compare their relative efficiencies.

There are two major steps that must be executed in a call of the binary search algorithm:

(1) If the lower index is less than or equal to the upper index, then the middle of the file is located and its key is compared to the value that you are searching for.
(2) In the worst case, the algorithm must be executed with a file that is roughly half as large as in the previous execution.

If we assume that Step 1 takes one time unit and $T(n)$ is the worst-case time for a file of n records, then

$$T(n) = 1 + T([n/2]). \qquad (8.4a)$$

For simplicity, we will assume that

$$T(0) = 0, \qquad (8.4b)$$

even though the conditions of Step 1 must be evaluated as false if $n = 0$. You might wonder why $n/2$ is truncated in 8.4a. If n is odd, then $n = 2k + 1$ for some $k \geq 0$, the middle of the file will be the $(k + 1)$st record, and no matter what half of the file the search is directed to, the reduced file will have $k = \lfloor n/2 \rfloor$ records. On the other hand, if n is even, then $n = 2k$ for $k \geq 1$. The middle of the file will be the kth record, and the worst case will occur if we are directed to the k records that come after the middle (the $(k + 1)$st through $(2k)$th records). Again the reduced file has $\lfloor n/2 \rfloor$ records.

Solution of 8.4a and 8.4b. To determine $T(n)$, the easiest case is when n is a power of two. If we compute $T(2^m)$, $m \geq 0$, by iteration, our results are

$$T(1) = 1 + T(0) = 1$$
$$T(2) = 1 + T(1) = 2$$
$$T(4) = 1 + T(2) = 3$$
$$T(8) = 1 + T(4) = 4.$$

The pattern that is established makes it clear that $T(2^m) = m + 1$. This result would seem to indicate that every time you double the size of your file, the search time increases by only one unit.

A more complete solution can be obtained if we represent n in binary form. For each $n \geq 1$, there exists a non-negative integer r such that

$$2^{r-1} \leq n < 2^r \qquad\qquad (8.4c)$$

For example, if $n = 21$, $2^4 \leq 21 < 2^5$; therefore, $r = 5$. If n satisfies 8.4c, its binary representation requires r digits. For example, $21_{\text{TEN}} = 10101_{\text{TWO}}$.

In general, $n = (a_1 a_2 \cdots a_r)_{\text{TWO}}$, where $a_1 = 1$. Note that in this form, $\lfloor n/2 \rfloor$ is easy to describe: it is the $r - 1$ digit binary number $a_1 a_2 \cdots a_{r-1}$. Therefore,

$$
\begin{aligned}
T(n) &= T(a_1 \cdots a_{r-1} a_r) \\
&= 1 + T(a_1 \cdots a_{r-1}) \\
&= 1 + (1 + T(a_1 \cdots a_{r-2})) \\
&= 2 + T(a_1 \cdots a_{r-2}) \\
&\quad\vdots \\
&= (r - 1) + T(a_1) \\
&= (r - 1) + 1 \qquad (\text{since } T(1) = 1) \\
&= r.
\end{aligned}
$$

From the pattern that we've just established, $T(n)$ reduces to r. A formal inductive proof of this statement is possible. However, we expect that most readers would be satisfied with the argument above. Any skeptics are invited to provide the inductive proof. If $n = 21$:

$$
\begin{aligned}
T(21) &= T(10101) = 1 + T(1010) \\
&= 1 + (1 + T(101)) \\
&= 1 + (1 + (1 + T(10))) \\
&= 1 + (1 + (1 + T(1)))) \\
&= 1 + (1 + (1 + (1 + T(0))))) \\
&= 5.
\end{aligned}
$$

Conclusion: The solution to 8.4a and 8.4b is $T(n) = r$, where $2^{r-1} \leq n < 2^r$, $n \geq 1$.

A less cumbersome statement of this fact is that for $n \geq 1$, $T(n) = \lfloor \log_2 n \rfloor + 1$. For example, $T(21) = \lfloor \log_2 21 \rfloor + 1 = 4 + 1 = 5$.

REVIEW OF LOGARITHMS

Any discussion of logarithms must start by establishing a base, which can be any positive number other than 1. With the exception of Theorem 8.4.1, our base will be 2. We will see that the use of a different base (10 and

$e \approx 2.171828$ are the other common ones) only has the effect of multiplying each logarithm by a constant. Therefore, the base that you use really isn't very important. Our choice of base 2 logarithms is convenient for the problems that we are considering.

The base 2 logarithm of a positive number represents an exponent and is defined by $\log_2 a = x \Leftrightarrow 2^x = a, a > 0$. For example, $\log_2 8 = 3$ because $2^3 = 8$ and $\log_2 1.414 \approx 0.5$ because $2^{0.5} \approx 1.414$. A graph of the function $f(x) = \log_2 x$ in Figure 8.4.1a shows that if $a < b$, the $\log_2 a < \log_2 b$; i.e., when x is increased, $\log_2 x$ is increased. However, if we move x from

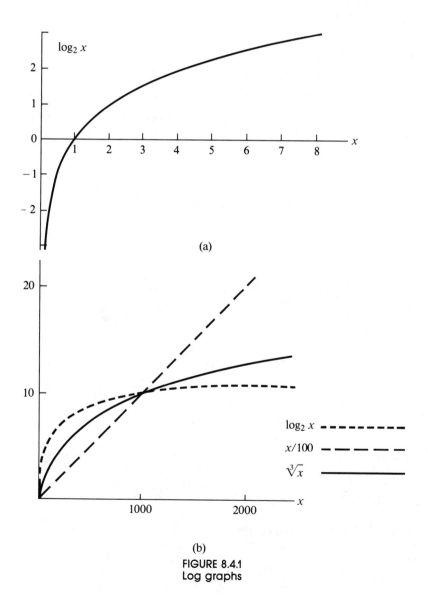

(a)

(b)
FIGURE 8.4.1
Log graphs

$2^{10} = 1024$ to $2^{11} = 2048$, $\log_2 x$ only changes from 10 to 11. This slow rate of increase of the logarithm function is an important point to remember. An algorithm acting on n pieces of data that can be executed in $\log_2 n$ time units can handle significantly larger sets of data than an algorithm that can be executed in $n/100$ or even $\sqrt[3]{n}$ time units (see Figure 8.4.1b). The graph of $T(n) = \lfloor \log_2 n \rfloor + 1$ would show the same behavior.

Theorem 8.4.1. Fundamental Properties of Logarithms. *A few more properties that we will use in subsequent discussions involving logarithms are*

$$\log_2 1 = 0 \qquad\qquad\qquad\qquad\qquad\qquad (8.4d)$$
$$\log_2 (ab) = \log_2 a + \log_2 b \qquad a, b > 0 \qquad (8.4c)$$
$$\log_2 (a/b) = \log_2 a - \log_2 b \qquad a, b > 0 \qquad (8.4d)$$
$$\log_2 a^r = r \log_2 a \qquad\qquad\quad a > 0, r \in \mathbf{R} \quad (8.4e)$$
$$2^{\log_2 a} = a \qquad\qquad\qquad\qquad a > 0 \qquad\qquad (8.4f)$$

We can derive the final expression for $T(n)$ using the properties of logarithms.

$$T(n) = r \Leftrightarrow 2^{r-1} \le n < 2^r$$
$$\Leftrightarrow \log_2 2^{r-1} \le \log_2 n < \log_2 2^r,$$

since the logarithm, base 2, is increasing with its argument.

$$\Leftrightarrow r - 1 \le \log_2 n < r$$
$$\Leftrightarrow r - 1 = \lfloor \log_2 n \rfloor$$
$$\Leftrightarrow r = \lfloor \log_2 n \rfloor + 1$$

We can apply several of these properties of logarithms to get an alternate expression for $T(n)$:

$$\lfloor \log_2 n \rfloor + 1 = \lfloor (\log_2 n) + 1 \rfloor$$
$$= \lfloor \log_2 n + \log_2 2 \rfloor$$
$$= \lfloor \log_2 2n \rfloor.$$

Definition: *Logarithms for base b, $b > 0$, $b \ne 1$:*

$$\log_b a = x \Leftrightarrow b^x = a, a > 0.$$

Theorem 8.4.2. *Let $b > 0$, $b \ne 1$. Then for all $a > 0$, $\log_b a = \log_2 a / \log_2 b$. Therefore, if $b > 1$, base b logarithms can be obtained from base 2 logarithms by dividing by the positive scaling factor $\log_2 b$. If $b < 1$, this scaling factor is negative.*

Proof: By an analogue of 8.4f, $a = b^{\log_b a}$. Therefore, if we take the base 2 logarithm of both sides of this equality we obtain:

$$\log_2 a = \log_2(b^{\log_b a})$$
$$= \log_b a \; \log_2 b.$$

To obtain the desired result, divide both sides of this last equation by $\log_2 b$. #

Note: $\log_2 10 = 3.32192$ and $\log_2 e = 1.55269$.

If the time that was assigned to Step 1 of the binary search algorithm is changed, we wouldn't expect the form of the solution to be very different. If $T(n) = a + T(\lfloor n/2 \rfloor)$ with $T(0) = c$, then $T(n) = c + a \lfloor \log_2 2n \rfloor$.

A further generalization would be to add a coefficient to $T(\lfloor n/2 \rfloor)$: $T(n) = a + bT(\lfloor n/2 \rfloor)$ with $T(0) = c$, where $a, b, c \in R$, and $b \neq 1$ is not quite as simple to derive. First, if we consider values of n that are powers of 2:

$$T(1) = a + b\,T(0) = a + bc$$
$$T(2) = a + b(a + bc) = a + ab + b_{2c}$$
$$T(4) = a + b(a + ba + b^2c) = a + ab + ab^2 + b^3c$$

.

.

.

$$T(2^r) = a + ab + ab^2 + \cdots + ab^r + b^{r+1}c.$$

If n is not a power of 2, by reasoning that is identical to what we used to solve 8.4a and 8.4b,

$$T(n) = \sum_{k=0}^{r} ab^k + b^{r+1}c,$$

where $r = \lfloor \log_2 n \rfloor$.

The first term of this expression can be written in closed form. Let x be that sum:

$$x = a + ab + ab^2 + \cdots + ab^r$$
$$bx = ab + ab^2 + \cdots + ab^r + ab^{r+1}$$

We've multiplied each term of x by b and aligned the identical terms in x and bx. Now if we subtract, $x - bx = a - ab^{r+1}$. Therefore,

$$x(1 - b) = a(1 - b^{r+1}), \text{ and}$$
$$x = a\frac{1 - b^{r+1}}{1 - b} = a\frac{b^{r+1} - 1}{b - 1}.$$

A closed form expression for $T(n)$ is

$$T(n) = a\frac{b^{r+1} - 1}{b - 1} + b^{r+1}c,$$

where $r = \lfloor \log_2 n \rfloor$.

Example 8.4.3. The efficiency of any search algorithm such as the binary search relies on the fact that the search file is sorted according to a key value and that the search is based on the key value. There are several methods for sorting a file. One example is the bubble sort. You might be familiar with this one since it is a popular "first sorting algorithm." A time analysis of the algorithm shows that if $B(n)$ is the worst-case time needed to complete the bubble sort on n records, then $B(n) = (n-1) + B(n-1)$ and $B(1) = 0$. The solution of this relation is a quadratic function $B(n) = \frac{1}{2}(n^2 - n)$. The growth rate of a quadratic function such as this one is controlled by its squared term. Any other terms are dwarfed by it as n gets large. For the bubble sort, this means that if we double the size of the file that we are to sort, n changes to $2n$ and so n^2 becomes $4n^2$. Therefore, the time needed to do a bubble sort is quadrupled.

One alternative to the bubble sort is the merge sort. Here is a simple version of this algorithm for sorting $F:r(1), \ldots, r(n), n \geq 1$. If $n = 1$, the file is sorted trivially. If $n \geq 2$ then:

(1) Divide F into $F1:r(1), \ldots, r(\lfloor n/2 \rfloor)$ and $F2:r(\lfloor n/2 \rfloor + 1), \ldots, r(n)$.
(2) Sort $F1$ and $F2$ using a merge sort.
(3) Merge the sorted files $F1$ and $F2$ into one sorted file. If the sort is to be done in descending order of key values, you continue to choose the higher key value from the fronts of $F1$ and $F2$ and place them in the back of F.

Note that $F1$ will always have $\lfloor n/2 \rfloor$ records and $F2$ will have $\lceil n/2 \rceil$ records; thus, if n is odd, $F2$ gets one more record then $F1$. We will assume that the time required to perform Step 1 of the algorithm is insignificant compared to the other steps; therefore, we will assign a time value of zero to this step. Step 3 requires roughly n comparisons and n movements of records from $F1$ and $F2$ to F; thus, its time is proportional to n. For this reason, we will assume that Step 3 takes n time units. Since Step 2 requires $T(\lfloor n/2 \rfloor) + T(\lceil n/2 \rceil)$ time units,

$$T(n) = n + T(\lfloor n/2 \rfloor) + T(\lceil n/2 \rceil), \qquad (8.4g)$$

with the initial condition

$$T(1) = 0. \qquad (8.4h)$$

Instead of an exact solution of 8.4g and 8.4h, we will be content with an estimate for $T(n)$. First, consider the case of $n = 2^r, r \geq 1$:

$$T(2^1) = T(2) = 2 + T(1) + T(1) = 2 = 1 \quad 2$$
$$T(2^2) = T(4) = 4 + T(2) + T(2) = 8 = 2 \quad 4$$
$$T(2^3) = T(8) = 8 + T(4) + T(4) = 24 = 3 \quad 8$$

$$\cdot$$

$$\cdot$$

$$\cdot$$

$$T(2^r) = r2^r = (\log_2 2^r)2^r.$$

TABLE 8.4.1
Comparison of Times for Bubble Sort and Merge Sort

n	$B(n)$	$T_e(n)$
10	45	34
50	1225	283
100	4950	665
500	124750	4483
1000	499500	9966

Thus, if n is a power of 2, $T(n) = n \log_2 n$. Now if, for some $r \geq 2$, $2^{r-1} < n < 2^r$, then $(r - 1)2^{r-1} \leq T(n) < r2^r$. (This can be proven by induction on r.) As n increases from 2^{r-1} to 2^r, $T(n)$ increases from $(r - 1)2^{r-1}$ to $r2^r$ and is slightly larger than $\lceil n \log_2 n \rceil$. The discrepancy is small enough so that $T_e(n) = \lceil n \log_2 n \rceil$ can be considered a solution of 8.4g and 8.4h for the purposes of comparing the merge sort with other algorithms. Table 8.4.1 compares $B(n)$ with $T_e(n)$ for selected values of n.

One word of warning: This analysis is not complete. For example, it doesn't take into account the memory that is needed to perform the algorithm.

Definition: Derangement. *A derangement of a set A is a permutation of A (i.e., A bijection from A into A) such that $f(a) \neq a$ for all $a \in A$.*

Example 8.4.4. If $A = \{1, 2, \ldots, n)$, an interesting question might be "How many derangements are there of A?" We know that our answer is bounded above by $n!$. We can also expect our answer to be quite a bit smaller than $n!$ since n is the image of itself for $(n-1)!$ of the permutations of A.

Let $D(n)$ be the number of derangements of $\{1, 2, \ldots, n)$. Our answer will come from discovering a recurrence relation on D.

Suppose that $n \geq 3$. If we are to construct a derangement of $\{1, 2, \ldots, n)$, f, then $f(n) = k \neq n$. Thus, the image of n can be selected in $n - 1$ different ways. No matter which of the $n - 1$ choices we make, we can complete the definition of f in one of two ways. First, we can decide to make $f(k) = n$, leaving $D(n - 2)$ ways of completing the definition of f, since f will be a derangement of $\{1, 2, \ldots, n\} - \{n, k\}$. Second, if we decide to select $f(k) \neq n$, each of the $D(n - 1)$ derangements of $\{1, 2, \ldots, n - 1\}$ can be used to define f. If g is a derangement of $\{1, 2, \ldots, n - 1\}$ such that $g(p) = k$, then define f by

$$f(j) = \begin{cases} n & \text{if} \quad j = p \\ k & \text{if} \quad j = n \\ g(j) & \text{otherwise} \end{cases}$$

Note that with our second construction of f, $f(f(n)) = f(k) \neq n$, while in the first construction, $f(f(n)) = f(k) = n$. Therefore, no derangement of $\{1, \ldots, n\}$ with $f(n) = k$ can be constructed by both methods.

N	D(N)	N!	D(N)/(N!)
1	0	1	0.000000000000000
2	1	2	0.500000000000000
3	2	6	0.333333333333334
4	9	24	0.375000000000000
5	44	120	0.366666666666667
6	265	720	0.368055555555555
7	1854	5040	0.367857142857142
8	14833	40320	0.367881944444443
9	133496	362880	0.367879188712521
10	1334961	3628800	0.367879464285714
11	14684570	39916800	0.367879439233606
12	176214841	479001600	0.367879441321282
13	2290792932	6227020800	0.367879441160690
14	32071101049	87178291200	0.367879441172162
15	481066515734	1307674368000	0.367879441171397
16	7697064251745	20922789888000	0.367879441171445

$D(16)/(16!) =$ 2.718281828459027

FIGURE 8.4.2
$D(n)$ compared to $n!$

To recap our result, we see that f is determined by first choosing one of $n - 1$ images of n and then constructing the remainder of f in one of $D(n - 2) + D(n - 1)$ ways. Therefore,

$$D(n) = (n - 1)(D(n - 2) + D(n - 1)). \qquad (8.4i)$$

This homogeneous second-order linear relation with variable coefficients, together with the initial conditions $D(1) = 0$ and $D(2) = 1$, completely define D. Instead of deriving a solution of this relation by analytical methods, we will give an empirical derivation of an approximation of $D(n)$. Since the derangements of $\{1, \ldots, n\}$ are drawn from a pool of $n!$ permutations, we will see what percentage of these permutations are derangements by listing the values of $n!$, $D(n)$ and $D(n)/n!$. To avoid having to write a complicated program, we restrict our maximum value of n to the largest integer n for which $n! \leq$ MAXINT.

On our computer, that value is 16. The results we obtained (see Figure 8.4.2) indicate that as n grows, $D(n)/n!$ hardly changes at all. If this quotient is computed to 8 decimal places, for $n \geq 12$, $D(n)/n! = 0.36787944$. The reciprocal of this number, which $D(n)/n!$ seems to be tending towards, is, to eight places, 2.71828182. This number appears in so many places in mathematics that it has its own name, e. An approximate solution of our recurrence relation on D is then $D(n) \approx n!/e$.

EXERCISES FOR SECTION 8.4

A Exercises

1. Solve the following recurrence relations. Indicate whether your solution is an improvement over iteration.

(a) $nS(n) - S(n - 1) = 0, S(0) = 1.$
(b) $T(k) + 3kT(k - 1) = 0, T(0) = 1.$
(c) $U(k) - ((k - 1)/k) U(k - 1) = 0, k \geq 2, U(1) = 1.$
(d) $S(n) + (3^n + (-1)^n 2) S(n - 1) = 0, S(0) = 0.2.$

2. Solve as completely as possible:
    (a) $T(n) = 3 + T(\lfloor n/2 \rfloor), \quad T(0) = 0.$
    (b) $T(n) = 1 + \frac{1}{2}(\lfloor n/2 \rfloor), \quad T(0) = 2.$
    (c) $V(n) = 1 + V(\lfloor n/8 \rfloor), \quad V(0) = 0.$ (Hint: Write n in octal form.)

3. Prove by induction that if $T(n) = 1 + T(\lfloor n/2 \rfloor), T(0) = 0$ and $2^{r-1} \leq n < 2^r, r \geq 1$, then $T(n) = r.$

4. Prove that if $n \geq 0, \lfloor n/2 \rfloor + \lceil n/2 \rceil = n.$ (Hint: Consider the cases n odd and n even separately.)

    B Exercises

5. Solve: $Q(n) = 1 + Q(\lfloor \sqrt{n} \rfloor), n \geq 2, \quad Q(1) = 0.$

6. Solve: $R(n) = n + R(\lfloor n/2 \rfloor), n \geq 1, \quad R(0) = 0.$

7. Suppose that Step 1 of the mergesort algorithm did take a significant amount of time. Assume that it takes 0.1 time units, independent of the value of n:

    (a) write out a new recurrence relation for $T(n)$ that takes this factor into account,
    (b) solve for $T(2^r), r \geq 0,$
    (c) assuming the solution for powers of 2 is a good estimate for all n, compare your result to the solution in the text. As n gets large, is there really much difference?

8.5 Generating Functions

This section contains an introduction to the topic of generating functions and how they are used to solve recurrence relations, among other problems. Methods that employ generating functions are based on the concept that you can take a problem involving sequences and translate it into a problem involving generating functions. Once you've solved the new problem, a translation back to sequences gives you a solution of the original problem.

This section covers:

(a) Definition of a generating function
(b) Solution of a recurrence relation using generating functions to identify the skills needed to use generating functions
(c) An introduction and/or review of the skills identified in Point b
(d) Some applications of generating functions

Definition: Generating Function of a Sequence. *The generating function of a sequence S with terms $S_0, S_1, S_2, \ldots$, is the infinite sum*

$$G(S;z) = \sum_{n=0}^{\infty} S_n z^n = S_0 + S_1 z + S_2 z^2 + S_3 z^3 + \cdots$$

The domain and codomain of generating functions will not be of any concern to us since we will only be performing algebraic operations on them.

Example 8.5.1.

(a) If $S_n = 3^n$, $n \geq 0$, then

$$G(S;z) = 1 + 3z + 9z^2 + 27z^3 + \cdots$$

$$= \sum_{n=0}^{\infty} 3^n z^n = \sum_{n=0}^{\infty} (3z)^n.$$

We can obtain a closed form expression for $G(S;z)$ by observing that $G(S;z) - 3zG(S;z) = 1$. Therefore,

$$G(S;z) = 1/(1 - 3z).$$

(b) Finite sequences have generating functions. For example, the sequence of binomial coefficients $C(n, 0)$, $C(n, 1)$, . . . , $C(n, n)$, $n \geq 1$ has generating function

$$G(C(n,);z) = C(n, 0) + C(n, 1)z + \cdots + C(n, n)z^n$$

$$= \sum_{k=0}^{n} C(n, k)z^k$$

$$= (1 + z)^n$$

by application of the binomial formula.

(c) If $Q(n) = n^2$,

$$G(Q;z) = \sum_{n=0}^{\infty} n^2 z^n = \sum_{k=0}^{\infty} k^2 z^k.$$

Note that the index that is used in the summation has no significance. Also, note that the lower limit of the summation could start at 1 since $Q(0) = 0$.

SOLUTION OF A RECURRENCE RELATION USING GENERATING FUNCTIONS

Problem: Solve $S(n) - 2S(n - 1) - 3S(n - 2) = 0$, $n \geq 2$, with $S(0) = 3$ and $S(1) = 1$.

(1) Translate the recurrence relation into an equation about generating functions.
   In our example, let $V(n) = S(n) - 2S(n - 1) - 3S(n - 2)$, $n \geq 2$, with $V(0) = 0$ and $V(1) = 0$. Therefore,

$$G(V;z) = 0 + 0z + \sum_{n=2}^{\infty} (S(n) - 2S(n-1) - 3S(n-2))z^n = 0$$

(2) Solve for the generating function of the unknown sequence,

$$G(S;z) = \sum_{n=0}^{\infty} S(n)z^n.$$

$$0 = \sum_{n=2}^{\infty} (S(n) - 2S(n-1) - 3S(n-2))z^n$$

$$= \sum_{n=2}^{\infty} S(n)z^n - 2\left(\sum_{n=2}^{\infty} S(n-1)z^n\right) - 3\left(\sum_{n=2}^{\infty} S(n-2)z^n\right).$$

Close examination of the three sums above shows:

$$\sum_{n=2}^{\infty} S(n)z^n = S(0) + S(1)z + \left(\sum_{n=2}^{\infty} S(n)z^n\right) - S(0) - S(1)z$$

$$= G(S;z) - 3 - z,$$

since $S(0) = 3$ and $S(1) = 1$.

$$\sum_{n=2}^{\infty} S(n-1)z^n = z\left(\sum_{n=2}^{\infty} S(n-1)z^{n-1}\right)$$

$$= z\left(\sum_{n=1}^{\infty} S(n)z^n\right)$$

$$= z\left(S(0) + \sum_{n=1}^{\infty} S(n)z^n - S(0)\right)$$

$$= z(G(S;z) - 3)$$

$$\sum_{n=2}^{\infty} S(n-2)z^n = z^2\left(\sum_{n=2}^{\infty} S(n-2)z^{n-2}\right)$$

$$= z^2 G(S; Z).$$

Therefore,

$$(G(S;z) - 3 - z) - 2z(G(S;z) - 3) - 3z^2 G(S;z) = 0$$

$$G(S;z) - 2zG(S;z) - 3z^2 G(S;z) = 3 - 5z$$

$$G(S;z) = \frac{3 - 5z}{1 - 2z - 3z^2}.$$

(3) Determine the sequence whose generating function is the one obtained in Step 2.

For our example, we need to know one general fact about the closed form expression of an exponential sequence (a proof will be given later):

$$[T(n) = ba^n, n \geq 0] \Leftrightarrow G(T;z) = \frac{b}{1 - az} \qquad (8.5a)$$

Now, in order to recognize S in our example, we must write our closed form expression for $G(S;z)$ as a sum of terms like $G(T;z)$ above. Note that the denominator of $G(S;z)$ can be factored:

$$G(S;z) = \frac{3 - 5z}{1 - 2z - 3z^2} = \frac{3 - 5z}{(1 - 3z)(1 + z)}.$$

If you look at this last expression for $G(S;z)$ closely, you can imagine how it could be the result of addition of two fractions,

$$\frac{3 - 5z}{(1 - 3z)(1 + z)} = \frac{A}{1 - 3z} + \frac{B}{1 + z}, \qquad (8.5b)$$

where A and B are two real numbers that must be determined. Starting on the right of 8.5b, it should be clear that the sum, for any A and B, would look like the left-hand side. The process of finding values of A and B that make 8.5b true is called the *partial fractions decomposition of the left-hand side*:

$$\frac{A}{1 - 3z} + \frac{B}{1 + z} = \frac{A(1 + z)}{(1 - 3z)(1 + z)} + \frac{B(1 - 3z)}{(1 - 3z)(1 + z)}$$

$$= \frac{(A + B) + (A - 3B)z}{(1 - 3z)(1 + z)}.$$

Therefore,

$$\begin{Bmatrix} A + B = 3 \\ A - 3B = -5 \end{Bmatrix} \Rightarrow \begin{Bmatrix} A = 1 \\ B = 2 \end{Bmatrix}$$

and

$$G(S;z) = \frac{1}{1 - 3z} + \frac{2}{1 + z}.$$

We can apply 8.5a to each term of $G(S;z)$:

$\dfrac{1}{1 - 3z}$ is the generating function for $S_1(n) = 1(3^n) = 3^n$, and

$\dfrac{2}{1 + z}$ is the generating function for $S_2(n) = 2(-1)^n$.

Therefore, $S(n) = 3^n + 2(-1)^n$.

From this example, we see that there are several skills that must be mastered in order to work with generating functions. You must be able to:

(a) manipulate summation expressions and their indices (in Step 2);
(b) solve algebraic equations and manipulate algebraic expressions, including partial function decompositions (Steps 2 and 3); and
(c) identify sequences with their generating functions (Steps 1 and 3).

We will concentrate on the last skill first, a proficiency in the other skills is a product of doing as many exercises and reading as many examples as possible.

First, we must identify the operations on sequences and on generating functions.

Operations on Sequences: Let S and T be sequences of numbers and let c be a real number. Define the *sum* $S + T$, the *scalar product* cS, the *product ST*, the *convolution* $S*T$, the *pop operation* $S \uparrow$ (read "S pop"), and the *push operation* $S \downarrow$ (read "S push") termwise for $k \geq 0$ by

$$(S + T)(k) = S(k) + T(k)$$
$$(cS)(k) = cS(k)$$
$$(ST)(k) = S(k)T(k)$$
$$(S*T)(k) = \sum_{j=0}^{k} S(j)T(k - j)$$
$$(S \uparrow)(k) = S(k + 1)$$

and

$$(S \downarrow)(k) = \begin{cases} 0 & \text{if } k = 0 \\ S(k - 1) & \text{if } k > 0. \end{cases}$$

If one imagines a sequence to be a matrix with one row and an infinite number of columns, $S + T$ and cS are exactly as in matrix addition and scalar multiplication. There is no obvious similarity between the other operations and matrix operations.

The pop and push operations can be understood by imagining a sequence to be an infinite stack of numbers with $S(0)$ at the top, $S(1)$ next, etc., as in Figure 8.5.1a. The sequence $S \uparrow$ is obtained by "popping" $S(0)$ from the

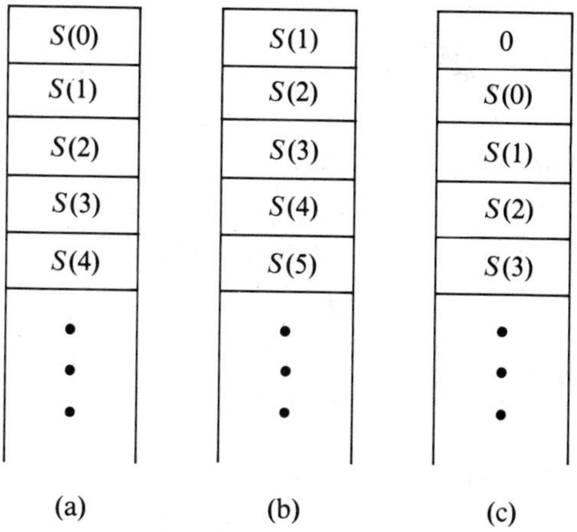

$S(0)$	$S(1)$	0
$S(1)$	$S(2)$	$S(0)$
$S(2)$	$S(3)$	$S(1)$
$S(3)$	$S(4)$	$S(2)$
$S(4)$	$S(5)$	$S(3)$
•	•	•
•	•	•
•	•	•
(a)	(b)	(c)

FIGURE 8.5.1
Stack interpretation of pop and push operations.

stack, leaving a stack as in Figure 8.5.1b, with $S(1)$ at the top, $S(2)$ next, etc. The sequence $S\downarrow$ is obtained by placing a zero at the top of stack, resulting in a stack as in Figure 8.5.1c. Keep these figures in mind when we discuss the pop and push operations.

Example 8.5.2. If $S(n) = n$, $T(n) = n^2$, $U(n) = 2^n$, and $R(n) = n2^n$,

(a) $(S + T)(n) = n + n^2$

(b) $(U + R)(n) = 2^n + n2^n = (1 + n)2^n$

(c) $(2U)(n) = 2 \cdot 2^n = 2^{n+1} = (U\uparrow)(n)$

(d) $(\frac{1}{2}R)(n) = \frac{1}{2}n2^n = n2^{n-1}$

(e) $(ST)(n) = n\,n^2 = n^3$

(f) $(S*T)(n) = \sum_{j=0}^{n} S(j)\,T(n-j) = \sum_{j=0}^{n} j(n-j)^2$

$$= \sum_{j=0}^{n} (jn^2 - 2nj^2 + j^3)$$

$$= n^2 \sum_{j=0}^{n} j - 2n \sum_{j=0}^{n} j^2 + \sum_{j=0}^{n} j^3$$

(by Exercise 14 of Section 8.3)

$$= n^2\left(n(n+1)/2\right) - 2n\left((2n+1)(n+1)/6\right)$$
$$+ \left((n(n+1)/2)\right)^2$$

$$= n^2(n+1)(n-1)/12$$

(g) $(U*U)(n) = \sum_{j=0}^{n} U(j)U(n-j) = \sum_{j=0}^{n} 2^j 2^{n-j}$

$$= (n+1)2^n$$

(h) $(S\uparrow)(n) = n + 1)$

(i) $(S\downarrow)(n) = max\{0, n-1\}$

(j) $((S\downarrow)\downarrow)(n) = max\{0, n-2\}$

(k) $(U\downarrow)(n) = 2^{n-1}$ if $n > 0$, $(U\downarrow)(0) = 0$

(l) $((U\downarrow)\uparrow)(n) = (U\downarrow)(n+1) = 2^n = U(n)$

(m) $((U\uparrow)\downarrow)(n) = \begin{cases} 0 \text{ if } n = 0 \\ U\uparrow(n-1) \text{ if } n > 0 \end{cases} = \begin{cases} 0 \text{ if } n = 0 \\ U(n) \text{ if } n > 0 \end{cases}$

Note that $(U\downarrow)\uparrow \neq (U\uparrow)\downarrow$.

Definition: If S is a sequence of numbers, define
$$S \uparrow p = (S \uparrow (p - 1)) \uparrow \text{ if } p > 1 \text{ and } S \uparrow 1 = S \uparrow.$$

Similarly, define
$$S \downarrow p = (S \downarrow (p - 1)) \downarrow \text{ and } S \downarrow 1 = S \downarrow.$$

Note that
$$(S \uparrow 2)(k) = ((S \uparrow) \uparrow)(k) = (S \uparrow)(k + 1) = S(k + 2).$$

In general,
$$(S \uparrow p)(k) = S(p + k), \text{ and}$$

$$(S \downarrow p)(k) = \begin{cases} 0 & \text{if } k < p \\ S(k - p) & \text{if } k \geq p. \end{cases}$$

Operations on Generating Functions: If
$$G(z) = \sum_{k=0}^{\infty} a_k z^k \text{ and } H(z) = \sum_{k=0}^{\infty} b_k z^k$$

are generating functions and c is a real number, then the *sum $G + H$, scalar product cG, product GH,* and *monomial product $z^p G$, $p \geq 1$* are generating functions, where

$$(G + H)(z) = \sum_{k=0}^{\infty} (a_k + b_k) z^k$$

$$(cG)(z) = c \sum_{k=0}^{\infty} a_k z^k = \sum_{k=0}^{\infty} (ca_k) z^k$$

$$(GH)(z) = \sum_{k=0}^{\infty} c_k z^k, \text{ where } c_k = \sum_{j=0}^{k} a_j b_{k-j}$$

$$(z^p G)(Z) = z^p \sum_{k=0}^{\infty} a_k z^k = \sum_{k=0}^{\infty} a_k z^{k+p}$$

$$= \sum_{n=p}^{\infty} a_{n-p} z^n.$$

The last sum is obtained by substituting $n - p$ for k in the previous sum.

Example 8.5.3. If

$$D(z) = \sum_{k=0}^{\infty} kz^k \text{ and } H(z) = \sum_{k=0}^{\infty} 2^k z^k,$$

$$(D + H)(z) = \sum_{k=0}^{\infty} (k + 2^k)z^k$$

$$(2H)(z) = \sum_{k=0}^{\infty} (2 \; 2^k)z^k = \sum_{k=0}^{\infty} 2^{k+1}z^k = (H(z) - 1)/z$$

$$(zD)(z) = z \sum_{k=0}^{\infty} kz^k = \sum_{k=0}^{\infty} kz^{k+1} = \sum_{k=1}^{\infty} (k - 1)z^k$$

$$= D(z) - \sum_{k=1}^{\infty} z^k.$$

$$(DH)(z) = \sum_{k=0}^{\infty} \left(\sum_{j=0}^{k} j2^{k-j} \right) z^k$$

$$(HH)(z) = \sum_{k=0}^{\infty} \left(\sum_{j=0}^{k} 2^j 2^{k-j} z^k \right) = \sum_{k=0}^{\infty} (k + 1)2^k z^k.$$

Note: $D(z) = G(S;z)$, $H(z) = G(U;z)$ from Example 8.5.2.

Now we establish the connection between the operations on sequences and generating functions. Let S and T be sequences and let c be a real number:

$$G(S + T;z) = G(S;z) + G(T;z) \tag{8.5c}$$

$$G(cS;z) = cG(S;z) \tag{8.5d}$$

$$G(S*T;z) = G(S;z) \, G(T;z) \tag{8.5e}$$

$$G(S \uparrow ;z) = (G(S;z) - S(0))/z \tag{8.5f}$$

$$G(S \downarrow ;z) = zG(S;z) \tag{8.5g}$$

In words, 8.5c says that the generating function of the sum of two sequences equals the sum of the generating functions of those sequences. Take the time to write out the other four identities in your own words. From the previous examples, these identities should be fairly obvious, with the possible exception of the last two. We will prove 8.5f as part of the next theorem and leave the proof of 8.5g to the interested reader. Note that there is no operation on generating functions that is related to sequence multiplication; that is, $G(ST;z)$ cannot be simplified.

Theorem 8.5.1. *If* $p \geq 1$,

(a) $G(S \uparrow p;z) = \left((G(S;z) - \sum_{k=0}^{p-1} S(k)z^k) \right)/z^p.$

(b) $G(S \downarrow p;z) = z^p \, G(S;z).$

Proof of Part a by induction.

Basis: $S \uparrow 1 = S \uparrow$.

$$G(S \uparrow, z) = \sum_{k=0}^{\infty} S(k + 1)z^k = \sum_{k=1}^{\infty} S(k)z^{k-1}$$

$$= \left(\sum_{k=1}^{\infty} S(k)z^k \right) / z$$

$$= \left(S(0) + \sum_{k=1}^{\infty} S(k)z^k - S(0) \right) / z$$

$$= (G(S;z) - S(0)) / z.$$

Therefore, Part a is true for $p = 1$.

Induction. Suppose that for some $r \geq 1$, a is true:

$$G(S \uparrow (r + 1); z) = G((S \uparrow r) \uparrow ; z)$$

$$= (G(S \uparrow r; z) - (S \uparrow r)(0)) / z \text{ by the basis}$$

$$= \frac{\left(\dfrac{G(S;z) - \sum_{k=0}^{r-1} S(k)z^k}{z^r} \right) - S(r)}{z}$$

by the induction hypothesis.
Now write $S(r)$ as $(S(r)z^r)/z^r$ and obtain

$$= \left(\left(G(S;z) - \sum_{k=0}^{r} S(k)z^k \right) / z^r \right) / z$$

$$= \left(G(S;z) - \sum_{k=0}^{r} S(k)z^k \right) / z^{r+1} \cdot \#$$

CLOSED FORM EXPRESSIONS FOR GENERATING FUNCTIONS

The most basic tool used to express generating functions in closed form is the closed form expression for the *geometric series,* which is an expression of the form $a + ar + ar^2 + \cdots$. It can either be terminated or extended infinitely.

Finite Geometric Series: $a + ar^n + \cdots + ar = a \left(\dfrac{1 - r^{n+1}}{1 - r} \right)$ (8.5h)

Infinite Geometric Series: $a + ar + ar^2 + \cdots = \dfrac{a}{1 - r}$ (8.5i)

Restrictions: If a and r represent numbers:

(1) r must   not   equal   1   in   the   finite   case.   Note   that
    $a + ar + \cdots + ar^n = (n + 1)a$ if $r = 1$.

(2) In the infinite case, the absolute value of r must be less than 1.

These restrictions don't come into play with generating functions. We could derive 8.5h by noting that if $S(n) = a + ar + \cdots + ar^n$, $n \geq 0$, then $S(n) = rS(n - 1) + a$, $n \geq 1$ (See Exercise 10 of Section 8.3). An alternative derivation was used in Section 8.4. We will take the same steps to derive 8.5i. Let

$$x = a + ar + ar^2 + \cdots$$
$$rx = ar + ar^2 + \cdots$$
$$x - rx = a$$

Therefore, $x = a/(1 - r)$.

Example 8.5.4.

(a) If $S(n) = 9 \cdot 5^n$, $n \geq 0$, $G(S;z)$ is an infinite geometric series with $a = 9$ and $r = 5z$. Therefore, $G(S;z) = 9/(1 - 5z)$.
(b) If $T(n) = 4$, $n \geq 0$, then $G(S;z) = 4/(1 - z)$.
(c) If $U(n) = 3(-1)^n$, then $G(U,z) = 3/(1 + z)$.
(d) *Let* $C(n) = S(n) + T(n) + U(n) = 9 \cdot 5^n + 4 + 3(-1)^n$.

$$G(C;z) = G(S;z) + G(T;z) + G(U;z)$$
$$= 9/(1 - 5z) + 4/(1 - z) + 3/(1 + z)$$
$$= (16 - 34z - 11z^2)/(1 - 5z - z^2 + 5z^3)$$

Given a choice between the last form of $G(C;z)$ and the previous sum of three fractions, we would prefer leaving it as a sum of three functions. As we saw in an earlier example, a partial fractions decomposition of a fraction such as the last expression requires some time.

(e) If $G(Q;z) = 34/(2 - 3z)$, then Q can be determined by multiplying the numerator and denominator by $1/2$ to obtain $17/(1 - 1.5z)$. We recognize this fraction as the sum of the infinite geometric series with $a = 17$ and $r = 1.5z$. Therefore $Q(n) = 17(3/2)^n$.
(f) If $G(A;z) = (1 + z)^3$, then we expand $(1 + z)^3$ to $1 + 3z + 3z^2 + z^3$. Therefore $A(0) = 1$, $A(1) = 3$ $A(2) = 3$, $A(3) = 1$, and, since there are no higher-powered terms, $A(n) = 0$, $n \geq 4$. A more concise way of describing A is $A(k) = C(3;k)$, since $C(n, k)$ is usually interpreted as 0 of $k > n$.

TABLE 8.5.1
Closed form Expression for Some Generating Functions

Sequence	Generating Function
$S(k) = ba^k$	$G(S; z) = \dfrac{b}{1 - az}$
$S(k) = k$	$G(S; z) = \dfrac{z}{(1 - z)^2}$
$S(k) = bka^k$	$G(S; z) = \dfrac{abz}{(1 - az)^2}$
$S(k) = \dfrac{1}{n!}$	$G(S; z) = e^z \ (e \approx 2.71828)$
$S(k) = \begin{cases} C(n, k) & 0 \le k \le n \\ 0 & k > n \end{cases}$	$G(S; z) = (1 + z)^n$

Table 8.5.1 contains some closed form expressions for the generating functions of some common sequences.

Example 8.5.5. Solve $S(k) + 3S(k - 1) - 4S(k - 2) = 0$, $k \ge 2$, with $S(0) = 3$ and $S(1) = -2$. The solution will be obtained using the same steps that were used earlier in this section, with one variation.

(1) Translate to an equation about generating functions. First, we change the index of the recurrence relation by substituting $n + 2$ for k. The result is $S(n + 2) + 3S(n + 1) - 4S(n) = 0$, $n \ge 0$. Now, if $V(n) = S(n + 2) + 3S(n + 1) - 4S(n)$, then V is the zero sequence, which has a zero generating function. Furthermore, $V = S \uparrow 2 + 3(S \uparrow) - 4S$. Therefore,

$$0 = G(V;z)$$
$$= G(S \uparrow 2;z) + G(3(S \uparrow);z) - G(4S;z)$$
$$= (G(S;z) - S(0) - S(1)z)/z^2 + 3(G(S;z) - S(0))/z - 4G(S;z).$$

(2) Solve for $G(S;z)$:

$$\frac{G(S;z) - 3 + 2z}{z^2} + 3\frac{G(S;z) - 3}{z} - 4\,G(S;z) = 0.$$

Multiply by z^2:

$$G(S;z) - 3 + 2z + 3z(G(S;z) - 3) - 4z^2G(S;z) = 0$$
$$G(S;z) + 3zG(S;z) - 4z^2G(S;z) = 3 + 7z$$
$$(1 + 3z - 4z^2)G(S;z) = 3 + 7z.$$

Therefore,

$$G(S;z) = \frac{3 + 7z}{1 + 3z - 4z^2}.$$

(3) Determine S from its generating function. $1 + 3z - 4z^2 = (1 + 4z)$
$\times (1 - z)$, thus a partial fraction decomposition of $G(S;z)$ would be:

$$\frac{A}{1 + 4z} + \frac{B}{1 - z} = \frac{A(1 - z) + B(1 + 4z)}{(1 + 4z)(1 - z)}$$

$$= \frac{(A + B) + (4B - A)z}{(1 + 4z)(1 - z)}.$$

Therefore, $A + B = 3$ and $4B - A = 7$. A solution of this set of equations
is $A = 1$, $B = 2$.

$$G(S;z) = \frac{1}{1 + 4z} + \frac{2}{1 - z}.$$

$\dfrac{1}{1 + 4z}$ is the generating function of $S_1(n) = (-4)^n$, and

$\dfrac{2}{1 - z}$ is the generating function of $S_2(n) = 2(1)^n = 2$.

In conclusion, since $G(S;z) = G(S_1,z) + G(S_2,z)$, $S(n) = 2 + (-4)^n$.

Example 8.5.6.   Let $A = \{a, b, c, d, e\}$ and let A^* be the set of all strings
of length zero or more that can be made using each of the elements of A zero
or more times. By the generalized rule of products, there are 5^n such strings
that have length n, $n \geq 0$. Suppose that X_n is the set of strings of length n with
the property that all of the a's and b's precede all of the c's, d's, and e's. Thus
$aaabde \in X_6$, but $abcabc \notin X_n$. Let $R(n) = \#X_n$. A closed form expression
for R can be obtained by recognizing R as the convolution of two sequences.
To illustrate our point, we will consider the calculation of $R(6)$.

Note that if a string belongs to X_6, it starts with k characters from $\{a, b\}$ and
is followed by $6 - k$ characters from $\{c, d, e\}$. Let $S(k)$ be the number of
strings of a's and b's with length k and let $T(k)$ be the number of strings of
c's, d's, and e's with length k. By the generalized rule of products, $S(k) = 2^k$
and $T(k) = 3^k$. Among the strings in X_6 are the ones that start with two a's
and b's and end with $4c$'s, d's and e's. There are $S(2) T(4)$ such strings. By
the   law   of   addition, $\#X_6 = R(6) = S(0)T(6) + S(1)T(5) + \cdots$
$+ S(5)T(1) + S(6)T(0)$. Note that the sixth term of R is the sixth term of the
convolution of S with T, $S*T$. Think about the general situation for a while
and it should be clear that $R = S*T$. Now, our course of action will be to:

(a) determine the generating functions of S and T,
(b) multiply $G(S;z)$ and $G(T;z)$ to obtain $G(S*T;z) = G(R;z)$ (by 10.5e), and
(c) determine R on the basis of $G(R;z)$.

(a) $G(S;z) = \sum\limits_{k=0}^{\infty} 2^k z^k = \dfrac{1}{1 - 2z}$, and $G(T;z) = \sum\limits_{k=0}^{\infty} 3^k z^k = \dfrac{1}{1 - 3z}$.

(b) $G(R;z) = G(S;z)G(T;z) = \dfrac{1}{(1 - 2z)(1 - 3z)}$.

(c) To recognize R from $G(R;z)$, we must do a partial fractions decomposition:

$$\frac{1}{(1 - 2z)(1 - 3z)} = \frac{A}{(1 - 2z)} + \frac{B}{(1 - 3z)} = \frac{(A + B) + (-3A - 2B)z}{(1 - 2z)(1 - 3z)}.$$

Therefore, $A + B = 1$ and $-3A - 2B = 0$. The solution of this pair of equations yields $A = -2$ and $B = 3$.
Since

$$G(R;z) = \frac{-2}{1 - 2z} + \frac{3}{1 - 3z},$$

which is the sum of the generating functions of $-2(2^k)$ and $3(3^k)$.

$$R(k) = -2(2^k) + 3(3^k)$$
$$= 3^{k+1} - 2^{k+1}.$$

For example, $R(6) = 3^7 - 2^7 = 2187 - 128 = 2059$. Naturally, this equals the sum that we obtained for $(S*T)(6)$.

EXTRA FOR EXPERTS

The remainder of this section is intended for readers who have had, or who intend to take, a course in combinatorics. We do not advise that it be included in a typical course. The method that was used in Example 8.5.6 is a very powerful one and can be used to solve many problems in combinatorics. We close this section with a general description of the problems that can be solved in this way, followed by some examples.

Consider the situation in which $P_1, P_2, \ldots, P_m$ are n actions that must be taken, each of which results in a well-defined outcome. For each $k = 1, 2, \ldots, m$ define X_k to be the set of possible outcomes of P_k. We will assume that each outcome can be quantified in some way and that the quantification of the elements of X_k is defined by the function $Q_k : X_k \rightarrow \{0, 1, 2, \ldots\}$. Thus, each outcome has a non-negative integer associated with it. Finally, define a frequency function $F_k : \{0, 1, 2, \ldots\} \rightarrow \{0, 1, 2, \ldots\}$ such that $F_k(n)$ is the number of elements of X_k that have a quantification of n.

Now, based on these assumptions, we can define the problems that can be solved. If a process P is defined as a sequence of processes $P_1, P_2, \ldots, P_m$ as above, and if the outcome of P, which would be an element of $X_1 \times X_2 \times \cdots \times X_m$, is quantified by $Q(a_1, a_2, \ldots, a_m) = Q_1(a_1) + \cdots + Q_m(a_m)$, then the frequency function, F, for P is the convolution of the frequency functions for $P_1, P_2, \ldots, P_m$, which has a generating function equal to the product of the generating functions of $F_1, F_2, \ldots, F_m$. I.e.:

$$G(F;z) = G(F_1;z)G(F_2;z) \cdots G(F_m;z) \qquad (8.5j)$$

Example 8.5.7. Suppose that you roll a die two times and add up the numbers on the top face for each roll. Since the faces on the die represent the integers 1 through 6, the sum must be between 2 and 12. How many ways can any one of these sums be obtained? Obviously 2 can be obtained only one way, with two 1's. There are two sequences that yield a sum of 3: 1—2 and 2—1. To obtain all of the frequencies with which the numbers 2 through 12 can be obtained, we set up the situation as follows. For $j = 1, 2; P_j$ is the rolling of the die for the jth time. $X_j = \{1, 2, \ldots, 6\}$ and $Q_j : X_j \rightarrow \{0, 1, 2, \ldots\}$ is defined by $Q_j(x) = x$. Since each number appears on a die exactly once, the frequency function is $F_j(k) = 1$ if $1 \leq k \leq 6$, and $F_j(k) = 0$ otherwise. The process of rolling the die two times is quantified by adding up the Q_j's; that is, $Q(a_1, a_2) = Q_1(a_1) + Q_2(a_2) = a_1 + a_2$. The frequency function for the process of rolling the die two times is then

$$\begin{aligned} G(F;z) &= G(F_1;z)G(F_2;z) \\ &= (z + z^2 + z^3 + z^4 + z^5 + z^6)^2 \\ &= z^2 + 2z^3 + 3z^4 + 4z^5 + 5z^6 + 6z^7 \\ &\quad + 5z^8 + 4z^9 + 3z^{10} + 2z^{11} + z^{12} \end{aligned}$$

Now, to obtain $F(k)$, just read the coefficient of z^k.

To apply this method, the crucial step is to decompose a large process in the proper way so that it fits into the general situation that we've described.

Example 8.5.8. Suppose that an organization is divided into three geographic sections, A, B, and C. Suppose that an executive committee of 11 members must be selected so that no more than 5 members from any one section are on the committee and that Sections A, B, and C must have minimums of 3, 2, and 2 members, respectively, on the committee. Looking at only the number of members from each section on the committee, how many ways can the committee be made up? One example of a valid committee would be 4 A's, 4 B's, and 3 C's.

Let P_A be the process of deciding *how many* members (not who) from Section A will serve on the committee. $X_A = \{3, 4, 5\}$ and $Q_A(k) = k$. The frequency function, F_A, is defined by $F_A(k) = 1$ if $k \in X_A$, with $F_A(k) = 0$

otherwise. $G(F_A;z)$ is then $z^3 + z^4 + z^5$. Similarly, $G(F_B;z) = z^2 + z^3 + z^4 + z^5 = G(F_C;z)$. Since the committee must have 11 members, our answer will be the coefficient of z^{11} in $G(F_A;z)G(F_B;z)G(F_C;z)$, which is 10.

EXERCISES FOR SECTION 8.5

A Exercises

1. For the following expressions, obtain partial fraction decompositions and identify the sequence having the expression as a generating function.

   (a) $\dfrac{5 + 2z}{1 - 4z^2}$

   (b) $\dfrac{6 - 29z}{1 - 11z + 30z^2}$

   (c) $\dfrac{32 - 22z}{2 - 3z + z^2}$

2. If $P(k) - 6P(k - 1) + 5P(k - 2) = 0$, $P(0) = 2$ and $P(1) = 2$, what is the generating function of P, $G(P;z)$?

3. Find the generating function of the Fibonacci sequence. (Hint: Start with the equation $F(k + 2) = F(k + 1) + F(k)$, $k \geq 0$, with $F(0) = F(1) = 1$.)

C Exercises

In the next two exercises, describe your answer in terms of the coefficient of a generating function, unless you have access to a program that will mutiply functions.

4. A game is played by rolling a die five times. For the kth roll, one point is added to your score if you roll a number higher than k. Otherwise, your score is zero for that roll. For example, the sequence of rolls 2—3—4—1—2 gives you a total score of three, while a sequence of 1—2—3—4—5 gives you a score of zero. Of the $6^5 = 7776$ possible sequences of rolls, how many give you a score of zero? of one? . . . of five?

5. Suppose that you roll a die ten times in a row and record the square of each number that you roll. How many ways could the sum of the squares of your rolls equal 40?

chapter

9

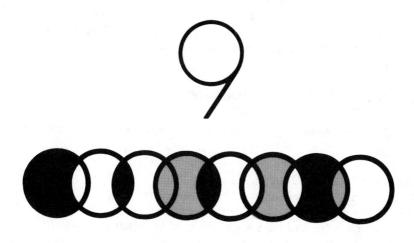

GRAPH THEORY

GOALS

This chapter has three principle goals. First, we will identify the basic components of a graph and some of the optional features that many graphs have. Second, we will discuss some of the questions that are most commonly asked of graphs. Third, we want to make the reader aware of how graphs are used to model different situations.

In Section 9.1, we will discuss these topics in general and in later sections we will take a closer look at selected topics in the theory of graphs.

Chapter 10 will continue our discussion with an examination of trees, a special type of graph.

9.1 Graphs—A General Introduction

Recall that the following definition of a directed graph was introduced in Chapter 6.

Definition: Directed Graph. *A directed graph consists of a set of vertices, V, and a set of edges, E, connecting certain elements of V. Each element of E is an ordered pair (i.e., an element of V × V). The first entry is the initial vertex of the edge and the second entry is the terminal vertex. In certain cases, there will be more than one edge between two vertices, in which case the different edges are identified with labels.*

Despite the set terminology in this definition, we usually think of a graph as a picture, an aid in visualizing a situation. In Chapter 6, we introduced this concept to help understand relations on sets. Although those relations were principally of a mathematical nature, it remains true that when we see a graph, it tells us how the elements of a set are related to one another.

Definition: Simple Graph and Multigraph. *A simple graph is one for which there is no more than one edge directed from any one vertex to any other vertex. All other graphs are called multigraphs.*

To illustrate the points that we will make in this chapter, we will introduce the following examples of graphs.

Example 9.1.1. A Directed Graph. Figure 9.1.1 is an example of a simple directed graph. In set terms, this graph is (V, E), where $V = \{s, a, b\}$ and $E = \{(s, a), (s, b), (a, b), (b, a), (b, b)\}$. Note how each edge is labeled either 0 or 1. There are often reasons for labeling even simple graphs. Some labels are to help make a graph easier to discuss; others are more significant. We will discuss the significance of the labels on this graph later.

Example 9.1.2. An Undirected Graph. A network of computers can be easily described using a graph. Figure 9.1.2 describes a network of five

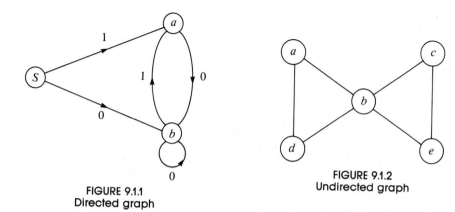

FIGURE 9.1.1
Directed graph

FIGURE 9.1.2
Undirected graph

computers, a, b, c, d, and e. An edge between any two vertices indicates that direct two-way communication is possible between the two computers. Note that the edges of this graph are not directed. This is due to the fact that the relation that is being displayed is symmetric (i.e., if X can communicate with Y, then Y can communicate with X). Although directed edges could be used here, it would simply clutter the graph.

There are several other situations for which this graph can serve as a model. One of them is to interpret the vertices as cities and the edges as roads, an abstraction of a map such as the one in Figure 9.1.3. Another interpretation is as an abstraction of the floor plan of a house (Figure 9.1.4). Vertex a represents the outside of the house; all others represent rooms. Two vertices are connected if there is a door between them.

Definition: Undirected Graph. *An undirected graph consists of a set V, called a vertex set, and a set E of two-element subsets of V, called the edge set. The two-element subsets are drawn as lines connecting the vertices.*

The undirected graph of Figure 9.1.2 is $V = \{a, b, c, d, e\}$ and $E = \{\{a, b,\}, \{a, d\}, \{b, c\}, \{b, d\}, \{c, e\}\}$. A directed graph can be turned into an undirected graph under the condition that if an edge exists from vertex v to

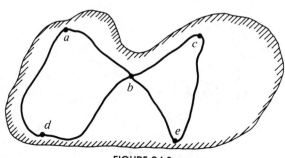

FIGURE 9.1.3
Road map

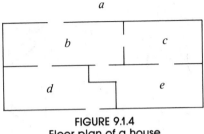

FIGURE 9.1.4
Floor plan of a house

vertex w, then an edge exists from vertex w to vertex v. In other words, the relation "is connected by an edge to" is a symmetric relation on the vertices.

Definition: Complete Undirected Graph. *A complete undirected graph of n vertices is an undirected graph with the property that each pair of distinct vertices are connected to one another. Such a graph is usually denoted by K_n.*

Example 9.1.3. A Multigraph. A common occurrence of a multigraph is a road map. The cities and towns on the map can be thought of as vertices, while the roads are the edges. It is not uncommon to have more than one road connecting two cities. In order to give clear travel directions, we name or number roads so that there is no ambiguity. We use the same method to describe the edges of the multigraph in Figure 9.1.5. There is no question what e_3 is; however, referring to the edge $(2, 3)$ would be ambiguous.

Example 9.1.4. A flowchart is a common example of a simple graph that requires labels for its vertices and some of its edges. Figure 9.1.6 is one such example which illustrates how many problems are solved. At the start of the

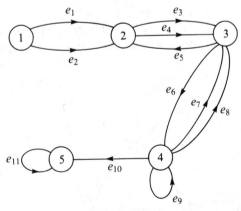

FIGURE 9.1.5
Multigraph

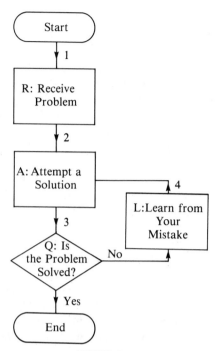

FIGURE 9.1.6
Flowchart for the problem-solving process

problem-solving process, we are at the vertex labeled "Start" and at the end (if we are lucky enough to have solved the problem) we will be at the vertex labeled "End." The sequence of vertices that we pass through as we move from "Start" to "End" is called a *path*. The "Start" vertex is called the *initial vertex of the path,* while the "End" is called the *final, or terminal, vertex.* Suppose that the problem is solved after two attempts; then the path that was taken is Start, R, A, Q, L, A, Q, End. An alternate path description would be to list the edges that were used: 1, 2, 3, No, 4, 3, Yes. This second method of describing a path has the advantage of being applicable for multigraphs. On the graph in Figure 9.1.5, the vertex list 1, 2, 3, 4, 3 does not clearly describe a path between 1 and 3, but e_1, e_4, e_6, e_7 is unambiguous.

A SUMMARY OF PATH NOTATION AND TERMINOLOGY

If x and y are two vertices of a graph, then a *path* between x and y describes a motion from x to y along edges of the graph. Vertex x is called the *initial vertex* of the path and y is called the *terminal vertex*. A path between x and y can always be described by its edge list, the list of edges that were used: $(e_1, e_2, \ldots, e_n)$, where : (1) the initial vertex of e_1 is x; (2) the terminal vertex of e_i is the initial vertex of e_{i+}, $i = 1, 2, \ldots, n - 1$; and (3) the terminal vertex of e_n is y. The number of edges in the edge list is the *path length*. A path on a simple graph can also be described by a vertex list. A path of length n will have a list of $n + 1$ vertices $(v_0, v_1, \ldots, v_n)$, where, for $k = 0$, $1, \ldots, n - 1$, (v_k, v_{k+1}) is an edge on the graph. A *circuit* is a path that terminates at its initial vertex.

Suppose that a path between two vertices has an edge list $(e_1, e_2, \ldots, e_n)$. A *subpath* of this graph is any portion of the path described by one or more consecutive edges in the edge list. For example, (3, no, 4) is a subpath of (1, 2, 3, no, 4, 3, yes). Any path is its own subpath; however, we call it an *improper subpath* of itself. All other subpaths are called *proper subpaths*.

A path or circuit is *simple* if it contains no proper subpath that is a circuit. This is the same as saying that a path or circuit is simple if it does not visit any vertex more than once except for the common initial and terminal vertex in a circuit. In the problem-solving method described in Figure 9.1.6, the path that you take is simple only if you reach a solution on the first try.

Example 9.1.5. The leadership structure of a corporation is often represented with a graph as in Figure 9.1.7. The principle behind such a structure is that everyone but the president has a single immediate supervisor. Any action that anyone takes can reach the president only through a unique "chain of command". This chain-of-command property is characteristic of a special type of graph called a *tree*. Note that the edges of this graph are not directed, but, as in a Hasse diagram, the relation between two connecting vertices is clear. That is, the top vertex is the supervisor of the lower vertex.

The process of structured (or top-down) problem solving results in a graph that is similar to this tree. Starting with the top of the tree, which would represent the whole problem, the problem is divided into a sequence of

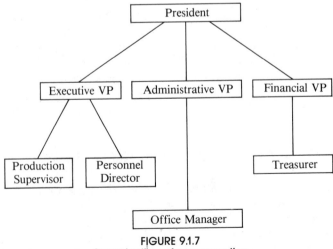

FIGURE 9.1.7
Organization of a corporation

separate subproblems. Each subproblem is divided further into smaller sub-problems in the same way until the solutions of the lowest problems are easy enough to recognize.

From these examples, we can see that although a graph can be defined, in short, as a collection of vertices and edges, an integral part of most graphs is the labeling of the vertices and edges that allows us to interpret the graph as a model for some situation.

Example 9.1.6. A Graph as a Model for a Set of Strings. Suppose that you would like to mechanically describe the set of strings of 0's and 1's having no consecutive 1's. One way to visualize a string of this kind is with the graph in Figure 9.1.1. Consider any path starting at vertex s. If the label on each graph is considered to be the output to a printer, then the output will have no consecutive 1's. For example, the path that is described by the vertex list $(s, a, b, b, a, b, b, a, b)$ would result in an output of 10010010.

Example 9.1.7. A Tournament Graph. Suppose that four teams compete in a round-robin sporting event; i.e., each team meets every other team once, and each game is played until a winner is determined. If the teams are named, A, B, C, and D, we can define the relation b on the set of teams by XbY if X beat Y. For one set of results, the graph of b might look like Figure 9.1.8.

Definition: Tournament Graph.

(a) *A tournament graph is a directed graph with the property that no edge connects a vertex to itself, and between any two vertices there is at most one edge.*

(b) *A complete (or round-robin) tournament graph is a tournament graph with the property that between any two distinct vertices there is exactly one edge.*

(c) *A single-elimination tournament graph is a tournament graph with the properties that: (i) one vertex (the champion) has no edge terminating*

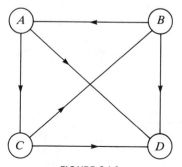

FIGURE 9.1.8
Round-robin graph with four vertices

at it and at least one edge initiating from it; (ii) every other vertex is the terminal vertex of exactly one edge; and (iii) there is a path from the champion vertex to every other vertex.

Example 9.1.8. The major league baseball championship is decided with with a single-elimination tournament. The two divisional champions in the American League (East and West) compete in a series of games. The loser is eliminated and the winner competes against the winner of the National League series (which is decided as in the American League). The tournament graph of the 1983 championship is in Figure 9.1.9.

The question "Once you have a graph, what do you do with it"? might come to mind. The following list of common questions and comments about graphs is a partial list that will give you an overview of the remainder of the chapter.

Question 1. How can a graph be represented as a data structure for use on a computer? We will discuss some common Pascal data structures that are used to represent graphs in Section 9.2.

Question 2. Given two vertices in a graph, does there exist a path between them? The existence of a path between any or all pairs of vertices in a graph will be discussed in Section 9.3. A related question is: How many paths of a certain type or length are there between two vertices?

Question 3. Is there a path (or circuit) that passes through every vertex (or uses every edge) exactly once? Paths of this kind are called *traversals*. We will discuss traversals in Section 9.4.

Question 4. Suppose that a cost is associated with the use of each vertex and/or edge in a path. What is the "cheapest" path, circuit, or traversal of a given kind? Problems of this kind will be discussed in Section 9.5.

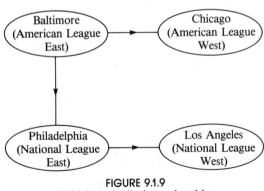

FIGURE 9.1.9
1983 baseball championship

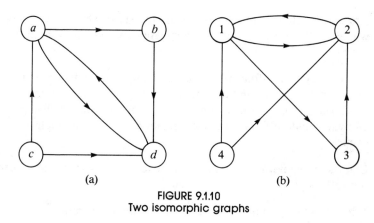

FIGURE 9.1.10
Two isomorphic graphs

Question 5.   Given the specifications of a graph, or the graph itself, what is the best way to draw the graph? The desire for neatness makes this a reasonable question. Another goal might be to avoid having edges of the graph cross one another.

ISOMORPHIC GRAPHS

We will close this section by establishing the relation "is isomorphic to," a form of equality on graphs. The graphs in Figure 9.1.10 obviously share some similarities, such as the number of vertices and the number of edges. It happens that they are even more similar than just that. If the letters a, b, c, and d in (a) are replaced with the numbers 1, 3, 2 and 4, respectively, and they are moved around so that they appear as in (b), you obtain (b).

Definition: Isomorphic Graphs. *Let (V, E) and (V', E') be two graphs. They are isomorphic if there exists a bijection, f, from V into V' such that $(v_i, v_j) \in E$ if and only if $(f(v_i), f(v_j)) \in E'$. For multigraphs, we add that the number of edges connecting v_i to v_j must equal the number of edges from $f(v_i)$ to $f(v_j)$.*

EXERCISES FOR SECTION 9.1

A Exercises

1.   What is the significance of the fact that there is a path connecting vertex B with every other vertex in Figure 9.1.2, as it applies to various situations that it models?

2.   Draw a graph similar to Figure 9.1.1 that represents the set of strings of 0's and 1's containing no more than two consecutive 1's.

3. Draw a directed graph which models the set of strings of 0's and 1's where all of the 1's must appear consecutively. Assume that the starting vertex has indegree 0.

4. In the NCAA final-four basketball tournament, the East champion plays the West champion, and the champions from the Mideast and Midwest play. The winners of the two games play for the national championship. Draw the eight different single-elimination tournament graphs that could occur.

5. Can a simple undirected graph of eight vertices have forty edges excluding self-loops?

6. Which of the graphs in Figure 9.1.11 are isomorphic? What is the correspondence between their vertices?

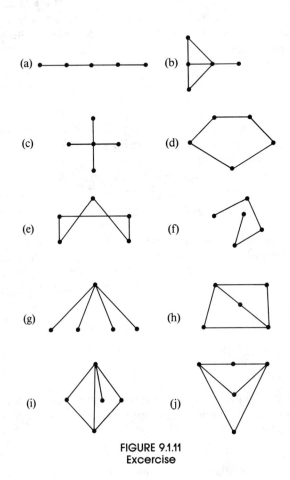

FIGURE 9.1.11
Excercise

7. (a) How many edges does a complete tournament graph with *n* vertices have?

   (b) How many edges does a single-elimination tournament graph with *n* vertices have?

8. Draw a complete undirected graph with 1, 2, 3, 4, and 5 vertices. How many edges does a K_n have?

9.2   Data Structures for Graphs (Optional)   OMIT

In this section, we will present two methods for representing a graph as a data structure. The structure that you choose for a particular graph depends on the number of vertices and edges and on what you intend to do with the graph. We will give general guidelines for making this decision.

General Assumptions.   We will assume that we have a graph with N vertices that can be indexed in some way (possibly arbitrarily) by the subrange $1 \,.\,.\, N$. In addition, we will assume that each vertex contains information that can be represented as a record of type VInfo. One common field of this record would contain the name of the vertex. For both of our data structures, we will start with V:ARRAY[$1 \,.\,.\, N$] OF VInfo. For simplicity, we are separating edge information from this array; however, the edge information could be included in V in all cases.

Data Structure 1:   *Adjacency Matrix.* As we saw in Chapter 6, the information about edges in a graph can be summarized with an adjacency matrix. The variable declaration

```
Edge: ARRAY[1..N, 1..N] OF BOOLEAN
```

is sufficient if no additional information is needed for the edges. To initialize Edge, we would first set each component to FALSE, then input pairs of integer values from $1 \,.\,.\, N$ that correspond to the edges of the graph. Note that this is the same as the adjacency matrix for a relation, with the exception that we used 0 and 1 for entries instead of FALSE and TRUE.

Data Structure 2:   *Linked Lists of Edges.* Note that the initializing procedure for an adjacency matrix presumes that a list of edges for the graph exists. This data structure maintains this list form. For each vertex in our graph, there will be a list of edges that initiate at that vertex. This can be accomplished using either Pascal pointer variables or an array of records with integer pointers. For simplicity, we will arrange the edges in a linear linked list; however, in extremely large graphs, the reader should be aware that faster access could be made to the edges with more sophisticated data structures.

Using Pascal pointer variables, we could use the declarations

```
Link = ↑Edge;
Edge = RECORD
         TermVertex: 1..N;
         Info: EdgeInfo; (*Optional*)
         Next: Link
       END;
EList: ARRAY[1..N] OF Link
```

In this structure, EList[k] represents a linear linked list of edges that initiate at vertex k.

As an alternative to pointer variables, particularly in other languages, you can use integer pointers. The declarations that would be used in this case are

```
MaxEdges = (estimate of the maximum number of edges that will
            be needed);
Link = 0..MaxEdges;(*0 = No link*)
EdgePool: ARRAY[1..MaxEdges] of
          RECORD
               TermVertex: 1..N;
               Info: EdgeInfo;
               Next: Link
          END;
EList: ARRAY[1..N] OF LINK
```

In this structure, EList[k] is the index of EdgePool that contains the first edge in the list of edges that initiate from vertex k. If EList[k] = 0, then no edges initiate at vertex k.

A natural question to ask is: Which data structure should be used in a given situation? There are several factors that determine the ideal structure for a given graph.

(a) *Memory Needed.* In general, for small graphs, the adjacency matrix structure is the most efficient. This is due to the fact that setting up a list of edges requires a certain amount of overhead memory that is not justified for small graphs. To be more precise, let n be the number of vertices in our graph, e = the number of edges in the graph (here an undirected edge counts for two edges, one in each direction), b = the number of bytes needed to store a Boolean variable (one bit is sufficient, but many systems will use full byte), z = the number of bytes needed to store an integer, p = the number of bytes needed to store a pointer, and i = the number of bytes needed to store additional information for an edge. Under these assumptions, an adjacency matrix will use $M_A = n^2(b + i)$ bytes, and the edge list structure will require $M_E = np + e(z + i + p)$ bytes.

Sparse and Dense Graphs. If a simple graph has n vertices, the maximum number of edges that it can have is n^2. If e is small in relation to n^2, then the graph is called *sparse*. If e is nearly n^2, the graph is called *dense*. The terms *small* and *nearly* are vague, but a reasonable interpretation is e is small if it is less than $n^2/4$, and it is nearly n^2 if e is greater than $3n^2/4$. If a graph is sparse, the term $e(2 + 1 + p)$ in M_E is small, and M_E is likely to be less than M_A. On the other hand, if a graph is dense, $e(z + i + p)$ is large and the adjacency matrix structure is likely to be more efficient.

(b) How will you be accessing the edges? If you will be wanting to know whether an edge between vertices exists, you can access this information very quickly with an adjacency matrix. To obtain the same information with edge lists, you must examine a list, which could take some time. On the other hand, if you will frequently need a list of all edges that initiate at a given vertex, this can be obtained more efficiently using an edge list as opposed to an adjacency matrix.

(c) Multigraphs can be represented with edge lists, while the only way to represent a multigraph with an adjacency matrix is by making part of each element of the matrix a list.

Example 9.2.1. Consider the graph in Figure 9.1.2 with the interpretation that it represents a floor plan for a house. We might use the following declarations to represent the graph:

RoomType =   (Living, Kitchen, Bedroom, Bath, Outside);

```
V: ARRAY[1..5] OF RECORD
                  Name: Char;
                  RType: RoomType;
               END
```

The elements of V could be assigned the values

k	V[k].Name	V[k].RType
1	'A'	Outside
2	'B'	Living
3	'C'	Bath
4	'D'	Kitchen
5	'E'	Bedroom

Which data structure should we use for representing the edges of this graph? Suppose that an integer requires two bytes ($z = 2$), a Pascal pointer requires one byte ($p = 1$), a Boolean variable requires one byte ($b = 1$), and no additional information is needed for each edge ($i = 0$). The adjacency matrix structure requires $M_A = 5^2(1 + 0) = 25$ bytes, while the edge list structure requires $M_E = 5 \times 2 + 12(2 + 0 + 2) = 58$ bytes with integer

	A	B	C	D	E
A	False	True	False	True	False
B	True	False	True	True	True
C	False	True	False	False	True
D	True	True	False	False	False
E	False	True	True	False	False

FIGURE 9.2.1
Adjacency matrix for Figure 9.1.2

pointers. With Pascal pointers, this can be reduced to $M_E = 5 \times 2 + 12 (2 + 0 + 1) = 46$ bytes. Using the memory criterion alone, the adjacency matrix structure is preferred. Of course, the memory amounts are so small in this case that other factors might contribute to our decision. For this reason, we will describe both structures.

The adjacency matrix for this graph appears in Figure 9.2.1.

To use integer pointers, we could declare

```
MaxEdges = 12;
Link = 0..MaxEdges;
EdgePool: ARRAY[1..12] OF RECORD
                              TermVert: 1..5;
                              Next: Link
                        END;
EList: ARRAY[1..5] OF LINK
```

The contents of EList and EdgePool is not unique. One possibility is

EdgePool[k]

j	EList[j]		k	TermVert	Next
1	4		1	3	5
2	1		2	1	10
3	9		3	5	0
4	2		4	2	6
5	12		5	1	7
			6	4	0
			7	4	8
			8	5	0
			9	2	3
			10	2	0
			11	3	0
			12	2	11

The linked list that this set of values represents appears in Figure 9.2.2.

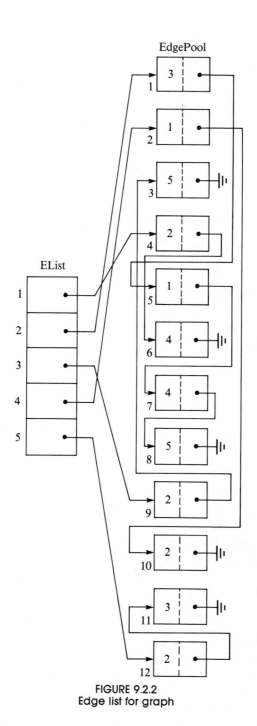

FIGURE 9.2.2
Edge list for graph

189

```
PROCEDURE AddEdge (I,J: 1..5);
VAR  T: Link;
BEGIN
   NEW(T);
   T↑.TermVert := J;
   T↑.Next := EList[I];
   EList[I] := T
END
```

FIGURE 9.2.3
Procedure for inserting an edge

Note that we have come full circle with this figure, which is a graph that helps us picture the data structure that in turn represents a graph.

If you prefer to use Pascal pointer variables, then the declarations would be

```
Link = ↑Edge;
Edge = RECORD
            TermVert:  1 . . 5;
            Next:  Link;
       END
```

In this case, we need not specify the number of edges, so that if we cut an extra door between two rooms, the edge that we would create in the graph could easily be accommodated. If the door connects room i with room j, we could simply create new edge records and add them to the edge lists of vertex i and vertex j. This can be coded most efficiently by defining the procedure AddEdge (Figure 9.2.3) and then executing AddEdge (i, j) and AddEdge (j, i).

Example 9.2.2. Consider an undirected graph represented by the top 200 college football teams in the United States for a given year. If two of the teams play one another, then imagine an edge connecting those two teams. Since the average team plays 11 games, most of which will be against other "Top 200" teams, we could expect up to 11 × 200 edges in our graph ($e = 2200$). Since 2200 is much smaller than 200^2, we could call this a sparse graph and would expect to find the edge list structure more efficient.

Example 9.2.3. Consider a timesharing system that consists of a processor and 8 terminals. At any one time, there can be several different terminals competing for processor time. A graph such as in Figure 9.2.4 can be used to visualize the current state of the system. Vertex P represents the processor and, unless no terminals are in use, there will be exactly one edge from P that terminates at the terminal that currently has the processor's "attention." All active terminals are arranged in a circular pattern. If an edge (T_i, T_j) is labeled F (forward), then T_j precedes T_i in the waiting line for processor time. If an

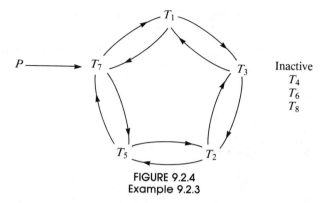

FIGURE 9.2.4
Example 9.2.3

edge is labeled B (backward), then the first vertex precedes the second in the waiting line.

A declaration for the vertices might be

```
V:  ARRAY[0 . . 8] of VInfo,
```

where $V[0]$ represents the processor. We can use a variation on the adjacency matrix to represent the graph:

```
EType = (Forward, Backward, None);

Edge = ARRAY[0 . . 8, 1 . . 8] OF EType
```

The situation in Figure 9.2.4 would be represented as

$$
\text{Edge} = \begin{array}{c} \\ 0 \\ 1 \\ 2 \\ 3 \\ 4 \\ 5 \\ 6 \\ 7 \\ 8 \end{array}
\begin{array}{cccccccc}
1 & 2 & 3 & 4 & 5 & 6 & 7 & 8 \\
\left[\begin{array}{cccccccc}
N & N & N & N & N & N & F & N \\
N & N & F & N & N & N & B & N \\
N & N & B & N & F & N & N & N \\
B & F & N & N & N & N & N & N \\
N & N & N & N & N & N & N & N \\
N & B & N & N & N & N & F & N \\
N & N & N & N & N & N & N & N \\
F & N & N & N & B & N & N & N \\
N & N & N & N & N & N & N & N
\end{array}\right]
\end{array}
$$

A second representation of this graph would be with a linked list of edges; however, we can use the fact that the maximum number of edges that can initiate from a vertex is two. The edge information can be included in the vertex array by including a backward and foreword pointer:

```
V:  ARRAY[0 . . 8] OF RECORD
                    Info:  VInfo;
                    Forward, Backward:  0 . . 8
          END
```

Again, a pointer value of 0 represents "no edge." The situation in Figure 9.2.4 would be represented by the values

I	V[I].Forward	V[I].Backward
0	7	0
1	3	7
2	5	3
3	2	1
4	0	0
5	7	2
6	0	0
7	1	5
8	0	0

EXERCISES FOR SECTION 9.2

A Exercises

1. Describe appropriate data structures to represent the following graphs. Include information fields that you feel are appropriate.

   (a) Vertices: Cities of the world that are served by at least one airline. Edges: Pairs of cities that are connected by a regular direct flight.

   (b) Vertices: The top 20 microcomputer systems (in terms of sales). Edges: An edge connects C_A to C_B if C_A has been evaluated to be "as good as" C_B by an independent evaluating group.

   (c) Vertices: All English words. Edges: An edge connects word x to word y if x is a prefix of y.

   (d) Vertices: ASCII characters. Edges: Edges connect characters that differ in their binary code by exactly two bits.

2. Each edge of a graph is colored with one of the four colors red, blue, yellow, or green. How could you represent the edges in this graph using a variation of the adjacency matrix structure?

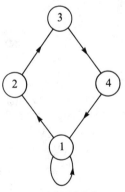

FIGURE 9.2.5
Exercise

3. Let G be the graph in Figure 9.2.5. What is the adjacency matrix for the graph H which has vertex set $\{1, 2, 3, 4\}$ and has an edge from i to j if G has a path of length 2 from i to j?

C Exercise

4. Write a Boolean-valued Pascal function to determine whether vertex I is connected to vertex J by an edge, where the edge information is stored in edge lists.

9.3 Connectivity

This section is devoted to a question which, when posed in relation to the graphs that we have examined, seems trivial. That question is: Given two vertices, s and t, of a graph, is there a path from s to t? If $s = t$, this question is interpreted as asking whether there is a circuit of positive length starting at s. Of course, for the graphs that we have seen up to now, this question can be answered after a brief examination.

There are two situations under which a question of this kind is non-trivial. One is where the graph is very large and an "examination" of the graph could take a considerable amount of time. Anyone who has tried to solve a maze may have run into a similar problem. The second interesting situation is when we want to pose the question to a machine. If only the information on edges between the vertices is part of the data structure for the graph, how can you put that information together to determine whether two vertices can be connected by a path?

Connectivity Terminology. Let v and w be vertices of a directed graph. Vertex v is *connected* to vertex w if there is a path from v to w. Two vertices are *strongly connected* if they are connected in both directions to one another. A graph is connected if, for each pair of distinct vertices, v and w, v is connected to w or w is connected to v. A graph is strongly connected if every pair of its vertices is strongly connected. For an undirected graph, in which edges can be used in either direction, the notions of strongly connected and connected are the same.

Theorem 9.3.1. *If a graph has n vertices and vertex v is connected to vertex w, then there exists a path from v to w of length no more than n.*

Proof (Indirect): Suppose that v is connected to w, but the shortest path from v to w has length m, where m is greater than n. A vertex list for a path of length m will have $m + 1$ vertices. This path can be represented as $(v_0, v_1, \ldots, v_m)$, where $v_0 = v$ and $v_m = w$. Note that since there are only n vertices and m vertices are listed in the path after v_0, there must be some duplication in the last m vertices of the vertex list, which represents a circuit in the path. This means that our path of minimum length can be reduced, which is a contradiction. #

Method 1: Adjacency Matrix Method. Suppose that the information about edges in a graph is stored (or can be stored) in an adjacency matrix, E. The relation, r, that E defines is vrw if there is an edge connecting v to w. Recall that the composition of r with itself, r^2, is defined by vr^2w if there exists a vertex y such that vry and yrw; i.e., v is connected to w by a path of length 2. We could prove by induction that the relation r^k, $k \geq 1$, is defined by vr^kw if there is a path of length k from v to w. Since the transitive closure, r^+, is the union of r, r^2, r^3, . . . , we can answer our connectivity question by determining the transitive closure of r, which can be done most easily by keeping our relation in matrix form. Theorem 9.3.1 is significant in our calculations because it tells us that we need only go as far as E^n to determine the matrix of the transitive closure. This was a feature of Warshall's algorithm, which we left unproven in Chapter 6.

The main advantage of the adjacency matrix method is that the transitive closure matrix can answer all questions about the existence of paths between vertices. Vertex v_i is connected to v_j if $E_{ij}^+ = 1$ (or True). A directed graph is connected if $E_{ij}^+ = 1$ or $E_{ji}^+ = 1$ for each $i \neq j$. A directed graph is strongly connected if its transitive closure matrix has no zeros.

One disadvantage to answering connectivity questions by the adjacency matrix method is that several matrix multiplications may be needed. If the number of vertices is large, this can be quite time consuming. This is tempered by the fact that adjacency matrices are most often used to represent either small or dense graphs. In a dense graph, the minimum path length between two vertices tends to be small. A second disadvantage is that the transitive closure matrix tells us whether a path exists, but not what the path is.

Method 2: Broadcasting. We will describe this method first with an example.

Example 9.3.1. The football team at Mediocre State University (MSU) has had a bad year, 2 wins and 9 losses. Thirty days after the end of the football season, the athletic board is meeting to decide whether to rehire the head coach; things look bad for him. However, on the day of the meeting, the coach releases the following list of results from the past year:

Mediocre State defeated Local A&M.

Local A&M defeated City College.

City College defeated Corn State U.

. . . (25 results later)

Tough Tech defeated Enormous State University (ESU).

But ESU went on to win the national championship.

The athletic board was so impressed that they hired the coach with a raise in pay! How did the coach come up with such a list?

In reality, such lists exist occasionally and appear in newspapers from time to time. Of course they really don't prove anything since each team that defeated MSU in our example above can produce a similar chain of results. Since college football records are readily available, the Coach could have found this list by trial and error. All that he needed to start with was that his team won at least one game. Since ESU lost one game, there was some hope of producing the chain.

The problem of finding this list is equivalent to finding a path in the tournament graph for last year's football season that initiates at MSU and ends at ESU (we will ignore tied games). Such a graph is far from complete and would be represented using edge lists. To make the coach's problem interesting, let's imagine that only the winner of a game remembers the result of the game. The coach's problem has now taken on the flavor of a maze. To reach ESU, he must communicate with the various teams along the path. One way that the coach could have discovered his list in time is by sending the following letter to the two teams that MSU defeated during the season:

Dear Football Coach:
    Please follow these directions exactly.
    (1) If you are the coach at ESU, call the coach at MSU now and tell him who sent you this letter.
    (2) If you are not the coach at ESU and this is the first letter of this type that you have received, then:
        (a) Remember who you received this letter from.
        (b) Send copies of this letter, signed by you, to each of the coaches whose teams you defeated during the past year.
    (3) Ignore this letter if you have received one like it already.

Signed,

Coach of MSU

Observations: From the conditions of this letter, it should be clear that if everyone cooperates and if mail can be sent and received in one day:

(a) If a path of length n exists from MSU to ESU, then the coach will know about it in n days.
(b) By making a series of phone calls, the coach can obtain the path that he wants by first calling the coach who defeated ESU (the person who sent ESU's coach that letter). This coach will know who sent him a letter, etc. Therefore, the vertex list of the desired path is obtained in reverse order.
(c) If a total of M football games were played, no more than M letters will be sent out.
(d) If a day passes without any letter being sent out, no path from MSU to ESU exists.

(e) This method could be extended to obtain a list of all teams that a given
team can be connected to. Simply imagine a series of letters like the one
above sent by each football coach and targeted at every other coach.

The general problem of finding a path between two vertices in a graph, if
one exists, can be solved exactly as we solved the problem above. The
following algorithm is commonly called the *Breadth-First Search*.

Algorithm 9.3.1. A Broadcasting Algorithm for Finding a Path between
Vertex i and Vertex j of a Graph Having n Vertices. The records $V[k]$,
$k = 1, \ldots, n$, consist of a Boolean field $V[k]$.found and an integer field
$V[k]$.from. The sets $D_1, D_2, \ldots$, called depth sets, have the property that
if k is in D_r, then the shortest path from vertex i to vertex k is of length r. In
Step 5, a stack is used to put the vertex list for the path from the vertex i to
vertex j in the proper order.

```
1.  For k:=1 to n Do V[k].found:=False
2.  r:=0
3.  D₀:={i}
4.  While (NOT V[j].found) and Dᵣ ≠ Φ DO
    4.1  Dᵣ₊₁:= Φ
    4.2  For each k in Dᵣ DO
            For each edge (k, t) DO
                If   V[t].found = False then
                     V[t].found: = True;
                     V[t].from: = k;
                     Dᵣ₊₁:= Dᵣ₊₁ U {t}
    4.3  r:= r + 1
5.  If V[j].found = True Then
    5.1  S:= Empty Stack
    5.2  k:=j
    5.3  While V[k].from ≠ i DO
         Push k into S
         k:= V[k].from
```

Notes on Algorithm 9.3.1:

(a) This algorithm will produce one path from vertex i to vertex j, if one
exists, and that path will be as short as possible. If more than one path
of this length exists, then the one that is produced depends on the order
in which the edges are examined and the order in which the elements
of D_r are examined in Step 4.

(b) The condition $D_r \neq \Phi$ is analogous to the condition that if no mail is
sent in a given day, then MSU cannot be connected to ESU.

(c) This algorithm can be easily revised to find paths to all vertices that can
be reached from vertex i. Step 5 would be put off until a specific path

to a vertex is needed since the information in V contains an efficient list of all paths. The algorithm can also be extended further to find paths between any two vertices.

Example 9.3.2. Consider the graph in Figure 9.3.1. The existence of a path from vertex 2 to vertex 3 is not difficult to determine by examination. After a few seconds, you should be able to find two paths of length four. Algorithm 9.3.1 will produce one of them.

Suppose that the edges from each vertex are sorted in ascending order by terminal vertex. For example the edges from vertex 3 would be in the order (3, 1), (3, 4), (3, 5). In addition, assume that in the body of Step 4 of the algorithm, the elements of D_r are used in ascending order. Then at the end of Step 4, the value of V will be

```
k               1  2  3  4  5  6
V[k].found      T  T  T  T  T  T
V[k].from       2  4  6  1  1  4
Depth set       1  3  4  2  2  3   (value of r for which k ∈ Dr)
```

Therefore, the path (2, 1, 4, 6, 3) is produced by the algorithm. Note that if we wanted a path from 2 to 5, the information in V produces the path (2, 1, 5) since $V[5].\text{from} = 1$ and $V[1].\text{from} = 2$. A shortest circuit that initiates at vertex 2 is also available by noting that $V[2].\text{from} = 4$, $V[4].\text{from} = 1$, and $V[1].\text{from} = 2$; thus the circuit (2, 1, 4, 2) is obtained.

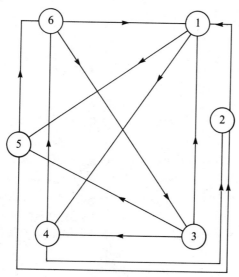

FIGURE 9.3.1
Example 9.3.2

EXERCISES FOR SECTION 9.3

A Exercises   *use Dikstra's algorithm*

1. Apply Algorithm 9.3.1 to find a path from 5 to 1 in Figure 9.3.1. What would be the final value of V? Assume that the terminal vertices in edge lists and elements of the depth sets are put into ascending order, as we assumed in Example 3.9.1.

2. Apply Algorithm 9.3.1 to find a path from the bedroom to outside using the edge list data structure in Example 9.2.1. Assume that the elements of the depth sets are put into ascending order.

$K_n| = \dfrac{n(n-1)}{2}$

3. In a simple undirected graph with no self-loops, what is the maximum number of edges that you can have keeping the graph unconnected. What is the minimum number of edges that will assure that the graph is connected?

4. Use a broadcasting algorithm to determine the shortest path from vertex a to vertex i in the graphs shown in Figure 9.3.2. List the depth sets and the stack that is obtained.

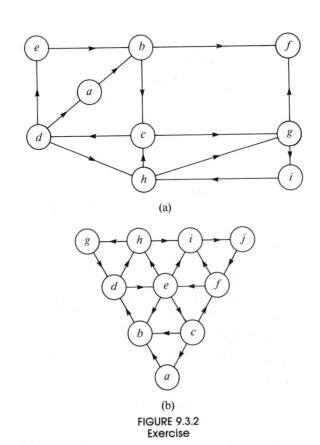

(a)

(b)

FIGURE 9.3.2
Exercise

B Exercise

5. Prove (by induction on k) that if the relation r on vertices of a graph is defined by vrw if there is an edge connecting v to w, then r^k, $k \geq 1$, is defined by $vr^k w$ if there is a path of length k from v to w.

C Exercise

6. Write a Pascal procedure that implements Algorithm 9.3.1, in which edge information is stored in an adjacency matrix.

9.4 Traversals: Eulerian and Hamiltonian Graphs

The subject of graph traversals has a long history. In fact, the solution by Leonhard Euler (Switzerland, 1707–83) of the Konigsberg Bridge Problem is considered by many to represent the birth of graph theory.

THE KONIGSBERG BRIDGE PROBLEM AND EULERIAN GRAPHS

A map of the Prussian city of Konigsberg (circa 1735) in Figure 9.4.1 shows that there were seven bridges connecting the four land masses that made up the city. The legend of this problem states that the citizens of Konigsberg searched in vain for a walking tour that passed over each bridge exactly once. No one could design such a tour and the search was abruptly abandoned with the publication of Euler's Theorem.

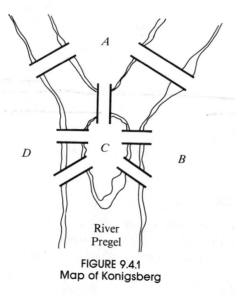

FIGURE 9.4.1
Map of Konigsberg

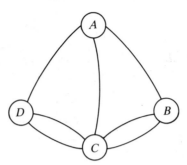

FIGURE 9.4.2
Multigraph representation of Konigsberg

Theorem 9.4.1. Euler's Theorem—Konigsberg Case. *No walking tour of Konigsberg can be designed so that each bridge is used exactly once.*

Proof: The map of Konigsberg can be represented as an undirected multigraph, as in Figure 9.4.2. The four land masses are the vertices and each edge represents a bridge. The desired tour is then a path that uses each edge once and only once. Since the path can start and end at two different vertices, there are two remaining vertices that must be intermediate vertices in the path. If x is an intermediate vertex, then every time that you visit x, you must use two edges, one to enter and one to exit. Therefore, there must be an even number of edges connecting x to the other vertices. Since every vertex in the Konigsberg graph has an odd number of edges, no tour of the type that is desired is possible. #

As is typical of most mathematicians, Euler wasn't satisfied with solving only the Konigsberg problem. His original theorem, which is paraphrased below, concerned the existence of paths and circuits like those sought in Konigsberg. These paths and circuits have become associated with Euler's name.

Definitions: Eulerian Paths, Circuits, Graphs. *A Eulerian path through a graph is a path whose edge list contains each edge of the graph exactly once. If the path is a circuit, then it is called a Eulerian circuit. A Eulerian graph is a graph that possesses a Eulerian path.*

In order to state the general theorem efficiently, we must define the degree of a vertex.

Definition: Degree.

  (a) *Let v be a vertex of an undirected graph. The degree of v ($deg(v)$) is the number of edges that connect v to the other vertices in the graph.*
  (b) *If v is a vertex of a directed graph, then the outdegree of v ($outdeg(v)$)*

is the number of edges of the graph that initiate at v. The indegree of v (indeg(v)) is the number of edges that terminate at v.

Example 9.4.1.

(a) The degrees of A, B, C, and D in Figure 9.4.2 are 3, 3, 5, and 3, respectively.

(b) In a tournament graph, *outdeg(v)* is the number of wins for v and *indeg(v)* is the number of losses. In a complete (round-robin) tournament graph with n vertices, *outdeg(v)* + *indeg(v)* = $n - 1$ for each vertex.

Theorem 9.4.2. Euler's Theorem—General Case. *An undirected graph is Eulerian if and only if it is connected and has either zero or two vertices with an odd degree. If no vertex has an odd degree, then the graph has a Eulerian circuit.*

Proof: It can be proven by induction that the number of vertices in an undirected graph that have an odd degree must be even. We will leave the proof of this fact to the reader as an exercise. The necessity of having either zero or two vertices of odd degree is clear from the proof of the Konigsberg case of this theorem. Therefore, we will concentrate on proving that this condition is sufficient to ensure that a graph is Eulerian. Let k be the number of vertices with odd degree.

Phase 1. If $k = 0$, start at any vertex, v_0, and travel along any path, not using any edge twice. Since each vertex has an even degree, this path can always be continued past each vertex that you reach except v_0. The result is a circuit that includes v_0. If $k = 2$, let v_0 be either one of the vertices of odd degree. Trace any path starting at v_0 using up edges until you can go no further, as in the $k = 0$ case. This time, the path that you obtain must end at the other vertex of odd degree.

At the end of Phase 1, we have an initial path that may or may not be Eulerian. If it is not Eulerian, Phase 2 can be repeated until all of the edges have been used. Since the number of unused edges is decreased in Phase 2, a Eulerian path must be obtained in a finite number of steps.

Phase 2. As we enter this phase, we have constructed a path that uses a proper subset of the edges in our graph. We will refer to this path as the *current path*. Let V be the vertices of our graph, E the edges, and E_u the edges that have been used in the current path. Consider the graph $G' = (V, E - E_u)$. Note that every vertex in G' has an even degree. Select any edge, e, from G'. Let v_a and v_b be the vertices that e connects. Trace a new path starting at v_a whose first edge is e. Be sure that at least one vertex of the new path is also in the current path. This will always be possible since (V, E) is connected. Starting at v_a, there exists a path in (V, E) to any vertex in the

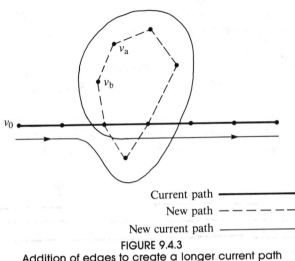

Current path ————————
New path — — — — — —
New current path ————————

FIGURE 9.4.3
Addition of edges to create a longer current path

current path. At some point along this path, which we can consider the start of the new path, we will have intersected the current path. Since the degree of each vertex in G' is even, any path that we start at v_a can be continued until it is a circuit. Now, we simply augment the current path with the new path. As we travel along the current path, the first time that we intersect the new path, we travel along it (see Figure 9.4.3). Once we complete the circuit that is the new path, we resume the traversal of the current path.

If the result of this phase is a Eulerian path, then we are finished; otherwise, repeat this phase. #

Example 9.4.2. The complete undirected graphs K_2 and K_{2i+1}, $i = 1, 2, 3,$ ..., are Eulerian. If i is greater than one, then K_{2i} is not Eulerian.

HAMILTONIAN GRAPHS

To search for a path that uses every vertex of a graph exactly once seems to be a natural next problem after you have considered Eulerian graphs. The Irish mathematician Sir William Hamilton (1805–65) is given credit for first defining such paths.

Definitions: Hamiltonian Paths, Circuits, Graphs. *A Hamiltonian path through a graph is a path whose vertex list contains each vertex of the graph exactly once, except if the path is a circuit, in which case the initial vertex appears a second time as the terminal vertex. If the path is a circuit, then it is called a Hamiltonian circuit. A Hamiltonian graph is a graph that possesses a Hamiltonian path.*

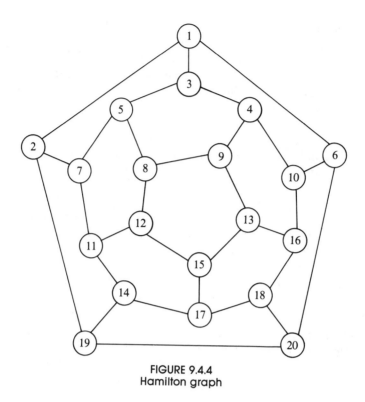

FIGURE 9.4.4
Hamilton graph

Example 9.4.3. Figure 9.4.4 contains a graph that is Hamiltonian. In fact, it is the graph that Hamilton used as an example to pose the question of existence of Hamiltonian paths in 1859. In its original form, the puzzle that was posed to readers was called "Around the World." The vertices were labeled with names of major cities of the world and the object was to complete a tour of these cities.

Unfortunately, a simple condition doesn't exist that characterizes a Hamiltonian graph. An obvious necessary condition is that the graph be connected; however, there is a connected undirected graph with four vertices that is not Hamiltonian. Can you draw such a graph? Hint: Y.

A Note on What Is Possible and What Is Impossible.   The search for a Hamiltonian path in a graph is typical of many simple-sounding problems in graph theory that have proven to be very difficult to solve. Although there are simple algorithms for conducting the search, they are impractical for large problems because they take such a long time to complete as graph size increases. Currently, every algorithm to search for a Hamiltonian path in a graph takes exponential time to complete. That is, if $T(n)$ is the time it takes to search a graph of n vertices, then there is a positive real number a, $a > 1$, such that $T(n) > a^n$ for all but possibly a finite number of positive values for

n. No matter how close to one we can make a, a^n will grow at such a fast rate that the algorithm will not be feasible for large values of n. For a given algorithm, the value of a depends on the relative times that are assigned to the steps, but in the search for Hamiltonian paths, the actual execution time for known algorithms is large with 20 vertices. For 1000 vertices, no algorithm is likely to be practical, and for 10,000 vertices, no currently known algorithm could be executed.

It is an unproven but widely held belief that no faster algorithm exists to search for Hamiltonian paths. A faster algorithm would have to be one that takes only polynomial time; that is, $T(n) < p(n)$, for some polynomial sequence p.

To sum up, the problem of determining whether a graph is Hamiltonian is certainly possible; however, for large graphs we consider the problem impossible. Most of the problems that we will discuss in the next section, particularly the Traveling Salesman Problem, are thought to be impossible in the same sense.

The following graphs, which appear in many applications, are all Hamiltonian.

Definition: The n-cube. *Let n be greater than or equal to 1. Let B^n be the set of strings of 0's and 1's with length n. The n-cube is the undirected graph with a vertex for each string in B^n and an edge connecting each pair of strings that differ in exactly one position. The 1-cube, 2-cube, 3-cube and 4-cube are pictured in Figure 9.4.5.*

A Hamiltonian circuit of the n-cube can be described recursively. The circuit itself, called the *Gray Code*, is not the only Hamiltonian circuit of the n-cube, but it is the easiest to describe. The standard way to write the Gray Code is as a column of strings, where the last string is followed by the first string to complete the circuit.

Basis ($n = 1$): The Gray Code for the 1-cube is $G_1 = \begin{pmatrix} 0 \\ 1 \end{pmatrix}$. Note that the edge between 0 and 1 is used twice in this circuit.

Recursion: Given the Gray Code for the n-cube, $n \geq 1$, then G_{n+1} is obtained by (1) listing G_n with each string prefixed with 0, and then (2) reversing the list of strings in G_n with each string prefixed with 1. Symbolically, the recursion can be expressed as

$$G_{n+1} = \begin{pmatrix} 0G_n \\ 1G_n^r \end{pmatrix}.$$

where G_r^r is the reverse of list G_n.

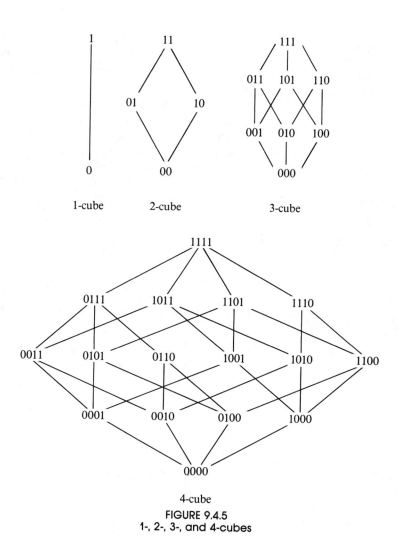

4-cube
FIGURE 9.4.5
1-, 2-, 3-, and 4-cubes

The Gray Codes for the 2-cube and 3-cube are

$$G_2 = \begin{pmatrix} 00 \\ 01 \\ 11 \\ 10 \end{pmatrix} \qquad\qquad G_3 = \begin{pmatrix} 000 \\ 001 \\ 011 \\ 010 \\ 110 \\ 111 \\ 101 \\ 100 \end{pmatrix}$$

An Application of the Gray Code. In a statistical analysis, there is often a variable that depends on several factors, but exactly which factors are significant may not be obvious. For each subset of factors, there would be certain quantities to be calculated. One such quantity is the multiple correlation coefficient for a subset. If the correlation coefficient for a given subset, A, is known, then the value for any subset that is obtained by either deleting or adding an element to A can be obtained quickly. To calculate the correlation coefficient for each set, we simply travel along G_n, where n is the number of factors being studied. The first vertex will always be the string of 0's, which represents the empty set. For each vertex that you visit, the set that it corresponds to contains the kth factor if the kth character is a 1.

EXERCISES FOR SECTION 9.4

A Exercises

1. Locate a map of New York City and draw a graph that represents its land masses and bridges. Is there a Eulerian path through New York City? You can do the same with any other city that has at least two land masses.

2. Which of the drawings in Figure 9.4.6 can be drawn without removing your pencil from the paper and without drawing any line twice?

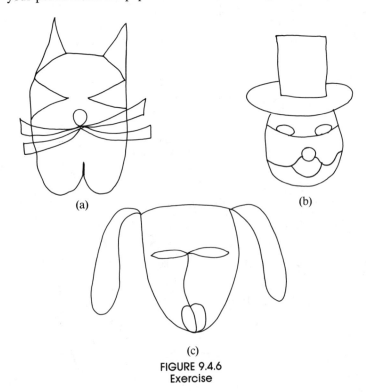

(a)

(b)

(c)

FIGURE 9.4.6
Exercise

3. Write out the Gray Code for the 4-cube.

4. Find a Hamiltonian circuit for the graph in Figure 9.4.4.

5. The Euler Construction Company has been contracted to construct an extra bridge in Konigsberg so that a Eulerian path through the town exists. Can this be done, and, if so, where should the bridge be built?

6. (a) How can you quickly decide which of the graphs in Figure 9.4.7 has a Eulerian path?

   (b) Find a Eulerian path for the graph which has one.

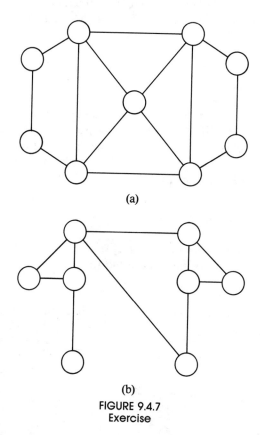

(a)

(b)

FIGURE 9.4.7
Exercise

B Exercises

7. Formulate Euler's theorem for directed graphs.

8. Prove that the number of vertices in an undirected graph with odd degree must be even. (Hint: Prove by induction on the number of edges.)

9. Under what condition will a round-robin tournament graph be Eulerian? Hamiltonian?

9.5 Graph Optimization

The common thread that connects all of the problems in this section is the desire to optimize (maximize or minimize) a quantity that is associated with a graph. We will concentrate most of our attention on two of these problems, the Traveling Salesman Problem and the Maximum Flow Problem. At the close of this section, we will discuss some other common optimization problems.

Definition: Weighted Graph. *A weighted graph, (V, E, w), is a graph (V, E) together with a weight function $w: E \to \mathbf{R}$. If $e \in E$, $w(e)$ is the weight on edge e.*

As you will see in our examples, $w(e)$ is usually a cost associated with the edge e; therefore, most weights will be positive.

Example 9.5.1. Let V be the set of six capital cities in New England: Boston, Augusta, Hartford, Providence, Concord, and Montpelier. Let E be $\{\{a, b\} \in V : a \neq b\}$; i.e., (V, E) is a complete unordered graph. A weight function on this graph is $w(c_1, c_2) = $ the distance from c_1 to c_2. Many road maps define distance functions as in Figure 9.5.1.

THE TRAVELING SALESMAN PROBLEM

Given a weighted graph, find a circuit $(e_1, e_2, \ldots, e_n)$ that visits every vertex at least once and minimizes the sum of the weights,

$$\sum_{i=1}^{n} w(e_1).$$

Any such circuit is called an *optimal path* and is a solution to the *Traveling Salesman Problem.*

	Augusta, ME	Boston, MA	Concord, NH	Hartford, CT	Montpelier, VT	Providence, RI
Augusta, ME	—	165	148	266	190	208
Boston, MA	165	—	75	103	192	43
Concord, NH	148	75	—	142	117	109
Hartford, CT	266	103	142	—	204	70
Montpelier, VT	190	192	117	204	—	223
Providence, RI	208	43	109	70	223	—

FIGURE 9.5.1
Distance between capital cities of New England

Notes.

(a) Some statements of the Traveling Salesman Problem require that the circuit be Hamiltonian. In many applications, the graph in question will be complete and this restriction presents no problem.

(b) If the weight on each edge is constant, e.g., $w(e) = 1$, then the solution to the Traveling Salesman Problem will be any Hamiltonian circuit, if one exists.

Example 9.5.2. The Traveling Salesman Problem gets its name from the situation of a salesman who wants to minimize the number of miles that he travels in visiting his customers. For example, if a salesman from Boston must visit the other capital cities of New England, then the problem is to find a circuit in the weighted graph of Example 9.5.1. Note that distance and cost are clearly related in this case.

The search for an efficient algorithm that solves the Traveling Salesman Problem has occupied researchers for years. If the graph in question is complete, then there are $(n - 1)!$ different circuits. As n gets large, it is inefficient to check every possible circuit. The most efficient algorithms for solving the Traveling Salesman Problem take an amount of time that is proportional to $n2^n$. Since this quantity grows so quickly, we can't expect to have the time to solve the Traveling Salesman Problem for large values of n.

Most of the useful algorithms that have been developed have to be heuristic; that is, they find a circuit that should be close to the optimal one. One such algorithm is the "closest neighbor" algorithm, one of the earliest attempts at solving the Traveling Salesman Problem. The general idea behind this algorithm is, starting at any vertex, to visit the closest neighbor to the starting point. At each vertex, the next vertex that is visited is the closest one that has not been reached. This shortsighted approach typifies heuristic algorithms called *greedy algorithms*, which attempt to solve a minimization (maximization) problem by minimizing (maximizing) the quantity associated with only the first step.

Algorithm 9.5.1. The Closest Neighbor Algorithm. Let $G = (V, E, w)$ be a complete weighted graph with $\#V = n$. The *closest neighbor circuit* through G starting at v_1 is $(v_1, v_2, \ldots, v_n)$, defined by the steps:

1. $V_1 := V - \{v_1\}$.
2. For $k := 2$ to $n - 1$ do:
   2.1 $v_k :=$ the closest vertex in V_{k-1} to v_{k-1};
       i.e., $w(v_{k-1}, v_k) = \min \{w(v_{k-1}, v) : v \in V_{k-1}\}$.
       In case of a tie for closest, v_k may be chosen arbitrarily.
   2.2 $V_k := V_{k-1} - \{v_k\}$.
3. $v_n :=$ the only element in V_n.

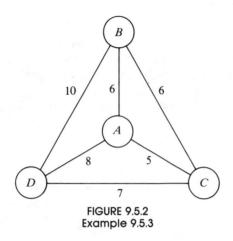

FIGURE 9.5.2
Example 9.5.3

The cost of the closest neighbor circuit is

$$\sum_{k=1}^{n-1} w(v_k, v_{k+1}) + w(v_n, v_1).$$

Example 9.5.3. The closest neighbor circuit starting at A in Figure 9.5.2 is (A, C, B, D, A), with a cost of 29. The optimal (or cheapest) path is (A, B, C, D, A), with a cost of 27.

Although the closest neighbor circuit is often not optimal, we may be satisfied if it is close to optimal. If C_{opt} and C_{cn} are the costs of optimal and closest neighbor circuits in a graph, then it is always the case that $C_{opt} \leq C_{cn}$, or $C_{cn}/C_{opt} \geq 1$. We can assess how good the closest neighbor algorithm is by determining how small the quantity C_{cn}/C_{opt} gets. If it is always near 1, then the algorithm is good. However, if there are graphs for which it is large, then the algorithm may be discarded. Note that in Example 9.5.3, $C_{cn}/C_{opt} = 29/27$, which would be considered good.

Example 9.5.4. A salesman must make stops at vertices A, B, and C, which are all on the same one-way street. The graph in Figure 9.5.3 is weighted by the function $w(i, j) = $ the time it takes to drive from vertex i to vertex j. Note that if j is down the street from i, then $w(i, j)$ is less than $w(j, i)$. The values of C_{opt} and C_{cn} are 20 and 32. Verify that C_{cn} is 32 by using the closest neighbor algorithm. The value of $C_{cn}/C_{opt} = 1.6$ is significant in this case since our salesman would spend 60 percent more time on the road if he used the closest neighbor algorithm.

A more general result relating to the closest neighbor algorithm presumes that the graph in question is complete and that the weight function satisfies the conditions (1) $w(x, y) = w(y, x)$ for all x, y in the vertex set, and (2) $w(x, y) + w(y, z) \geq w(x, z)$ for all x, y, z in the vertex set. The first condition is called the *symmetry condition* and the second is the *triangle inequality*.

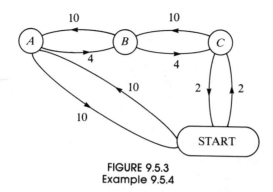

FIGURE 9.5.3
Example 9.5.4

Theorem 9.5.1. *If (V, E, w) is a complete weighted graph that satisfies the symmetry and triangle inequality conditions, then*

$$C_{cn}/C_{opt} \leq \lceil \log_2 (2n) \rceil / 2. \qquad (9.5a)$$

Proof: See Liu, pages 105–109.

Notes.

(a) If $\#V$ is 8, then this theorem says that C_{cn} can be no larger than twice the size of C_{opt}; however, it doesn't say that the closest neighbor circuit will necessarily be that far from an optimal circuit. The quantity $\lceil \log_2 (2n) \rceil / 2$ is called an *upper bound* for the ratio C_{cn}/C_{opt}. It tells us only that things can't be any worse than the upper bound. Certainly, there are many graphs with 8 vertices such that the optimal and closest neighbor circuits are the same. What is left unstated in this theorem is whether there are graphs for which the quantities in 9.5a are equal. If there are such graphs, we say that the upper bound is *sharp*.
(b) The value of C_{cn}/C_{opt} in Example 9.5.4 is 1.6, which is greater than $\lceil \log_2 (8) \rceil / 2 = 1.5$; however, the weight function in this example does not satisfy the conditions of the theorem.

TRAVELING SALESMAN PROBLEM—UNIT SQUARE VERSION

Example 9.5.5. A robot is programmed to weld joints on square metal plates. Each plate must be welded at prescribed points on the square. To minimize the time it takes to complete the job, the total distance that the robot's arm moves should be minimized. Let $d (P, Q)$ be the distance between P and Q. Assume that before each plate can be welded, the arm must be positioned at a certain point P_0. Given a list of n points, we want to put them in order so that

$$d(P_0, P_1) + d(P_1, P_2) + \cdots + d(P_n, P_0)$$

is as small as possible.

The problem that is outlined in Example 9.5.5 is of such obvious importance that it is probably the most studied version of the Traveling Salesman Problem. What follows is the usual statement of the problem. If $[0, 1] = \{x \text{ in } \mathbf{R} : 0 \leq x \leq 1\}$, let $S = [0, 1]^2$, the *unit square*. Given n pairs of real numbers (x_1, y_1), (x_2, y_2), ... and (x_n, y_n) in S that represent the n vertices of a K_n, find a circuit of the graph that minimizes the sum of the distances traveled in traversing the circuit.

Since the problem calls for a circuit, it doesn't matter which vertex we start at; assume that we will start at (x_1, y_1). Once the problem is solved, we can always change our starting position. A function can most efficiently describe a circuit in this problem. Every bijection $f: \{1, \ldots, n\} \rightarrow \{1, \ldots, n\}$ with $f(1) = 1$ describes a circuit $((x_1, y_1), (x_{f(2)}, y_{f(n)}), \ldots, (x_{f(n)}, y_{f(n)}))$. Since there are $(n - 1)!$ such bijections, an examination of all possible circuits is not feasible for large values of n.

One hueristic algorithm that is popular is the *strip algorithm*:

Algorithm 9.5.2. Given n points in the unit square,

Phase 1:

(1.1) Divide the square into $\lceil \sqrt{n/2} \rceil$ vertical strips, as in Figure 9.5.4. Let d be the width of each strip. If a point lies on a boundary between two strips, consider it part of the left-hand strip.

(1.2) Starting from the left, find the first strip that contains one of the points. Locate your starting point by selecting the first point that is encountered in that strip as you travel from bottom to top. We will assume that that point is (x_1, y_1).

(1.3) Alternate traveling up and down the strips that contain vertices until all of the vertices have been reached.

(1.4) Return to the starting point.

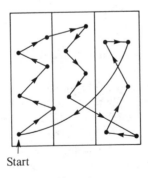 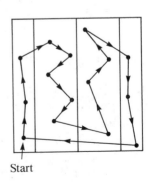

Start   Start

Phase 1   Phase 2

FIGURE 9.5.4
Strip algorithm

Phase 2:

(2.1) Shift all strips $d/2$ units to the right (creating a small strip on the left).
(2.2) Repeat Steps 1.2 through 1.4 of Phase 1 with the new strips.

When the two phases are complete, choose the shorter of the two circuits obtained.

Step 1.3 needs a bit more explanation. How do you travel up or down a strip? In most cases, the vertices in a strip will be vertically distributed so that the order in which they are visited is obvious. In some cases, however, the order might not be clear as in the third strip in Figure 9.5.4. Within a strip, the order in which you visit the points (if you are going up the strip) is determined thusly: (x_1, y_1) precedes (x_2, y_2) if $y_1 < y_2$ or if $y_1 = y_2$ and $x_1 < x_2$. In traveling down a strip, replace $y_1 < y_2$ with $y_2 < y_1$.

The selection of $\lceil \sqrt{n/2} \rceil$ strips was made in a 1959 paper by Beardwood, Halton, and Hammersley. It balances the problems that arise if the number of strips is too small or too large. If the square is divided into too few strips, some strips may be packed with vertices so that visiting them would require excessive horizontal motion. If too many strips are used, excessive vertical motion tends to be the result. An update on what is known about this algorithm is contained in the paper by K. J. Supowit, E. M. Reingold, and D. A. Plaisted.

Since the construction of a circuit in the square consists of sorting the given points, it should come as no surprise that the strip algorithm requires a time that is roughly a multiple of $n \log (n)$ time units when n points are to be visited.

The worst case that has been encountered with this algorithm is one in which the circuit obtained has a total distance of approximately $\sqrt{2n}$ (see Sopowit et al).

NETWORKS AND THE MAXIMUM FLOW PROBLEM

A *network* is a simple weighted directed graph that contains two distinguished vertices called the *source* and *sink*. An example of a real situation that can be represented by a network is a city's water system. A reservoir would be the source, while a distribution point in the city to all of the users would be the sink. The system of pumps and pipes that carries the water from source to sink makes up the remaining network. We can assume that the water that passes through a pipe in one minute is controlled by a pump and the maximum rate is determined by the size of the pipe and the strength of the pump. This maximum rate of flow through a pipe is called its *capacity* and is the information that the weight function of network contains.

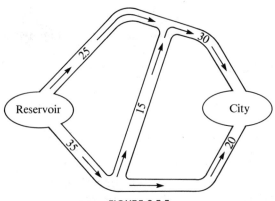

FIGURE 9.5.5
Diagram of a city's water system.

Example 9.5.6. Consider the system that is illustrated in Figure 9.5.5. The numbers that appear next to each pipe indicate the capacity of that pipe in thousands of gallons per minute. This map can be drawn in the form of a network, as in Figure 9.5.6.

Although the material passing through this network is water, networks can also represent the flow of other materials, such as automobiles, electricity, and telephone calls.

The Maximum Flow Problem is derived from the objective of moving the maximum amount of water or other material from the source to the sink. To measure this amount, we define a *flow* as a function $f : E \rightarrow \mathbf{R}$ such that (1) the flow of material through any edge is non-negative and no larger than its capacity: $0 \le f(e) \le w(e)$, for all e in E; and (2) for each vertex other than the source and sink, the total amount of material that is directed into a vertex is equal to the total amount that is directed out:

$$\sum_{(x,\, v) \in E} f(x, v) = \sum_{(v,\, y) \in E} f(v, y)$$

Flow into v = Flow out of v \hfill (9.5b)

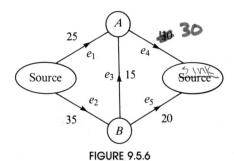

FIGURE 9.5.6
Example 9.5.6

The summation notation on the left of 9.5b represents the sum of the flows through each edge in E that has v as a terminal vertex. The right-hand side indicates that you should add all of the flows through edges that initiate at v.

Theorem 9.5.2. If f is a flow, then

$$\sum_{(\text{source},\ v)\in E} f(\text{source}, v)$$

and

$$\sum_{(v,\ \text{sink})\in E} f(v, \text{sink})$$

have a common value, which is called the *value of the flow*. We will denote the value of a flow f by $V(f)$. The value of a flow represents the amount of material that passes through the network with that flow.

Proof.   Subtract the right-hand side of 9.5b from the left-hand side. The result is: Flow into v $-$ Flow out of $v = 0$. Now sum up these differences for each vertex in $V' = V - \{\text{source}, \text{sink}\}$. The result is

$$\sum_{v\in V'}\left(\sum_{(x,\ v)\in E} f(x, v) - \sum_{(v,\ y)\in E} f(v, y)\right) = 0 \qquad (9.5c)$$

Now observe that if an edge connects two vertices in V', its flow appears as both a positive and a negative term in 9.5c. This means that the only positive terms that are not cancelled out are the flows out of the source. In addition, the only negative terms that remain are the flows into the sink. Therefore,

$$\sum_{(\text{source},\ x)\in E} f(\text{source}, x) - \sum_{(y,\ \text{sink})\in E} f(y, \text{sink}) = 0. \quad \#$$

MAXIMAL FLOWS

Since the Maximum Flow Problem consists of maximizing the amount of material that passes through a given network, it is equivalent to finding a flow with the largest possible value. Any such flow is called a maximal flow.

For the network in Figure 9.5.6, one flow is f_1, defined by $f_1(e_1) = 25$, $f_1(e_2) = 20$, $f_1(e_3) = 0$, $f_1(e_4) = 25$, and $f_1(e_5) = 20$. The value of f_1, $V(f_1)$, is 45. Since the total flow into the sink can be no larger than 50 ($w(e_4) + w(e_5) = 30 + 20$), we can tell that f_1 is not very far from the solution. Can you improve on f_1 at all? The sum of the capacities into the sink can't always be obtained by a flow. The same is true for the sum of the capacities out of the source. In this case, the sum of the capacities out of the source is 60, which obviously can't be reached in this network.

A solution of the Maximum Flow Problem for this network is the maximal flow $f_2 : f_2(e_1) = 25$, $f_2(e_2) = 25$, $f_2(e_3) = 5$, $f_2(e_4) = 30$, and $f_2(e_5) = 20$, with $V(f_2) = 50$. This solution is not unique. In fact, there is an infinite number of maximal flows for this problem.

There have been several algorithms developed to solve the Maximal Flow Problem. One of these is the Ford and Funkerson Algorithm (FFA). The FFA consists of repeatedly finding paths in a network called *flow augmenting paths* until no impovement can be made in the flow that has been obtained.

Definition: Flow Augmenting Path. *Given a flow f in a network (V, E), a flow augmenting path with respect to f is a simple path from the source to the sink using edges both in their forward and their reverse directions such that for each edge e in the path, $w(e) - f(e) > 0$ if e is used in its forward direction and $f(e) > 0$ if e is used in the reverse direction.*

Example 9.5.7. For f_1 in Example 9.5.6, a flow augmenting path would be (e_2, e_3, e_4) since

$$w(e_2) - f_1(e_2) = 15,$$
$$w(e_3) - f_1(e_3) = 15, \text{ and}$$
$$w(e_4) - f_1(e_4) = 5.$$

These positive differences represent unused capacities, and the smallest value represents the amount of flow that can be added to each edge in the path. Note that by adding 5 to each edge in our path, we obtain f_2, which is maximal. If an edge with a positive flow is used in its reverse direction, it is contributing a movement of material that is counterproductive to the objective of maximizing flow. This is why the algorithm directs us to decrease the flow through that edge.

Algorithm 9.5.3. The Ford and Funkerson Algorithm.

(1) Define the flow function f_0 by $f_0(e) = 0$ for each edge $e \in E$.
(2) $i := 0$.
(3) Repeat:
    (3.1) If possible, find a flow augmenting path with respect to f.
    (3.2) If a flow augmenting path exists, then:
        (3.2.1) Determine
$$d = \min \{\{w(e) - f_i(e) : e \text{ is used in the forward direction}\},$$
$$\{f_i(e) : e \text{ is used in the reverse direction}\}\}.$$
        (3.2.2) Define f_{i+1} by

$$f_{i+1}(e) = \begin{cases} f_i(e) \text{ if } e \text{ is not used in the path;} \\ f_i(e) + d \text{ if } e \text{ is used in the forward direction;} \\ f_i(e) - d \text{ if } e \text{ is used in the reverse direction.} \end{cases}$$

        (3.2.3) $i := i + 1$.
    until no flow augmenting path exists.
(4) Terminate with a maximal flow f_i.

Notes:

(a) It should be clear that every flow augmenting path leads to a flow of increased value and that none of the capacities of the network can be violated.

(b) The depth-first search should be used to find flow augmenting paths since it is far more efficient than the breadth-first search in this situation. The depth-first search differs from the broadcasting algorithm (a variation of the breadth-first search) in that you sequentially visit vertices until you reach a "dead end" and then backtrack.

A Depth-First Search for the Sink Initiating at the Source: Let E' be the set of directed edges that can be used in producing a flow augmenting path. Add to the network a vertex called start and the edge (start, source).

(1) $S :=$ Vertex set of the network.
(2.1) $p :=$ start.
(2.2) $p :=$ Source (Move p along the edge (start, source).)
(3) WHILE p is not equal to start or sink DO.
   If an edge in E' exists that takes you from p to another vertex in S, then set p to be that next vertex and delete the edge from E'. Otherwise, reassign p to be the vertex that p was reached from (i.e., backtrack).
(4) If $p =$ start, no flow augmenting path exists.
   If $p =$ sink, you have found a flow augmenting path.

(c) There have been networks discovered for which the FFA does not terminate in a finite number of steps. These examples all have irrational capacities. It has been proven that if all capacities are positive integers, the FFA terminates in a finite number of steps. See Ford and Funkerson, Even, or Berge for details.

(d) When you use the FFA to solve the Maximum Flow Problem by hand it is convenient to label each edge of the network with the fraction $f_i(e)/w(e)$.

Example 9.5.8. Consider the network in Figure 9.5.7, where the current flow, f, is indicated by a labeling of the edges. The path (source, v_2, v_1, v_3, sink) is a flow augmenting path. Note that (v_1, v_3) is used in the reverse direction, which is allowed because $f(v_1, v_3) > 0$. The value of the new flow that we obtain is 8. This flow must be maximal since the capacities out of the source add up to 8. This maximal flow is defined by the labeling of Figure 9.5.8.

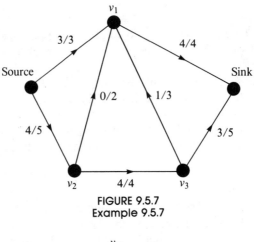

FIGURE 9.5.7
Example 9.5.7

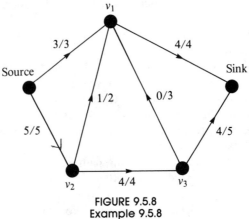

FIGURE 9.5.8
Example 9.5.8

OTHER GRAPH OPTIMIZATION PROBLEMS

(a) *The Minimum Spanning Tree Problem:* Given a weighted graph, (V, E, w), find a subset E' of E with the properties that (V, E') is connected and the sum of the weights of edges in E' are as small as possible. We will discuss this problem in Chapter 10.

(b) *The Minimum Matching Problem:* Given an undirected weighted graph, (V, E, w), with an even number of vertices, pair up the vertices so that each pair is connected by an edge and the sum of these edges is as small as possible. A unit square version of this problem has been studied extensively. See the paper by K. J. Supowit, E. M. Reingold, and D. A. Plaisted for details on what is known about this version of the problem.

(c) *The Graph Center Problem.* Given a connected, undirected, weighted graph, find a vertex (the center) in the graph with the property that the distance from the center to every other vertex is as small as possible. "As small as possible" could be interpreted either as minimizing the sum of the distances to each vertex or minimizing the maximum distance from the center to a vertex.

EXERCISES FOR SECTION 9.5

A Exercises

1. Find the closest neighbor circuit through the six capitals of New England starting at Boston. Does it matter which city you start in?

2. Is Theorem 9.5.1 sharp for $n = 3$? For $n = 4$?

3. Given the following sets of points in the unit square, find the shortest circuit that visits all of the points and find the circuit that is obtained with the strip algorithm.
    (a) $\{(0.1k, 0.1k) : k = 0, 1, 2, \ldots, 10\}$
    (b) $\{(0.1, 0.3), (0.3, 0.8), (0.5, 0.3), (0.7, 0.9), (0.9, 0.1)\}$
    (c) $\{(0.0, 0.5), (0.5, 0.0), (0.5, 1.0), (1.0, 0.5)\}$
    (d) $\{(0,0), (0.2, 0.6), (0.4, 0.1), (0.6, 0.8), (0.7, 0.5)\}$

4. For $n = 4, 5$, and 6, locate n points in the unit square for which the strip algorithm works poorly.

5. (a) [easy] Find two maximal flows for the network in Figure 9.5.6 other than the one in the text.
    (b) [harder] Describe the set of all maximal flows for the same network.
    (c) [hardest] Prove that if a network has two maximal flows, then it has an infinite number of maximal flows.

6. Find maximal flows for the networks in Figure 9.5.9.

B Exercise

7. Discuss reasons that the closest neighbor algorithm is not used in the unit square version of the Traveling Salesman Problem. (Hint: Count the number of comparisons of distances that must be done.)

C Exercises

8. Devise a "closest neighbor" algorithm for matching points in the unit square.

9. Explore the possibility of solving the traveling salesman problem in the "unit box": $[0, 1]^3$.

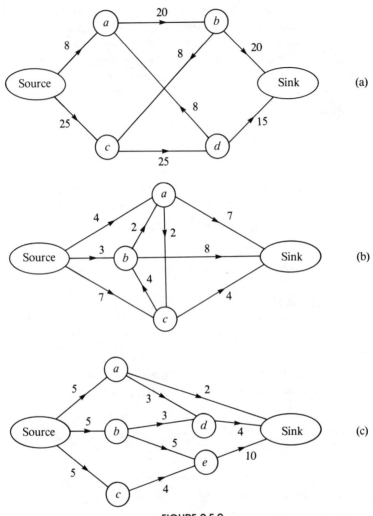

FIGURE 9.5.9
Exercise to find maximal flows for the network

9.6 Planarity and Colorings

The topics in this section are related to how graphs are drawn.

Planarity—Can a given graph be drawn in an plane so that no edges intersect? Certainly, it is natural to avoid intersections, but up to now we haven't gone out of our way to do so.

Colorings—Suppose that each vertex in an undirected graph is to be colored so that no two vertices that are connected by an edge have the same color. How many colors are needed? This question is motivated by the problem of drawing a map so that no two bordering countries are colored the same. A similar question can be asked for coloring edges.

Definition: Planar Graph. *A graph is planar if it can be drawn in a plane (e.g., a sheet of paper) so that no edges cross.*

Example 9.6.1. The graph in Figure 6.2.1 is planar, as is demonstrated in Figure 6.2.2.

Notes:

(a) *In discussing planarity, we need only consider simple undirected graphs with no self-loops.* All other graphs can be treated as such since all of the edges that relate any two vertices can be considered as one "package" that clearly can be drawn in a plane.
(b) Can you think of a graph that is not planar? How would you prove that it isn't planar? Proving the nonexistence of something is usually more difficult than proving its existence. This case is no exception. Intuitively, we would expect that sparse graphs would be planar and dense graphs would be non-planar. Theorem 9.6.2 will verify that dense graphs are indeed non-planar.
(c) The topic of planarity is a result of trying to restrict a graph to two dimensions. Is there an analogous topic for three dimensions? What graphs can be drawn in one dimension?
   Answer to Note c: If a graph has only a finite number of vertices, it can always be drawn in three dimensions. This is not true for all graphs with an infinite number of vertices. The only "one-dimensional" graphs are the ones that consist of a finite number of chains, as in Figure 9.6.1, with one or more vertex in each chain.

Example 9.6.2. A discussion of planarity is not complete without mentioning the famous Three Utilities Puzzle. The object of the puzzle is to supply three houses, A, B, and C, with the three utilities, gas, electric, and water. The constraint that makes this puzzle impossible to solve is that no utility lines may intersect. There is a "solution" if you allow one of the lines to run under one house on its way to a second house. This solution is invalid when the puzzle is posed in the form of a graph, which is to draw the graph in Figure 9.6.2 so that no two edges intersect. This graph is one of the simplest non-planar graphs.

A planar graph divides the plane into one or more *regions*. Two points on the plane lie in the same region if you can draw a curve connecting the two

FIGURE 9.6.1
Chains of length one, two, and three

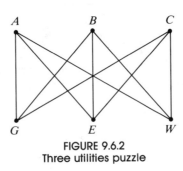

FIGURE 9.6.2
Three utilities puzzle

points that does not pass through an edge. One of these regions will be of
infinite area. Each point on the plane is either a vertex, a point on an edge,
or a point in a region. A remarkable fact about the geography of planar graphs
is the following theorem that is attributed to Euler.

Theorem 9.6.1. *If $G = (V, E)$ is a connected planar graph with r regions,*
$v = \#V$ *and* $e = \#E$, *then*

$$v + r - e = 2. \tag{9.6a}$$

Experiment: Jot down a graph right now and count the number of vertices, regions, and edges that you have. If $v + r - e$ is not 2, then your graph
is either not planar or not connected.

Proof of Theorem 9.6.1 by Induction on e, for $e \geq 0$.

Basis: If $e = 0$, then G must be a graph with one vertex, $v = 1$, and there
is one infinite region, $r = 1$.

$$v + r - e = 1 + 1 - 0 = 2.$$

Therefore, the basis is complete.

Induction: Suppose that G has k edges, $k \geq 1$, and that all connected
planar graphs with less than k edges satisfy 9.6a. If G has a vertex with degree
1, identify one such vertex as v_1 and the edge that connects it to the rest of
G by e_1. If all of the vertices of G have degree 2 or more, select any edge that
is part of the boundary of the infinite region and call it e_1. We will consider
the two cases where v_1 exists and v_1 does not exist separately. The latter case
will be broken into two subcases.
    Start by drawing all of G except v_1 (if it exists) and e_1. We will call this
graph G'. Figure 9.6.3 illustrates the different possibilities that we must
consider. If G' is connected (as in Graphs a and b), the induction hypothesis
states that if G' has v' vertices and r' regions, then $v' + r' - (k - 1) = 2$.

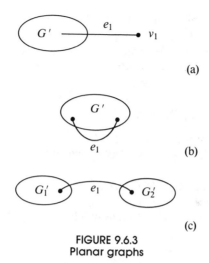

(a)

(b)

(c)

FIGURE 9.6.3
Planar graphs

Case 1:   If v_1 exists, then the addition of v_1 and e_1 to G' completes G. By doing this, we increase the number of vertices to $v = v' + 1$ and the number of edges to $e = (k - 1) + 1 = k$. The number of regions stays the same, $r = r'$. Then

$$v + r - e = (v' + 1) + r' - k$$
$$= v' + r' - (k - 1)$$
$$= 2$$

by the induction hypothesis, and 9.6a is true in the first case.

Case 2:   If no v_1 exists, then either G' is connected or it isn't. We will complete the subcase where G' is connected and leave the other subcase to the reader. When we add e_1 to G', as in Figure 9.6.3b, the number of vertices doesn't increase ($v = v'$), but the number of edges and regions both increase by 1 ($r = r' + 1$ and $e = (k - 1) + 1 = k$). Then

$$v + r - e = v' + (r' + 1) - k$$
$$= v' + r' - (k - 1)$$
$$= 2.$$

Hence, 9.6a is true for the first subcase of Case 2.

Hint for Second Subcase: If G' is not connected, then G' is made up of two components that are planar graphs with less than k edges. They share one common region, the infinite one. #

Theorem 9.6.2.   *If $G = (V, E)$ is a connected planar graph with v vertices and e edges, then if v is greater than or equal to 3,*

$$e \leq 3v - 6. \tag{9.6b}$$

Remark: One implication of 9.6b is that the number of edges in a connected planar graph will never be larger than three times its number of vertices (as long as it has at least three vertices). Since the maximum number of edges in a graph with v vertices is a quadratic function of v, as v increases, planar graphs are more and more sparse.

Outline of a Proof of Theorem 9.6.2.

(a) Let r be the number of regions in G. For each region, count the number of edges that comprise its border. The sum of these counts must be at least $3r$.
(b) The number of edges in G must be at least $(3r)/2$.
(c) $e \geq (3r)/2 \Rightarrow r \leq 2e/3$.
(d) Substitute $2e/3$ for r in 9.6a to obtain an inequality that is equivalent to 9.6b.   #

The following theorem will be useful later.

Theorem 9.6.3.   If G is a connected planar graph, then it has a vertex with degree 5 or less.

Proof by Contradiction. We can assume that G has at least seven vertices, for otherwise the degree of any vertex is at most 5. Suppose that G is a connected planar graph and each vertex has a degree of 6 or more. Then, since each edge contributes to the degree of two vertices, $e \geq 6v/2 = 3v$. However, Theorem 9.6.2 states that the $e \leq 3v - 6 < 3v$, which is a contradiction.   #

GRAPH COLORING

The map of Euler Island in Figure 9.6.4 shows that there are seven towns on the island. Suppose that a cartographer must produce a colored map in which no two towns that share a boundary have the same color. To keep costs down,

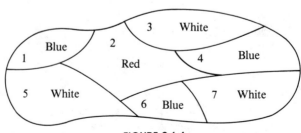

FIGURE 9.6.4
Map of Euler Island

she wants to minimize the number of different colors that appear on the map. How many colors are sufficient? For Euler Island, the answer is three. This problem motivates a more general problem.

   The Graph Coloring Problem.   Given an undirected graph $G = (V, E)$, find a "coloring function" f from V into a set of colors H such that (v_i, v_j) in $E \Rightarrow f(v_i) \neq f(v_j)$ and H has the smallest possible cardinality. The cardinality of H is called the *chromatic number of G, $c(G)$.*

   Notes:

   (a) A coloring function into an n element set is called an *n-coloring*.
   (b) In terms of this general problem, the chromatic number of the graph of Euler Island is three. To see that no more than three colors are needed, we need only display a 3-coloring: $f(1) = f(4) = f(6) =$ blue, $f(2) =$ red, and $f(3) = f(5) = f(7) =$ white. This coloring is not unique. The next smallest set of colors would be of two colors, and you should be able to convince yourself that no 2-coloring exists for this graph.

   In the mid-nineteenth century, it became clear that the typical planar graph had a chromatic number of no more than 4. At that point, mathematicians attacked the Four-Color Conjecture, which is that if G is any planar graph, then its chromatic number is no more than 4. Although the conjecture is quite easy to state, it took over 100 years, until 1976, to prove the conjecture in the affirmative.

Theorem 9.6.3.   The Four-Color Theorem.   *If G is any planar graph, then $c(G) \leq 4$.*

   A proof of the Four-Color Theorem is far beyond the scope of this text, but we can prove a theorem that is only 25 percent inferior.

Theorem 9.6.4.   The Five-Color Theorem.   *If G is a planar graph, then $c(G) \leq 5$.*
   The number 5 is not a sharp upper bound for $c(G)$ by the Four-Color Theorem.

Proof by Induction on the Number of Vertices in the Graph.

Basis: Clearly, a graph with one vertex has chromatic number of 1.

Induction: Assume that all planar graphs with $n - 1$ vertices have a chromatic number of 5 or less. Let G be a planar graph. By Theorem 9.6.2, there exists a vertex v with $deg\ (v) \leq 5$. Let $G - v$ be the planar graph obtained by deleting v and all edges that connect v to other vertices in G. By the induction hypothesis, $G - v$ has a 5-coloring. Assume that the colors

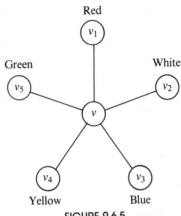

FIGURE 9.6.5

used are red, white, blue, green, and yellow. If *deg* $(v) < 5$, then we can produce a 5-coloring of G by selecting a color that is not used in coloring the vertices that are connected to v with an edge in G.

If *deg* $(v) = 5$, then we can use the same approach if the five vertices that are adjacent to v are not all colored differently. We are now left with the possibility that v_1, v_2, v_3, v_4, and v_5 are all connected to v by an edge and they are all colored differently. Assume that they are colored red, white, blue, yellow, and green, respectively, as in Figure 9.6.5. Suppose that v_1 and v_3 are not connected to one another using only red and blue vertices in $G - v$. If we consider all paths that start at v_1 and go through only red and blue vertices, then we cannot reach v_3. If we exchange the colors of the vertices in these paths, including v_1, we still have a 5-coloring of $G - v$. Since v_1 is now blue, we can color v red.

Now suppose that v_1 is connected to v_3 using only red and blue vertices. Then a path from v_1 to v_3 by using red and blue vertices followed by the edges (v_3, v) and (v, v_1) completes a circuit that either encloses v_2 or encloses v_4 and v_5. Therefore, no path from v_2 to v_4 exists using only white and yellow vertices. We can then repeat the same process as in the previous paragraph with v_2 and v_4, which will alow us to color v white. #

Definition: Bipartite Graph. *A bipartite graph is a graph that has a 2-coloring. Equivalently, a graph is bipartite if its vertices can be partitioned into two non-empty subsets so that each edge connects a vertex from each subset.*

Example 9.6.3.

(a) The graph of the Three Utilities Puzzle is bipartite. The vertices are partitioned into the utilities and the homes. Of course a 2-coloring of the graph is to color the utilities red and the homes blue.

(b) For $n \geq 1$, the n-cube is bipartite. The sets of strings with odd and with even numbers of 1's are "colored" differently.

(c) Let V be a set of 64 vertices, one for each square on a chess board. We can index the elements of V by v_{ij} = square on the ith row, j^{th} column. Connect vertices in V according to whether or not you can move a knight from one square to another. Using our indexing of V,

$$(v_{ij}, v_{kl}) \text{ in } E \text{ iff } |i - k| + |j - l| = 3$$

and

$$|i - k| \cdot |j - l| = 2.$$

(V, E) is a bipartite graph. The usual coloring of a chessboard is the desired 2-coloring.

How can you recognize whether a graph is bipartite? Unlike planarity, there is a nice equivalent condition for a graph to be bipartite.

Theorem 9.6.5. *An undirected graph is bipartite if and only if it has no circuit of odd length.*

Proof. ($\Rightarrow$) Let $G = (V, E)$ be a bipartite graph that is partitioned into two sets, $R(ed)$ and $B(lue)$ that define a 2-coloring. Suppose that there does exist a circuit of odd length. Since it is odd, the numbers of red and blue vertices in the circuit must be unequal. On the other hand, we can specify a direction in the circuit and define for the vertices of the circuit by $f(v) =$ the next vertex in the circuit after v. Note that f is a bijection. Hence the number of red vertices in the circuit equals the number of blue vertices, a contradiction. #

($\Leftarrow$) Assume that G has no circuit of odd length greater than or equal to 3. For each component of G, select any vertex, w and color it red. Then for every other vertex v in the component, find the path of shortest distance from w to v. If the length of the path is odd, color v blue, and if it is even, color v red. We claim that this method defines a 2-coloring of G. Suppose that it does not define a 2-coloring. Then let v_a and v_b be two vertices with identical colors that are connected with an edge. By the way that we colored G, neither v_a nor v_b is the w of its component. We can now construct a circuit with an odd length in G. First, we start at the w vertex for the component and follow the shortest path to v_a. Then follow the edge (v_a, v_b), and finally, follow the reverse of a shortest path from w to v_b. Since v_a and v_b have the same color, the first and third segments of this circuit have lengths that are both odd or even, and their sum must be even. The addition of the single edge (v_a, v_b) shows us that this circuit has an odd length. This contradicts our premise. #

Note: An efficient algorithm for finding a 2-coloring of a graph can be designed using the method that is used in the second part of the proof above.

EXERCISES FOR SECTION 9.6

A Exercises

1. Complete the proof of Theorem 9.6.1.

2. Use the outline of a proof of Theorem 9.6.2 to write a complete proof. Be sure to point out where the premise $v \geq 3$ is essential.

3. Apply theorem 9.6.2 to prove that once n gets to a certain size, a K_n is non-planar. What is the largest complete planar graph?

4. Can you apply Theorem 9.6.2 to prove that the Three Utilities Puzzle can't be solved?

5. What are the chromatic numbers of the following graphs?

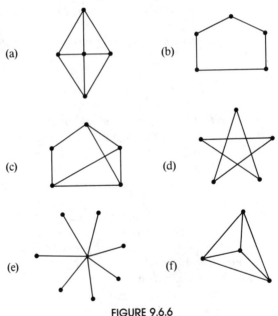

FIGURE 9.6.6
Exercise on chromatic number

6. Prove that if an undirected graph has a subgraph that is a K_3, then its chromatic number is at least 3.

7. What is the chromatic number of the United States?

8. What is $c(K_n)$, $n \geq 1$?

B Exercises

9. Design an algorithm to determine whether a graph is bipartite.

10. Let $G = (V, E)$ with $\#V \geq 11$, and let U be the set of all undirected edges between distinct vertices in V. Prove that either G or $G' = (V, E^c)$ is non-planar.

C Exercises

11. Prove that any graph with a finite number of vertices can be drawn in three dimensions so that no edges intersect.

12. Suppose that you had to color the edges of an undirected graph so that for each vertex, the edges that it is connected to have different colors. How can this problem be transformed into a vertex coloring problem?

13. (a) Suppose that the edges of a K_6 are colored either red or blue. Prove that there will be either a "red K_3" (a subset of the vertex set with three vertices connected by red edges) or a "blue K_3."

(b) Suppose that six persons are selected at random. Prove that either there exists a subset of three of them with the property that each can speak to the other in a common language or there exist three persons, no two of which can speak together in a common language.

chapter

10

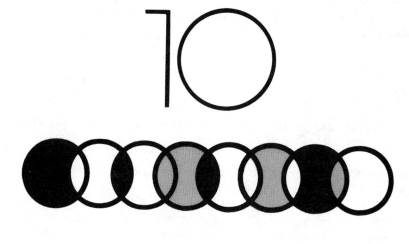

TREES

10 GOALS

In this chapter we will study the class of graphs called *trees*. Trees are frequently used in both mathematics and the sciences. Our solution of Example 2.1 is one simple instance. Since they are often used to illustrate or prove other concepts, a poor background in trees can be a serious handicap. For this reason, our ultimate goals are to: (1) define the various common types of trees, (2) identify some basic properties of trees, and (3) discuss some of the common applications of trees.

10.1   What is a Tree?

What distinguishes all trees from other types of graphs is the absence of certain paths called *cycles*.

Definition: Cycle.   *A cycle is a circuit whose edge list contains no duplicates.*

The simplest example of a cycle in an undirected graph is a pair of vertices with two edges connecting them. Since trees are cycle-free, we can rule out all multigraphs from consideration as trees.

Trees can either be undirected or directed graphs. We will concentrate on the undirected variety for now.

Definition: Tree.   *An undirected graph is a tree if it is connected and contains no cycles or self-loops.*

Example 10.1.1.

(a) Graphs a, b, and c in Figure 10.1.1 are all trees, while graphs d, e, and f are not trees.
(b) A K_2 is a tree. However, if $n \geq 3$, a K_n is not a tree.
(c) In a loose sense, a botanical tree is a mathematical tree. There are no cycles in the branch structure of a botanical tree.
(d) The structure of some chemical compounds are modeled by a tree. For example, butane (Figure 10.1.2a) consists of four carbon molecules and ten hydrogen molecules, where an edge between two molecules represents a bond between them. A bond is a force that keeps two molecules together. The same set of molecules can be linked together in a different tree structure to give us the compound isobutane (Figure 10.1.2b). There are some compounds whose graphs are not trees. One example is benzene (Figure 10.1.2c).
One type of graph that is not a tree, but is closely related, is a forest.

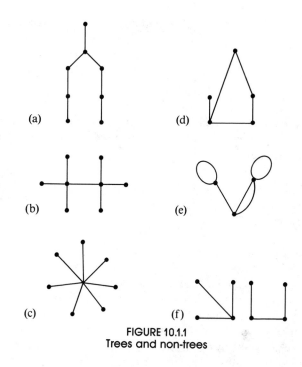

(a)

(b)

(c)

(d)

(e)

(f)

FIGURE 10.1.1
Trees and non-trees

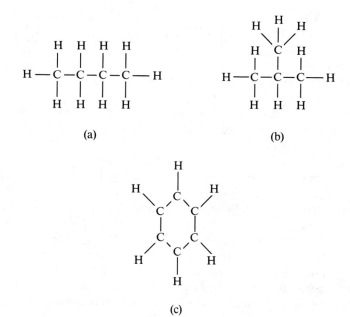

(a)

(b)

(c)

FIGURE 10.1.2
Simple organic compounds

233

Definition: Forest.   *A forest is an undirected graph whose components are all trees.*

Example 10.1.2.   The left-hand side of Figure 10.1.1 can be viewed as a forest of three trees.

We will now examine several conditions that are equivalent to the one that defines a tree. The following theorem will be used as a tool in proving that the conditions are equivalent.

Theorem 10.1.1.   *Let $G = (V, E)$ be an undirected graph with no self-loops, and let v_a, $v_b \in V$. If two different simple paths exist between v_a and v_b, then there exists a cycle in G.*

   Proof: Let $p_1 = (e_1, e_2, \ldots , e_m)$ and $p_2 = (f_1, f_2, \ldots , f_n)$ be two different simple paths from v_a to v_b. The first step that we will take is to delete from p_1 and p_2 the initial edges that are identical. This is, if $e_1 = f_1$, $e_2 = f_2$, $\ldots , e_j = f_j$, and $e_{j+1} \neq f_{j-1}$, delete the first j edges of both paths. Once this is done, both paths start at the same vertex, call it v_c, and still both end at v_b. Now we construct a cycle by starting at v_c and following what is left of p_1 until we first meet what is left of p_2. If this first meeting occurs at vertex v_d, then the remainder of the cycle is completed by following the portion of p_2 that starts at v_d and ends at v_c. #

Theorem 10.1.2.   *Let $G = (V, E)$ be an undirected graph with no self-loops and $\#V = n$. The following are all equivalent:*

(1) G is a tree.
(2) For each pair of distinct vertices in V, there exists a unique simple path between them.
(3) G is connected and if $e \in E$, then $(V, E-\{e\})$ is disconnected.
(4) G contains no cycles, but by adding one edge, you create a cycle.
(5) G is connected and $\#E = n - 1$.

   Proof: Strategy. Most of this theorem can be proven by proving the following chain of implications: $(1) \Rightarrow (2)$, $(2) \Rightarrow (3)$, $(3) \Rightarrow (4)$, and $(4) \Rightarrow (1)$. Once these implications have been demonstrated, the transitive closure of $\Rightarrow$ on $\{1, 2, 3, 4\}$ establishes the equivalence of the first four conditions. The proof that Statement 5 is equivalent to the first four can be done by induction, which we will leave to the interested reader.
   $(1) \Rightarrow (2)$ (Indirect). Assume that G is a tree and that there exists a pair of vertices, v_1 and v_2, between which there is either no path or there are at least two distinct paths. Both of these possibilities contradict the premise that G is a tree. If no path exists, G is disconnected, and if two paths exist, a cycle can be obtained by Theorem 10.1.1.

(2) $\Rightarrow$ (3). We now use Statement 2 as a premise. Since each pair of vertices in V are connected by one path, G is connected. Now if we select any edge e in E, it connects two vertices, v_1 and v_2. By (2), there is no simple path connecting v_1 to v_2 other than (e). In addition, since any path can be reduced to a simple path by deleting some of its edges, no path at all can exist between v_1 and v_2 in $(V, E - \{e\})$. Hence $(V, E - \{e\})$ is disconnected.

(3) $\Rightarrow$ (4). Now we will assume that Statement 3 is true. We must show that G has no cycles and that adding an edge to G creates a cycle. We will use an indirect proof for this part. Since (4) is a conjunction, by DeMorgan's Law its negation is a disjunction and we must consider two cases. First, suppose that G has a cycle. Then the deletion of any edge in the cycle keeps the graph connected, which contradicts (3). The second case is that the addition of an edge to G does not create a cycle. Then there are two distinct paths between the vertices that the new edge connects. By Theorem 10.1.1, a cycle can then be created, which is a contradiction.

(4) $\Rightarrow$ (1). Assume that G contains no cycles and that the addition of an edge creates a cycle. All that we need to prove to verify that G is a tree is that G is connected. If it is not connected, then select any two vertices that are not connected. If we add an edge to connect them, the fact that a cycle is created implies that a second path between the two vertices can be found which is in the original graph, which is a contradiction. #

The usual definition of a directed tree is based on whether the associated undirected graph, which is obtained by "erasing" its directional arrows, is a tree. In Section 10.3 we will introduce the rooted tree, which is a special type of directed tree.

EXERCISES FOR SECTION 10.1

A Exercises

1. Given the following vertex sets, draw all possible undirected trees that connect them.
   (a) $V_a = \{\text{right, left}\}$
   (b) $V_b = \{+, -, 0\}$
   (c) $V_c = \{\text{north, south, east, west}\}$.

2. Are all trees planar? If they are, can you explain why? If they are not, you should be able to find a non-planar tree.

3. Prove that if G is a simple undirected graph with no self-loops, then G is a tree if and only if G is connected and $\#E = \#V - 1$.

4. (a) Prove that if $G = (V, E)$ is a tree and $e \in E$, then $(V, E - \{e\})$ is a forest of two trees.
   (b) Prove that if (V_1, E_1) and (V_2, E_2) are disjoint trees and e is an edge that connects a vertex in V_1 to a vertex in V_2, then $(V_1 \cup V_2, E_1 \cup E_2 \cup \{e\})$ is a tree.

5. (a) Prove that any tree with at least two vertices has at least two vertices of degree 1.
    (b) Prove that if a tree is not a chain, then it has at least three vertices of degree 1.

10.2. Spanning Trees

The topic of spanning trees is motivated by a graph-optimization problem.

Example 10.2.1. A map of Atlantis (Figure 10.2.1) shows that there are four campuses in its university system. A new phone system is being installed and the objective is to allow for communication between any two campuses. to achieve this objective, the university must buy direct lines between certain pairs of campuses. Let G be the graph with a vertex for each campus and an edge for each direct phone line. Total communication is equivalent to G being a connected graph. This is due to the fact that two campuses can communicate over any number of lines. To minimize costs, the university wants to buy a minimum number of lines.

The solutions to this problem are all trees. Any graph that satisfies the requirements of the university must be connected, and if a cycle does exist, any line in the cycle can be deleted, reducing the cost. Each of the sixteen trees that can be drawn to connect the vertices North, South, East, and West (see Exercise 1c of Section 10.1) solves the problem as it is stated. Note that in each case, three direct lines must be purchased. There are two considerations that can help reduce the number of solutions that would be considered.

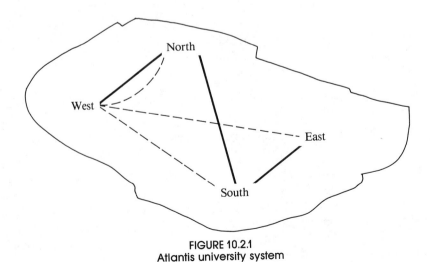

FIGURE 10.2.1
Atlantis university system

Objective 1:   Given that the cost of each line depends on certain factors, such as the distance between the campuses, select a tree whose cost is as low as possible.

Objective 2:   Suppose that communication over multiple lines is noisier as the number of lines increases. Select a tree with the property that the maximum number of lines that any pair of campuses must use to communicate with is as small as possible.

Typically, these objectives are not compatible; that is, you cannot always simultaneously achieve these objectives in all cases. In the case of the Atlantis university system, the solution with respect to Objective 1 is indicated with solid lines in Figure 10.2.1. There are four solutions to the problem with respect to Objective 2: any tree in which one campus is directly connected to the other three. One solution with respect to Objective 2 is indicated with dotted lines in Figure 10.2.1. After satisfying the conditions of Objective 2, it would seem reasonable to select the cheapest of the four trees.

Definitions.

(a) Spanning Set.   *Let G = (V, E) be a connected undirected graph. A spanning set for G is a subset E' of E such that (V, E') is connected.*
(b) Spanning Tree.   *If E' is a spanning set for G and T = (V, E') is a tree, then T is called a spanning tree for G.*

Notes

(a) If (V, E') is a spanning tree, $\#E' = \#V - 1$.
(b) The significance of a spanning tree is that it is a minimal spanning set. A smaller set would not span the graph, while a larger set would have a cycle, which has an edge that is superfluous.

For the remainder of this section, we will discuss two of the many topics that relate to spanning trees. The first is the Minimal Spanning Tree Problem, which is a generalization of Objective 1 above. The second is the Minimum Diameter Spanning Tree Problem, which is a generalization of Objective 2.

THE MINIMAL SPANNING TREE PROBLEM

Given a weighted connected undirected graph $G = (V, E, w)$, find a spanning tree (V, E') for which $\sum_{e \in E'} w(e)$ is as small as possible.

Unlike many of the graph-optimization problems that we've examined, a solution to this problem can be obtained efficiently. It is a situation in which a greedy algorithm works.

Definition: Bridge.   *Let G = (V, E) be an undirected graph and let {L, R} be a partition of V. A bridge between L and R is an edge in E that connects a vertex in L to a vertex in R.*

Theorem 10.2.1. *Let $G = (V, E, w)$ be a weighted connected undirected graph. Let V be partitioned into two sets L and R. If $e*$ is a bridge of least weight between L and R, then there exists a minimal spanning tree for G that includes $e*$.*

Proof (Indirect):   Suppose that no minimal spanning tree including $e*$ exists. Let $T = (V, E')$ be a minimal spanning tree. If we add $e*$ to T, a cycle is created, and this cycle must contain another bridge, e, between L and R. Since $w(e*) \leq w(e)$, we can delete e and the new tree, which includes $e*$, must also be a minimal spanning tree. #

Example 10.2.2.   The bridges between the vertex sets $\{a, b, c\}$ and $\{d, e\}$ in Figure 10.2.2 are the edges $\{b, d\}$ and $\{c, e\}$. According to the theorem, a minimal spanning tree that includes (b, d) exists. By examination, you should be able to see that this is true. Is it true that only the bridges of minimal weight can be part of a minimal spanning tree? The answer is no.

Theorem 10.2.1 essentially tells us that a minimal spanning tree can be constructed recursively by continually adding minimally weighted bridges to a set of edges.

Algorithm 10.2.1.   Greedy Algorithm for Finding a Minimal Spanning Tree.   Let $G = (V, E, w)$ be a connected, weighted, undirected graph, and let v_1 be an arbitrary vertex in V. The following steps lead to a minimal spanning tree for G. L and R will be sets of vertices and E' is a set of edges.

(1) (Initialize) $L := V - \{v_1\}$, $R := \{v_1\}$, $E := \emptyset$.
(2) (Build the tree)   While $L \neq \emptyset$ do:
    (2.1) Find $e* = (\{v_L, v_R\})$, a bridge of minimum weight between L and R.
    (2.2) (Update sets) $R := R \cup \{v_L\}, L := L - \{v_L\}$ and $E' := E' \cup \{e*\}$.
(3) Terminate with a minimal spanning tree (V, E').

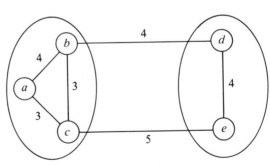

FIGURE 10.2.2

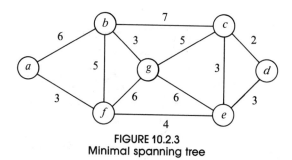

FIGURE 10.2.3
Minimal spanning tree

Notes:

(a) If more than one minimal spanning tree exists, then the one that is obtained depends on v_1 and the means by which $e*$ is selected in Step 2.1 if two minimally weighted bridges exist.
(b) Warning: If two minimally weighted bridges exist between L and R, do not try to speed up the algorithm by adding all of them to E'.
(c) That Algorithm 10.2.1 yields a minimal spanning tree can be proven by induction with the use of Theorem 10.2.1.
(d) If it is not known whether G is connected, Algorithm 10.2.1 can be revised to handle this possibility. The key change (in Step 2.1) would be to determine whether any bridge at all exists between L and R. The condition of the while loop in Step 2 must also be changed somewhat.

Example 10.2.3. Consider the graph in Figure 10.2.3. If we apply Algorithm 10.2.1 starting at a, we obtain the following edge list in the order given: $\{a, f\}$, $\{f, e\}$, $\{e, c\}$, $\{c, d\}$, $\{f, b\}$, $\{b, g\}$. The total of the weights of these edges is 20. The method that we have used (in Step 2.1) to select a bridge when more than one minimally weighted bridge exists is to order all bridges alphabetically by the vertex in L and then, if further ties exist, by the vertex in R. The first vertex in that order is selected in Step 2.1 of the algorithm.

THE MINIMUM DIAMETER SPANNING TREE PROBLEM

Given a connected undirected graph $G = (V, E)$, find a spanning tree $T = (V, E')$ of G such that the longest path in T is as short as possible.

Example 10.2.4. The Minimum Diameter Spanning Tree Problem is easy to solve in a K_n. Select any vertex v_1 and construct the spanning tree whose edge set is the set of edges that connect v_1 to the other vertices in the K_n. Figure 10.2.4 illustrates a solution for a K_5.

For incomplete graphs, a two-stage algorithm is needed. In short, the first step is to locate a "center" of the graph. The maximum distance from a center to any other vertex is as small as possible. Once a center is located, a

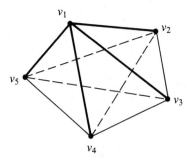

FIGURE 10.2.4
Minimum diameter spanning tree for K_5

breadth-first search of the graph is used to construct the spanning tree. The breadth-first search is essentially the broadcasting algorithm that we discussed in Section 9.4 applied to undirected graphs.

EXERCISES FOR SECTION 10.2

A Exercises

1. Suppose that after Atlantis U.'s phone system is in place, a fifth campus is established and that a transmission line can be bought to connect the new campus to any old campus. Is this larger system the most economical one possible with respect to Objective 1? Can you always satisfy Objective 2?

2. Construct a minimal spanning tree for the capital cities in New England (See Figure 10.1.1).

3. Show that the answer to the question posed in Example 10.2.2 is no.

4. Find a minimal spanning tree for the graphs in Figure 10.2.5.

5. Find a minimum diameter spanning tree for the graphs in Figure 10.2.6.

10.3 Rooted Trees

In the next two sections, we will discuss rooted trees. Our primary focuses will be on general rooted trees and on a special case, ordered binary trees.

Informal Definition and Terminology: What differentiates rooted trees from undirected trees is that a *rooted tree* contains a distinguished vertex, called the *root*. Consider the tree in Figure 10.3.1. Vertex *A* has been designated the root of the tree. If we choose any other vertex in the tree, such as

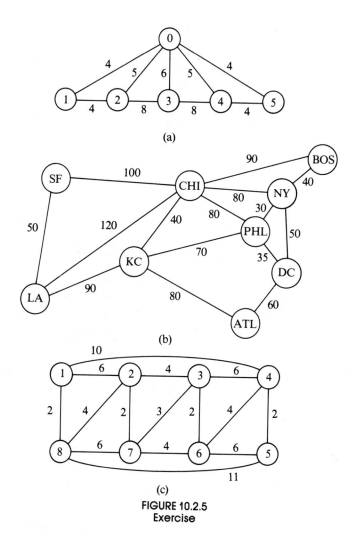

(a)

(b)

(c)

FIGURE 10.2.5
Exercise

M, we know that there is a unique path from A to M. The vertices on this path, (A, D, K, M), are described in genealogical terms:

M is a *child* of K (so is L).

K is $M's$ *parent*.

A, D, and K are $M's$ *ancestors*.

D, K, and M are *descendants* of A.

These genealogical relationships are often easier to visualize if the tree is rewritten so that children are positioned below their parents, as in Figure 10.3.2.

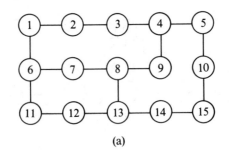

(a)

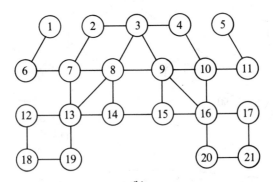

(b)

FIGURE 10.2.6
Exercise

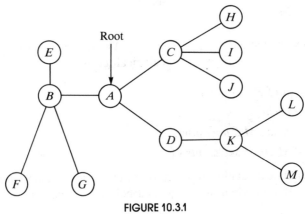

FIGURE 10.3.1
Rooted tree

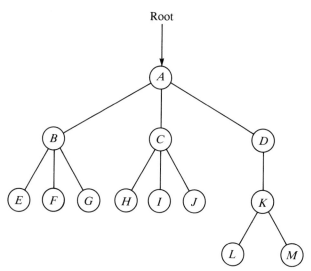

Root

FIGURE 10.3.2
Rooted tree of Figure 10.3.1, redrawn

With this format, it is easy to see that each vertex in the tree can be thought of as the root of a tree that contains, in addition to itself, all of its descendants. For example, D is the root of a tree that contains D, K, L, and M. Furthermore, K is the root of a tree that contains K, L, and M. Finally, L and M are roots of trees that contain only themselves. From this observation, we can give a formal definition of a rooted tree.

Definition: Rooted Tree.

(a) Basis:
  (i) A tree with no vertices is a rooted tree (the empty tree).
  (ii) A single vertex with no children is a rooted tree.
(b) Recursion:
  Let T_1, T_2, . . . , T_r, $r \geq 1$, be disjoint rooted trees with roots v_1, v_2, . . . , v_r, respectively, and v a vertex that does not belong to any of these trees. Then a rooted tree, rooted at v, is obtained by making v the parent of the vertices v_1, v_2, . . . , and v_r. We call T_1, T_2, . . . , T_r subtrees of the large tree.

The *level* of a vertex of a rooted tree is the number of edges that separate the vertex from the root. The level of the root is zero. The *depth* of a tree is the maximum level of the vertices in the tree. The depth of a tree in Figure 10.3.2 is three, which is the level of the vertices L and M. The vertices E, F, G, H, I, J, and K have level two. B, C, and D are at level one and A has level zero.

Example 10.3.1. Figure 2.1 is a rooted tree, with start as the root. It is an example of what is called a *decision tree*.

Example 10.3.2: Pascal Note on Record Structure. Frequently, the fields of a record are themselves records. For example, student information can be stored in a Pascal record:

```
Student = RECORD
          Name:  RECORD
                    First:  String;
                    Middle: Char;
                    Last:   String
                 END;
          Major: String;
          SAT:   RECORD
                    Math,Verbal:  0..800
          END
        END.
```

The organization of this data can be visualized with a rooted tree such as the one in Figure 10.3.3.

In general, you can represent a record type, T, as a rooted tree with T as the root and a subtree for each field. Those fields that are records are roots of further subtrees, while the non-records (often called *elementary fields*) have no further children in the tree.

KRUSKAL'S ALGORITHM

An alternate algorithm for constructing a minimal spanning tree uses a forest of rooted trees. First we will describe the algorithm in its simplest terms. Afterward, we will describe how rooted trees are used to implement the

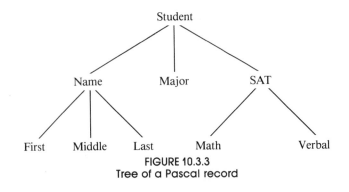

FIGURE 10.3.3
Tree of a Pascal record

algorithm. Finally, we will describe a simple data type and operations that make the algorithm quite easy to program. In all versions of this algorithm, assume that $G = (V, E, w)$ is a weighted undirected graph with $\#V = m$ and $\#E = n$.

Kruskal's Algorithm—Version 1.

(1) Sort the edges of G in ascending order according to weight. That is, $i \leq j \Leftrightarrow w(e_i) \leq w(e_j)$.

(2) Go down the list obtained in Step 1 and add edges to a spanning set (initially empty) of edges so that the set of edges does not form a cycle. When an edge that would create a cycle is encountered, ignore it. Continue examining edges until either $m - 1$ edges have been selected or you have come to the end of the edge list. If $m - 1$ edges are selected, these edges make up a minimal spanning tree for G. If fewer than $m - 1$ edges are selected, G is not connected.

Note: Step 1 can be accomplished using one of any number of standard sorting routines. Using the most efficient sorting routine, the time required to perform this step is proportional to $n \log(n)$. The second step of the algorithm, also of $n \log(n)$ time complexity, is the one that uses a forest of rooted trees to test for whether an edge should be added to the spanning set.

Kruskal's Algorithm—Version 2.

(1) Sort the edges as in Version 1.

(2) (2.1) Initialize each vertex in V to be the root of its own rooted tree.

(2.2) Go down the list of edges until either a spanning tree is completed or the edge list has been exhausted. For each edge $e = \{v_1, v_2\}$ we can determine whether e can be added to the spanning set without forming a cycle by determining whether the root of v_1's tree is equal to the root of v_2's tree. If the two roots are equal, then ignore e. If the roots are different, then we can add e to the spanning set. In addition, we merge the trees that v_1 and v_2 belong to. This is accomplished by either making v_1's root the parent of v_2's root or vice versa.

Notes:

(a) Since we start the algorithm with m trees and each addition of an edge decreases the number of trees by one, we end the algorithm with one rooted tree, provided a spanning tree exists.

(b) The rooted tree that we obtain is not the spanning tree itself.

(c) Suppose that we define the operation merge(r_1, r_2), where r_1 and r_2 are roots of trees, to create a new tree, rooted at r_1, so that the parent of

r_2 is r_1. To minimize the time it takes to find the root of a given tree, we want to keep the depths of our trees as small as possible. Suppose that the depth of r_1's tree is j and that the depth of r_2's tree is k. If $j = k$, it doesn't matter which way we merge the trees. However, if j is less than k, merge (r_1, r_2) will give us a tree of depth $k + 1$, while merge (r_2, r_1) gives us a tree of depth k. For this reason, we should, if possible, always merge smaller trees in larger trees; that is, the smaller tree's root should be the child of the larger tree's root.

A Data Structure for Kruskal's Algorithm. The important information that we need to maintain about the rooted trees in this algorithm is the identity of each vertex's parent. This can be accomplished quite easily using an array of integers, Parent.

```
Parent: ARRAY [1..m] of 0..m
```

We interpret the elements of Parent as follows:

If Parent[K] = 0 then v_k is the root of a tree.

If Parent[K] = j then v_j is the parent of v_k.

Assume that we have an array of E that contains the information on the edges of the graph.

```
E: ARRAY [1..n] OF RECORD
                    v1, v2: 1..m;
                    wgt: Real;
                    span: Boolean
          END
```

We will use the span field to keep track of whether an edge is part of the spanning tree that is being constructed. This field is initialized to False at the start of the algorithm.

The following function returns the root of the tree containing vertex k.

```
FUNCTION Root(K: 1..m) : 1..m;
BEGIN
If Parent[K] = 0 THEN Root := K
                 ELSE Root := Root(Parent[K])
END;
```

Merge is a very simple procedure that merges two trees as required in the algorithm. Note that our version does not minimize the depth of the resulting tree.

```
PROCEDURE Merge (VAR P: ARRAY[1..m]
                      of Integer; j, K:1..m);
BEGIN
   P[K]:=j
END;
```

Kruskal's Algorithm—Version 3.

(1) Sort the edges in E in ascending order according to the weight field.
(2) For $i := 1$ TO m DO Parent[i]: $= 0$;
(3) For $j := 1$ TO n DO $E[j]$.span:$=$ False;
(4) $r := 0$ (*number of edges in the current spanning set*)
(5) $q: = 1$ (*index of the current edge to be examined*)
(6) While $r \leq (m - 2)$ and $q \leq n$ DO
    (6.1) $j := \text{Root}(E[q].v1)$
    (6.2) $k := \text{Root}(E[q].v2)$
    (6.3) IF $j = k$

> THEN $q : = q + 1$
> ELSE BEGIN
> > Merge(parent, j, k);
> > $E[q]$.span $: =$ True;
> > $r : = r + 1$;
> > $q : = q + 1$
> END

The algorithm terminates with a minimum spanning tree if $r = m - 1$; otherwise, the graph is not connected.

EXERCISES FOR SECTION 10.3

A Exercises

1. Suppose that an undirected tree has diameter d and that you would like to select a vertex of the tree as a root so that the resulting rooted tree has the smallest depth possible. How would such a root be selected and what would be the depth of the tree (in terms of d)?

2. Use Kruskal's algorithm to find a minimal spanning tree for the graphs in Figure 10.3.4. What is the final rooted tree? Merge trees so that the depth of the merged tree is minimized. If you must merge two trees of equal depth, make the root with the lower number the root of the new tree.

C Exercises

3. (a) Suppose that information on buildings is arranged in records with five fields: the name of the building, its location, its owner, its height, and its floor space. The location and owner fields are records that include all of the information that you would expect, such as street, city, and state, together with the owner's name (first, middle, last) in the owner field. Draw a rooted tree to describe this type of record.

   (b) Write a Pascal type of definition for information on a book that corresponds to the rooted tree in Figure 10.3.5.

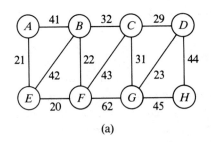

(a)

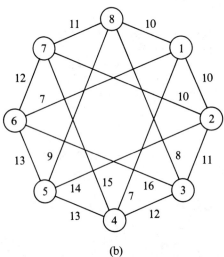

(b)

FIGURE 10.3.4
Exercise

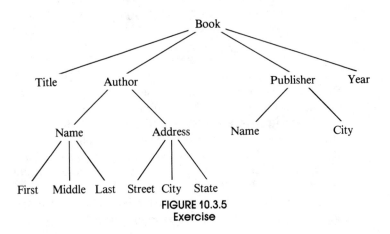

FIGURE 10.3.5
Exercise

4. Instead of keeping the value of Parent[k] = 0 when v_k is a root, you can hold the depth of the tree by making Parent[k] = $-1*$ (depth of the tree rooted at v_k). Revise Kruskal's algorithm so that the depth of the trees are kept as small as possible.

10.4 Binary Trees

An *ordered rooted tree* is a rooted tree whose subtrees are put into a definite order and are, themselves, ordered rooted trees. An empty tree and a single vertex with no descendants (no subtrees) are ordered rooted trees.

Example 10.4.1. The trees in Figure 10.4.1 are identical rooted trees, but as ordered trees, they are different.

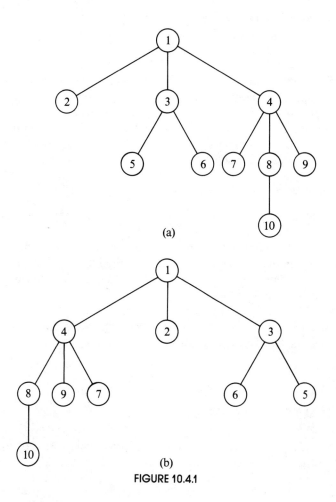

(a)

(b)

FIGURE 10.4.1

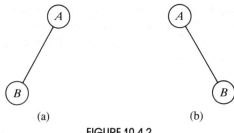

(a)   (b)

FIGURE 10.4.2

If a tree rooted at v has p subtrees, we would refer to them as the first, second, ... , p^{th} subtrees. If we restrict the number of subtrees of each vertex to be less than or equal to two, we have a binary (ordered) tree.

Definition: Binary Tree. *A binary tree is*

(a) *a tree consisting of no vertices (the empty tree), or*
(b) *a vertex with two subtrees that are both binary trees. The subtrees are called the left and right subtrees.*

The difference between binary trees and ordered trees is that every vertex of a binary tree has exactly two subtrees (one or both of which may be empty), while a vertex of an ordered tree may have any number of subtrees. The two trees in Figure 10.4.2 would be considered identical as ordered trees; however, they are different binary trees. Tree a has an empty right subtree and Tree b has an empty left subtree.

Terminology and General Facts:

(a) A vertex of a binary tree with two empty subtrees is called a *leaf*. All other vertices are called *internal vertices*.
(b) The number of leaves in a binary tree can vary from one up to roughly half the number of vertices in the tree (see Exercise 4 of this section).
(c) The maximum number of vertices at level k of a binary tree is 2^k, $k \geq 0$ (see Exercise 5 of this section).
(d) A *full binary tree* is a tree in which each vertex has either zero or two empty subtrees. In other words, each vertex has either two or zero children. See Exercise 6 of this section for a general fact about full binary trees.

TRAVERSALS OF BINARY TREES

The traversal of a binary tree consists of visiting each vertex of the tree in some prescribed order. Unlike graph traversals, the consecutive vertices that are visited are not always connected with an edge. The most common binary tree traversals are differentiated by the order in which the root and its subtrees are visited. The three traversals are best described recursively and are:

(1) Preorder Traversal:
   (a) Visit the root of the tree.
   (b) Preorder traverse the left subtree.
   (c) Preorder traverse the right subtree.
(2) Inorder Traversal:
   (a) Inorder traverse the left subtree.
   (b) Visit the root of the tree.
   (c) Inorder traverse the right subtree.
(3) Postorder Traversal:
   (a) Postorder traverse the left subtree.
   (b) Postorder traverse the right subtree.
   (c) Visit the root of the tree.

Any traversal of an empty tree consists of doing nothing.

Example 10.4.2. For the tree in Figure 10.4.3, the orders in which the vertices are visited are:

A-B-D-E-C-F-G, for the preorder traversal,
D-B-E-A-F-C-G, for the inorder traversal, and
D-E-B-F-G-C-A, for the ~~preorder~~ postorder traversal.

Example 10.4.3. Binary Tree Sort. Given a collection of integers (or other objects than can be ordered), one technique for sorting is a binary tree sort. If the integers are $a_1, a_2, \ldots, a_n$, $n \geq 1$, we first execute the following algorithm that creates a binary tree:

(1) Insert a_1 into the root of the tree.
(2) For $k := 2$ to n (*insert a_k into the tree*)
   (2.1) $r := a_1$
   (2.2) inserted := false

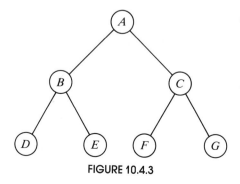

FIGURE 10.4.3

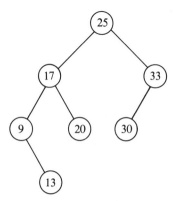

FIGURE 10.4.4

(2.3) While not inserted do
  If $a_k < r$ then
    if r has a left child
      then $r :=$ left child of r
      else make a_k the left child of r and
        inserted $:=$ true
    else $(*a_k \geq r*)$
      if r has a right child
        then r $:=$ right child of r
        else make a_k the right child of r and
          inserted $:=$ true

If the integers to be sorted are 25, 17, 9, 20, 33, 13, and 30, then the tree that is created is the one in Figure 10.4.4. The inorder traversal of this tree is 9, 13, 17, 20, 25, 30, 33, the integers in ascending order. In general, the inorder traversal of the tree that is constructed in the algorithm above will produce a sorted list. The preorder and postorder traversals of the tree have no meaning.

EXPRESSION TREES

A convenient way to represent an algebraic expression is by its expression tree. Consider the expression $X = a*b - c/d + e$. Since it is customary to put a precedence on multiplication/divisions, X is evaluated as $((a*b) - (c/d)) + e$. Consecutive multiplication/divisions or addition/subtractions are evaluated from left to right. We can analyze X further by noting that it is the sum of two simpler expressions $(a*b) - (c/d)$ and e. The first of these expressions can be broken down further into the difference of the expressions $a*b$ and c/d. When we decompose any expression into (left expression) (operation) (right expression), the expression tree of that expression is the

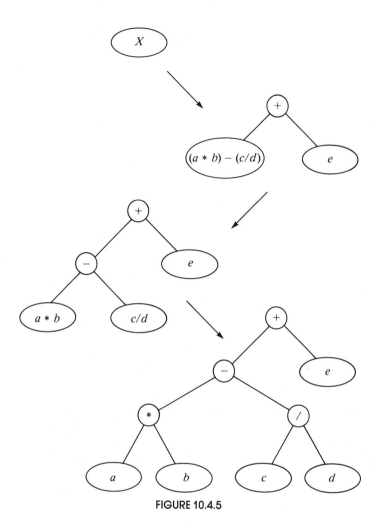

FIGURE 10.4.5

binary tree whose root contains the operation and whose left and right subtrees are the trees of the left and right expressions, respectively. Additionally, a simple variable or a number has an expression tree which is a single vertex containing the variable or number. The evolution of the expression tree for expression X appears in Figure 10.4.5.

Example 10.4.2.

(a) If we intend to apply the addition and subtraction operations in X first, we would parenthesize the expression to $a*(b - c)/(d + e)$. Its expression tree appears in Figure 10.4.6a.

(b) The expression trees for $a^2 - b^2$ and for $(a + b)*(a - b)$ appear in Figures 10.4.6b and 10.4.6c.

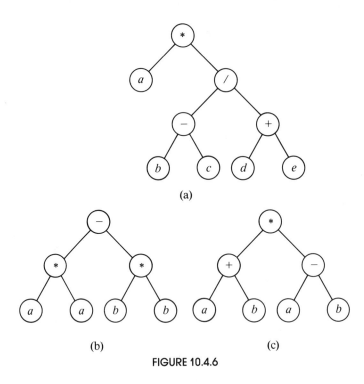

(a)

(b)   (c)

FIGURE 10.4.6

The three traversals of an operation tree are all significant. A binary operation applied to a pair of numbers can be written in three ways. One is the familiar *infix* form, such as $a + b$ for the sum of a and b. Another form is *prefix*, in which the same sum is written $+ab$. The final form is *postfix*, in which the sum is written $ab+$. Algebraic expressions involving the four standard arithmetic operations ($+$, $-$, $*$, and $/$) in prefix and postfix form are defined as follows:

Prefix: (a) A variable or number is a prefix expression.
           (b) Any operation followed by a pair of prefix expressions is a prefix expression.

Postfix: (a) A variable or number is a postfix expression.
           (b) Any pair of postfix expressions followed by an operation is a postfix expression.

The connection between traversals of an expression tree and these forms is simple:

(a) The preorder traversal of an expression tree will result in the prefix form of the expression.
(b) The postorder traversal of an expression tree will result in the postfix form of the expression.
(c) The inorder traversal of an operation tree will not, in general, yield the

proper infix form of the expression. If an expression requires parentheses in infix form, an inorder traversal of its expression tree has the effect of removing the parentheses.

Example 10.4.5. The preorder traversal of the tree in Figure 10.4.5 is $+-*ab/cde$, which is the prefix version of expression X. The postfix traversal is $ab*cd/-e+$. Note that since the original form of X needed no parentheses, the inorder traversal, $a*b-c/d+e$, is the correct infix version.

COUNTING BINARY TREES

We close this section with a formula for the number of different binary trees with n vertices. The formula is derived using generating functions. Although the complete details are beyond the scope of this text, we will supply an overview of the derivation in order to illustrate how generating functions are used in advanced combinatorics.

Let $B(n)$ be the number of different binary trees of size n (n vertices), $n \geq 0$. By our definition of a binary tree, $B(0) = 1$. Now consider any positive integer $n + 1$, $n \geq 0$. A binary tree of size $n + 1$ has two subtrees, the sizes of which add up to n. The possibilities can be broken down into $n + 1$ cases:

Case 0:   Left subtree has size 0; right subtree has size n.
Case 1:   Left subtree has size 1; right subtree has size $n - 1$.
.
.
.
Case k: Left subtree has size k; right subtree has size $n - k$.
.
.
.
Case n: Left subtree has size n; right subtree has size 0.

In Case k, we can count the number of possibilities by multiplying the number of ways that the left subtree can be filled ($B(k)$) by the number of ways that the right subtree can be filled ($B(n - k)$). Since the sum of these products equals $B(n + 1)$, we obtain recurrence relation:

$$B(n + 1) = B(0)B(n) + B(1)B(n - 1) + \cdots + B(n)B(0)$$
$$= \sum_{k=0}^{n} B(k)B(n - k) \qquad n \geq 0.$$

Now take the generating function of both sides of this recurrence relation:

$$\sum_{n=0}^{\infty} B(n + 1)z^n = \sum_{n=0}^{\infty} \left[\sum_{k=0}^{n} B(k)B(n - k) \right] z^n$$

or $G(B\uparrow; z) = G(B*B; z) = G(B; z)^2$. Therefore, if we abbreviate $G(B; z)$ to G, we obtain $(G - B(0))/z = G^2$ or $G - 1 = zG^2$, since $B(0) = 1$. We can write this final equation in the standard form of a quadratic equation in the unknown G: $zG^2 - G + 1 = 0$. The two solutions of this equation are

$$G_1 = (1 + \sqrt{1 - 4z})/2z$$

and

$$G_2 = (1 - \sqrt{1 - 4z})/2z.$$

The gap in our deviation occurs here. By applying calculus techniques, we can eliminate G_2 and so $G(B; z) = G_1$. After expanding G_1 by something called its Taylor series, we can obtain a closed form expression for $B(n)$, which is simplified to

$$B(n) = \frac{1}{n+1}\binom{2n}{n}.$$

Many advanced texts such as Standish have a complete derivation of this solution.

EXERCISES FOR SECTION 10.4

A Exercises

1. Draw the expression trees for the following expressions:
   (a) $a(b + c)$
   (b) $(ab) + c$
   (c) $ab + ac$
   (d) $bb - 4ac$
   (e) $((a_3x + a_2)x + a_1)x + a_0$

2. Draw the expression trees for
   (a) $\dfrac{x^2 + 1}{x - 1}$
   (b) $xy + xz + yz$

3. Write out the preorder, inorder, and postorder traversals of the trees in Exercise 1 above.

4. Verify the formula for $B(k)$, $0 \le k \le 3$ by drawing all binary trees with three or fewer vertices.

5. (a) Draw a binary tree with seven vertices and only one leaf.
   (b) Draw a binary tree with seven vertices and as many leaves as possible.

B Exercises

6. Prove that the maximum number of vertices at level k of a binary tree is 2^k and that a tree with that many vertices at level k must have at least $2^{k+1} - 1$ vertices.

7. Prove that if T is a full binary tree, then the number of leaves of T is one more than the number of internal vertices (non-leaves).

chapter

11

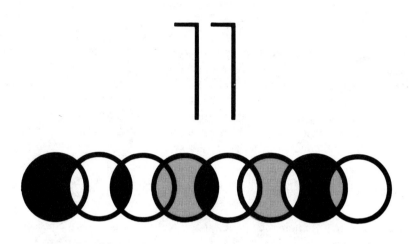

ALGEBRAIC SYSTEMS

GOALS

The primary goal of this chapter is to make the reader aware of what an algebraic system is and how algebraic systems can be studied at different levels of abstraction. After describing the concrete, axiomatic, and universal levels, we will introduce one of the most important algebraic systems at the axiomatic level, the group. In this chapter, group theory will be a vehicle for introducing the universal concepts of isomorphism, direct product, subsystem, and generating set. These concepts can be applied to all algebraic systems. The simplicity of group theory will help the reader obtain a good intuitive understanding of these concepts. In Chapter 15, we will introduce some additional concepts and applications of group theory. We will close the chapter with a discussion of how some computer hardware and software systems use the concept of an algebraic system.

11.1 Operations

One of the first mathematical skills that we all learn is how to add a pair of positive integers. A young child soon recognizes that something is wrong if a sum has two values, particularly if his or her sum is different from the teacher's. In addition, it is unlikely that a child would consider assigning a non-positive value to the sum of two positive integers. In other words, at an early age we probably know that the sum of two positive integers is unique and belongs to the set of positive integers. This is what characterizes all operations on a set.

Definition: Binary Operation. *Let S be a non-empty set. A binary operation on S is a rule that assigns to each ordered pair of elements of S a unique element of S. In other words, a binary operation is a function from $S \times S$ into S.*

Example 11.1.1. Union and intersection are both binary operations on the power set of any universe. Addition and multiplication are binary operators on the natural numbers. Addition and multiplication are binary operations on the set of 2 by 2 real matrices, $M_{2 \times 2}(\mathbf{R})$. Division is a binary operation on some sets of numbers, like the positive reals. But on the integers $(1 \div 2 \notin \mathbf{Z})$ and even on the reals $(1 \div 0$ is not defined), division is not a binary operation.

Notes:

(a) We stress that the image of each ordered pair must be in S. This requirement disqualifies subtraction on the natural numbers from consideration as a binary operation, since $1 - 2$ is not a natural number. Subtraction is a binary operation on the integers.

(b) *On Notation*. Despite the fact that a binary operation is a function, symbols, not letters, are used to name them. The most commonly used symbol for a binary operation is an asterisk, $*$. We will also use $ when a second symbol is needed.

(c) If $*$ is a binary operation on S and $a, b \in S$, there are three common ways of denoting the image of the pair (a, b). They are:

$* \, ab$	$a * b$	$ab \, *$
Prefix Form	Infix Form	Postfix Form

We are all familiar with infix form. For example, $2 + 3$ is how everyone is taught to write the sum of 2 and 3. But notice how $2 + 3$ was just described in the previous sentence! The word *sum* preceded 2 and 3. Orally, prefix form is quite natural to us. The prefix and postfix forms are superior to infix form in some respects. In Chapter 10, we saw that algebraic expressions with more than one operation didn't need parentheses if they were in prefix or postfix form. Due to our familiarity with infix form, we will use it throughout the remainder of this book.

Some operations, such as negation of numbers and complementation of sets, are not binary, but unary operators.

Definition: Unary Operation. *Let S be a non-empty set. A unary operator on S is a rule that assigns to each element of S a unique element of S. In other words, a unary operator is a function from S into S.*

COMMON PROPERTIES OF OPERATIONS

Whenever an operation on a set is encountered, there are several properties that should immediately come to mind. To effectively make use of an operation, you should know which of these properties it has. By now, you should be familiar with most of these properties. We will list the most common ones here to refresh your memory and define them for the first time in a general setting. Let S be any set and $*$ a binary operation on S.

Properties that apply to a single binary operation:

$*$ is *commutative* if $a * b = b * a$ for all $a, b \in S$.

$*$ is *associative* if $(a * b) * c = a * (b * c)$ for all $a, b, c \in S$.

$*$ has an *identity* if there exists an element, e, in S such that $a * e = e * a = a$ for all $a \in S$.

$*$ has the *inverse* property if for each $a \in S$, there exists an element b in S such that $a * b = b * a = e$. b is called *an inverse of* a.

$*$ is *idempotent* if $a * a = a$ for all $a \in S$.

Properties that apply to two binary operations:

Let $ be a second binary operation on S.

$ is *left distributive* over $*$ if $a\,\$\,(b\,*\,c) = (a\,\$\,b)\,*\,(a\,\$\,c)$ for all $a, b, c \in S$.

$ is *right distributive* over $*$ if $(b\,*\,c)\,\$\,a = (b\,\$\,a)\,*\,(c\,\$\,a)$ for all $a, b, c \in S$.

$ is *distributive over* $*$ if $ is both left and right distributive over $*$.

A common property of unary operations is involution. A unary operation $-$ on S has the *involution property* if $-(-a) = a$ for all $a \in S$.

If T is a subset of S, we say that T is *closed under* $*$ if $a, b \in T$ implies that $a\,*\,b \in T$. In other words, by operating on elements of T with $*$, you can't obtain new elements that are outside of T.

Example 11.1.2.

(a) The odd integers are closed under multiplication, but not under addition.

(b) Let p be a proposition over U and let A be the set of propositions over U that imply p. That is; $q \in A$ if $q \Rightarrow p$. Then A is closed under both conjunction and disjunction.

(c) The set of positive multiples of 5 is closed under both addition and multiplication.

Note: It is important to realize that the properties listed above depend on both the set and the operation(s).

OPERATION TABLES

If the set on which an operation is defined is small, a table is often a good way of describing the operation. For example, we might want to define $\oplus$ on $\{0, 1, 2\}$ by $a \oplus b = a + b$ if $a + b \le 2$ and $a \oplus b = a + b - 3$ if $a + b > 2$. The table for $\oplus$ is

$\oplus$	0	1	2
0	0	1	2
1	1	2	0
2	2	0	1

The top row and left column are the column and row headings, respectively. To determine $a \oplus b$, find the entry in Row a and Column b. The following operation table serves to define $*$ on $\{i, j, k\}$.

*	i	j	k
i	i	i	i
j	j	j	j
k	k	k	k

Note that $j * k = j$, yet $k * j = k$. Thus, $*$ is not commutative. Commutivity is easy to identify in a table: the table must be symmetric with respect to the diagonal going from the top left to lower right.

EXERCISES FOR SECTION 11.1

A Exercises

1. Determine the properties that the following operations have. They are all operations on the positive integers.
   - (a) addition
   - (b) multiplication
   - (c) M defined by a M b = larger of a and b
   - (d) m defined by a m b = smaller of a and b
   - (e) @ defined by a @ $b = a^b$

2. Which pairs of operations in Exercise 1 are distributive over one another?

3. Let $*$ be an operation on S and $A, B \subseteq S$. Prove that if A and B are both closed under $*$, then $A \cap B$ is also closed under $*$, but $A \cup B$ need not be.

4. How can you pick out the identity of an operation from its table?

5. Define $a * b$ by $|a - b|$ (the absolute value of $a - b$). Which properties does $*$ have on the set of non-negative integers?

2 Algebraic Systems

An *algebraic system* is a mathematical system consisting of a set called the *domain* and one or more operations on the domain. If V is the domain and $*_1, \ldots, *_n$ are the operations, $[V; *_1, \ldots, *_n]$ is used to denote the mathematical system. If the context is clear, this notation is abbreviated to V.

Example 11.2.1.

(a) Let B* be the set of all finite strings of 0's and 1's including the null (or empty) string, λ. An algebraic system is obtained by adding the operation of *concatenation*. The concatenation of two strings is simply the linking of the two strings together in the order indicated. The concatenation of a with b is often denoted (a) (b) with no operation symbol. For example, $(01101)\,(101) = 01101101$ and $(\lambda)\,(100) = 100$. Note that concatenation is an associative operation and that λ is the identity for concatenation.

(b) Let M be any non-empty set and let $*$ be any operation on M that is associative and for which an identity exists in M.

Our second example might seem strange, but we include it to illustrate a point. The algebraic system B^* is a special case of $[M; *\,]$. Most of us are much more comfortable with B^* than with M. No doubt the reason is that the elements in B^* are more concrete. We know what they look like and exactly how they are combined. The description of M is so vague that we don't even know what the elements are, much less how they are combined. Why would anyone want to study M? The reason is related to this question: What theorems are of interest in an algebraic system? Answering this question is one of our main objectives in this chapter. Certain properties of algebraic systems are called *algebraic properties*, and any theorem that says something about the algebraic properties of a system would be of interest. The ability to identify what is algebraic and what isn't is one of the skills that you should learn from this chapter.

Now, back to the question of why we study M. Our answer is to illustrate the usefulness of M with a theorem about M.

Theorem 11.2.1. *If a, b are elements of M and $a * b = b * a$, then $(a * b) * (a * b) = (a * a) * (b * b)$.*

Proof:

$$
\begin{aligned}
(a * b) * (a * b) &= a * (b * (a * b)) && \text{Why?}\\
&= a * ((b * a) * b) && \text{Why?}\\
&= a * ((a * b) * b) && \text{Why?}\\
&= a * (a * (b * b)) && \text{Why?}\\
&= (a * a) * (b * b) && \text{Why? \#}
\end{aligned}
$$

The power of this theorem is that it can be applied to any algebraic system that M describes. Since B^* is one such system, we can apply Theorem 11.2.1 to any two strings that commute—for example, 01 and 0101. Although a special case of this theorem could have been proven for B^*, it would not have been any easier to prove, and it would not have given us any insight into other special cases of M.

Example 11.2.2. Consider the set of two by two real matrices, $M_{2\times2}(\mathbf{R})$, with the operation of matrix multiplication. In this context, Theorem 11.2.1 can be interpreted as saying that if $AB = BA$, $(AB)^2 = A^2B^2$. One pair of matrices that this theorem applies to is

$$\begin{bmatrix} 2 & 1 \\ 1 & 2 \end{bmatrix} \text{ and } \begin{bmatrix} 3 & -4 \\ -4 & 3 \end{bmatrix}.$$

LEVELS OF ABSTRACTION

One of the fundamental tools in mathematics is abstraction. There are three levels of abstraction that we will identify for algebraic systems: concrete, axiomatic, and universal.

Concrete Level. Almost all of the mathematics that you have done in the past was at the concrete level. As a rule, if you can give examples of a few typical elements of the domain and describe how the operations act on them, you are describing a concrete algebraic system. Two examples of concrete systems are $B*$ and $M_{2\times2}(\mathbf{R})$. A few others are:

(a) The integers with addition. Of course, addition isn't the only standard operation that we could include. Technically, if we were to add multiplication, we would have a different system.
(b) The subsets of the natural numbers, with union, intersection, and complementation.
(c) The complex numbers with addition and multiplication.

Axiomatic Level. The next level of abstraction is the axiomatic level. At this level, the elements of the domain are not specified, but certain axioms are stated about the number of operations and their properties. The system that we called M is an axiomatic sytem. Some combinations of axioms are so common that a name is given to any algebraic system that they apply to. Any system with the properties of M is called a *monoid*. The study of M would be called *monoid theory*. The assumptions that we made about M, associativity and the existence of an identity, are called the *monoid axioms*. One of your few brushes with the axiomatic level may have been in your elementary algebra course. Many algebra texts identify the properties of the real numbers with addition and multiplication as the field axioms. As we will see in Chapter 16, Rings and Fields, the real numbers share these axioms with other concrete systems, all of which are called *fields*.

Universal Level. The final level of abstraction is the universal level. There are certain concepts, called *universal algebra concepts*, that can be applied to the study of all algebraic systems. Although a purely universal approach to algebra would be much too abstract for our purposes, defining concepts at this level should make it easier to organize the various algebraic theories in your own mind. In this chapter, we will consider the concepts of isomorphism, subsystem, and direct product.

GROUPS

To illustrate the axiomatic level and the universal concepts, we will consider yet another kind of axiomatic system, the group. In Chapter 5 we noted that the simplest equation in matrix algebra that we are often called upon to solve is $AX = B$, where A and B are known square matrices and X is an unknown matrix. To solve this equation, we need the associative, identity, and inverse laws. We call the systems that have these properties *groups*.

Definition: Group. *A group consists of a non-empty set G and an operation $*$ on G satisfying the properties.*

(a) $*$ is associative on G.
(b) There exists an identity element, e, in G such that $a * e = e * a = a$ for all a in G.
(c) For all $a \in G$, there exists an inverse, a^{-1}, in G such that $a * a^{-1} = a^{-1} * a = e$.

A group is denoted by its set's name, G, or occasionally by $[G; *]$ to emphasize the operation. At the concrete level, most sets have a standard operation associated with them that will form a group. As we will see below, the integers with addition is a group. Therefore, in group theory, $\mathbf{Z}$ always stands for $[\mathbf{Z}; +]$.

Generic Symbols.   At the axiomatic and universal levels, there are often symbols that have a special meaning attached to them. In group theory, the letter e is used to denote the identity element of whatever group is being discussed. The inverse of an element a is usually denoted a^{-1} and is read "a inverse". When a concrete group is discussed, these symbols are dropped in favor of concrete symbols. These concrete symbols may or may not be similar to the generic symbols. For example, the identity element of the group of integers is 0, and the inverse of n is denoted by $-n$, the additive inverse of n.

The asterisk could also be considered a generic symbol since it is used to denote operations on the axiomatic level.

Example 11.2.3.

(a) The integers with addition is a group. We know that addition is associative. Zero is the identity for addition: $0 + n = n + 0 = n$ for all integers n. The additive inverse of any integer is obtained by multiplying by -1. Thus the inverse of n is $-n$.

(b) The integers with multiplication is not a group. Although multiplication is associative and 1 is the identity, not all integers have a multiplicative inverse *in* $\mathbf{Z}$. For example, the multiplicative inverse of 10 is 0.1, but 0.1 is not an integer.

(c) The power set of any set U with the operation of symmetric difference, $\oplus$, is a group. If A and B are sets, then $A \oplus B$ is $(A \cup B) - (A \cap B)$.

We will leave it to the reader to prove that $\oplus$ is associative over $\mathcal{P}(U)$. The identity is the empty set: $A \oplus \varnothing = A$. Every set is its own inverse since $A \oplus A = \varnothing$. Note that $\mathcal{P}(U)$ is not a group with union or intersection.

Definition: Abelian Group. *A group is abelian if its operation is commutative. Most of the groups that we will discuss in this book will be abelian. The term abelian is used to honor the mathematician N. Abel (1802–29), who helped develop group theory.*

EXERCISES FOR SECTION 11.2

A Exercises

1. Discuss the analogy between the terms *generic* and *concrete* for algebraic systems and the terms *generic* and *trade* for prescription drugs.

2. Discuss the connection between groups and monoids. Is every monoid a group? Is every group a monoid?

3. Which of the following are groups?
   (a) B^* with concatenation (Example 11.2.1a)
   (b) $M_{2\times 3}(\mathbf{R})$ with matrix addition
   (c) $M_{2\times 2}(\mathbf{R})$ with matrix multiplication
   (d) the positive real numbers with multiplication
   (e) the non-zero real numbers with multiplication
   (f) $\{1, -1\}$ with multiplication
   (g) the positive integers with the operation M defined by $a\,M\,b = $ larger of a and b

4. Prove that, $\oplus$, defined by $A \oplus B = (A \cup B) - (A \cap B)$ is an associative operation on $\mathcal{P}(U)$.

5. The following problem supplies an example of a non-abelian group. A rook matrix is a matrix that has only 0's and 1's as entries such that each row has exactly one 1 and each column has exactly one 1. The term *rook matrix* is derived from the fact that each rook matrix represents the placement of n rooks on an $n \times n$ chessboard such that none of the rooks can attack one another. A rook in chess can move only vertically or horizontally, but not diagonally. Let R_n be the set of $n \times n$ rook matrices. There are six 3×3 rook matrices:

$$I = \begin{bmatrix} 1 & 0 & 0 \\ 0 & 1 & 0 \\ 0 & 0 & 1 \end{bmatrix} \quad R_1 = \begin{bmatrix} 0 & 1 & 0 \\ 0 & 0 & 1 \\ 1 & 0 & 0 \end{bmatrix} \quad R_2 = \begin{bmatrix} 0 & 0 & 1 \\ 1 & 0 & 0 \\ 0 & 1 & 0 \end{bmatrix}$$

$$F_1 = \begin{bmatrix} 1 & 0 & 0 \\ 0 & 0 & 1 \\ 0 & 1 & 0 \end{bmatrix} \quad F_2 = \begin{bmatrix} 0 & 0 & 1 \\ 0 & 1 & 0 \\ 1 & 0 & 0 \end{bmatrix} \quad F_3 = \begin{bmatrix} 0 & 1 & 0 \\ 1 & 0 & 0 \\ 0 & 0 & 1 \end{bmatrix}$$

(a) List the 2 × 2 rook matrices. They form a group under matrix multiplication. Write out the multiplication table. Is the group abelian?

(b) Write out the multiplication table for R_3. This is another group. Is it abelian?

(c) How many 4 × 4 rook matrices are there? How many $n \times n$ rook matrices are there?

6. For each of the following sets, identify the standard operation that results in a group. What is the identity of each group?

(a) The set of non-zero real numbers.

(b) The set of 2 by 3 matrices with rational entries.

B Exercise

7. Let $V = \{e, a, b, c\}$. Let $*$ be defined (partially) by $x * x = e$ for all $x \in V$. Write a complete table for $*$ so that $[V; *]$ is a group.

11.3 Some General Properties of Groups

In this section, we will present some of the most basic theorems of group theory. Keep in mind that each of these theorems tells us something about every group. We will illustrate this point at the close of the section.

Theorem 11.3.1. *The identity of a group is unique.*

One difficulty that students often encounter is how to get started in proving a theorem like this. The difficulty is certainly not in the theorem's complexity. Before actually starting the proof, we rephrase the theorem so that the implication it states is clear.

Theorem 11.3.1 (Rephrased). *If $G = [G; *]$ is a group and e is an identity of G, then no other element of G is an identity of G.*

Proof (Indirect):   Suppose that $f \in G, f \neq e$, and f is an identity of G. We will show that $f = e$, a contradiction, which completes the proof:

$$f = f * e \qquad \text{Since } e \text{ is an identity.}$$
$$= e. \qquad \text{Since } f \text{ is an identity. \#}$$

Theorem 11.3.2. *The inverse of any element of a group is unique.*

The same problem is encountered here as in the previous theorem. We will leave it to the reader to rephrase this theorem. The proof is also left to the reader to write out in detail. Here is a hint: If b and c are both inverses of a, then you can prove that $b = c$. If you have difficulty with this proof, note that we have already proven it in a concrete setting in Chapter 5.

The significance of Theorem 11.3.2 is that we can refer to the inverse of an element without ambiguity. The notation for the inverse of a is usually a^{-1} (note the exception below).

Example 11.3.1.

(a) e^{-1} is the inverse of the identity e, which always is e.
(b) $(a^{-1})^{-1}$ is the inverse of a^{-1}, which is always equal to a (see Theorem 11.3.3).
(c) $(x * y * z)^{-1}$ is the inverse of $(x * y * z)$.
(d) In a concrete group with an operation that is addition or is similar to addition, the inverse of a is usually written $-a$. For example, the inverse of $x - 3$ in the group $[\mathbf{Z}; +]$ is written $-(x - 3)$. In the group of 2 by 2 matrices over the real numbers, the inverse of

$$\begin{bmatrix} 4 & 1 \\ 1 & -3 \end{bmatrix} \text{ is written } -\begin{bmatrix} 4 & 1 \\ 1 & -3 \end{bmatrix}.$$

Theorem 11.3.3. *If a is an element of group G, then $(a^{-1})^{-1} = a$.*

Theorem 11.3.3 (Rephrased). *If a has inverse b and b has inverse c, then $a = c$.*

Proof:

$$
\begin{aligned}
a &= a * (b * c) && \text{Because } c \text{ is the inverse of } b \\
  &= (a * b) * c && \text{Why?} \\
  &= e * c && \text{Why?} \\
  &= c. && \text{\#}
\end{aligned}
$$

Theorem 11.3.4. *If a and b are elements of group G, then $(a * b)^{-1} = b^{-1} * a^{-1}$.*

Note: This theorem simply gives you a formula for the inverse of $a * b$. This formula should be familiar. In Chapter 5 we saw that if A and B are invertible matrices, then $(AB)^{-1} = B^{-1}A^{-1}$.

Proof: Let $x = b^{-1} * a^{-1}$. We will prove that x is the inverse of $a * b$.

$$
\begin{aligned}
(a * b) * x &= (a * b) * (b^{-1} * a^{-1}) \\
            &= a * (b * (b^{-1} * a^{-1})) && \text{Why?} \\
            &= a * ((b * b^{-1}) * a^{-1}) && \text{Why?} \\
            &= a * (e * a^{-1}) && \text{Why?} \\
            &= a * a^{-1} && \text{Why?} \\
            &= e.
\end{aligned}
$$

Similarly, $x * (a * b) = e$; therefore, $(a * b)^{-1} = x = b^{-1} * a^{-1}$. #

Theorem 11.3.5: Cancellation Laws. *If a, b, and c are elements of group G, both $a * b = a * c$ and $b * a = c * a$ imply that $b = c$.*

Proof: Since $a * b = a * c$, we can operate on both $a * b$ and $a * c$ on the left with a^{-1}:

$$a^{-1} * (a * b) = a^{-1} * (a * c)$$
$$(a^{-1} * a) * b = (a^{-1} * a) * c$$
$$e * b = e * c$$
$$b = c.$$

This completes the proof of the left cancellation law. The right law can be proven in exactly the same way. #

Theorem 11.3.6: Linear Equations in a Group. *If G is a group and a, $b \in G$, the equation $a * x = b$ has a unique solution, $x = a^{-1} * b$. In addition, the equation $x * a = b$ has a unique solution, $x = b * a^{-1}$.*

Proof (for $a * x = b$): $a * x = b = (a * a^{-1}) * b = a * (a^{-1} * b)$. By the cancellation law, we can conclude that $x = a^{-1} * b$. #
  If c and d are two solutions of the equation $a * x = b$, then $a * c = b = a * d$ and, by the cancellation law, $c = d$. This verifies that $a^{-1} * b$ is the only solution of $a * x = b$.

Note: Our proof of Theorem 11.3.6 was analogous to solving $4x = 9$ in the following way: $4x = 9 = ((4)(0.25))9 = 4((0.25)(9))$; therefore, by cancelling 4, $x = (0.25)(9) = 2.25$.
  If a is an element of a group G, then we establish the notation that $a * a = a^2, a * a * a = a^3, \ldots$.
  For $n \geq 0$, define a^n recursively by $a^0 = e$ and $a^n = a^{n-1} * a$ if $n \geq 1$.
  In addition, negative exponents are allowed: for example, a^{-3} is defined as $(a^3)^{-1}$. In general, if $n \geq 1$, then $(a^{-n}) = (a^n)^{-1}$.

Example 8.3.2.

(a) In the group of positive real numbers with multiplication, $5^3 = 5^2 \cdot 5 = (5^1 \cdot 5) \cdot 5 = ((5^0 \cdot 5)5) \cdot 5 = 125$.
$5^{-3} = (5^3)^{-1} = 125^{-1} = 1/125$.
(b) In a group with addition, we use a different form of notation. For example, in $[\mathbf{Z}; +]$, $a + a$ is written as $2a$, $a + a + a$ is written as $3a$, etc. The inverse of a multiple of a such as $-(a + a + a + a + a) = -(5a)$ is written as $-5a$.

Based on the definitions for exponentiation above, there are several properties that can be proven. They are all identical to the exponentiation properties from elementary algebra. We will leave the proofs of these properties to the interested reader.

Theorem 11.3.7:   Properties of Exponentiation. *If a is an element of a group G, and n and m are integers,*

(a) $(a^n)^{-1} = (a^{-1})^n$;
(b) $a^n * a^m = a^{n+m}$;
(c) $(a^n)^m = a^{nm}$.

Our final theorem is the only one that contains a hypothesis about the group in question. The theorem only applies to finite groups.

Theorem 11.3.8.   *If G is a finite group, #G = n, and a is an element of G, then there exists a positive integer m such that $a^m = e$ and $m \leq n$.*

Proof: Consider the list $a, a^2, a^3, \ldots, a^{n+1}$. Since there are $n + 1$ elements of G in this list, there must be some duplication. Suppose that $a^p = a^q$, with $p < q$. Let $m = q - p$. Then $a^m = a^{q-p} = a^q * (a^q)^{-1} = e$. Furthermore, since $1 \leq p < q \leq n + 1$, $q - p = m \leq n$. #
Consider the concrete group $[\mathbf{Z}; +]$. All of the theorems that we have stated in this section except for the last one say something about $\mathbf{Z}$. Among the facts that we conclude from the theorems about $\mathbf{Z}$ are:

Since the inverse of 5 is -5, the inverse of -5 is 5.
The inverse of $-6 + 71$ is $-(71) + -(-6) = -71 + 6$.
The solution of $12 + x = 22$ is $x = -12 + 22$.
$-4(6) + 2(6) = (-4 + 2)(6) = -2(6) = -(2(6))$.
$7(4(3)) = (7 \; 4)(3) = 28(3)$ (twenty-eight 3's).

EXERCISES FOR SECTION 11.3

A Exercises

1.   Let $[G; *]$ be a group and a be an element of G. Define $f: G \to G$ by $f(x) = a * x$.

   (a)   Prove that f is a bijection.
   (b)   On the basis of Part a, describe a set of bijections on the set of integers.

2.   Rephrase Theorem 11.3.2 and write out a clear proof.

3.   Prove by induction on n that if $a_1, a_2, \ldots, a_n$ are elements of a group G, $n \geq 2$, then $(a_1 * a_2 * \cdots * a_n)^{-1} = a_n^{-1} * a_{n-1}^{-1} * \cdots * a_1^{-1}$. Interpret this result in terms of $[\mathbf{Z}; +]$ and $[\mathbf{R}; \bullet]$.

4.   True or false? If a, b, c are elements of a group G, and $a * b = c * a$, then $b = c$. Explain your answer.

5.   Prove Theorem 11.3.7.

6. Each of the following facts can be derived by identifying a certain group and then applying one of the theorems of this section to it. For each fact, list the group and the theorem that is used.

   (a) $(1/3)5$ is the only solution of $3x = 5$.
   (b) $-(-(-18))) = -18$.
   (c) If A, B, C are 3×3 matrices over the real numbers, with $A + B = A + C$, then $A = C$.
   (d) There is only one subset of the natural numbers for which $K \oplus A = A$ for every $A \subseteq N$.

11.4 Z_n, the Integers Modulo n

In this section we introduce a collection of concrete groups, one for each positive integer, that will provide us with a wealth of examples and applications.

The Division Property for Integers. If m, $n \in \mathbf{Z}$, $n > 0$, then there exist two unique integers, q (quotient) and r (remainder), such that $m = nq + r$ and $0 \le r < n$.

Note: The division property says that if m is divided by n, you will obtain a quotient and a remainder, where the remainder is less than n. This is a fact that most elementary school students learn when they are introduced to long division. In performing the division $1986 \div 97$, you obtain a quotient of 20 and a remainder of 46; i.e., $1986 = (97)(20) + 46$.

If two numbers, a and b, share the same remainder after dividing by n, we say that they are *congruent modulo n*, denoted $a = b(mod\ n)$. For example, $13 = 38(mod\ 5)$ because $13 = (5)(2) + 3$ and $38 = (5)(7) + 3$.

Modular Arithmetic. If n is a positive integer, we define the operations of addition modulo n $(+_n)$ and multiplication modulo n $(\times_n)$ as follows. If a, $b \in \mathbf{Z}$,

$a +_n b$ = remainder after $a + b$ is divided by n, and
$a \times_n b$ = remainder after $a \times b$ is divided by n.

Notes:

(a) The result of doing arithmetic modulo n is always an integer between 0 and $n-1$ (Why?). This observation implies that $\{0, 1, \ldots, n-1\}$ is closed under modulo n arithmetic.
(b) It is always true that $a +_n b = (a + b)\ (mod\ n)$ and $a \times_n b = (a \times b)$ $(mod\ n)$. For example, $4 +_7 5 = 2 = 9\ (mod\ 7)$ and $4 \times_7 5 = 6 = 20\ (mod\ 7)$.
(c) We will use the notation $\mathbf{Z}_n$ to denote the set $\{0, 1, 2, \ldots, n-1\}$.

PROPERTIES OF MODULAR ARITHMETIC ON $\mathbf{Z}_n$

Addition modulo n is always commutative and associative; 0 is the identity for $+_n$ and every element of $\mathbf{Z}_n$ has an additive inverse.

Multiplication modulo n is always commutative and associative, and 1 is the identity for $\times_n$.

Multiplication modulo n is distributive (both right and left) over addition modulo n.

Theorem 11.4.1. *If $a \in \mathbf{Z}_n$, $a \neq 0$, then $-a = n - a$.*

Proof: $a +_n (n - a) = (a + (n - a))(mod\ n) = n(mod\ n)$. The only element of $\mathbf{Z}_n$ that is congruent to n is 0. #

Note: The algebraic properties of $+_n$ and $\times_n$ on $\mathbf{Z}_n$ are identical to the properties of addition and multiplication on $\mathbf{Z}$.

The group $\mathbf{Z}_n$. For each $n \geq 1$, $[\mathbf{Z}_n; +_n]$ is a group. Henceforth, we will use the notation $\mathbf{Z}_n$ when referring to this group. Figure 11.4.1 contains the tables for $\mathbf{Z}_1$ through $\mathbf{Z}_5$.

Example 11.4.1.

(a) We are all somewhat familiar with $\mathbf{Z}_{12}$ since the hours of the day are counted using this group, except for the fact that 12 is used in place of 0. If someone started a four-hour trip at 10 o'clock, she would arrive at $10 +_{12} 4 = 2$ o'clock. If a satellite orbits the earth every two hours and starts its first orbit at 5 o'clock, it would end its first orbit at

$+_1$	0
0	0

$+_2$	0	1
0	0	1
1	1	0

$+_3$	0	1	2
0	0	1	2
1	1	2	0
2	2	0	1

$+_4$	0	1	2	3
0	0	1	2	3
1	1	2	3	0
2	2	3	0	1
3	3	0	1	0

$+_5$	0	1	2	3	4
0	0	1	2	3	4
1	1	2	3	4	0
2	2	3	4	0	1
3	3	4	0	1	2
4	4	0	1	2	3

FIGURE 11.4.1
Addition tables for $\mathbf{Z}_n$, $1 \leq n \leq 5$

$5 +_{12} 2 = 7$ o'clock. Its seventh orbit would end at $5 +_{12} 7(2) = 7$ o'clock.

(b) Virtually all computers represent unsigned integers in binary form with a fixed number of digits. A very small computer might reserve seven bits to store the value of an integer. There are only 2^7 different values that can be stored in seven bits. Since the smallest value is 0, represented as 0000000, the maximum value will be $2^7 - 1 = 127$, repesented as 1111111. When a command is given to add two integer values, and the two values have a sum of 128 or more, overflow occurs. For example, if we try to add 56 and 95, the sum is an eight-digit binary integer 10010111. One common procedure is to retain the seven lowest-ordered digits. The result of adding 56 and 95 would be $0010111_{two} = 23 = (56 + 95) \ (mod \ 128)$. Integer arithmetic with this computer would actually be modulo 128 arithmetic.

PASCAL NOTE

With a slight adjustment, the MOD operation in Pascal can be used to do modular arithmetic. If N is a positive integer, A and B have integer values, and $A + B$ has a nonnegative value, then the expression $(A + B)$ MOD N will have a value equal to $A +_N B$. For example, $(4 + 2)$ MOD 5 has a value of 1. If $A + B$ has a negative value, the value of $(A + B)$ MOD N will often be $A +_N B - N$. For example, $(3 + (-6))$ MOD 10 will have a value of -3; while the value of $3 +_{10} (-6)$ is 7, since $-3 = 10(-1) + 7$, (where $q = -1, r = 7$). Before using the MOD operation, test it with a few calculations. Some compilers define MOD differently.

Programs to check the algebraic properties of modular arithmetic can easily be written. See Figure 11.4.2 for associativity of addition. The generality of a mathematical proof is far superior to such a program, but the program serves as a reminder of what needs to be done in the absence of a better proof.

EXERCISES FOR SECTION 11.4

A Exercises

1. Calculate:
   (a) $7 +_8 3$
   (b) $7 \times_8 3$
   (c) $4 \times_8 4$
   (d) $10 +_{12} 2$
   (e) $6 \times_8 2 +_8 6 \times_8 5$
   (f) $6 \times_8 (2 +_8 5)$
   (g) $3 \times_5 3 \times_5 3 \times_5 3 = 3^4 \ (mod \ 5)$
   (h) $2 \times_{11} 7$

```
FUNCTION Associative(n: Integer): Boolean;
(* Tests for the truth of "+ mod n is associative." *)

Var  i,j,k: Integer;    a: Boolean;

BEGIN
    i:=0; a := True;
    WHILE ( i < n ) AND a  DO
        BEGIN
            j:= 0;
            WHILE ( j < n ) AND a  DO
                BEGIN
                    k:=0;
                    WHILE ( k < n ) AND a  DO
                        IF ((((i + j) MOD n) + k ) MOD n)<>
                            ((i + ((j + k) MOD n)) MOD n)
                                THEN a := False
                                ELSE k := k + 1;
                    j := j + 1
                END;
            i := i + 1
        END;
    Associative := a
END;
```

FIGURE 11.4.2
Pascal function to test for associativity of $+_n$

2. List the additive inverses of the following elements:
   (a) 4, 6, 9 in Z_{10}
   (b) 16, 25, 40 in Z_{50}

3. In the group Z_{11}, what are:
   (a) 3(4)
   (b) 36(4)
   (c) How could you efficiently compute $m(4)$, $m \in Z$?

4. Prove the division property by induction on m. (See hint in the Solutions section.)

5. Write a Pascal program that determines, for any positive n, whether $\times_n$ is distributive over $+_n$.

6. Prove that $\{1, 2, 3, 4\}$ is a group under the operation $\times_5$.

B Exercises

7. A student is asked to solve the following equations under the requirement that he or she work solely in Z_2. List all solutions.
   (a) $x^2 + 1 = 0$.
   (b) $x^2 + x + 1 = 0$.

8. Determine the solutions of the equations in Exercise 7 in Z_5.

C Exercise

9. Write a Pascal function which determines the remainder obtained by
    dividing a by b where $b \neq 0$. Do this by:
    (a) using the MOD operation
    (b) using only addition and subtraction

11.5 Subsystems

The subsystem is a fundamental concept of algebra at the universal level.

Definition: Subsystem. *If* $[V; *_1, \ldots, *_n]$ *is an algebraic system of a
certain kind and W is a subset of V, then W is a subsystem of V if*
$[W; *_1, \ldots, *_n]$ *is an algebraic system of the same kind as V. The usual
notation for "W is a subsystem of V" is* $W \leq V$.

Since the definition of a subsystem is at the universal level, we can cite
examples of the concept of subsystem at both the axiomatic and concrete
level.

Example 11.5.1.

(a) (Axiomatic) If $[G; *]$ is a group, and H is a subset of G, then H is a
    subgroup of G if $[H; *]$ is a group.
(b) (Concrete) $U = \{-1, 1\}$ is a subgroup of $[\mathbf{R}^*; .]$. Take the time now
    to write out the multiplication table of U and convince yourself that
    $[U; .]$ is a group.
(c) (Concrete) The even integers, $2\mathbf{Z} = \{2k : k \text{ is an integer}\}$ is a subgroup
    of $[\mathbf{Z}; +]$. Convince yourself of this fact.
(d) (Concrete) The set of nonnegative integers is not a subgroup of $[\mathbf{Z}; +]$.
    All of the group axioms are true for this subset except one: no positive
    integer has a positive additive inverse. Therefore, the inverse property
    is not true. Note that every group axiom must be true for a subset to
    be a subgroup.
(e) (Axiomatic) If M is a monoid and P is a subset of M, then P is a
    submonoid of M if P is a monoid.
(f) (Concrete) If B^* is the set of strings of 0's and 1's of length zero or
    more with the operation of concatenation, then two submonoids of S
    are: (i) the set of strings of even length, and (ii) the set of strings that
    contain no 0's. The set of strings of length less than 100 is not a
    submonoid because it isn't closed under concatenation. Why isn't the
    set of strings of length 100 or more not a submonoid of B^*?

For the remainder of this section, we will concentrate on the
properties of subgroups. The first order of business is to establish a

systematic way of determining whether a subset of a group is a subgroup.

Theorem/Algorithm 11.5.1. *To determine whether H, a subset of group* $[G; *]$, *is a subgroup, it is sufficient to prove:*

(a) *H is closed under* $*$; *i.e., a, b in H implies that* $a * b$ *is also in H;*
(b) *H contains the identity element for* $*$; *and*
(c) *H contains the inverse of every one of its elements; i.e., if a is in H, then* a^{-1} *is also in H.*

Proof: Our proof consists of verifying that if the three properties above are true, then all the axioms of a group are true for $[H; *]$. By Condition a, $*$ can be considered an operation on H. The associative, identity, and inverse properties are the axioms that are needed. The identity and inverse properties are true by Conditions b and c, respectively, leaving only the associative property. Since, $[G; *]$ is a group, $a * (b * c) = (a * b) * c$ for all a, b, c in G. Certainly, if this equation is true for all choices of three elements from G, it will be true for all choices of three elements from H, since H is a subset of G. #

For every group with at least two elements, there are at least two subgroups: they are the whole group and $\{e\}$. Since these two are automatic, they are not considered very interesting and are called the *improper subgroups* of the group. All other subgroups, if there are any, are called *proper subgroups*.

We can apply Theorem 11.5.1 at both the concrete and axiomatic levels.

Examples 11.5.2.

(a) (Concrete) We can verify that $2\mathbf{Z} \leq \mathbf{Z}$, as stated in Example 11.5.1. Whenever you want to discuss a subset, you must find some convenient way of describing its elements. An element of $2\mathbf{Z}$ can be described as 2 times an integer; i.e., $a \in 2\mathbf{Z}$ is equivalent to $(\exists k)_{\mathbf{Z}}(a = 2k)$. Now we can verify that the three conditions of Theorem 11.5.1 are true for $2\mathbf{Z}$. First, if $a, b \in 2\mathbf{Z}$, then there exist $j, k \in \mathbf{Z}$ such that $a = 2j$ and $b = 2k$. Then $a + b = 2j + 2k = 2(j + k)$. Since $j + k$ is an integer, $a + b$ must be an element of $2\mathbf{Z}$. Second, zero belongs to $2\mathbf{Z}$ $(0 = 2(0))$. Finally, if $a \in 2\mathbf{Z}$ and $a = 2k$, $-a = -(2k) = 2(-k)$; therefore, $-a \in 2\mathbf{Z}$. By theorem 11.5.1, $2\mathbf{Z} \leq \mathbf{Z}$.

How would this argument change if you were asked to prove that $3\mathbf{Z} \leq \mathbf{Z}$? Or $n\mathbf{Z} \leq \mathbf{Z}$, $n \geq 2$?

(b) (Concrete) We can prove that $H = \{0, 3, 6, 9\}$ is a subgroup of Z_{12}. First, for each ordered pair (a, b) in $H \times H$, $a +_{12} b$ is in H. This can be checked without too much trouble since $\#(H \times H) = 16$. Thus we can conclude that H is closed under $+_{12}$. Second, $0 \in H$. Third, $-0 = 0$, $-3 = 9$, $-6 = 6$, and $-9 = 3$. Therefore, the inverse of each element of H belongs to H.

(c) (Axiomatic) If H and K are both subgroups of a group G, then $H \cap K$ is a subgroup of G. To justify this statement, we have no concrete information to work with, only the facts that $H \leq G$ and $K \leq G$. Our proof that $H \cap K \leq G$ reflects this and is an exercise in applying the definitions of intersection and subgroup. (i) If a and b are elements of $H \cap K$, then a and b both belong to H, and since $H \leq G$, $a * b$ must be an element of H. Similarly, $a * b \in K$; therefore, $a * b \in H \cap K$. (ii) The identity of G must belong to both H and K; hence it belongs to $H \cap K$. (iii) If $a \in H \cap K$, then $a \in H$, and since $H \leq G$, $a^{-1} \in K$, which implies that a^{-1} is an element of $H \cap K$. Again, by the theorem, $H \cap K \leq G$.

Now that this fact has been established, we can apply it to any pair of subgroups of any group. For example, since $2\mathbf{Z}$ and $3\mathbf{Z}$ are both subgroups of $[\mathbf{Z};+]$, $2\mathbf{Z} \cap 3\mathbf{Z}$ is also a subgroup of $\mathbf{Z}$. Note that if $a \in 2\mathbf{Z} \cap 3\mathbf{Z}$, then a must have a factor of 3; i.e., there exists $k \in \mathbf{Z}$ such that $a = 3k$. In addition, a must be even, therefore k must be even. There exists $j \in \mathbf{Z}$ such that $k = 2j$, therefore $a = 3(2j)$. This indicates that $2\mathbf{Z} \cap 3\mathbf{Z} \subseteq 6\mathbf{Z}$. The opposite containment can easily be established; therefore, $2\mathbf{Z} \cap 3\mathbf{Z} = 6\mathbf{Z}$.

Given a finite group, we can apply Theorem 11.3.7 to obtain a simpler condition for a subset to be a subgroup.

Theorem/Algorithm 11.5.2. *If $[G; *]$ is a finite group, H is a non-empty subset of G, and you can verify that H is closed under $*$, then H is a subgroup of G.*

Proof: In this proof, we demonstrate that Conditions b and c of Theorem 11.5.1 follow from the closure of H under $*$, which is Condition a. First, select any element of H; call it a. The powers of a : $a^1, a^2, \ldots$, are all in H by the closure property. Therefore, by Theorem 11.3.7, there exists m, $1 \leq m \leq \#G$ such that $a^m = e$; hence $e \in H$. To prove that c is true, we let a be any element of H. If $a = e$, then a^{-1} is in H since $e^{-1} = e$. If $a \neq e$, $a^m = e$ for some m between 2 and $\#G$. $e = a^m = a^{m-1} * a$. Therefore, $a^{-1} = a^{m-1}$, which belongs to H since $m - 1 \geq 1$. #

Example 11.5.3.   To determine whether $H_1 = \{0, 5, 10\}$ and $H_2 = \{0, 4, 8, 12\}$ are subgroups of $\mathbf{Z}_{15}$, we need only write out the addition tables (modulo 15) for these sets:

$+_{15}$	0	5	10
0	0	5	10
5	5	10	0
10	10	0	5

$+_{15}$	0	4	8	12
0	0	4	8	12
4	4	8	12	1
8	8	12	1	5
12	12	1	5	9

Note that H_1 is a subgroup of $\mathbf{Z}_{15}$. Since the interior of the addition table for H_2 contains elements that are outside of H_2, H_2 is not a subgroup of $\mathbf{Z}_{15}$.

One kind of subgroup that merits special mention due to its simplicity is the *cyclic subgroup*.

Definition: Cyclic Subgroup Generated by an Element. *If G is a group and* $a \in G$, *the cyclic subgroup generated by a, (a), is the set of powers of a and their inverses:*

$$(a) = \{a^n : n \in \mathbf{Z}\}$$

A subgroup H is cyclic if there exists $a \in H$ such that $H = (a)$.

Notes:

(a) If $(a) = G$, then we say that G is generated by a, a generates G, and G is cyclic. We will study cyclic groups in more detail in Chapter 15.
(b) If the operation on G is additive, then $(a) = \{na : n \in \mathbf{Z}\}$.

Example 11.5.4.

(a) In $[\mathbf{R}^+; .]$, $(2) = \{2^n : n \in Z\} = \{. . . , 0.125, 0.25, 0.5, 1, 2, 4, 8, 16, . . .\}$.
(b) In $\mathbf{Z}_{15}$, $(6) = \{0, 3, 6, 9, 12\}$. If G is finite, you need list only the positive powers of a up to the first occurrence of the identity to obtain all of (a). In $\mathbf{Z}_{15}$, the multiples of 6 are 6, $2(6) = 12$, $3(6) = 3$, $4(6) = 9$, and $5(6) = 0$. Note that $\{0, 3, 6, 9, 12\}$ is also (3), (9), and (12). This shows that a cyclic subgroup can have many different generators.

   If you want to list the cyclic subgroups of a group, the following theorem can save you some time.

Theorem 11.5.3.   *If a is an element of group G, then* $(a) = (a^{-1})$.
This is an easy way of seeing that (9) in $\mathbf{Z}_{15}$ equals (6), since $-6 = 9$.

EXERCISES FOR SECTION 11.5

A Exercises

1. Which of the following subsets of the real numbers is a subgroup of $[\mathbf{R}; +]$?

   (a) the rational numbers
   (b) the positive real numbers
   (c) $\{k/2 : k$ is an integer$\}$
   (d) $\{2^k : k$ is an integer$\}$
   (e) $\{x : -100 \leq x \leq 100\}$

2. Describe in simpler terms the following subgroups of **Z**:
   (a) $5\mathbf{Z} \cap 4\mathbf{Z}$
   (b) $4\mathbf{Z} \cap 6\mathbf{Z}$ (be careful)
   (c) the only finite subgroup of **Z**

3. Find at least two proper subgroups of R_3, the set of 3 by 3 rook matrices (see Exercise 4 of Section 11.2).

4. Where should you place the following in Figure 11.5.1?
   (a) e
   (b) a^{-1}
   (c) xy

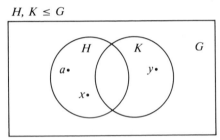

$H, K \leq G$

FIGURE 11.5.1
Exercise

5. (a) List the cyclic subgroups of $\mathbf{Z}_6$ and draw an ordering diagram for the relation "is a subset of" on these subgroups.
   (b) Do the same for $\mathbf{Z}_{12}$.
   (c) Do the same for $\mathbf{Z}_8$.
   (d) On the basis of your results in Parts a, b, and c, what would you expect if you did the same with $\mathbf{Z}_{24}$?

B Exercises

6. Subgroups generated by subsets of a group. The concept of a cyclic subgroup is a special case of the concept that we will discuss here. Let $[G; *]$ be a group and S a nonempty subset of G. Define the set (S) recursively by: (i) if a is in S, then a is in (S); (ii) if $a, b \in (S)$, then $a * b \in (S)$; and (iii) if $a \in (S)$, then $a^{-1} \in (S)$.
   (a) Prove that the identity of G is in (S). By its definition, it should be clear that $(S) \leq G$; however, the identity of G isn't in (S) by the definition.
   (b) What is $(\{9, 15\})$ in $[\mathbf{Z}; +]$?
   (c) Prove that if $H \leq G$ and $S \subseteq H$, then $(S) \leq H$. This proves that (S) is contained in every subgroup of G that contains S; i.e., $(S) = \cap \{H : S \subseteq H \text{ and } H \leq G\}$.

(d) Describe $(\{0.5, 3\})$ in $[\mathbf{R}^+; .]$; in $[\mathbf{R}; +]$.

(e) If $j, k \in \mathbf{Z}$, $(\{j, k\})$ is a cyclic subgroup of $\mathbf{Z}$. In terms of j and k, what is a generator of $(\{j, k\})$?

7. Prove that if $H, K \leq G$ and $H \cup K = G$, then $H = G$ or $K = G$. (Hint: Use an indirect argument.)

11.6 Direct Products

Our second universal algebraic concept lets us look in the opposite direction from subsystems. Direct products allow us to create larger systems. In the following definition, we avoid complicating the notation by not specifying how many operations the systems have.

Definition: Direct product. *If* $[V_1; *_1, \#_1, \ldots], [V_2; *_2, \#_2, \ldots], \ldots, [V_n; *_n, \#_n, \ldots]$ *are algebraic systems of the same kind, then the direct product of these systems is* $V = V_1 \times V_2 \times \cdots \times V_n$, *with operations defined below. The elements of V are n-tuples of the form* $(a_1, \ldots, a_n)$, *where* $a_k \in V_k, k = 1, \ldots, n$. *The systems* $V_1, V_2, \ldots, V_n$ *are called the factors of V. There are as many operations on V as there are on the factors. Each of these operations is defined componentwise. For example, if*

$(a_1, \ldots, a_n), (b_1, \ldots, b_n) \in V$,

$(a_1, \ldots, a_n) * (b_1, \ldots, b_n) = (a_1 *_1 b_1, \ldots, a_n *_n b_n)$

$(a_1, \ldots, a_n) \# (b_1, \ldots, b_n) = (a_1 \#_1 b_1, \ldots, a_n \#_n b_n), \ldots$

Example 11.6.1. Consider the monoids $\mathbf{N}$ (the set of natural numbers with addition) and $B*$ (the set of finite strings of 0's and 1's with concatenation). The direct product of $\mathbf{N}$ with $B*$ is a monoid. We illustrate its operation, which we will denote by $*$, with examples:

$(4, 001) * (3, 11) = (4 + 3, (001)(11)) = (7, 00111)$

$(0, 11010) * (3, 01) = (3, 1101001)$

$(0, \lambda) * (129, 00011) = (0 + 129, (\lambda)(00011)) = (129, 00011)$

$(\lambda = \text{null string})$

$(2, 01) * (8, 10) = (10, 0110)$, and

$(8, 10) * (2, 01) = (10, 1001)$.

Note that our new monoid is not commutative. What is the identity for $*$?

Notes:

(a) On notation. If two or more consecutive factors in a direct product are identical, it is common to combine them using exponential notation.

For example, $\mathbf{Z} \times \mathbf{Z} \times \mathbf{R}$ can be written $\mathbf{Z}^2 \times \mathbf{R}$, and $\mathbf{R} \times \mathbf{R} \times \mathbf{R}$ $\times \mathbf{R}$ can be written $\mathbf{R}^4$. This is purely a notational convenience; no exponentiation is really taking place.

(b) In our definition of a direct product, the operations are called *componentwise operations*, and they are indeed operations on V. Consider $*$ above. If two n-tuples, a and b, are selected from V, the first components of a and b, a_1 and b_1, are operated on with $*_1$ to obtain $a_1 *_1 b_1$, the first component of $a * b$. Note that since $*_1$ is an operation on V_1, $a_1 *_1 b_1$ is an element of V_1. Similarly, all the components of $a * b$ as they are defined belong to their proper sets.

One significant fact about componentwise operations is that the components of the result can all be computed at the same time (concurrently). The time required to compute in a direct product can be reduced to a length of time that is not much longer than the maximum amount of time needed to compute in the factors (see Figure 11.6.1).

(c) A direct product of algebraic systems is not always an algebraic system of the same type as its factors. This is due to the fact that certain axioms that are true for the factors may not be true for the set of n-tuples. This situation does not occur with groups, however. You will find that whenever a new type of algebraic system is introduced, call it type T, one of the first theorems that is usually proven, if possible, is that the direct product of two or more of systems of type T is a system of type T.

Theorem 11.6.1.   *The direct product of two or more groups is a group; i.e., the algebraic properties of a system obtained by taking the direct product of two or more groups includes the group axioms.*

We will only present the proof of this theorem for the direct product of two groups. Some slight revisions can be made to obtain a proof for any number of factors.

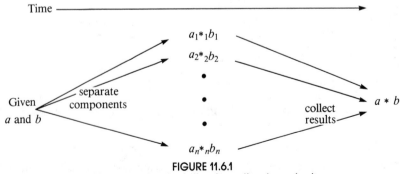

FIGURE 11.6.1
Concurrent calculation in a direct product

Proof: Stating that the direct product of two groups is a group is a short way of saying that if $[G_1; *_1]$ and $[G_1; *_1]$ are groups, then $[G_1 \times G_2; *]$ is also a group, where $*$ is the componentwise operation on $G_1 \times G_2$.

Associativity of $*$: If $a, b, c \in G_1 \times G_2$,

$$
\begin{aligned}
a * (b * c) &= (a_1, a_2) * ((b_1, b_2) * (c_1, c_2)) \\
&= (a_1, a_2) * (b_1 *_1 c_1, b_2 *_2 c_2) \\
&= (a_1 *_1 (b_1 *_1 c_1), a_2 *_2 (b_2 *_2 c_2)) \\
&= ((a_1 *_1 b_1) *_1 c_1, (a_2 *_2 b_2) *_2 c_2) \\
&= (a_1 *_1 b_1, a_2 *_2 b_2) * (c_1, c_2) \\
&= ((a_1, a_2) * (b_1, b_2)) * (c_1, c_2) \\
&= (a * b) * c.
\end{aligned}
$$

An identity for $*$: As you might expect, if e_1 and e_2 are identities for G_1 and G_2, respectively, then $e = (e_1, e_2)$ is the identity for $G_1 \times G_2$. If $a \in G_1 \times G_2$,

$$
\begin{aligned}
a * e &= (a_1, a_2) * (e_1, e_2) \\
&= (a_1 *_1 e_1, a_2 *_2 e_2) \\
&= (a_1, a_2) \\
&= a.
\end{aligned}
$$

Similarly, $e * a = a$.

Inverses in $G_1 \times G_2$: The inverse of an element is determined componentwise: $a^{-1} = (a_1, a_2)^{-1} = (a_1^{-1}, a_2^{-1})$. To verify, we compute $a * a^{-1}$:

$$
\begin{aligned}
a * a^{-1} &= (a_1, a_2) * (a_1^{-1}, a_2^{-1}) \\
&= (a_1 *_1 a_1^{-1}, a_2 *_2 a_2^{-1}) \\
&= (e_1, e_2) \\
&= e.
\end{aligned}
$$

Similarly, $a^{-1} * a = e$. #

Example 11.6.2.

(a) If $n \geq 2$, $\mathbf{Z}_2^n$, the direct product of n factors of $\mathbf{Z}_2$ is a group with 2^n elements. We will take a closer look at $\mathbf{Z}_2^3 = \mathbf{Z}_2 \times \mathbf{Z}_2 \times \mathbf{Z}_2$. The elements of this group are triples of zeros and ones. Since the operation on $\mathbf{Z}_2$ is $+_2$, we will use the symbol $+$ for the operation on $\mathbf{Z}_2^3$. Two of the eight triples in the group are $a = (1, 0, 1)$ and $b = (0, 0, 1)$. Their "sum" is $a + b = (1 +_2 0, 0 +_2 0, 1 +_2 1) = (1, 0, 0)$. One interesting fact about this group is that each element is its own inverse. For example $a + a = (1, 0, 1) + (1, 0, 1) = (0, 0, 0)$; therefore $-a = a$. We use the additive notation for the inverse of a because we are using a form of addition. Note that $\{(0, 0, 0), (1, 0, 1)\}$ is a subgroup of $\mathbf{Z}_2^3$. Write out the "addition" table for this set and apply Theorem 11.5.2. The same can be said for any set consisting of $(0, 0, 0)$ and another element of $\mathbf{Z}_2^3$.

(b) The direct product of the positive real numbers with the integers modulo 4, $\mathbf{R}^+ \times \mathbf{Z}_4$ is an infinite group since one of its factors is infinite. The operations on the factors are multiplication and modular addition, so we will select the neutral symbol # for the operation on $\mathbf{R}^+ \times \mathbf{Z}_4$. If $a = (4, 3)$ and $b = (0.5, 2)$, then a # b = $(4, 3)$ # $(0.5, 2)$ = $(4 \times 0.5, 3 +_4 2)$ = $(2, 1)$, $b^2 = b$ # b = $(0.5, 2)$ # $(0.5, 2)$ = $(0.25, 0)$, $a^{-1} = (4^{-1}, -3) = (0.25, 1)$ and $b^{-1} = (0.5^{-1}, -2) = (2, 2)$.

It would be incorrect to say that $\mathbf{Z}_4$ is a subgroup of $\mathbf{R}^+ \times \mathbf{Z}_4$, but there is a subgroup of the direct product that closely resembles $\mathbf{Z}_4$. It is $\{(1, 0), (1, 1), (1, 2), (1, 3)\}$. It's table is

#	(1, 0)	(1, 1)	(1, 2)	(1, 3)
(1, 0)	(1, 0)	(1, 1)	(1, 2)	(1, 3)
(1, 1)	(1, 1)	(1, 2)	(1, 3)	(1, 0)
(1, 2)	(1, 2)	(1, 3)	(1, 0)	(1, 1)
(1, 3)	(1, 3)	(1, 0)	(1, 1)	(1, 2)

Imagine erasing (1,) throughout the table and writing $+_4$ in place of #. What would you get? We will explore this phenomenon in detail in the next section.

The whole direct product could be visualized as four parallel half-lines labeled 0, 1, 2, and 3 (Figure 11.6.2). On the kth line, the point that lies x units to the right of the zero mark would be (x, k). The set $\{(2^n, n(1)); n \in \mathbf{Z}\}$, which is plotted on Figure 11.6.2, is a subgroup of $\mathbf{R}^+ \times \mathbf{Z}_4$. What cyclic subgroup is it? The answer: $((2, 1))$ or $((\frac{1}{2}, 3))$.

A more conventional direct product is $\mathbf{R}^2$, the direct product of two factors of $[\mathbf{R}; +]$. The operation on $\mathbf{R}^2$ is componentwise addition; hence we will

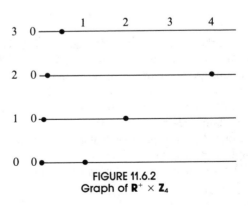

FIGURE 11.6.2
Graph of $\mathbf{R}^+ \times \mathbf{Z}_4$

use + as the operation symbol for this group. You should be familiar with this operation, since it is identical to addition of 2×1 matrices. The Cartesian coordinate system can be used to visualize $\mathbf{R}^2$ geometrically. We plot the pair (s, t) on the plane in the usual way: s units along the x axis and t units along the y axis. There is a variety of different subgroups of $\mathbf{R}^2$, a few of which are: (1) $\{(x, 0) : x \in R\}$, all of the points on the x axis; (2) $\{(x, y) : 2x - y = 0\}$, all of the points that are on the line $2x - y = 0$; (3) If a, $b \in \mathbf{R}$, $\{(x, y) : ax + by = 0\}$ (the first two subgroups are special cases of this one, which represents any line that passes through the origin); (4) $\{(x, y) : 2x - y = k, k \in \mathbf{Z}\}$, a set of lines that are parallel to $2x - y = 0$; and (5) $\{(n, 3n) : n \in \mathbf{Z}\}$, which is the only countable subgroup that we have listed.

We will leave it to the reader to verify that these sets are subgroups. We will only point out how the fourth example, call it H, is closed under "addition". If $a = (p, q)$ and $b = (s, t)$ and both belong to H, then $2p - q = j$ and $2s - 1 = k$, where both j and k are integers. $a + b = (p, q) + (s, t) = (p + s, q + t)$. We can determine whether $a + b$ belongs to H by deciding whether or not $2(p + s) - (q + t)$ is an integer: $2(p + s) - (q + t) = 2p + 2s - q - t = (2p - q) + (2s - t) = j + k$, which is an integer. This completes a proof that H is closed under the operation of $\mathbf{R}^2$.

Several useful facts can be stated in regards to the direct product of two or more groups. We will combine them into one theorem, which we will present with no proof. Parts a and c were derived for $n = 2$ in the proof of Theorem 11.6.1.

Theorem 11.6.2. *If $G = G_1 \times G_2 \times \cdots \times G_n$ is a direct product of n groups and $(a_1, a_2, \ldots, a_n) \in G$, then:*

(a) $(a_1, a_2, \ldots, a_n)^{-1} = (a_1^{-1}, a_2^{-1}, \ldots, a_n^{-1})$.
(b) $(a_1, a_2, \ldots, a_n)^m = (a_1^m, a_2^m, \ldots, a_n^m)$.
(c) *The identity of G is $(e_1, e_2, \ldots, e_n)$, where e_j is the identity of G_j.*
(d) *G is abelian if and only if each of the factors $G_1, \ldots, G_n$ is abelian.*
(e) *If $H_1, H_2, \ldots, H_n$ are subgroups of the corresponding factors, then $H_1 \times H_2 \times \cdots \times H_n$ is a subgroup of G.*

Not all subgroups of a direct product are obtained as in Part c of Theorem 11.6.2. For example, $\{(n, n) : n \in \mathbf{Z}\}$ is a subgroup of $\mathbf{Z}^2$, but is not a direct product of two subgroups of $\mathbf{Z}$.

Example 11.6.3. Using the identity $(x + y) + x = y$ in $\mathbf{Z}_2^n$, we can devise a scheme for representing a symetrically linked list using only one link field. A symetrically linked list is a list in which each node contains a pointer to its immediate successor and its immediate predecessor (see Figure 11.6.3). If the pointers are n-digit binary addresses, then each pointer can be taken as an element of $\mathbf{Z}_2^n$. Lists of this type can be accomplished using cells with only one "address." In place of a left and a right pointer, the only "link" is the value of the sum (left link) + (right link). All standard list operations (merge,

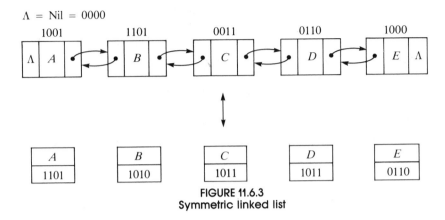

FIGURE 11.6.3
Symmetric linked list

insert, delete, traverse, etc.) are possible with this structure provided that you
know the value of the nil pointer and the address of the first (leftmost) cell.
Since first $\uparrow$ .left = nil, we can recover first $\uparrow$. right by adding the value of
nil: (nil + first $\uparrow$ .right) + nil = first $\uparrow$ .right. Now if we temporarily retain
the address of the first cell, we can recover the address of the third cell. The
link field of the second node contains the sum second $\uparrow$ .left + second $\uparrow$.
right = first + third. Therefore third = the link field of second + first. We
no longer need the address of the first cell, only the second and third, to
recover the fourth address, etc.

Algorithm 11.5.1.   Given a symmetric list represented as in Example 11.6.3,
a traversal of the list is accomplished as follows, where first is the address of
the first cell.

(1) yesterday :=nil
(2) today :=first
(3) While today $\neq$ nil do
    (3.1) Write(today)
    (3.2) tomorrow := today $\uparrow$ .link + yesterday
    (3.3) yesterday := today
    (3.4) today := tomorrow.

At any point in this algorithm it would be quite easy to insert a cell between
today and tomorrow. Can you discover how this would be accomplished?

EXERCISES FOR SECTION 11.6

A Exercises

1.   Write out the group table of $\mathbf{Z}_2 \times \mathbf{Z}_3$ and find the two proper subgroups
of this group.

2. List more examples of proper subgroups of $\mathbf{R}^2$ that are different from the ones in Example 11.6.2c.

3. Algebraic properties of the n-cube:
    (a) The four elements of $\mathbf{Z}_2^2$ can be visualized geometrically as the four corners of the 2-cube (see Figure 9.4.5). Algebraically describe the statements:
    (i) Corners a and b are adjacent.
    (ii) Corners a and b are diagonally opposite one another.
    (b) The eight elements of $\mathbf{Z}_2^3$ can be visualized as the eight corners of the 3-cube. One face contains $\mathbf{Z}_2 \times \mathbf{Z}_2 \times \{0\}$ and the opposite face contains the remaining four elements so that $(a, b, 1)$ is behind $(a, b, 0)$. As in Part a, describe Statements i and ii algebraically.
    (c) If you could imagine a geometric figure similar to the square or cube in n dimensions, and its corners were labeled by elements of $\mathbf{Z}_2^n$ as in Parts a and b, how would Statements i and ii be expressed algebraically?

4. (a) Suppose that you were to be given a group $[G; *]$ and asked to solve the equation $x * x = e$. Without knowing the group, can you anticipate how many solutions there will be?
    (b) Answer the same question as Part a for the equation $x * x = x$.

5. Which of the following sets are subgroups of $\mathbf{Z} \times \mathbf{Z}$? Give a reason for any negative answers.
    (a) $\{0\}$
    (b) $\{(2j, 2k) : j, k \in \mathbf{Z}\}$
    (c) $\{(2j + 1, 2k) : j, k \in \mathbf{Z}\}$
    (d) $\{(n, n^2) : n \in \mathbf{Z}\}$
    (e) $\{(j, k) : j + k \text{ is even}\}$

6. Determine the following values in group $\mathbf{Z}_3 \times \mathbf{R}^*$:
    (a) $(2, 1) * (1, 2)$
    (b) the identity element
    (c) $(1, 1/2)^{-1}$

11.7 Isomorphisms

The following informal definition of isomorphic systems should be memorized. No matter how technical a discussion about isomorphic systems becomes, keep in mind that this is the essence of the concept.

Definition: Isomorphic Systems/Isomorphism. *Two algebraic systems are isomorphic if there exists a translation rule between them so that any true statement in one system can be translated to a true statement in the other system. The translation rule is called an isomorphism.*

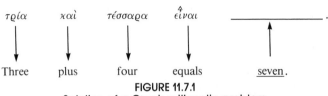

FIGURE 11.7.1
Solution of a Greek arithmetic problem

Example 11.7.1.   Imagine that you are an eight-year-old child who has been reared in an English-speaking family, has moved to Greece, and has been placed in a Greek school. Suppose that your new teacher asks the class to do an addition problem that has been written out in Greek, such as the one in Figure 11.7.1. The natural thing for you to do is to pull out your Greek-English/English-Greek dictionary and translate the Greek words to English. After you've solved the problem, you can consult the same dictionary to obtain the proper Greek word that the teacher wants. Although this is not the recommended method of learning a foreign language, it will surely yield the correct answer to the problem. Mathematically, we may say that the system of Greek integers with addition (kai) is isomorphic to English integers with addition (plus). The problem of translation between natural languages is more difficult than this though, because two complete natural languages are not isomorphic, or at least the isomorphism between them is not contained in a simple dictionary.

Example 11.7.2:   Pascal Sets. In this example, we will describe how set type variables, such as the ones in Pascal, can be implemented on a computer. We will describe the two systems first and then describe the isomorphism between them.

System 1:   The Power Set of $\{1, 2, 3, 4, 5\}$ with the Operation of Union $\cup$. For simplicity, we will only discuss union. However, the other operations are implemented in a similar way.

System 2:   Strings of Five Bits of Computer Memory Together with An OR Gate. Individual bit values are either zero or one, so the elements of this system can be visualized as sequences of 5 zeros and ones. An OR gate, Figure 11.7.2, is a small piece of computer hardware that accepts two bit values at any one time and outputs either a zero or one, depending on the inputs. The output of an OR gate is one, except when the two bit values that it accepts are both zero, in which case the output is zero. The operation on this system actually consists of sequentially inputting the values of two bit strings into the OR gate. The result will be a new string of five zeros and ones. An alternate method of operating in this system is to use five OR gates and to input corresponding pairs of bits from the input strings into the gates concurrently.

System 1:

System 2:

$[\mathcal{P}(\{1,2,3,4,5\});\ \cup]$

Strings of 5 bits with OR

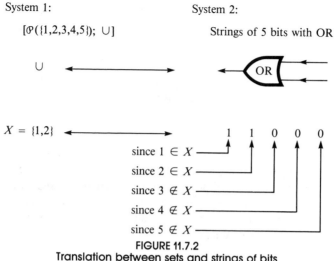

$X = \{1,2\}$

1 1 0 0 0

since $1 \in X$
since $2 \in X$
since $3 \notin X$
since $4 \notin X$
since $5 \notin X$

FIGURE 11.7.2
Translation between sets and strings of bits

The Isomorphism: Since each system has only one operation, it is clear that union and the OR gate translate into one another. The translation between sets and bit strings is easiest to describe by showing how to construct a set from a bit string. If $a_1 a_2 a_3 a_4 a_5$ is a bit string in System 2, the set that it translates to contains the number k if and only if a_k equals 1. For example, 10001 is translated to the set $\{1, 5\}$, while the set $\{1, 2\}$ is translated to 11000. Now imagine that your computer is like the child who knows English and must do a Greek problem. To execute a program that has code that includes the set expression $\{1, 2\} \cup \{1, 5\}$, it will follow the same procedure as the child to obtain the result, as shown in Figure 11.7.3.

Example 11.7.3:   Multiplying without Doing Multiplication. One isomorphism that has been used for years is the one between $[\mathbf{R}^{+}\,;\bullet]$ and $[\mathbf{R};+]$. Until the 1970s, when the price of calculators dropped, multiplication and exponentiation were performed with an isomorphism between these systems. The isomorphism ($\mathbf{R}^{+}$ to $\mathbf{R}$) between the two groups is that $\bullet$ is translated into $+$ and any positive real number a is translated to the logarithm of a. To translate from $\mathbf{R}$ to $\mathbf{R}^{+}$, you invert the logarithm function. If base ten logarithms are used, an element of $\mathbf{R}$, b, will be translated to 10^{b}. In precalculator days, the translation was done with a table of logarithms or with a slide rule. An example of how the isomorphism is used appears in Figure 11.7.4.

$\{1, 2\}$ $\cup$ $\{1, 5\}$ $=$ $\{1, 2, 5\}$

11000   OR   10001 $=$ 11001

FIGURE 11.7.3
Translation of a problem in set theory

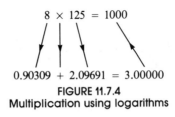

FIGURE 11.7.4
Multiplication using logarithms

The following definition of an isomorphism between two groups is a more formal one that appears in most abstract algebra texts. At first glance, it appears different, but we will point out how it is really a slight variation on the informal definition. It is the common definition because it is easy to apply; i.e., given a function, this definition tells you what to do to determine whether that function is an isomorphism.

PROCEDURE FOR SHOWING GROUPS ARE ISOMORPHIC

Definition: Group Isomorphism. *If $[G_1 ; *_1]$ and $[G_2 ; *_2]$ are groups, f: $G_1 \rightarrow G_2$ is an isomorphism from G_1 into G_2 if: (a) f is a bijection, and (b) $f(a *_1 b) = f(a) *_2 f(b)$ for all a, b in G_1. If such a function exists, then G_1 is isomorphic to G_2.*

Notes:

(a) There could be several different isomorphisms between the same pair of groups. Thus, if you are asked to demonstrate that two groups are isomorphic, your answer need not be unique.

(b) Any application of this definition requires a procedure outlined in Figure 11.7.5.

The first condition, that an isomorphism be a bijection, reflects the fact that every true statement in the first group should have exactly one corresponding true statement in the second group. This is exactly why we run into difficulty in translating between two natural languages. To see how Condition b of the formal definition is consistent with the informal definition, consider the function $L:\mathbf{R}^+ \rightarrow \mathbf{R}$ defined by $L(x) = \log_{10}x$. The translation diagram between $\mathbf{R}^+$ and $\mathbf{R}$ for the multiplication problem $a \cdot b$ appears in Figure 11.7.6. We arrive at the same result by computing $L^{-1}(L(a) + L(b))$ as we do by computing $a \cdot b$. If we apply the function L to the two results, we get the same image:

$$L(a \cdot b) = L(L^{-1}(L(a) + L(b))) = L(a) + L(b), \qquad (11.7a)$$

since $L(L^{-1}(x)) = x$. Note that 11.7a is exactly Condition b of the formal definition applied to the two groups $\mathbf{R}^+$ and $\mathbf{R}$.

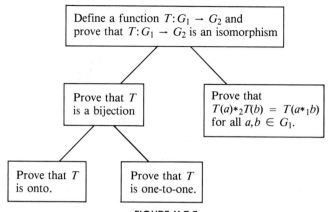

FIGURE 11.7.5
Steps in proving G_1 and G_2 are isomorphic

$$a \cdot b = L^{-1}(L(a \cdot b))$$

$$L(a) + L(b) = L(a \cdot b)$$

FIGURE 11.7.6
Multiplication using logarithms—general situation

Example 11.7.4. If

$$G = \left\{ \begin{bmatrix} 1 & a \\ 0 & 1 \end{bmatrix} : a \in \mathbf{R} \right\},$$

with matrix multiplication, $[\mathbf{R}; +]$ is isomorphic to G. Our translation rule is the function $f : \mathbf{R} \rightarrow G$ defined by

$$f(a) = \begin{bmatrix} 1 & a \\ 0 & 1 \end{bmatrix}.$$

Since groups have only one operation, there is no need to state explicitly that addition is translated to matrix multiplication. That f is a bijection is clear from its definition. If a and b are any real numbers,

$$f(a) f(b) = \begin{bmatrix} 1 & a \\ 0 & 1 \end{bmatrix} \begin{bmatrix} 1 & b \\ 0 & 1 \end{bmatrix}$$

$$= \begin{bmatrix} 1 & a + b \\ 0 & 1 \end{bmatrix}$$

$$= f(a + b).$$

We can apply this translation rule to determine the inverse of a matrix in G.

We know that $a + -a = 0$ is a true statement in **R**. Using f to translate this statement, we get

$$f(a)f(-a) = f(0) = \begin{bmatrix} 1 & 0 \\ 0 & 1 \end{bmatrix};$$

therefore

$$\begin{bmatrix} 1 & a \\ 0 & 1 \end{bmatrix}^{-1} = \begin{bmatrix} 1 & -a \\ 0 & 1 \end{bmatrix}.$$

Theorem 11.7.1 summarizes some of the general facts about group isomorphisms that are used most often in applications.

Theorem 11.7.1. *If $[G;*]$ and $[H;\#]$ are groups with identities e and e', respectively, and $T : G \to H$ is an isomorphism from G into H, then:*

(a) $T(e) = e'$,
(b) $T(a)^{-1} = T(a^{-1})$ for all $a \in G$, and
(c) If K is a subgroup of G, then $T(K) = \{T(a) : a \in K\}$ is a subgroup of H and is isomorphic to K.

"Is isomorphic to" is an equivalence relation on the set of all groups. Therefore, the set of all groups is partitioned into equivalence classes, each equivalence class containing groups that are isomorphic to one another.

PROCEDURES FOR SHOWING GROUPS ARE NOT ISOMORPHIC

How do you decide that two groups are not isomorphic to one another? The negation of "G and H are isomorphic" is that no translation rule between G and H exists. If G and H have different cardinalities, then no bijection from G into H can exist. Hence they are not isomorphic. Given that $\#G = \#H$, it is usually impractical to list all bijections from G into H and show that none of them satisfy Condition b of the formal definition. The best way to prove that two groups are not isomorphic is to find a true statement about one group that is not true about the other group. We illustrate this method in the following checklist that you can apply to most pairs of non-isomorphic groups in this book.

Assume that $[G;*]$ and $[H;\#]$ are groups. The following are reasons for G and H to be *not* isomorphic.

(a) G and H do not have the same cardinality. For example $\mathbf{Z}_{12} \times \mathbf{Z}_5$ can't be isomorphic to $\mathbf{Z}_{50}$ and $[\mathbf{R};+]$ can't be isomorphic to $[\mathbf{Q}^+; x]$.
(b) G is abelian and H is not abelian since $a * b = b * a$ is always true in G, but $T(a) \# T(b) = T(b) \# T(a)$ would not always be true. Two groups with six elements each are $\mathbf{Z}_6$ and the set of 3×3 rook matrices (see Exercise 4 in Section 11.2). The second group is non-abelian, therefore it can't be isomorphic to $\mathbf{Z}_6$.

(c) G has a certain kind of subgroup that H doesn't have. Theorem 11.7.1c states that this cannot happen if G is isomorphic to H. $[\mathbf{R^*};.]$ and $[\mathbf{R^+};.]$ are not isomorphic since $\mathbf{R^*}$ has a subgroup with two elements, $\{-1, 1\}$, while the proper subgroups of $\mathbf{R^+}$ are all infinite (convince yourself of this fact).

(d) The number of solutions of $x * x = e$ in G is not equal to the number of solutions of $y \# y = e'$ in H. $\mathbf{Z_8}$ is not isomorphic to $\mathbf{Z_2^3}$ since $x +_8 x = 0$ has two solutions, 0 and 4, while $y + y = (0, 0, 0)$ has eight solutions. If the operation in G is defined by a table, then the number of solutions of $x * x = e$ will be the number of occurrences of e in the main diagonal of the table. The equations $x^3 = e$, $x^4 = e, \ldots$, can also be used in the same way to identify non-isomorphic groups.

(e) One of the cyclic subgroups of G equals G (G is cyclic), while none of H's cyclic subgroups equals H (H is noncyclic). This is a special case of Condition c. $\mathbf{Z}$ and $\mathbf{Z} \times \mathbf{Z}$ are not isomorphic since $\mathbf{Z} = (1)$ and $\mathbf{Z} \times \mathbf{Z}$ is not cyclic.

EXERCISES FOR SECTION 11.7

A Exercises

1. State whether each pair of groups below is isomorphic. If it is, give an isomorphism; if it is not, give your reason.
    (a) $\mathbf{Z} \times \mathbf{R}$ and $\mathbf{R} \times \mathbf{Z}$
    (b) $\mathbf{Z_2} \times \mathbf{Z}$ and $\mathbf{Z} \times \mathbf{Z}$
    (c) $\mathbf{R}$ and $\mathbf{Q} \times \mathbf{Q}$
    (d) $\mathcal{P}(\{1, 2\})$ with symmetric difference and $\mathbf{Z_2^2}$
    (e) $\mathbf{Z_2^2}$ and $\mathbf{Z_4}$
    (f) $\mathbf{R^4}$ and $M_{2 \times 2}(\mathbf{R})$ with matrix addition
    (g) $\mathbf{R^2}$ and $\mathbf{R} \times \mathbf{R^+}$
    (h) $\mathbf{Z_2}$ and the 2×2 rook matrices
    (j) $\mathbf{Z_6}$ and $\mathbf{Z_2} \times \mathbf{Z_3}$

2. If you know two natural languages, show that they are not isomorphic.

3. Prove that the relation "is isomorphic to" on groups is transitive.

4. (a) Write out the tables for $G = [\{1, -1, i, -i\}; \cdot]$ where i is the complex number for which $i^2 = -1$. Show that G is isomorphic to $[\mathbf{Z_4}; +_4]$.
    (b) Solve $x^2 = -1$ in G by first translating to $\mathbf{Z_4}$, solving the equation in $\mathbf{Z_4}$, and then translating back to G.

B Exercises

5. It can be shown that there are five non-isomorphic groups of order eight. You should be able to describe three of them. Do so without use of tables. Be sure to explain why they are not isomorphic.

6. Prove Theorem 11.7.1.

7. Prove that all infinite cyclic groups are isomorphic to **Z**.

8. (a) Prove that **R*** is isomorphic to $\mathbf{Z}_2 \times \mathbf{R}$.
    (b) Describe how multiplication of non-zero real numbers can be accomplished doing only additions and translations.

11.8 Object-Oriented Programming

Some recent trends in computer software and hardware design make use of the concept of an algebraic structure. These new methods are classified under the topic of object-oriented programming. A feature of the language Ada is the package, which allows the programmer to combine data types and operations on them into one entity. This allows for the possibility of giving certain users access to data, but only in a certain context. For example, you might want to allow someone to read certain fields in records of a data file. A package that defines that file and a procedure that prints selected fields of a record could be used in this situation.

We could define a version of the group $\mathbf{Z}_{100}$ in Ada with the following package definition.

```
Package group is
     Type Z_100 is 0..99;
     Function sum(a,b: Z_100) return Z_100;
     zero : constant Z_100 := 0; --identity element
     function inverse (a: Z_100) return Z_100;
End group;
```

The definitions of functions in a package are contained in the body of that package. This allows a user to do certain things without knowing how they are done.

```
Package body group is
     Function sum (a,b: Z_100) return Z_100 is
          s: Z_100;
     Begin
          s := (a + b) mod 100;
          return s;
     End sum;
     Function inverse (a: Z_100) return Z_100 is
     Begin
          If a = zero then return a
                         else return 100 - a
     End inverse;
End group;
```

In Chapter 16, we will study a structure called a *ring*. One ring consists of the set $\mathbf{Z}_{100}$ with an expanded set of operations and constants. Its package definition would be

```
Package ring is
      Type Z_100 is 0..99;
      Function sum (a,b: Z_100) return Z_100;
      Function product (a,b: Z_100) return Z_100;
      zero : constant Z_100 := 0; --additive
      identity
      function inverse (a: Z_100) return Z_100;
      one : constant Z_100 := 1; --multiplicative
      identity
End ring;
```

Note how we are working with the same data type in both a group and a ring, but more can be done in a ring.

The next type of algebraic structure that we will study is a *vector space*. One such vector space consists of all ordered triples of real numbers; i.e., $\mathbf{R} \times \mathbf{R} \times \mathbf{R} = \mathbf{R}^3$. A package definition of that system would be

```
Package vector space is
      Type R_3 is Array (1..3) of Real;
      Function Sum (v,w: R_3) return R_3;
      Function Scalar_Product (c:real, v:R_3)
      return R_3;
      zero : constant R_3 := (0, 0, 0);
End vector space;
```

Object-oriented programming has also been used to coordinate hardware systems that employ multiple processors. The article by Kinder describes how this method has been used in a real system.

EXERCISES FOR SECTION 11.8

C Exercises

1. Write out package descriptions in pseudocode or Ada (if you are familiar with the language) for
   (a) the monoid of integers with multiplication
   (b) the direct product of groups $[\mathbf{Z}_4; +_4]$ and $[\mathbf{R}; +]$
   (c) the group $\mathbf{Z}_2^{10}$

2. Write out package bodies for the systems in Exercise 1 above.

3. What is the value of the expression sum(a, sum(b, c)) = sum(sum(a, b), c) in the group package? What group property does this expression refer to?

chapter

12

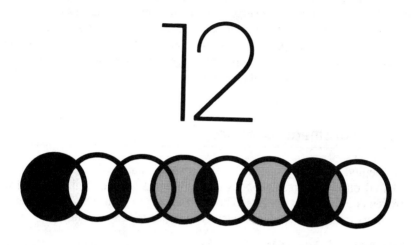

MORE MATRIX ALGEBRA

12 GOALS

In Chapter 5 we studied matrix operations and the algebra of sets and logic. We also made note of the strong resemblance of matrix algebra to elementary algebra. The reader should briefly review this material. In this chapter we shall look at a powerful matrix tool in the applied sciences—namely, a technique for solving systems of linear equations. We will then use this process for determining the inverse of $n \times n$ matrices, $n \geq 2$, when they exist. We conclude by a development of the diagonalization process, with a discussion of several of its applications.

12.1 Systems of Linear Equations

The method of solving systems of equations by matrices that we will look at is based on procedures involving equations that we are familiar with from previous mathematics courses. The main idea is to reduce a given system of equations to another simpler system which has the same solutions.

Definition: Solution Set. *Given a system of equations involving real variables* $x_1, \ldots , x_n$, *the solution set of the system is the set of n-tuples in* $\mathbf{R}^n$, $(a_1, \ldots , a_n)$ *such that the substitutions* $x_1 = a_1, \ldots , x_n = a_n$ *make all the equations true.*

If the variables are from a set S, then the solution set will be a subset of S^n.

Definition: Equivalent Systems of Equations. *Two systems of linear equations are called equivalent if they have the same set of solutions.*

Example 12.1.1. The previous definition tells us that if we know that the system

$$\begin{cases} 4x_1 + 2x_2 + x_3 = 1 \\ 2x_1 + x_2 + x_3 = 4 \\ 2x_1 + 2x_2 + x_3 = 3 \end{cases}$$

is equivalent to the system

$$\begin{cases} x_1 + 0x_2 + 0x_3 = -1 \\ 0x_1 + x_2 + 0x_3 = -1 \\ 0x_1 + 0x_2 + 1x_3 = 7 \end{cases}$$

then both systems have the solution set $\{(-1, -1, 7)\}$; i.e.,

$$x_1 = -1, x_2 = -1, x_3 = 7.$$

Theorem 12.1.1: Elementary Operations on Equations. *If any sequence of the following operations is performed on a system of equations, the resulting system is equivalent to the original system:*

(1) *Interchange any two equations in the system.*
(2) *Multiply both sides of any equation by a non-zero constant.*
(3) *Multiply both sides of any equation by a non-zero constant and add the result to a second equation in the system, with the sum replacing the latter equation.*

Let us now use the above theorem to work out the details of Example 12.1.1.

Step 1. We will first change the coefficient of x_1 in Equation 1 to a 1 and then use it as a pivot to obtain 0's for the coefficients of x_1 in Equations 2 and 3.

$$(1.1) \begin{cases} 4x_1 + 2x_2 + x_3 = 1 \\ 2x_1 + x_x + x_3 = 4 \\ 2x_1 + 2x_2 + x_3 = 3 \end{cases}$$ Multiply Equation 1 by 1/4 to obtain

$$(1.2) \begin{cases} x_1 + \frac{1}{2}x_2 + \frac{1}{4}x_3 = \frac{1}{4} \\ 2x_1 + x_2 + x_3 = 4 \\ 2x_1 + 2x_2 + x_3 = 3 \end{cases}$$ Multiply Equation 1 by -2 and add the result to Equation 2 to obtain

$$(1.3) \begin{cases} x_1 + \frac{1}{2}x_2 + \frac{1}{4}x_3 = \frac{1}{4} \\ 0x_1 + 0x_2 + \frac{1}{2}x_3 = 7/2 \\ 2x_1 + 2x_2 + x_3 = 3 \end{cases}$$ Multiply Equation 1 by -2 and add the result to Equation 3 to obtain

$$(1.4) \begin{cases} x_1 + \frac{1}{2}x_2 + \frac{1}{4}x_3 = \frac{1}{4} \\ 0x_1 + 0x_2 + \frac{1}{2}x_3 = 7/2 \\ 0x_1 + x_2 + \frac{1}{2}x_3 = 5/2 \end{cases}$$

Step 2. We would now like to proceed in a fashion analogous to Step 1—namely, multiply the coefficient of x_2 in the second equation by a suitable number so that the result is a 1. Then use it as a pivot to obtain 0's as coefficients for x_2 in the first and third equations. This is clearly impossible (Why?), so we will first interchange Equations 2 and 3 and proceed as outlined above.

$$(2.1) \begin{cases} x_1 + \frac{1}{2}x_2 + \frac{1}{4}x_3 = \frac{1}{4} \\ 0x_1 + 0x_2 + \frac{1}{2}x_3 = 7/2 \\ 0x_1 + x_2 + \frac{1}{2}x_3 = 5/2 \end{cases}$$ Interchange Equations 2 and 3 to obtain

(2.2) $\begin{cases} x_1 + \frac{1}{2}x_2 + \frac{1}{4}x_3 = \frac{1}{4} \\ 0x_1 + x_2 + \frac{1}{2}x_3 = 5/2 \\ 0x_1 + 0x_2 + \frac{1}{2}x_3 = 7/2 \end{cases}$ Multiply Equation 2 by $-\frac{1}{2}$ and add the result to Equation 1 to obtain

(2.3) $\begin{cases} x_1 + 0x_2 + 0x_3 = -1 \\ 0x_1 + x_2 + \frac{1}{2}x_3 = 5/2 \\ 0x_1 + 0x_2 + \frac{1}{2}x_3 = 7/2 \end{cases}$

Step 3. Next, we will change the coefficient of x_3 in the third equation to a 1 and then use it as a pivot to obtain 0's for the coefficients of x_3 in Equations 1 and 2.

(3.1) $\begin{cases} x_1 + 0x_2 + 0x_3 = -1 \\ 0x_1 + x_2 + \frac{1}{2}x_3 = 5/2 \\ 0x_1 + 0x_2 + \frac{1}{2}x_3 = 7/2 \end{cases}$ Multiply Equation 3 by 2 to obtain

(3.2) $\begin{cases} x_1 + 0x_2 + 0x_3 = -1 \\ 0x_1 + x_2 + \frac{1}{2}x_3 = 5/2 \\ 0x_1 + 0x_2 + x_3 = 7 \end{cases}$ Multiply Equation 3 by $-\frac{1}{2}$ and add the result to Equation 2 to obtain

(3.3) $\begin{cases} x_1 = 0x_2 + 0x_3 = -1 \\ 0x_1 + x_2 + 0x_3 = -1 \\ 0x_1 + 0x_2 + x_3 = 7 \end{cases}$

From the system of equations in Step 3.3, we see that the solution to the original system (Step 1.1) is $x_1 = -1$, $x_2 = -1$ and $x_3 = 7$.

In the above sequence of steps, we note that the variables serve the sole purpose of keeping the coefficients in the appropriate location. This we can effect by using matrices. The matrix of the system given in Step 1.1 is

$$\begin{bmatrix} 4 & 2 & 1 & 1 \\ 2 & 1 & 1 & 4 \\ 2 & 2 & 1 & 3 \end{bmatrix},$$

where the matrix of the first three columns is called the *coefficient matrix* and the complete matrix is referred to as the *augmented matrix*. Since we are now using matrices to solve the system, we will translate the definition for elementary row operations into matrix language.

Definition: Elementary Row Operations. *The following operations on a matrix are called elementary row operations:*

(1) Interchange any two rows of the matrix.
(2) Multiply any row of the matrix by a non-zero constant.

(3)   *Multiply any row of the matrix by a non-zero constant and add the result to a second row, with the sum replacing the latter row.*

Definition: Row Equivalent.   *Two matrices, A and B, are said to be row-equivalent if one can be obtained from the other by any one elementary row operation or by any sequence of elementary row operations.*

If we use the convention R_i to stand for Row i of a matrix and $\rightarrow$ to stand for row equivalent, then

$$A \xrightarrow{\quad c\,R_i + R_j \quad} B$$

means that the matrix B is obtained from the matrix A by multiplying the ith row of A by c and adding the result to Row j. The matrix notation for the system given in Step 1.1 with the subsequent steps are:

$$
\begin{bmatrix} 4 & 2 & 1 & 1 \\ 2 & 1 & 1 & 4 \\ 2 & 2 & 1 & 3 \end{bmatrix}
\xrightarrow{\tfrac{1}{4}R_1}
\begin{bmatrix} 1 & \tfrac{1}{2} & \tfrac{1}{4} & \tfrac{1}{4} \\ 2 & 1 & 1 & 4 \\ 2 & 2 & 1 & 3 \end{bmatrix}
$$

$$
\xrightarrow{-2R_1 + R_2}
\begin{bmatrix} 1 & \tfrac{1}{2} & \tfrac{1}{4} & \tfrac{1}{4} \\ 0 & 0 & \tfrac{1}{2} & 7/2 \\ 2 & 2 & 1 & 3 \end{bmatrix}
$$

$$
\xrightarrow{-2R_1 + R_3}
\begin{bmatrix} 1 & \tfrac{1}{2} & \tfrac{1}{4} & \tfrac{1}{4} \\ 0 & 0 & \tfrac{1}{2} & 7/2 \\ 0 & 1 & \tfrac{1}{2} & 5/2 \end{bmatrix}
$$

$$
\xrightarrow[R_2 \text{ and } R_3]{\text{Interchange}}
\begin{bmatrix} 1 & \tfrac{1}{2} & \tfrac{1}{4} & \tfrac{1}{4} \\ 0 & 1 & \tfrac{1}{2} & 5/2 \\ 0 & 0 & \tfrac{1}{2} & 7/2 \end{bmatrix}
$$

$$
\xrightarrow{-\tfrac{1}{2}R_2 + R_1}
\begin{bmatrix} 1 & 0 & 0 & -1 \\ 0 & 1 & \tfrac{1}{2} & 5/2 \\ 0 & 0 & \tfrac{1}{2} & 7/2 \end{bmatrix}
$$

$$
\xrightarrow{2R_3}
\begin{bmatrix} 1 & 0 & 0 & -1 \\ 0 & 1 & \tfrac{1}{2} & 5/2 \\ 0 & 0 & 1 & 7 \end{bmatrix}
$$

$$
\xrightarrow{-\tfrac{1}{2}R_3 + R_2}
\begin{bmatrix} 1 & 0 & 0 & -1 \\ 0 & 1 & 0 & -1 \\ 0 & 0 & 1 & 7 \end{bmatrix}
$$

This again gives us the solution. This procedure is called the *Gauss-Jordan elimination method*.

It is important to remember when solving any system of equations via this or any similar approach that at any step in the procedure we can rewrite the matrix in "equation format" to help us to interpret the meaning of the augmented matrix.

In Example 12.1.1 we obtained a unique solution, only one triple, namely $(-1, -1, 7)$, which satisfies all three equations. For a system involving three unknowns, are there any other possible results? To answer this question, let's review some basic facts from analytic geometry.

The graph of a linear equation in three-dimensional space is a plane. So geometrically we can visualize the three linear equations as three planes in three-space. Certainly the three planes can intersect in a unique point, as in Example 12.1.1, or the planes could be parallel. If they are parallel, there are no common points of intersection; that is, there are no real numbers x_1, x_2, and x_3 which will satisfy all three equations. Also, the three planes could intersect along a common axis or line. In this case, there would be an infinite number of real numbers triples in $\mathbf{R}^3$ which would satisfy all three equations. We generalize:

In a system of n linear equations, n unknowns, there can be:

(1) a unique solution,
(2) no solution, or
(3) an infinite number of solutions.

To illustrate these points, consider the following examples:

Example 12.1.2.   Find all solutions to the system

$$\begin{cases} 2x_1 + 3x_2 + x_3 = 2 \\ x_1 + x_2 + 5x_3 = 4 \\ 2x_1 + 2x_2 + 10x_3 = 6 \end{cases}$$

The reader can verify that

$$\begin{bmatrix} 1 & 3 & 1 & 2 \\ 1 & 1 & 5 & 4 \\ 2 & 2 & 10 & 6 \end{bmatrix}$$

reduces to

$$\begin{bmatrix} 2 & 3 & 1 & 2 \\ 1 & 1 & 5 & 4 \\ 0 & 0 & 0 & -2 \end{bmatrix}.$$

(See Exercise 2 of this section.)

We can row-reduce this matrix further if we wish. However, any further row-reduction will not substantially change the last row, which, in equation form, is $0x_1 + 0x_2 + 0x_3 = -2$. It is clear that we cannot find real numbers x_1, x_2, x_3 which will satisfy this one equation, hence we cannot find real numbers x_1, x_2, x_3 which will satisfy all three original equations simultaneously. When this occurs, we say that the system has no solution, or the solution set is empty.

Example 12.1.3. Next let's attempt to find all of the solutions to:

$$\begin{cases} x_1 + 6x_2 + 2x_3 = 1 \\ 2x_1 + x_2 + 3x_3 = 2 \\ 4x_1 + 2x_2 + 6x_3 = 4 \end{cases}$$

The matrix

$$\begin{bmatrix} 1 & 6 & 2 & 1 \\ 2 & 1 & 3 & 2 \\ 4 & 2 & 6 & 4 \end{bmatrix}$$

reduces to

$$\begin{bmatrix} 1 & 0 & 16/11 & 1 \\ 0 & 1 & 1/11 & 0 \\ 0 & 0 & 0 & 0 \end{bmatrix}.$$

If we apply additional elementary row operations to this matrix, it will only become more complicated. In particular, we cannot obtain a 1 in the third row, third column. Since the matrix is in simplest form, we will express it in equation format to help us determine the solution set.

$$x_1 + 0x_2 + 16/11x_3 = 1$$
$$0x_1 + 1x_2 + 1/11x_3 = 0$$
$$0x_1 + 0x_2 + 0x_3 = 0$$

Any real numbers x_1, x_2, and x_3 will satisfy the last equation. However, the first equation can be rewritten as $x_1 = 1 - 16/11x_3$, which describes the coordinate x_1 in terms of x_3. Similarly, the second equation gives x_2. A convenient way of listing the solutions of this system is to use set notation. If we call the solution set of the system S, then $S = \{(1 - 16/11x_3, -1/11x_3, x_3) | x_3 \in \mathbf{R}\}$. What this means is that if we wanted to list all solutions, we would replace x_3 by all possible numbers. Clearly, there is an infinite number of solutions, two of which are $(1, 0, 0)$ and $(-15, -1, 11)$.

A Word of Caution: Frequently we may obtain "different-looking" answers to the same problem when a system has an infinite number of answers. Assume a student's solutions set to Example 12.1.3 is $A = \{(1 + 16x_2, x_2, -11x_2) | x_2$ is a real number$\}$. Certainly the result described by S looks different from that described by A. To see whether they indeed describe the same set, we wish to determine whether every solution produced in S can be generated in A. The solution generated by S, $(-15, -1, 11)$, can be produced by A by taking $x_2 = -1$. We must prove that every solution described in S is described in A and, conversely, that every solution described in A is described in S. (See Exercise 3 of this section.)

To summarize the procedure in the Gauss-Jordan technique for solving systems of equations, we attempt to obtain 1's along the main diagonal of the coefficient matrix with 0's above and below the diagonal, as in Example 12.1.1. We may find in attempting this that the closest we can come is to put the coefficient matrix in "simplest" form, as in Example 12.1.3, or we may find that the situation of Example 12.1.1 evolves as part of the process. In this latter case, we can terminate the process and state that the system has no solutions. The final matrix forms of Examples 12.1.1 and 12.1.3 are called *echelon forms*.

In practice, systems of linear equations are solved using computers. Generally, the Gauss-Jordan algorithm is the most useful; however, slight variations of this algorithm are also used. The different approaches share many of the same advantages and disadvantages. The two major concerns of all methods are:

(1) minimizing inaccuracies due to rounding off errors and
(2) minimizing computer time.

The accuracy of the Gauss-Jordan method can be improved by always choosing the element with the largest absolute value as the pivot element, as in the following algorithm.

Algorithm 12.1.1. Given a matrix equation $Ax = b$, where A is $n \times n$, let C be the augmented matrix $[A \mid b]$. A solution, if one exists, can be obtained by performing the following algorithm ($C_i = i^{th}$ row of C):

(1) Singular := false
(2) $S := \{1, 2, \ldots, n\}$
(3) For col := 1 to n do
    (3.1) Select row so that $C[\text{row, col}] = \max \{|C[i, \text{col}]| : i \in S\}$
    (3.2) $S := S - \{\text{row}\}$
    (3.3) If $C[\text{row, col}] \neq 0$
        then
        (3.3.1) $C_{\text{row}} := (1/C[\text{row, col}])*C_{\text{row}}$
        (3.3.2) For $i := 1$ to n DO
                if $i \neq$ row then $C_i := C_i - C[i, \text{row}]*C_{\text{row}}$
        else singular := true

At the end of this algorithm:
If singular = false, then the solution of $Ax = b$ is $x = (C[1, n + 1], \ldots, C[n, n + 1])$.

If singular = true, then either $Ax = b$ has no solution or an infinite number of solutions.

Warning: Actual implementation on a computer requires that the condition of Step 3.3 be changed to $|C[\text{row, col}]| > E$, where E is a small positive number that would depend on the machine in use.

An introductory discussion with several Pascal programs for solutions of systems of equations (and therefore for finding inverses of matrices) can be found in Miller.

EXERCISES FOR SECTION 12.1

A Exercises

1. Solve the following systems:

(a) $\begin{cases} 2x_1 + x_2 = 3 \\ x_1 - x_2 = 1 \end{cases}$

(b) $\begin{cases} x_1 + x_2 + 2x_3 = 1 \\ x_1 + 2x_2 - x_3 = -1 \\ x_1 + 3x_2 + x_3 = 5 \end{cases}$

(c) $\begin{cases} 2x_1 + 2x_2 + 4x_3 = 2 \\ 2x_1 + x_2 + 4x_3 = 0 \\ 3x_1 + 5x_2 + x_3 = 0 \end{cases}$

(d) $\begin{cases} 6x_1 + 7x_2 + 2x_3 = 3 \\ 4x_1 + 2x_2 + x_3 = -2 \\ 6x_1 + x_2 + x_3 = 1 \end{cases}$

2. (a)   Write out the details of Example 12.1.2.
    (b)   Write out the details of Example 12.1.3.

3. (a)   Use the solution set S of Example 12.1.3 to list three different solutions to the given system. Then show that each of these solutions can be described by the set A of Example 12.1.3.
    (b)   Prove that $S = A$.

4. Solve the following systems by describing the solution sets completely:

(a) $\begin{cases} 2x_1 + x_2 + 3x_3 = 2 \\ 4x_1 + x_2 + 2x_3 = -1 \\ 8x_1 + 2x_2 + 4x_3 = 4 \end{cases}$

(b) $\begin{cases} 2x_1 + x_2 + 3x_3 = 5 \\ 4x_1 + x_2 + 2x_3 = -1 \\ 8x_1 + 2x_2 + 4x_3 = -2 \end{cases}$

(c) $\begin{cases} x_1 - x_2 + 3x_3 = 7 \\ x_1 + 3x_2 + x_3 = 4 \end{cases}$

(d) $\begin{cases} x_1 + x_2 - x_3 + 2x_4 = 1 \\ x_1 + 2x_2 + 3x_3 + x_4 = 5 \\ x_1 + 3x_2 + 2x_3 - x_4 = -1 \end{cases}$

(e) $\begin{cases} x_1 + x_2 + 2x_3 + x_4 = 3 \\ x_1 - x_2 + 3x_3 - x_4 = -2 \\ 3x_1 + 3x_2 + 6x_3 + 3x_4 = 9 \end{cases}$

B Exercise

5. Given a system of n linear equations in n unknowns in matrix form
    $Ax = b$, prove that if b is a matrix of all 0's, then the solution set of
    $Ax = b$ is a subgroup of $\mathbf{R}^n$.

12.2 Matrix Inversion

In Chapter 5 we defined the inverse of an $n \times n$ matrix. We noted that not
all matrices have inverses, but when the inverse of a matrix exists, it is
unique. This enables us to define the inverse of an $n \times n$ matrix A as the
unique matrix B such that $AB = BA = I$, where I is the $n \times n$ identity
matrix. In order to obtain some practical experience, we developed a formula
which allowed us to determine the inverse of invertible 2×2 matrices. We
will now use the Gauss-Jordan procedure for solving systems of linear equations to compute the inverses, when they exist, of $n \times n$ matrices, $n \geq 2$.
The following procedure for a 3×3 matrix can be generalized for $n \times n$
matrices, $n \geq 2$.

Example 12.2.1. Given the matrix

$$A = \begin{bmatrix} 1 & 1 & 2 \\ 2 & 1 & 4 \\ 3 & 5 & 1 \end{bmatrix},$$

we want to find the matrix

$$B = \begin{bmatrix} x_{11} & x_{12} & x_{13} \\ x_{21} & x_{22} & x_{23} \\ x_{31} & x_{32} & x_{33} \end{bmatrix},$$

if it exists, such that (a) $AB = I$ and (b) $BA = I$. We will concentrate on
finding a matrix which satisfies Equation a and then verify that the matrix B
obtained satisfies Equation b.

$$\begin{bmatrix} 1 & 1 & 2 \\ 2 & 1 & 4 \\ 3 & 5 & 1 \end{bmatrix} \begin{bmatrix} x_{11} & x_{12} & x_{13} \\ x_{21} & x_{22} & x_{23} \\ x_{31} & x_{32} & x_{33} \end{bmatrix} = \begin{bmatrix} 1 & 0 & 0 \\ 0 & 1 & 0 \\ 0 & 0 & 1 \end{bmatrix}$$

$$\Leftrightarrow \begin{bmatrix} 1x_{11} + 1x_{21} + 2x_{31} & 1x_{12} + 1x_{22} + 2x_{32} & 1x_{13} + 1x_{23} + 2x_{33} \\ 2x_{11} + 1x_{21} + 4x_{31} & 2x_{12} + 1x_{22} + 4x_{32} & 2x_{13} + 1x_{23} + 4x_{33} \\ 3x_{11} + 5x_{21} + 1x_{31} & 3x_{12} + 4x_{22} + 1x_{32} & 3x_{13} + 5x_{23} + 1x_{33} \end{bmatrix}$$

$$\tag{12.2a}$$

$$= \begin{bmatrix} 1 & 0 & 0 \\ 0 & 1 & 0 \\ 0 & 0 & 1 \end{bmatrix}$$

By definition of equality of matrices, this gives us three systems of equations to solve. The matrix form of one of the 12.2a systems is:

$$\begin{bmatrix} 1 & 1 & 2 & 1 \\ 2 & 1 & 4 & 0 \\ 3 & 5 & 1 & 0 \end{bmatrix} \tag{12.2b}$$

Using the Gauss-Jordan technique of Section 12.1, we have:

$$\begin{bmatrix} 1 & 1 & 2 & 1 \\ 2 & 1 & 4 & 0 \\ 3 & 5 & 1 & 0 \end{bmatrix} \xrightarrow[\;-3R_1 + R_3\;]{-2R_1 + R_2 \text{ and}} \begin{bmatrix} 1 & 1 & 2 & 1 \\ 0 & -1 & 0 & -2 \\ 0 & 2 & -5 & -3 \end{bmatrix}$$

$$\xrightarrow[\;2R_2 + R_3\;]{R_2 + R_1 \text{ and}} \begin{bmatrix} 1 & 0 & 2 & -1 \\ 0 & -1 & 0 & -2 \\ 0 & 0 & -5 & -7 \end{bmatrix} \xrightarrow{2/5\,R_3 + R_1} \begin{bmatrix} 1 & 0 & 0 & -19/5 \\ 0 & -1 & 0 & -2 \\ 0 & 0 & -5 & -7 \end{bmatrix}$$

$$\xrightarrow[\;-1R_3\;]{-1R_2 \text{ and}} \begin{bmatrix} 1 & 0 & 0 & -19/5 \\ 0 & 1 & 0 & 2 \\ 0 & 0 & 1 & 7/5 \end{bmatrix}$$

So $x_{11} = -19/5$, $x_{21} = 2$, and $x_{31} = 7/5$, which gives us the first column of the matrix B. The matrix form of the system to obtain x_{12}, x_{22}, and x_{32}, the second column of B, is:

$$\begin{bmatrix} 1 & 1 & 2 & 0 \\ 2 & 1 & 4 & 1 \\ 3 & 5 & 1 & 0 \end{bmatrix}, \tag{12.2c}$$

which reduces to

$$\begin{bmatrix} 1 & 0 & 0 & 9/5 \\ 0 & 1 & 0 & -1 \\ 0 & 0 & 1 & -2/5 \end{bmatrix}. \tag{12.2d}$$

The critical idea to note here is that the coefficient matrix in 12.2c is the same as the matrix in 12.2b, hence the sequence of row operations that we used to reduce the matrix in 12.2b can be used to reduce the matrix in 12.2c. To determine the third column of B, we reduce

$$\begin{bmatrix} 1 & 1 & 2 & 0 \\ 2 & 1 & 4 & 0 \\ 3 & 5 & 1 & 1 \end{bmatrix}$$

to obtain $x_{13} = 2/5$, $x_{23} = 0$ and $x_{33} = -1/5$. Here again it is important to note that the sequence of row operations used to "solve" this system is exactly the same as those we used in the first system. Why not save ourselves a considerable amount of time and effort and solve all three systems simultaneously? This we can effect by augmenting the coefficient matrix by the identity matrix I. We then have

$$\begin{bmatrix} 1 & 1 & 2 & 1 & 0 & 0 \\ 2 & 1 & 4 & 0 & 1 & 0 \\ 3 & 5 & 1 & 0 & 0 & 1 \end{bmatrix}$$

[The same sequence of
row operations as above.]

$$\longrightarrow \quad \begin{bmatrix} 1 & 0 & 0 & -19/5 & 9/5 & 2/5 \\ 0 & 1 & 0 & 2 & -1 & 0 \\ 0 & 0 & 1 & 7/5 & -2/5 & -1/5 \end{bmatrix}.$$

So that

$$B = \begin{bmatrix} -19/5 & 9/5 & 2/5 \\ 2 & -1 & 0 \\ 7/5 & -2/5 & -1/5 \end{bmatrix}.$$

(The reader should verify that $BA = I$ so that $A^{-1} = B$.)

As the following theorem indicates, the verification that $BA = I$ is not necessary. The proof of the theorem is beyond the scope of this text. The interested reader can find it in most linear algebra texts.

Theorem 12.2.1.   *Let A be an n × n matrix. If a matrix B can be found such that AB = I, then BA = I, so that B = A⁻¹. In fact, to find A⁻¹, we need only find a matrix B that satisfies one of the two conditions AB = I or BA = I.*

It is clear from Chapter 5 and our discussions in this chapter that not all $n \times n$ matrices have inverses. How do we determine whether a matrix has an inverse using this method? The answer is quite simple: the technique we developed to compute inverses is a matrix approach to solving several systems of equations simultaneously.

Example 12.2.2.   The reader can verify that if

$$A = \begin{bmatrix} 1 & 2 & 1 \\ -1 & -2 & -1 \\ 0 & 1 & 2 \end{bmatrix},$$

then

$$\begin{bmatrix} 1 & 2 & 1 & 1 & 0 & 0 \\ -1 & -2 & -1 & 0 & 1 & 0 \\ 0 & 5 & 8 & 0 & 0 & 1 \end{bmatrix}$$

reduces to

$$\begin{bmatrix} 1 & 2 & 1 & 1 & 0 & 0 \\ 0 & 0 & 0 & 1 & 1 & 0 \\ 0 & 5 & 8 & 0 & 0 & 1 \end{bmatrix} \qquad (12.2e)$$

Although this matrix can be further row-reduced, it is not necessary to do so since in equation form we have:

(i) $\begin{cases} x_{11} + 2x_{21} + x_{31} = 1 \\ 0x_{11} + 0x_{21} + 0x_{31} = 1 \\ 0x_{11} + 5x_{21} + 8x_{31} = 0 \end{cases}$ (ii) $\begin{cases} x_{12} + 2x_{22} + x_{32} = 0 \\ 0x_{12} + 0x_{22} + 0x_{32} = 1 \\ 0x_{12} + 5x_{22} + 8x_{32} = 0 \end{cases}$

(iii) $\begin{cases} x_{13} + 2x_{23} + x_{33} = 0 \\ 0x_{13} + 0x_{23} + 0x_{33} = 0 \\ 0x_{13} + 5x_{23} + 8x_{33} = 1 \end{cases}$

Clearly, there is no solution to System i (or to System ii), therefore A^{-1} does not exist. From this discussion it should be obvious to the reader that the zero row of the coefficient matrix together with the non-zero entry in the fourth column of that row in matrix 12.2e tell us that A^{-1} does not exist.

EXERCISES FOR SECTION 12.2

A Exercises

1. In order to develop an understanding of the technique of this section, work out all the details of Example 12.2.1.

2. Use the method of this section to find A^{-1} whenever possible for the following matrices. If A^{-1} does not exist explain why.

(a) $A = \begin{bmatrix} 1 & 2 \\ -1 & 3 \end{bmatrix}$

(b) $A = \begin{bmatrix} 1 & 2 & 1 \\ -2 & -3 & -1 \\ 1 & 4 & 4 \end{bmatrix}$

(c) $A = \begin{bmatrix} 6 & 7 & 2 \\ 4 & 2 & 1 \\ 6 & 1 & 1 \end{bmatrix}$

(d) $A = \begin{bmatrix} 0 & 3 & 2 & 5 \\ 1 & -1 & 4 & 0 \\ 0 & 0 & 1 & 1 \\ 0 & 1 & 3 & -1 \end{bmatrix}$

(e) $A = \begin{bmatrix} 2 & 3 \\ 4 & 6 \end{bmatrix}$

(f) $A = \begin{bmatrix} 2 & 1 & 3 \\ 4 & 1 & 2 \\ 8 & 2 & 4 \end{bmatrix}$

3. (a) Find A^{-1} if

$$A = \begin{bmatrix} 2 & 0 & 0 \\ 0 & 3 & 0 \\ 0 & 0 & 5 \end{bmatrix} \qquad A = \begin{bmatrix} -1 & 0 & 0 & 0 \\ 0 & 5/2 & 0 & 0 \\ 0 & 0 & 1/7 & 0 \\ 0 & 0 & 0 & 3/4 \end{bmatrix}$$

   (b) If D is a diagonal matrix whose diagonal entries are non-zero, what is D^{-1}?

4. Express each system of equations in Exercise 1, Section 12.1, in the form $Ax = B$. Solve each system by first finding A^{-1}.

12.3 An Introduction to Vector Spaces and the Diagonalization Process

When we encountered various types of matrices in Chapter 5, it became apparent that a particular kind of matrix, the diagonal matrix, was much easier to use in computations. For example, if $A = \begin{bmatrix} 2 & 1 \\ 2 & 3 \end{bmatrix}$, then A^5 can be found. Its computation is tedious, however. If

$$A = \begin{bmatrix} 1 & 0 \\ 0 & 4 \end{bmatrix},$$

then

$$A^5 = \begin{bmatrix} 1^5 & 0 \\ 0 & 4^5 \end{bmatrix} = \begin{bmatrix} 1 & 0 \\ 0 & 1024 \end{bmatrix}.$$

In a variety of applications it is beneficial to be able to *diagonalize* a matrix. In this section we will investigate what this means and consider a few applications. In order to understand when the diagonalization process cn be performed, it is necessary to develop several of the underlying concepts of *linear algebra*.

By now, you realize that mathematicians tend to generalize. Once we have found a "good thing," something that is useful, we apply it to as many different concepts as possible. In doing so, we frequently find that the "different concepts" are not really different but only look different. Four sentences in four different languages might look dissimilar, but when they are translated into a common language, they might very well express the exact same idea.

Early in the development of mathematics, the concept of a vector led to a variety of applications in physics and engineering. We can certainly picture vectors, or "arrows," in the xy-plane and even in the three-dimensional space. Does it make sense to talk about vectors in four-dimensional space, in ten-dimensional space, or in any other mathematical situation? If so, what is the essence of a vector? Is it its shape or the properties, or rules, it follows? The shape in two- or three-space is just a picture, or geometric interpretation, of a vector. The essence is the rules, or properties, we wish vectors to follow so we can manipulate them algebraically. What follows is a definition of what is called a vector space, a "space of vectors." It is a list of all the essential properties of vectors, and it is the basic definition of what is called linear algebra.

Definition: Vector Space. *Let V be any non-empty set of objects. Define on V an operation, called addition, for any two elements $\vec{x}$ and $\vec{y} \in V$, and denote this operation by $\vec{x} + \vec{y}$. Let scalar multiplication be defined for a real number a from **R** and an element $\vec{x}$ from V denote this operation by $a\vec{x}$. The set V together with operations of addition and scalar multiplication is called a vector space over **R** if the following hold for all $\vec{x}$, $\vec{y}$, and $\vec{z}$ in V and for a and b in **R**:*

(1) $\vec{x} + \vec{y} \in V$.
(2) $\vec{x} + \vec{y} = \vec{y} + \vec{x}$.
(3) $(\vec{x} + \vec{y}) + \vec{z} = \vec{x} + (\vec{y} + \vec{z})$.
(4) There exists a vector in V, denoted by $\vec{0}$, such that $\vec{x} + \vec{0} = \vec{x}$.
(5) For each vector $\vec{x} \in V$, there exists a unique vector, called $-\vec{x} \in V$, such that $\vec{x} + (-\vec{x}) = \vec{0}$.

These are the main properties associated with the operation of addition. They can be summarized by saying that $[V; +]$ is an Abelian group.

The next five properties are associated with the operation of scalar multiplication:

(6) $a\vec{x} \in V$.
(7) $a(\vec{x} + \vec{y}) = a\vec{x} + a\vec{y}$.
(8) $(a + b)\vec{x} = a\vec{x} + b\vec{x}$.
(9) $a(b\vec{x}) = (ab)\vec{x}$.
(10) $1\vec{x} = \vec{x}$.

*In a vector space it is common to call the elements of V vectors and those from **R** scalars.*

Example 12.3.1. Let $V = M_{2\times3}(\mathbf{R})$ and let the operations of addition and scalar multiplication be the usual operations of addition and scalar multiplication for matrices. Then V together with these operations is a vector space over **R**. The reader is strongly encouraged to verify the definition for this example before proceeding further (see Exercise 3 of this section). Note we can call the elements of $M_{2\times3}(\mathbf{R})$ vectors even though they are not arrows.

Example 12.3.2. From Chapter 1, $\mathbf{R}^2 = \{(a_1, a_2) \mid a_1, a_2 \in \mathbf{R}\}$. If we define addition and scalar multiplication the natural way, that is, as we would on 1×2 matrices, then $\mathbf{R}^2$ is a vector space over $\mathbf{R}$. (See Exercise 4 of this section.)

In Example 12.3.2, we have the "bonus" that we can illustrate the algebraic concept geometrically. In mathematics, a "geometric bonus" does not always occur and, of course, is not necessary for the development or application of the concept. However, geometric illustrations are quite useful in helping us understand concepts and should be utilized whenever available.

Let's consider some illustrations of the vector space $\mathbf{R}^2$. Let $\overrightarrow{x} = (1, 4)$ and $\overrightarrow{y} = (3, 1)$.

If we illustrate the vector (x_1, x_2) as a directed line segment, or "arrow," from the point $(0, 0)$ to the point (x_1, x_2), the above vectors are as pictured in Figure 12.3.1, and $\overrightarrow{x} + \overrightarrow{y} = (1, 4) + (3, 1) = (4, 5)$, which also has the geometric representation as pictured in Figure 12.3.1. The vector $2\overrightarrow{x} = 2(1, 4) = (2, 8)$ is a vector in the same direction as $\overrightarrow{x}$ but twice its length.

Remarks:

(1) We will henceforth drop the arrow above a vector name and use the common convention that boldface letters toward the end of the alphabet are vectors, while letters early in the alphabet are scalars.

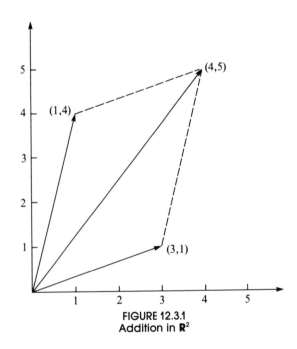

FIGURE 12.3.1
Addition in $\mathbf{R}^2$

(2) The vector $(a_1, a_2, \ldots, a_n) \in \mathbf{R}^n$ is referred to as an n-tuple.

(3) For those familiar with calculus, we are expressing the vector $\mathbf{x} = a_1\mathbf{i} + a_2\mathbf{j} + a_3\mathbf{k} \in \mathbf{R}^3$ as (a_1, a_2, a_3). This allows us to discuss vectors in $\mathbf{R}^n$ in much simpler notation.

In many situations a vector space V is given and we would like to describe the whole vector space by the smallest number of essential reference vectors. An example of this is the description of $\mathbf{R}^2$, the xy plane, via the x and y axes. Again our concepts must be algebraic in nature so we are not restricted solely to geometric considerations.

Definition: Linear Combination. *A vector* $\mathbf{y}$ *from a vector space V (over* $\mathbf{R}$*) is called a linear combination of the vectors* $\mathbf{x}_1, \mathbf{x}_2, \ldots, \mathbf{x}_n$ *if there exist scalars* $a_1, a_2, \ldots, a_n$ *from* $\mathbf{R}$ *such that* $\mathbf{y} = a_1\mathbf{x}_1 + a_2\mathbf{x}_2 + \cdots + a_n\mathbf{x}_n$.

Example 12.3.3. The vector $(2, 3)$ from $\mathbf{R}^2$ is a linear combination of the vectors $(1, 0)$ and $(0, 1)$, since $(2, 3) = 2(1, 0) + 3(0, 1)$.

Example 12.3.4. Prove that the vector $(5, 4)$ is a linear combination of the vectors $(4, 1)$ and $(1, 3)$. By the definition we must show that there exist scalars a_1 and a_2 such that:

$$(5, 4) = a_1(4, 1) + a_2(1, 3), \text{ which reduces to}$$
$$(5, 4) = (4a_1 + a_2, a_1 + 3a_2), \text{ which gives}$$
$$\begin{cases} 4a_1 + a_2 = 5, \\ a_1 + 3a_2 = 4 \end{cases} \text{ so that}$$
$$a_1 = 1 \text{ and } a_2 = 1.$$

Another way of looking at the above example is if we replace a_1 and a_2 both by 1, then the two vectors $(4, 1)$ and $(1, 3)$ produce, or generate, the vector $(5, 4)$. Of course, if we replace a_1 and a_2 by different scalars, we can generate more vectors from $\mathbf{R}^2$. If $a_1 = 3$ and $a_2 = -2$, then $a_1(4, 1) + a_2(1, 3) = 3(4, 1) - 2(1, 3) = (12-2, 3-3) = (10, 0)$. Will the vectors $(4, 1)$ and $(1, 3)$ generate all vectors in $\mathbf{R}^2$?

Example 12.3.5. To see if this is so, let (a, b) be an arbitrary vector in $\mathbf{R}^2$ and see if we can always find scalars a_1 and a_2 such that $a_1(4, 1) + a_2(1, 3) = (a, b)$. This is equivalent to solving the following system of equations:

$$4a_1 + a_2 = a$$
$$a_1 + 3a_2 = b,$$

which always has solutions for a_1 and a_2 regardless of the values of the real numbers a and b. Why? We formalize in a definition:

Definition: Generate. *Let* $\{x_1, x_2, \ldots, x_n\}$ *be a set of vectors in a vector space V over* **R**. *This set is said to generate, or span, V if, given any vector* $y \in V$, *we can always find scalars* $a_1, a_2, \ldots, a_n$ *such that* $y = a_1 x_1 + a_2 x_2 + \cdots + a_n x_n$. *A set that generates a vector space is called a generating set.*

We now give a geometric interpretation of the above.

We know that the standard coordinate system, x axis and y axis, were introduced in basic algebra in order to describe all points in the xy plane geometrically. It is also quite clear that to describe any point in the plane we need exactly two axes. Form a new coordinate system the following way:

Draw the vector $(4, 1)$ and an axis through $(4, 1)$ and label it the x' axis. Also draw the vector $(1, 3)$ and an axis through it to be labeled the y' axis. Draw the coordinate grid for the axis, that is, lines parallel, and let the unit lengths of this "new" plane be the lengths of the respective vectors, $(4, 1)$ and $(1, 3)$, so that we obtain Figure 12.3.2.

From Example 12.3.5 and Figure 12.3.2, we see that any position (point, vector) on the plane can be described (named) using the old (standard) axes or our new $x'y'$ axes. Hence the position which had the name $(1, 3)$ in reference to the standard axes has the name $(1, 0)$ with respect to the $x'y'$ axes, or, in the phraseology of linear algebra, the *coordinates* of the point $(1, 3)$ with respect to the $x'y'$ axes are $(1, 0)$.

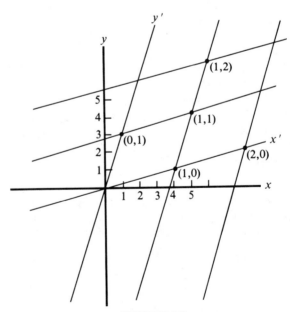

FIGURE 12.3.2

Example 12.3.6. From Example 12.3.4 we found that if we choose $a_1 = 1$ and $a_2 = 1$, then the two vectors $(4, 1)$ and $(1, 3)$ generate the vector $(5, 4)$. Another geometric interpretation of this problem is that the coordinates of the position $(5, 4)$ with respect to the $x'y'$ axes of Figure 12.3.1 is $(1, 1)$. In other words, a position in the plane has the name $(5, 4)$ in reference to the xy axes and the same position has the name $(1, 1)$ in reference to the $x'y'$ axes.

From the above, it is clear that we can use different axes to describe points or vectors in the plane. No matter what choice we use, we want to be able to describe each position in a unique manner. This is not the case in Figure 12.3.3. Any point in the plane could be described via the $x'y'$ axes, the $x'z'$ axes or the $y'z'$ axes. Therefore, in this case, a single point would have three different names, a very confusing situation.

We formalize the above discussion in two definitions and a theorem.

Definition: Linear Independence. *Let $\{x_1, x_2, \ldots, x_n\}$ be a set of vectors in a vector space V (over $\mathbf{R}$). This set is called linearly independent if the only solution to the equation $a_1x_1 + a_2x_2 + \cdots + a_nx_n = \mathbf{0}$ is that $a_1 = a_2 = \cdots = a_n = 0$. Otherwise the set is called a linearly dependent set.*

Definition: Basis. *A set of vectors $\{x_1, x_2, \ldots, x_n\}$ forms a basis for a vector space V (over $\mathbf{R}$) if:*

(1) the set generates V, and
(2) the set is linearly independent.

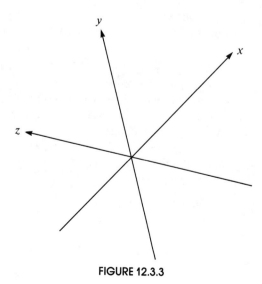

FIGURE 12.3.3

Theorem 12.3.1. *If* $\{x_1, x_2, \ldots, x_n\}$ *is a basis for a vector space V over* **R**, *then any vector* $y \in V$ *can be uniquely expressed as a linear combination of the* x_i's.

Proof: Assume that $\{x_1, x_2, \ldots, x_n\}$ is a basis for V over **R**. We must prove two facts:

    (1) each vector $y \in V$ can be expressed as a linear combination of the x_i's, and

    (2) each such expression is unique.

The proof of Part 1 is easy since the x_i's must generate V. (Why?) Therefore the proof of Part 1 is complete.

    The proof of Part 2 is also reasonable. We follow the standard approach for any uniqueness facts. Let y be any vector in V and assume that there are two different ways of expressing y, namely

$$y = a_1 x_1 + a_2 x_2 + \cdots + a_n x_n$$

and

$$y = b_1 x_1 + b_2 x_2 + \cdots + b_n x_n,$$

where at least one a_i is different from the corresponding b_i. Then $a_1 x_1 + a_2 x_2 + \cdots + a_n x_n = b_1 x_1 + b_2 x_2 + \cdots + b_n x_n$, so that $(a_1 - b_1) x_1 + (a_2 - b_2) x_2 + \cdots + (a_n - b_n) x_n = 0$. Now a crucial statement: since the x_i's form a linearly independent set, the only solution to the previous equation is that each of the coefficients must equal zero, so $a_i - b_i = 0$ for $i = 1, \ldots, n$; hence $a_i = b_i$ for all i. This contradicts the above assumption, so each vector $y \in V$ can only be expressed in one and only one way. #

    Theorem 12.3.1, together with the previous examples, gives us a clear insight into the meaning of linear independence, namely uniqueness.

Example 12.3.7. Prove that $\{(1, 1), (-1, 1)\}$ is a basis for $\mathbf{R}^2$ over **R** and explain what this means geometrically. First we must show that the vectors $(1, 1)$ and $(-1, 1)$ generate all of $\mathbf{R}^2$. This we can do by imitating Example 12.3.5 and leave it to the reader (see Exercise 10 of this section). Secondly, we must prove that the set is linearly independent.

    Let a_1 and a_2 be scalars such that $a_1(1, 1) + a_2(-1, 1) = (0, 0)$. We must prove that the only solution to the equation is that a_1 and a_2 must both equal zero. The above equation becomes:

$$(a_1 - a_2, a_1 + a_2) = (0, 0),$$

which gives

$$\begin{cases} a_1 - a_2 = 0 \text{ and} \\ a_1 + a_2 = 0. \end{cases}$$

so that $a_1 = 0$ and $a_2 = 0$ is the only solution to the system and therefore the set is linearly independent. To explain the results geometrically, note through Exercise 12, Part a, that the coordinates of each vector $y \in \mathbf{R}^2$ can be determined uniquely using the vectors $(1, 1)$ and $(-1, 1)$.

The concept of dimension is quite obvious for those vector spaces which have an immediate geometric interpretation. For example, the dimension of $\mathbf{R}^2$ is two and that of $\mathbf{R}^3$ is three. How can we define the concept of dimension algebraically so that the resulting definition correlates with that of $\mathbf{R}^2$ and $\mathbf{R}^3$? First we need a theorem which we will state without proof.

Theorem 12.3.2. *If V is a vector space with a basis containing exactly n elements, then all bases of V contain n elements.*

Definition: Dimension. *Let V be a vector space over $\mathbf{R}$ with basis $\{x_1, x_2, \ldots, x_n\}$. Then the dimension of V is n. Notation: dim $V = n$.*

EXERCISES FOR SECTION 12.3

A Exercises

1. If $a = 2$, $b = (-3)$,

$$A = \begin{bmatrix} 1 & 0 & -1 \\ 2 & 3 & 4 \end{bmatrix}, \quad B = \begin{bmatrix} 2 & -2 & 3 \\ 4 & 5 & 8 \end{bmatrix}, \quad \text{and} \quad C = \begin{bmatrix} 1 & 0 & 0 \\ 3 & 2 & -2 \end{bmatrix},$$

   verify that all ten properties of the definition of a vector space are true for $M_{2 \times 3}(\mathbf{R})$ with these values.

2. Let $a = 3$, $b = 4$, $x = (-1, 3)$, $y = (2, 3)$, and $z = (1, 0)$. Verify that all ten properties of the definition of a vector space are true for $\mathbf{R}^2$ for these values.

3. (a) Verify that $M_{2 \times 3}(\mathbf{R})$ is a vector space over $\mathbf{R}$.
   (b) Is $M_{m \times n}(\mathbf{R})$ a vector space over $\mathbf{R}$?

4. (a) Verify that $\mathbf{R}^2$ is a vector space over $\mathbf{R}$.
   (b) Is $\mathbf{R}^n$ a vector space over $\mathbf{R}$ for every positive integer n?

5. Let $P^3 = \{a_0 + a_1x + a_2x^2 + a_3x^3 \mid a_0, a_1, a_2, a_3 \in \mathbf{R}\}$; that is, P^3 is the set of all polynomials in x having real coefficients with degree ≤ 3. Verify that P^3 is a vector space over $\mathbf{R}$.

6. For each of the following, express the vector y as a linear combination of the vectors x_1 and x_2:
   (a) $y = (5, 6)$, $x_1 = (1, 0)$, and $x_2 = (0, 1)$
   (b) $y = (2, 1)$, $x_1 = (2, 1)$, and $x_2 = (1, 1)$
   (c) $y = (3, 4)$, $x_1 = (1, 1)$, and $x_2 = (-1, 1)$

7. Express the vector $\begin{bmatrix} 1 & 2 \\ -3 & 3 \end{bmatrix}$ as a linear combination of

$$\begin{bmatrix} 1 & 1 \\ 1 & 1 \end{bmatrix}, \quad \begin{bmatrix} -1 & 5 \\ 2 & 1 \end{bmatrix}, \quad \begin{bmatrix} 0 & 1 \\ 1 & 1 \end{bmatrix}, \quad \text{and} \quad \begin{bmatrix} 0 & 0 \\ 0 & 1 \end{bmatrix}.$$

8. Express the vector $x^3 - 4x^2 + 3$ as a linear combination of the vectors $1, x, x^2,$ and x^3.

9. (a) Show that the set $\{x_1, x_2\}$ generates R^2 for each of the parts in Exercise 6 of this section.
   (b) Show that $\{x_1, x_2, x_3\}$ generates R^2 where $x_1 = (1, 1), x_2 = (3, 4),$ and $x_3 = (-1, 5)$.
   (c) Create a set of four or more vectors which generates R^2.
   (d) What is the smallest number of vectors needed to generate R^2? R^n?
   (e) Show that

$$\begin{bmatrix} 1 & 0 \\ 0 & 0 \end{bmatrix}, \quad \begin{bmatrix} 0 & 1 \\ 0 & 0 \end{bmatrix}, \quad \begin{bmatrix} 0 & 0 \\ 1 & 0 \end{bmatrix}, \quad \begin{bmatrix} 0 & 0 \\ 0 & 1 \end{bmatrix}$$

   generates $M_{2\times2}(R)$.
   (f) Show that $\{1, x, x^2, x^3\}$ generates P^3.

10. Complete Example 12.3.7 by showing that $\{(1, 1), (-1, 1)\}$ generates R^2.

11. (a) Prove that $\{(4, 1), (1, 3)\}$ is a basis for R^2 over R.
    (b) Prove that $\{(1, 0), (3, 4)\}$ is a basis for R^2 over R.
    (c) Prove that $\{(1, 0, -1), (2, 1, 1), (1, -3, -1)\}$ is a basis for R^3 over R.
    (d) Prove that the sets in Exercise 9, Parts e and f, form bases of the respective vector spaces.

12. (a) Determine the coordinates of the points or vectors $(3, 4), (-1, 1),$ and $(1, 1)$ with respect to the basis $\{(1, 1), (-1, 1)\}$. Interpret your results geometrically.
    (b) Determine the coordinates of the point or vector $(3, 5, 6)$ with respect to the basis $\{(1, 0, 0), (0, 1, 0), (0, 0, 1)\}$. Explain why this basis is called the standard basis for R^3.

13. (a) Let $y_1 = (1, 3, 5, 9), y_2 = (5, 7, 6, 3),$ and $c = 2$. Find $y_1 + y_2$ and cy_1.
    (b) Let $f_1(x) = 1 + 3x + 5x^2 + 9x^3, f_2(x) = 5 + 7x + 6x^2 + 3x^3,$ and $c = 2$. Find $f_1(x) + f_2(x)$ and $cf_1(x)$.
    (c) Let $A = \begin{bmatrix} 1 & 3 \\ 5 & 9 \end{bmatrix}, B = \begin{bmatrix} 5 & 7 \\ 6 & 3 \end{bmatrix},$ and $c = 2$. Find $A + B$ and cA.
    (d) Are the vector spaces $R^4, P^3,$ and $M_{2\times2}(R)$ isomorphic to each other? Discuss with reference to Parts a, b, and c.

12.4   The Diagonalization Process

We now have the background to understand the main ideas behind the diagonalization process.

Definition: Eigenvalue, Eigenvector.   *Let A be an n × n matrix over* **R**. *λ is an eigenvalue of A if for some non-zero column vector* **X** ∈ **R**^{n} *we have* A**X** = λ**X**. **X** *is called an eigenvector corresponding to the eigenvalue* λ.

Example 12.4.1.   Find the eigenvalues and corresponding eigenvectors of the matrix

$$A = \begin{bmatrix} 2 & 1 \\ 2 & 3 \end{bmatrix}.$$

We want to find

$$\mathbf{X} = \begin{bmatrix} x_1 \\ x_2 \end{bmatrix},$$

and λ, such that A**X** = λ**X**

$$\Leftrightarrow \begin{bmatrix} 2 & 1 \\ 2 & 3 \end{bmatrix}\begin{bmatrix} x_1 \\ x_2 \end{bmatrix} = \lambda \begin{bmatrix} x_1 \\ x_2 \end{bmatrix}$$

$$\Leftrightarrow \begin{bmatrix} 2 & 1 \\ 2 & 3 \end{bmatrix}\begin{bmatrix} x_1 \\ x_2 \end{bmatrix} - \lambda \begin{bmatrix} x_1 \\ x_2 \end{bmatrix} = \begin{bmatrix} 0 \\ 0 \end{bmatrix}$$

$$\Leftrightarrow \begin{bmatrix} 2 & 1 \\ 2 & 3 \end{bmatrix} - \lambda \begin{bmatrix} 1 & 0 \\ 0 & 1 \end{bmatrix}\begin{bmatrix} x_1 \\ x_2 \end{bmatrix} = \begin{bmatrix} 0 \\ 0 \end{bmatrix}$$

$$\Leftrightarrow \begin{bmatrix} 2 - \lambda & 1 \\ 2 & 3 - \lambda \end{bmatrix}\begin{bmatrix} x_1 \\ x_2 \end{bmatrix} = \begin{bmatrix} 0 \\ 0 \end{bmatrix}. \qquad (12.4a)$$

The last matrix equation will have non-zero solutions if and only if

$$det \begin{bmatrix} 2 - \lambda & 1 \\ 2 & 3 - \lambda \end{bmatrix} = 0$$

or $(2 - \lambda)(3 - \lambda) - 2 = 0$, which becomes $\lambda^2 - 5\lambda + 4 = 0$; so λ = 1 or λ = 4. We now have to find the eigenvectors associated with each eigenvalue.

Case 1.   If λ = 1, then Equation 12.4a becomes:

$$\begin{bmatrix} 2 - 1 & 1 \\ 2 & 3 - 1 \end{bmatrix}\begin{bmatrix} x_1 \\ x_2 \end{bmatrix} = \begin{bmatrix} 0 \\ 0 \end{bmatrix}$$

$$\begin{bmatrix} 1 & 1 \\ 2 & 2 \end{bmatrix}\begin{bmatrix} x_1 \\ x_2 \end{bmatrix} = \begin{bmatrix} 0 \\ 0 \end{bmatrix},$$

which reduces to the single equation, $x_1 + x_2 = 0$. From this, $x_2 = -x_1$. This means the solution set of this equation is (in column notation)

$$E_1 = \left\{ \begin{bmatrix} x_1 \\ -x_1 \end{bmatrix} \,\middle|\, x_1 \in \mathbf{R} \right\}.$$

So any column vector of the form

$$\begin{bmatrix} x_1 \\ -x_1 \end{bmatrix},$$

where x_1 is any non-zero real number, is an eigenvector associated with the eigenvalue $\lambda = 1$. The reader should verify that, for example,

$$\begin{bmatrix} 2 & 1 \\ 2 & 3 \end{bmatrix} \begin{bmatrix} 2/3 \\ -2/3 \end{bmatrix} = 1 \begin{bmatrix} 2/3 \\ -2/3 \end{bmatrix},$$

so that $\begin{bmatrix} 2/3 \\ -2/3 \end{bmatrix}$ is an eigenvector of A with associated eigenvalue 1.

Case 2.   If $\lambda = 4$ then equation 12.4.1 becomes:

$$\begin{bmatrix} 2-4 & 1 \\ 2 & 3-4 \end{bmatrix} \begin{bmatrix} x_1 \\ x_2 \end{bmatrix} = \begin{bmatrix} 0 \\ 0 \end{bmatrix},$$

which reduces to the single equation $-2x_1 + x_2 = 0$, so that $x_2 = 2x_1$. The solution set of the equation is

$$E_2 = \left\{ \begin{bmatrix} x_1 \\ 2x_1 \end{bmatrix} \,\middle|\, x_1 \in \mathbf{R} \right\}.$$

Therefore, all eigenvectors of A associated with the eigenvalue $\lambda = 4$ are of the form $\begin{bmatrix} x_1 \\ 2x_1 \end{bmatrix}$, where x_1 can be any non-zero number.

The following theorems summarize the more important aspects of this example:

Theorem 12.4.1.   *Let A be any $n \times n$ matrix over* $\mathbf{R}$. *Then $\lambda \in \mathbf{R}$ is an eigenvalue of A if and only if $det(A - \lambda I) = 0$.*

The equation $det(A - \lambda I) = 0$ is called the *characteristic equation* and the left side of this equation is called the *characteristic polynomial* of A.

Theorem 12.4.2.   *Non-zero eigenvectors corresponding to distinct eigenvalues are linearly independent.*

The solution space of $(A - \lambda I)\mathbf{X} = \mathbf{0}$ is called the *eigenspace of A corresponding to* λ. This terminology is justified by Exercise 2 of this section.

We now consider the main aim of this section. Given an $n \times n$ (square) matrix A, we would like to "change" A into a diagonal matrix D, perform our tasks with the simpler matrix D, and then describe the results in terms of the given matrix A.

Definition: Diagonalizable. *An $n \times n$ matrix A is called diagonalizable if there exists an invertible $n \times n$ matrix P such that $P^{-1}AP$ is a diagonal matrix D. The matrix P is said to diagonalize the matrix A.*

Example 12.4.2. We will now diagonalize the matrix A of example 12.4.1. Form the matrix P as follows: Let $P^{(1)}$ be the first column of the matrix P. Choose for $P^{(1)}$ any eigenvector from E_1. We may as well choose a simple vector in E_1, so $P^{(1)} = \begin{bmatrix} 1 \\ -1 \end{bmatrix}$ is our candidate. Similarly, let $P^{(2)}$ be the second column of P, and choose for $P^{(2)}$ any eigenvector from E_2. The vector $P^{(2)} = \begin{bmatrix} 1 \\ 2 \end{bmatrix}$ is a reasonable choice, thus $P = \begin{bmatrix} 1 & 1 \\ -1 & 2 \end{bmatrix}$ and $P^{-1} = 1/3 \begin{bmatrix} 2 & -1 \\ 1 & 1 \end{bmatrix}$, so that $P^{-1}AP = 1/3 \begin{bmatrix} 2 & -1 \\ 1 & 1 \end{bmatrix}\begin{bmatrix} 2 & 1 \\ 2 & 3 \end{bmatrix}\begin{bmatrix} 1 & 1 \\ -1 & 2 \end{bmatrix} = \begin{bmatrix} 1 & 0 \\ 0 & 4 \end{bmatrix} = D.$

Note that the elements on the main diagonal of D are the eigenvalues of A, where D_{ii} is the eigenvalue corresponding to the eigenvector $P^{(i)}$.

Remarks:

(1) The first step in the diagonalization process is the determination of the eigenvalues. The ordering of the eigenvalues is purely arbitrary. If we choose $\lambda = 4$ first and $\lambda = 1$ second, the columns of P would be interchanged and D would be $\begin{bmatrix} 4 & 0 \\ 0 & 1 \end{bmatrix}$ (see Exercise 3c of this section). Nonetheless, the final outcome of the application to which we are applying the diagonalization process would be the same.

(2) If A is an $n \times n$ matrix with *distinct* eigenvalues, then P is also an $n \times n$ matrix whose columns $P^{(1)}, P^{(2)}, \ldots, P^{(n)}$ are n linearly independent vectors.

Example 12.4.3. Diagonalize the matrix

$$A = \begin{bmatrix} 1 & 0 & 0 \\ 0 & -5 & 3 \\ 0 & -6 & 4 \end{bmatrix}:$$

$$det(A - \lambda I) = det\begin{bmatrix} 1-\lambda & 0 & 0 \\ 0 & -5-\lambda & 3 \\ 0 & -6 & 4-\lambda \end{bmatrix}$$

$$= (1 - \lambda)\begin{vmatrix} -5-\lambda & 3 \\ -6 & 4-\lambda \end{vmatrix}$$

$$= (1 - \lambda)[(-5 - \lambda)(4 - \lambda) + 18]$$

$$= (1 - \lambda)(\lambda^2 + \lambda - 2).$$

Hence, the equation $det(A - \lambda I) = 0$ becomes

$$(1 - \lambda)(\lambda^2 + \lambda - 2) = 0 \Leftrightarrow (1 - \lambda)(\lambda - 1)(\lambda + 2) = 0$$

Therefore, our eigenvalues for the matrix A are $\lambda = 1$ and $\lambda = -2$. We note that we do not have three distinct eigenvalues, but we proceed as in the previous example to investigate the outcome of the problem.

Case 1. If $\lambda = -2$, then the equation $(A - \lambda I)\mathbf{X} = \mathbf{0}$ becomes

$$\begin{bmatrix} 3 & 0 & 0 \\ 0 & -3 & 3 \\ 0 & -6 & 6 \end{bmatrix} \begin{bmatrix} x_1 \\ x_2 \\ x_3 \end{bmatrix} = \begin{bmatrix} 0 \\ 0 \\ 0 \end{bmatrix},$$

which reduces to

$$\begin{cases} 3x_1 = 0 \\ -3x_2 + 3x_3 = 0 \end{cases} \qquad \text{or } x_1 = 0 \text{ and } x_2 = x_3.$$

Therefore, the solution, or eigenspace, corresponding to $\lambda = -2$ is

$$E_1 = \left\{ \begin{bmatrix} 0 \\ x_3 \\ x_3 \end{bmatrix} \,\middle|\, x_3 \in \mathbf{R} \right\},$$

so that

$$\begin{bmatrix} 0 \\ 1 \\ 1 \end{bmatrix}$$

is an eigenvector corresponding to the eigenvalue $\lambda = -2$, and can be used for our first column of P;

$$\text{take } P^{(1)} = \begin{bmatrix} 0 \\ 1 \\ 1 \end{bmatrix}.$$

Before we continue, we make the following useful observations:

(1) We could show, as in Exercise 2a of this section, that E_1 is a subspace of $\mathbf{R}^3$ with basis

$$\left\{ \begin{bmatrix} 0 \\ 1 \\ 1 \end{bmatrix} \right\},$$

or $\{(0, 1, 1)\}$, if you prefer row notation, so that dim $E_1 = 1$.

(2) Item 1 above can easily be seen by rewriting E_1 as

$$\left\{ x_3 \begin{bmatrix} 0 \\ 1 \\ 1 \end{bmatrix} \,\middle|\, x_3 \in \mathbf{R} \right\},$$

so every vector in E_1 can be uniquely expressed as a linear combination of the vector

$$\begin{bmatrix} 0 \\ 1 \\ 1 \end{bmatrix}.$$

The definition of basis tells us that

$$\left\{ \begin{bmatrix} 0 \\ 1 \\ 1 \end{bmatrix} \right\}$$

is a basis of E_1.

Case 2.   If $\lambda = 1$, then the equation $(A - \lambda I)\mathbf{X} = \mathbf{0}$ becomes

$$\begin{bmatrix} 0 & 0 & 0 \\ 0 & -6 & 3 \\ 0 & -6 & 3 \end{bmatrix} \begin{bmatrix} x_1 \\ x_2 \\ x_3 \end{bmatrix} = \begin{bmatrix} 0 \\ 0 \\ 0 \end{bmatrix},$$

which reduces to $-6x_2 + 3x_3 = 0$ or $-2x_2 + x_3 = 0$, so that $x_3 = 2x_2$. Hence, the eigenspace corresponding to $\lambda = 1$ is

$$E_2 = \left\{ \begin{bmatrix} x_1 \\ x_2 \\ 2x_2 \end{bmatrix} \middle| x_1, x_2 \in \mathbf{R} \right\}.$$

We note that the solution set contains two independent variables, x_1 and x_2. Further, note that we cannot express E_2 as a linear combination of a single vector as in Observation 2 above. However, after some experimentation with E_2, we see that we can rewrite E_2 as

$$E_2 = \left\{ x_1 \begin{bmatrix} 1 \\ 0 \\ 0 \end{bmatrix} + x_2 \begin{bmatrix} 0 \\ 1 \\ 2 \end{bmatrix} \middle| x_1 x_2 \in \mathbf{R} \right\},$$

which is a linear combination of two linearly independent vectors,

$$\begin{bmatrix} 1 \\ 0 \\ 0 \end{bmatrix} \quad \text{and} \quad \begin{bmatrix} 0 \\ 1 \\ 2 \end{bmatrix}.$$

Therefore, the eigenspace E_2 is a subspace of $\mathbf{R}^3$ with basis

$$\left\{ \begin{bmatrix} 1 \\ 0 \\ 0 \end{bmatrix}, \begin{bmatrix} 0 \\ 1 \\ 2 \end{bmatrix} \right\},$$

and *dim* $E_2 = 2$.

What this means with respect to the diagonalization process is that $\lambda = 1$ gives us both Column 2 and Column 3. The order is not important.

$$\text{Let } P^{(2)} = \begin{bmatrix} 1 \\ 0 \\ 0 \end{bmatrix} \text{ and } P^{(3)} = \begin{bmatrix} 0 \\ 1 \\ 2 \end{bmatrix}, \text{ so that } P = \begin{bmatrix} 0 & 1 & 0 \\ 1 & 0 & 1 \\ 1 & 0 & 2 \end{bmatrix}.$$

The reader can verify (see Exercise 5 of this section) that

$$P^{-1}AP = \begin{bmatrix} -2 & 0 & 0 \\ 0 & 1 & 0 \\ 0 & 0 & 1 \end{bmatrix}.$$

Remark: Assume the equation $det(A - I) = 0$ gives $\lambda^3 + 2\lambda^2 - \lambda - 2 = 0$. The solution to this equation becomes quite easy if we use the fact that all integer solutions (if they exist) of any polynomial with integral coefficients must be factors of the constant term. Hence, the above cubic may only have $\pm 1, \pm 2$ as integral solutions. We try $\lambda = 1$, and since substitution into the equation gives 0, $\lambda = 1$ is a solution. To determine the other (two) solutions, divide $\lambda^3 + 2\lambda^2 - \lambda - 2$ by $\lambda - 1$ to obtain $\lambda^2 + \lambda - 2$ so that

$$\lambda^3 + 2\lambda^2 - \lambda - 2 = 0$$
$$\Leftrightarrow (\lambda - 1)(\lambda^2 + \lambda - 2) = 0$$
$$\Leftrightarrow (\lambda - 1)^2(\lambda + 2) = 0,$$

giving $\lambda = 1$ and $\lambda = -2$.

In doing Example 12.4.3, the given 3×3 matrix A produced only two, not three, distinct eigenvalues, yet we were still able to diagonalize A. The reason we were able to diagonalize A was included in Example 12.4.3: that is, with only two different eigenvalues, we were able to generate three linearly independent eigenvectors. Again the main idea is to produce a matrix P. If A is an $n \times n$ matrix, P will be an $n \times n$ matrix, and its n columns must be linearly independent vectors. The main question in the study of diagonalizability is "When can it be done?" This is summarized in the following theorem.

Theorem 12.4.4. *Let A be an $n \times n$ matrix over R. Then A is diagonalizable if and only if A has n linearly independent eigenvectors.*

Proof: ($\Leftarrow$) Assume that A has linearly independent eigenvectors, $P^{(1)}$, $P^{(2)}, \ldots, P^{(n)}$, with corresponding eigenvalues, $\lambda_1, \lambda_2, \ldots, \lambda_n$. We want to prove that A is diagonalizable. The columns of the $n \times n$ matrix AP are $AP^{(1)}, AP^{(2)}, \ldots, AP^{(n)}$ (see Exercise 7 of this section). Then, since the $P^{(i)}$ are eigenvectors of P with associated eigenvalues λ_i, we have $AP^{(i)} = \lambda_i P^{(i)}$ for $i = 1, 2, \ldots, n$. But this means that $AP = PD$ (Why?), where

$$D = \begin{bmatrix} \lambda_1 & 0 & 0 & \cdots & 0 \\ 0 & \lambda_2 & 0 & \cdots & 0 \\ \vdots & & & & \vdots \\ 0 & & \cdots & & \lambda_n \end{bmatrix},$$

so that $P^{-1}AP = D$, which is the definition of the diagonalizability of A.

($\Rightarrow$) The proof in this direction involves a concept that is not covered in this text (rank of a matrix); however, it is not beyond the grasp of better students.

We now give an example of a matrix which is not diagonalizable.

Example 12.4.4. Let us attempt to diagonalize the matrix

$$A = \begin{bmatrix} 1 & 0 & 0 \\ 0 & 2 & 1 \\ 1 & -1 & 4 \end{bmatrix}.$$

$$det(A - \lambda I) = det \begin{bmatrix} 1-\lambda & 0 & 0 \\ 0 & 2-\lambda & 1 \\ 1 & -1 & 4-\lambda \end{bmatrix}$$

$$= (1 - \lambda) \begin{vmatrix} 2-\lambda & 1 \\ -1 & 4-\lambda \end{vmatrix}$$

$$= (1 - \lambda)((2 - \lambda)(4 - \lambda) + 1)$$

$$= (1 - \lambda)(\lambda - 3)^2,$$

and $(1 - \lambda)(\lambda - 3)^2 = 0$ gives $\lambda = 1$ and $\lambda = 3$.

Case 1. If $\lambda = 1$, then the equation $(A - \lambda I)X = \mathbf{0}$ becomes

$$\begin{bmatrix} 0 & 0 & 0 \\ 0 & 1 & 1 \\ 1 & -1 & 3 \end{bmatrix} \begin{bmatrix} x_1 \\ x_2 \\ x_3 \end{bmatrix} = \begin{bmatrix} 0 \\ 0 \\ 0 \end{bmatrix}$$

$$\Leftrightarrow \begin{cases} x_2 + x_3 = 0 \\ x_1 - x_2 + 3x_3 = 0. \end{cases}$$

This gives $x_2 = -x_3$ and $x_1 = -4x_3$, so that the eigenspace associated with $\lambda = 1$ is

$$E_1 = \left\{ \begin{bmatrix} 4x_3 \\ -x_3 \\ x_3 \end{bmatrix} \mid x_3 \in \mathbf{R} \right\},$$

which is a vector space with basis

$$\left\{ \begin{bmatrix} 4 \\ -1 \\ -1 \end{bmatrix} \right\},$$

so *dim* $E_1 = 1$. Let the first column of P, namely $P^{(1)}$, =

$$\begin{bmatrix} -4 \\ -1 \\ 1 \end{bmatrix}.$$

So far, this example is the same as the preceeding one.

Case 2.   If $\lambda = 3$, then the equation $(A - \lambda I)X = 0$ becomes

$$\begin{bmatrix} -2 & 0 & 0 \\ 0 & -1 & 1 \\ 1 & -1 & 1 \end{bmatrix} \begin{bmatrix} x_1 \\ x_2 \\ x_3 \end{bmatrix} = \begin{bmatrix} 0 \\ 0 \\ 0 \end{bmatrix},$$

giving

$$\begin{cases} -2x_1 = 0 \\ -x_2 + x_3 = 0 \\ x_1 - x_2 + x_3 = 0. \end{cases}$$

From the first equation, we obtain $x_1 = 0$, so that the eigenspace associated with $\lambda = 3$ is

$$E_2 = \left\{ \begin{bmatrix} 0 \\ x_3 \\ x_3 \end{bmatrix} \,\middle|\, x_3 \in \mathbf{R} \right\},$$

which is a vector space with basis

$$\left\{ \begin{bmatrix} 0 \\ 1 \\ 1 \end{bmatrix} \right\},$$

so *dim* $E_2 = 1$. So $\lambda = 3$ produces only one column of P. Since we began with only two eigenvalues, we had hoped that one of them would produce a vector space of dimension two, or, in matrix language, two linearly independent columns of P. Since A does not have three linearly independent eigenvectors by Theorem 12.4.4, A cannot be diagonalized.

EXERCISES FOR SECTION 12.4

A Exercises

1. List three different eigenvectors of the matrix A of Example 12.4.1 associated with the eigenvalues below. Verify your results.

    (a) $\lambda = 1$
    (b) $\lambda = 4$
    (c) Choose one of the three eigenvectors corresponding to $\lambda = 1$ and

one of the three eigenvectors corresponding to $\lambda = 4$, and show that the two chosen vectors are linearly independent.

2. (a) Prove that E_1 and E_2 (for Example 12.4.1) are vector spaces over **R**. Since they are also subsets of $\mathbf{R}^2$, they are called *sub-vector spaces*, or *subspaces* for short, of $\mathbf{R}^2$. Since these are subspaces consisting of eigenvectors, they are called eigenspaces.

   (b) Use the definition of dimension in the previous section to find *dim* E_1 and *dim* E_2. Note that *dim* E_1 + *dim* E_2 = *dim* $\mathbf{R}^2$. This is not a coincidence.

3. (a) Verify that $P^{-1}AP$ is indeed equal to $\begin{bmatrix} 1 & 0 \\ 0 & 4 \end{bmatrix}$, as indicated in Example 12.4.2.

   (b) Take any two non-zero eigenvectors of the matrix A of Example 12.4.2 and verify that $P^{-1}AP = \begin{bmatrix} 1 & 0 \\ 0 & 4 \end{bmatrix}$.

   (c) Choose $P^{(1)} = \begin{bmatrix} 1 \\ 2 \end{bmatrix}$ and $P^{(2)} = \begin{bmatrix} 1 \\ -1 \end{bmatrix}$, and find $P^{-1}AP$.

4. Diagonalize the following, if possible:

   (a) $\begin{bmatrix} 1 & 2 \\ 3 & 2 \end{bmatrix}$ (b) $\begin{bmatrix} -2 & 1 \\ -7 & 6 \end{bmatrix}$ (c) $\begin{bmatrix} 3 & 0 \\ 0 & 4 \end{bmatrix}$

   (d) $\begin{bmatrix} 0 & 1 \\ 1 & 1 \end{bmatrix}$ (e) $\begin{bmatrix} 2 & 1 \\ 4 & 2 \end{bmatrix}$ (f) $\begin{bmatrix} 2 & -1 \\ 1 & 0 \end{bmatrix}$

5. (a) Let

$$P = \begin{bmatrix} 0 & 1 & 0 \\ 1 & 0 & 1 \\ 1 & 0 & 2 \end{bmatrix},$$

   as in example 12.4.3, and verify that

$$P^{-1}AP = \begin{bmatrix} -2 & 0 & 0 \\ 0 & 1 & 0 \\ 0 & 0 & 1 \end{bmatrix}.$$

   (b) If you chose the columns of P in reverse order, what is $P^{-1}AP$?

6. Diagonalize, if possible, the following:

   (a) $\begin{bmatrix} 1 & -1 & 4 \\ 3 & 2 & -1 \\ 2 & 1 & -1 \end{bmatrix}$ (b) $\begin{bmatrix} 6 & 0 & 0 \\ 0 & 7 & -4 \\ 9 & 1 & 3 \end{bmatrix}$

   (c) $\begin{bmatrix} 1 & 3 & 6 \\ -3 & -5 & -6 \\ 3 & 3 & 4 \end{bmatrix}$ (d) $\begin{bmatrix} 1 & 1 & 0 \\ 1 & 0 & 1 \\ 0 & 1 & 1 \end{bmatrix}$

   (e) $\begin{bmatrix} 1 & -1 & 0 \\ -1 & 2 & -1 \\ 0 & -1 & 1 \end{bmatrix}$ (f) $\begin{bmatrix} 2 & -1 & 0 \\ -1 & 2 & -1 \\ 0 & -1 & 2 \end{bmatrix}$

B Exercise

7. (a) Let A and P be as in Example 12.4.3. Show that the columns of the matrix AP can be found by computing $AP^{(1)}$, $AP^{(2)}$, . . . , $AP^{(n)}$.
    (b) Let A and P be arbitrary $n \times n$ matrices. Prove that the columns of AP are $AP^{(1)}$, $AP^{(2)}$, . . . , $AP^{(n)}$.

12.5 Some Applications

A large and varied number of applications involve computations of powers of matrices. These applications can be found in science, the social sciences, economics, the analysis of relationships with groups, engineering, and, indeed, any area where mathematics is used and, therefore, where programs are to be developed. We will consider a few diverse examples here. A good introductory text for additional examples is the one by Williams.

To aid your understanding of the following examples, we develop a helpful technique to compute A^m for $m \geq 1$. If A can be diagonalized, then there is a matrix P such that $P^{-1}AP = D$, a diagonal matrix.

$$\Leftrightarrow \quad A = PDP^{-1}$$
$$\Leftrightarrow \quad A^m = PD^mP^{-1}$$

by Exercise 8a of Section 5.4. Of course, the last equation can also be derived, by the simple observation that

$$
\begin{aligned}
A^m &= (PDP^{-1})^m \\
&= (PDP^{-1})(PDP^{-1}) \cdots (PDP^{-1}) \quad (m \text{ factors}) \\
&= PD^mP^{-1}.
\end{aligned}
$$

Example 12.5.1: Recursion. Consider the Fibonacci sequence of Example 8.1.5:

$$F_0 = 1, F_1 = 1$$
$$F_k = F_{k-1} + F_{k-2}.$$

In order to formulate the problem in matrix form, we introduced the "dummy equation" $F_{k-1} = F_{k-1}$, so that now we have:

$$F_k = F_{k-1} + F_{k-2}$$
$$F_{k-1} = F_{k-1}.$$

Hence, in matrix form,

$$\begin{bmatrix} F_k \\ F_{k-1} \end{bmatrix} = \begin{bmatrix} 1 & 1 \\ 1 & 0 \end{bmatrix} \begin{bmatrix} F_{k-1} \\ F_{k-2} \end{bmatrix}$$

$$= A \begin{bmatrix} F_{k-1} \\ F_{k-2} \end{bmatrix} \quad \text{if } A = \begin{bmatrix} 1 & 1 \\ 1 & 0 \end{bmatrix}$$

$$= A^2 \begin{bmatrix} F_{k-2} \\ F_{k-3} \end{bmatrix} \qquad \text{Why?}$$

$$= A^3 \begin{bmatrix} F_{k-3} \\ F_{k-4} \end{bmatrix}$$

$$\vdots$$

$$= A^{k-1} \begin{bmatrix} F_1 \\ F_0 \end{bmatrix}$$

$$= A^{k-1} \begin{bmatrix} 1 \\ 1 \end{bmatrix} \qquad \text{Why?}$$

Next, by diagonalizing A and using the fact that $A^m = PD^mP^{-1}$, we can show that

$$F_k = \frac{1}{\sqrt{5}} \left[\left(\frac{1 + \sqrt{5}}{2} \right)^k - \left(\frac{1 - \sqrt{5}}{2} \right)^k \right]$$

(see Exercise 1a of this section).

Comments:

(1) An equation of the form $F_k = aF_{k-1} + bF_{k-2}$, where a and b are given constants, is sometimes referred to as a *homogeneous second-order* (to determine a term, we need to know the two predecessor terms) *linear difference equation*. The conditions $F_0 = c_0$ and $F_1 = c_1$ are called *initial conditions* and are given constants. Those of you who are familiar with differential equations may recognize that the above language parallels that in differential equations. Difference (recurrence) equations move forward discretely—that is, in a finite number of finite steps—while a differential equation moves continuously—that is, takes an infinite number of infinitesimal steps.

(2) A recurrence relationship of the form $F_k = aF_{k-1} + b$, where a and b are constants, is called a first-order difference equation. In order to write out the sequence, we need to know the (one) initial condition and a and b. Equations of this type can be solved similarly to the method outlined in Example 12.5.1 by introducing the superfluous equation $1 = 0 \cdot F_{k-1} + 1$ to obtain in matrix form:

$$\begin{bmatrix} F_k \\ 1 \end{bmatrix} = \begin{bmatrix} a & b \\ 0 & 1 \end{bmatrix} \begin{bmatrix} F_{k-1} \\ 1 \end{bmatrix}$$

$$= \begin{bmatrix} a & b \\ 0 & 1 \end{bmatrix}^k \begin{bmatrix} F_0 \\ 1 \end{bmatrix}. \qquad \text{Why?}$$

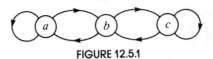

FIGURE 12.5.1

Example 12.5.2:   Graph Theory.   Consider the graph in Figure 12.5.1. From the procedures outlined in Section 6.4, the adjacency matrix of this graph is

$$A = \begin{bmatrix} 1 & 1 & 0 \\ 1 & 0 & 1 \\ 0 & 1 & 1 \end{bmatrix}.$$

Recall that A^k is the adjacency matrix of the relation r^k, where r is the relation $r = \{(a, a), (a, b), (b, a), (b, c), (c, b), (c, c)\}$ of the above graph. Also recall that in computing A^k, we used Boolean arithmetic. What happens if we use "regular" arithmetic? Then, for example,

$$A^2 = \begin{bmatrix} 2 & 1 & 1 \\ 1 & 2 & 1 \\ 1 & 1 & 2 \end{bmatrix}.$$

How can we interpret this? We note that $A_{33} = 2$ and that there are two paths of length two from node 3 (that is, c) to node 3 (c). Also, $A_{13} = 1$, and there is one path of length 2 from node a to node c. The reader should verify these claims from the graph in Figure 12.5.1.

Theorem 12.5.1.   *The entry $(A^k)_{ij}$ of A^k is the number of paths, or walks, of length k from node v_i to node v_j.*

How do we find A^k for possibly large values of k? From the discussion at the beginning of this section, we know that $A^k = PD^kP^{-1}$ if A is diagonalizable. We leave to the reader to show that $\lambda = 1, 2, -1$ are eigenvalues of A with eigenvectors

$$\begin{bmatrix} 1 \\ 0 \\ -1 \end{bmatrix}, \begin{bmatrix} 1 \\ 1 \\ 1 \end{bmatrix}, \begin{bmatrix} 1 \\ -2 \\ 1 \end{bmatrix}$$

respectively. So that

$$A^k = P \begin{bmatrix} 1 & 0 & 0 \\ 0 & 2^k & 0 \\ 0 & 0 & (-1)^k \end{bmatrix} P^{-1},$$

where

$$P^{-1} = \begin{bmatrix} 1/2 & 0 & -1/2 \\ 1/3 & 1/3 & 1/3 \\ 1/6 & -1/3 & 1/6 \end{bmatrix}.$$

See Exercise 4 of this section for the completion of this example.

Example 12.5.3: Matrix Calculus. Those who have studied calculus recall that the Maclurin series is a useful way of expressing many common functions. For example,

$$e^x = \sum_{k=0}^{\infty} \frac{x^k}{k!}.$$

Indeed, calculators and computers use these series for calculations. Given a polynomial $f(x)$, we defined the matrix-polynomial $f(A)$ for square matrices in Chapter 5. Hence, we are in a position to describe e^A for a $n \times n$ matrix A:

$$e^A = I + A + \frac{A^2}{2!} + \frac{A^3}{3!} + \cdots + \frac{A^n}{n!} + \cdots$$

$$= \sum_{k=0}^{\infty} \frac{A^k}{k!}.$$

Again we note the occurrence of high powers of a matrix.

Let A be an $n \times n$ diagonalizable matrix. Then there exists an invertible $n \times n$ matrix P such that $P^{-1}AP = D$, a diagonal matrix, so that

$$e^A = e^{PDP^{-1}}$$

$$= \sum_{k=0}^{\infty} \frac{(PDP^{-1})^k}{k!}$$

$$= P\left(\sum_{k=0}^{\infty} \frac{D^k}{k!}\right) P^{-1}. \qquad \text{Why?}$$

Comments on Example 12.5.3:

(1) Many of the ideas of calculus can be developed using matrices—differentiation, integration, etc., of matrix functions. For example, if

$$A(t) = \begin{bmatrix} t^3 & 3t^2 + 8t \\ e^t & 2 \end{bmatrix},$$

then

$$\frac{d(A(t))}{dt} = \begin{bmatrix} 3t^2 & 6t + 8 \\ e^t & 0 \end{bmatrix}.$$

(2) Many of the basic formulas in calculus are true in matrix calculus. For example,

$$\frac{d(A(t) + B(t))}{dt} = \frac{d(A(t))}{dt} + \frac{d(B(t))}{dt}, \qquad \frac{d(e^{At})}{dt} = Ae^{At}.$$

(3) Matrix calculus can be used to solve whole systems of differential equations, as opposed to "single" differential equations, quite similarly to the procedure used in ordinary differential equations.

(4) An introductory exploration of matrix calculus can be found in Bronson.

EXERCISES FOR SECTION 12.5

A Exercises

1. (a) Write out all the details of Example 12.5.1 to show that the formula for F_k given in the text is correct.

   (b) Use induction to prove the assertion made in Example 12.5.1 that

   $$\begin{bmatrix} F_k \\ F_{k-1} \end{bmatrix} = A^{k-1} \begin{bmatrix} 1 \\ 1 \end{bmatrix}.$$

2. (a) Do Example 8.3.8 of Chapter 8 using the method outlined in Example 12.5.1. Note that the terminology *characteristic equation, characteristic polynomial*, etc., introduced in Chapter 8, comes from the language of matrix algebra.

   (b) What is the significance of Algorithm 8.3.1, Part c, with respect to this section?

3. Solve $S(k) = 5S(k - 1) + 4$, with $S(0) = 0$, using the method of this section.

4. How many paths are there of length 6 between vertex 1 and vertex 3 in Figure 12.5.2? How many paths from vertex 2 to vertex 2 of length 6 are there? (Hint: The characteristic polynomial of the adjacency matrix is λ^4.)

5. Use the matrix A of Example 12.5.2 to:

   (a) Determine the number of paths of length 1 that exist from the vertex

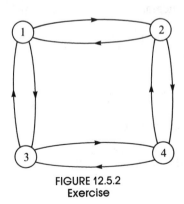

FIGURE 12.5.2
Exercise

      a to each of the vertices in Example 12.5.2. Verify using the graph. Do the same for vertices *b* and *c*.

(b)   Verify all the details of Example 12.5.2.

(c)   Use Example 12.5.2 to determine the number of paths of length 4 there are from each node in the graph of Figure 12.5.1 to every node in the graph. Verify your results using the graph.

6. (a)   Show that the following assertion is true:

$$e^A = \cdots = P\left(\sum_{k=0}^{\infty} \frac{D^k}{k!}\right) P^{-1}.$$

    (b)   Find e^A if $A = \begin{bmatrix} 2 & 1 \\ 2 & 3 \end{bmatrix}$.

    (c)   Find $\sin A$ if $A = \begin{bmatrix} 1 & 0 \\ 0 & 2 \end{bmatrix}$.

        (Hint: Recall $\sin x = \displaystyle\sum_{k=0}^{\infty} \frac{(-1)^k x^{2k+1}}{(2k+1)!}$.)

    (d)   Find $\sin A$ if $A = \begin{bmatrix} 3 & 0 \\ 0 & 4 \end{bmatrix}$.

7. We noted in Chapter 5 that since matrix algebra is not commutative under multiplication, certain difficulties arise. This problem still plagues us.

    (a)   Let $A = \begin{bmatrix} 1 & 1 \\ 0 & 0 \end{bmatrix}$ and $B = \begin{bmatrix} 0 & 0 \\ 0 & 2 \end{bmatrix}$. Find e^A, e^B, and e^{A+B}. Note that $e^A e^B \neq e^{A+B} \neq e^B e^A$.

    (b)   Show that $e^0 = I$ where 0 is the 2 × 2 0-matrix.

    (c)   Prove that for any matrix A, $(e^A)^{-1} = e^{-A}$. (Hint: Show that $(e^A)(e^{-A}) = I$ and use the uniqueness of inverse property.)

    (d)   Prove that if A and B commute, $e^A \cdot e^B = e^{A+B}$, thereby proving that $e^A e^B = e^B e^A$. How?

8. Another observation for adjacency matrices: For the matrix in Example 12.5.2, note that the sum of the elements in the row corresponding to the node *a* (that is, the first row) gives the outdegree (the number of edges

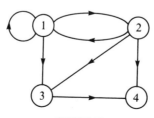

FIGURE 12.5.3

coming out from the node a) of a. Similarly, the sum of the elements in any given column gives the indegree of the node corresponding to that column.

(a) Using the matrix A of Example 12.5.2, find the outdegree and the indegree of each node. Verify by the graph.

(b) Repeat Part a for the directed graphs:

chapter

13

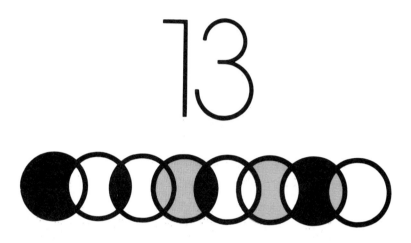

BOOLEAN ALGEBRA

13 GOALS

In this section we will develop an algebra which is particularly important to computer scientists, as it is the mathematical foundation of computer design, or switching theory. This algebra is called Boolean algebra after the mathematician George Boole (1815–64). The similarities of Boolean algebra and the algebra of sets and logic will be discussed, and we will discover special properties of finite Boolean algebras.

In order to achieve these goals, we will recall the basic ideas of posets introduced in Chapter 6 and develop the concept of a lattice, which has applications in finite state machines.

The reader should view the development of the topics of this chapter as another example of an algebraic system. Hence, we expect to define first the elements in the system, next the operations on the elements, and then the common properties of the operations in the system.

13.1 Posets Revisited

From Chapter 6, Section 3, we recall the following definition:

Definition: Poset. *A set L on which a partial ordering relation (reflexive, antisymmetric, and transitive) r is defined is called a partially ordered set, or poset, for short.*

We recall a few examples of posets:

(1) $L = \mathbf{R}$ and r is the relation $\leq$.
(2) $L = \mathcal{P}(A)$ where $A = \{a, b\}$ and r is the relation $\subseteq$.
(3) $L = \{1, 2, 3, 6\}$ and r is the relation $|$ (divides). We remind the reader that the pair (a, b) as an element of the relation r can be expressed as $(a, b) \in r$, or arb, depending on convenience and readability.

The posets we will concentrate on in this chapter will be those which have maxima and minima. These partial orderings resemble that of $\leq$ on $\mathbf{R}$, so the symbol $\leq$ is used to replace the symbol r in the definition of a partially ordered set. Hence, the definition of a poset becomes:

Definition: Poset. *A set on which a partial ordering, $\leq$, is defined is called a partially ordered set, or, in brief, a poset. Here, $\leq$ is a partial ordering on L if and only if for all a, b, c $\in$ L:*

(1) $a \leq a$ *(reflexivity)*,
(2) $a \leq b$ *and* $b \leq a \Rightarrow a = b$ *(antisymmetry), and*
(3) $a \leq b$ *and* $b \leq c \Rightarrow a \leq c$ *(transitivity).*

We now proceed to introduce maximum and minimum concepts. To do this, we will first define these concepts for two elements of the poset L, and then define the concepts over the whole poset L.

Definitions: Lower Bound, Upper Bound. *Let a, $b \in L$, a poset. Then $c \in L$ is a lower bound of a and b if $c \leq a$ and $c \leq b$. $d \in L$ is an upper bound of a and b if $a \leq d$ and $b \leq d$.*

Definitions: Greatest Lower Bound, Least Upper Bound. *Let L be a poset and $\leq$ be the partial ordering on L. Let a, $b \in L$, then:*

(1) *$g \in L$ is called the greatest lower bound (glb) of a and b if and only if $g \leq a$ and $g \leq b$, and further, if there exists an element $g' \in L$ such that if $g' \leq a$ and $g' \leq b$, then $g' \leq g$. This latter phrase says if g' is also a lower bound, then g is "greater" than g', so g is the greatest lower bound.*

(2) *An element $\ell \in L$ is called the least upper bound (lub) of a and b if and only if $a \leq \ell$ and $b \leq \ell$, and further, if there exists an element $\ell' \in L$ such that $a \leq \ell'$ and $b \leq \ell'$, then $\ell \leq \ell'$; that is, ℓ is the least of the upper bounds.*

Definitions: Greatest Element, Least Element. *An element m in a poset L is called the greatest (maximum) element if, for all $a \in L$, $a \leq m$. An element $s \in L$ is called the least (minimum) element if for all $a \in L$, $s \leq a$. The greatest and least elements, when they exist, are frequently denoted by 1 and 0 respectively.*

Example 13.1.1. Let $L = \{1, 3, 5, 7, 15, 21, 35, 105\}$ and let $\leq$ be the relation $|$ (divides) on L. Then L is a poset. To determine the *lub* of 3 qnd 7, we look for all $\ell \in L$, such that $3|\ell$ and $7|\ell$. Certainly, both $\ell = 21$ and $\ell = 105$ work. Next, since $21|105$, then $21 = lub$ of 3 and 7. Similarly, the *lub* of 3 and 5 is 15. The greatest element of L is 105 since $a|105$ for all $a \in L$. To find the *glb* of 15 and 35, we first consider all elements g of L such that $g \mid 15$ and $g \mid 35$. Certainly, both $g = 5$ and $g = 1$ satisfy these conditions. But since $1 \mid 5$, then $g = 5$ is the *glb* of 15 and 35. The least element of L is 1 since $1 \mid a$ for all $a \in L$.

Henceforth, for any positive integer n, D_n will denote the set of all positive integers which are divisors of n. For example, the set L of Example 13.1.1 is D_{105}.

Example 13.1.2. If $M = \mathcal{P}(A)$, where $A = \{a, b, c\}$, with the relation $\subseteq$ on M, then M is a poset. The *glb* of the elements $\{a, b\}$ and $\{a, c\}$ is $g = \{a\}$. For any other element g' of M which is subset of $\{a, b\}$ and $\{a, c\}$(there is only one; what is it?), $g' \subseteq g$. The least element of M is $\emptyset$ and the greatest element of M is $\{a, b, c\}$. The Hasse diagram of M is shown in Figure 13.1.1.

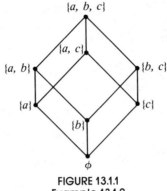

FIGURE 13.1.1
Example 13.1.2

With a little practice, it is quite easy to find the least upper bounds and greatest lower bounds of all possible pairs in M directly from the graph of M.

The previous examples and definitions indicate that the *lub* and *glb* are defined in terms of the partial ordering of the given poset. It is not yet clear whether all posets are such that every pair of elements has both a *lub* and a *glb*. Indeed, this is not the case (see Exercise 5).

EXERCISES FOR SECTION 13.1

A Exercises

1. Let $D_{30} = \{1, 2, 3, 5, 6, 10, 15, 30\}$ and let the relation $|$ be a partial ordering on D_{30}.

   (a) Find all lower bounds of 10 and 15.

   (b) Find the *glb* of 10 and 15.

   (c) Find all upper bounds of 10 and 15.

   (d) Determine the *lub* of 10 and 15.

   (e) Draw the Hasse diagram for D_{30} with $|$.

2. List the elements of the sets D_8, D_{50}, and D_{1001}.

3. Compare the Hasse diagram of Exercise 1 with that of Example 13.1.2. Note that the two diagrams are structurally the same.

4. (a) Determine the least upper bounds and greatest lower bounds of all pairs of elements of the poset M of Example 13.1.2 using Figure 13.1.1.

   (b) Verify the results in Part a using the definitions of least upper bound and greatest lower bound.

5. Figure 13.1.2 contains Hasse diagrams of posets.
   (a) Determine the *lub* and *glb* of all pairs of elements when they exist. Indicate those pairs which do not have a *lub* (or a *glb*).
   (b) Find the least and greatest elements of Figure 13.1.2, when they exist.

6. Prove: If a poset *L* has a least element, then this element is unique.

13.2 Lattices

In this section, we restrict our discussion to *lattices,* those posets where every pair of elements has a *lub* and a *glb*. We first introduce some notation.

Definitions: Join, Meet. *Let L be a poset under an ordering* $\leq$ *. Let a, b $\in$ L. We define:*

$a \vee b$ *(read "a join b") as lub of a and b, and*
$a \wedge b$ *(read "a meet b") as glb of a and b.*

Theorem 13.2.1. *Let L be a poset under* $\leq$ *and let a and b be elements in L, then*

(a) *If a and b have a lub, then this lub is unique.*
(b) *If a and b have a glb then this glb is unique.*

Proof of Part a: Assume to the contrary that *a* and *b* have two different *lub*'s and call them l_1 and l_2. Then, by definition of *lub*, we have:

$a \leq l_1, b \leq l_1$ and
$a \leq l_2, b \leq l_2$.

Further, since l_1 is a *lub*, then $l_1 \leq l_2$. But l_2 is also a *lub*, so $l_2 \leq l_1$. By the antisymmetric property of $\leq$, we have $l_1 = l_2$, which contradicts the original assumption.

Proof of Part b: See Exercise 1 of this section. #

Since both the *lub* and the *glb*:
   (1) act on pairs of elements in a poset *L* and
   (2) give unique results,
they can be regarded as binary operations, if they always exist.

Definition: Lattice. *A lattice is a poset L (under* $\leq$ *) in which every pair of elements has a lub and a glb. Since a lattice L is an algebraic system with binary operations* $\vee$ *and* $\wedge$ *, it is denoted by* $[L; \vee, \wedge]$.

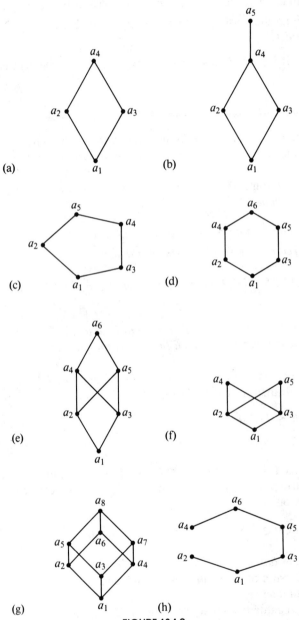

FIGURE 13.1.2
Figure to find the least and greatest elements

In example 13.1.2. the operation table for *lub* is easy, although admittedly, tedious to do, and we can observe that every pair of elements in this poset has a *lub*. See page 340.

The reader is encouraged to write out the operation table for $\wedge$ and to note that every pair of elements in this poset also has a *glb*, so that M together with these two operations is a lattice. We further observe that:

(1) $[\mathscr{P}(A); \vee , \wedge]$ is a lattice (under $\subseteq$) for any set A, and
(2) the join operation is the set operation of union and the meet operation is the operation intersection; i.e., $\vee = \cup$ and $\wedge = \cap$.

It can be shown (see the exercises) that the commutative laws, associative laws, idempotent laws, and absorption laws are all true for any lattice. An example of this is clearly $[\mathscr{P}(A); \cup , \cap]$, since these laws hold in the algebra of sets. This lattice is also distributive. This is not always the case for lattices in general, however.

Definition: Distributive Lattice. *Let* $[L; \vee , \wedge]$ *be a lattice (under* $\leq$ *).* $[L; \vee , \wedge]$ *is called a distributive lattice if and only if the distributive laws hold; that is, for all a, b, c* $\in L$*, we have:*

$$a \vee (b \wedge c) = (a \vee b) \wedge (a \vee c) \text{ and}$$
$$a \wedge (b \vee c) = (a \wedge b) \vee (a \wedge c).$$

Example 13.2.1. If A is any set, the lattice $[\mathscr{P}(A); \cup , \cap]$ (under $\subseteq$) is distributive.

Example 13.2.2. We now give an example of a lattice where the distributive laws do not hold. Let $L = \{1, 2, 3, 5, 30\}$. Then L is a poset under the relation divides. The operation tables for $\vee$ and $\wedge$ on L are:

$\vee$	1	2	3	5	30
1	1	2	3	5	30
2	2	2	30	30	30
3	3	30	3	30	30
5	5	30	30	5	30
30	30	30	30	30	30

$\wedge$	1	2	3	5	30
1	1	1	1	1	1
2	1	2	1	1	2
3	1	1	3	1	3
5	1	1	1	5	5
30	1	2	3	5	30

∨	∅	{a}	{b}	{c}	{a, b}	{a, c}	{b, c}	{a, b, c}
∅	∅	{a}	{b}	{c}	{a, b}	{a, c}	{b, c}	{a, b, c}
{a}	{a}	{a}	{a, b}	{a, c}	{a, b}	{a, c}	{a, b, c}	{a, b, c}
{b}	{b}	{a, b}	{b}	{b, c}	{a, b}	{a, b, c}	{b, c}	{a, b, c}
{c}	{c}	{a, c}	{b, c}	{c}	{a, b, c}	{a, c}	{b, c}	{a, b, c}
{a, b}	{a, b}	{a, b}	{a, b}	{a, b, c}	{a, b}	{a, b, c}	{a, b, c}	{a, b, c}
{a, c}	{a, c}	{a, c}	{a, b, c}	{a, c}	{a, b, c}	{a, c}	{a, b, c}	{a, b, c}
{b, c}	{b, c}	{a, b, c}	{b, c}	{b, c}	{a, b, c}	{a, b, c}	{b, c}	{a, b, c}
{a, b, c}	{a, b, c}	{a, b, c}	{a, b, c}	{a, b, c}	{a, b, c}	{a, b, c}	{a, b, c}	{a, b, c}

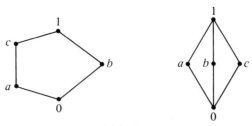

FIGURE 13.2.1
Non-distributive lattices

Since every pair of elements in L has both a join and a meet, $[L; \vee, \wedge]$ is a lattice (under $|$). Is this lattice distributive? We note that:

$$2 \vee (5 \wedge 3) = 2 \vee 1 = 2 \text{ and}$$
$$(2 \vee 5) \wedge (2 \vee 3) = 30 \wedge 30 = 30,$$

so that $a \vee (b \wedge c) \neq (a \vee b) \wedge (a \vee c)$ for some $a, b, c \in L$. Hence L is not a distributive lattice.

It can be shown that a lattice is nondistributive if and only if it contains a sublattice isomorphic to one of the lattices in Figure 13.2.1.

It is interesting to note that if $|$ is the relation "divides" on $\mathbf{P}$, then for a, $b \in \mathbf{P}$ we have:

$a \vee b = lcm(a, b)$, the least common multiple of a and b; that is, the smallest integer (in $\mathbf{P}$) which is divisible by both a and b; $a \wedge b = gcd(a, b)$, the greatest common divisor of a and b; that is, the largest integer which divides both a and b.

EXERCISES FOR SECTION 13.2

A Exercises

1. Prove Part b of Theorem 13.2.1.

2. Which of the posets in Exercise 5 of Section 13.1 are lattices? Which of the lattices are distributive?

B Exercises

3. (a) State the commutative laws, associative laws, idempotent laws, and absorption laws for lattices.
   (b) Prove these laws.

4. Let $[L; \vee, \wedge]$ be a lattice and $a, b, c \in L$. Prove:
   (a) $a \vee b \geq a$.
   (b) $a \vee b \geq b$.
   (c) $a \geq b$ and $a \geq c \Rightarrow a \geq b \vee c$.

13.3 Boolean Algebras

In order to define a Boolean algebra, we need the additional concept of complementation.

Definition: Complemented Lattice. *Let* $[L; \vee , \wedge]$ *be a lattice which contains a least element, denoted by* 0, *and a greatest element, denoted by* 1 *(such a lattice is called bounded).* $[L; \vee , \wedge]$ *is called a complemented lattice if and only if for every element a* $\in L$, *there exists an element* $\bar{a}$ *in L such that* $a \wedge \bar{a} = 0$ *and* $a \vee \bar{a} = 1$. *The element* $\bar{a}$ *is called a complement of the element a.*

Example 13.3.1.   Let $M = \mathcal{P}(A)$, where $A = \{a, b, c\}$. Then $[M; \cup , \cap]$ is a bounded lattice with $0 = \emptyset$ and $1 = A$. Then, to find if it exists, the complement, $\bar{B}$, of, say $B = \{a, b\} \in M$, we want $\bar{B}$ such that

$$\{a, b\} \cap \bar{B} = \emptyset \text{ and } \{a, b\} \cup \bar{B} = A.$$

Here, $\bar{B} = \{c\}$, and since it can be shown that each element of M has a complement (see Exercise 1), $[M; \cup , \cap]$ is a complemented lattice. Note that if A is any set and $M = \mathcal{P}(A)$, then $[M; \cup , \cap]$ is a complemented lattice where the complement of $B \in M$ is $\bar{B} = B^c = A - B$.
  In Example 13.3.1, we observe that the complement of each element of M is unique. Is this always the case? The answer is no. Consider the following.

Example 13.3.2.   Let $L = \{1, 2, 3, 5, 30\}$ and consider the lattice $[L; \vee , \wedge]$ (under "divides"). The least element of L is 1 and the greatest element is 30. Let us compute the complement of the element $a = 2$. We want to determine $\bar{a}$ such that $2 \wedge \bar{a} = 1$ and $2 \vee \bar{a} = 30$. Certainly, $\bar{a} = 3$ works, but so does $\bar{a} = 5$, so the complement of $a = 2$ in this lattice is not unique. However, $[L; \vee , \wedge]$ is still a complemented lattice since each element does have at least one complement.

  The following theorem gives us an insight into when uniqueness of complements occurs.

Theorem 13.3.1.   *If* $[L; \vee , \wedge]$ *is a complemented and distributive lattice, then the complement* $\bar{a}$ *of any element a* $\in L$ *is unique.*

  Proof: Let $a \in L$ and assume to the contrary that a has two complements, namely a_1 and a_2. Then by definition of complement,

$$a \wedge a_1 = 0 \text{ and } a \vee a_1 = 1,$$

Also,

$$a \wedge a_2 = 0 \text{ and } a \vee a_2 = 1.$$

So that

$$a_1 = a_1 \wedge 1 = a_1 \wedge (a \vee a_2)$$
$$= (a_1 \wedge a) \vee (a_1 \wedge a_2)$$
$$= 0 \vee (a_1 \wedge a_2)$$
$$= a_1 \wedge a_2.$$

On the other hand,

$$a_2 = a_2 \wedge 1 = a_2 \wedge (a \vee a_1)$$
$$= (a_2 \wedge a) \vee (a_2 \wedge a_1)$$
$$= 0 \vee (a_2 \wedge a_1)$$
$$= a_2 \wedge a_1.$$

Hence $a_1 = a_2$, which contradicts the assumption that a has two different complements, a_1 and a_2. #

Definition: Boolean Algebra. *A Boolean algebra is a lattice which contains a least element and a greatest element and which is both complemented and distributive.*

Since the complement of each element in a Boolean algebra is unique (by Theorem 13.3.1), complementation is a valid unary operation over the set under discussion, and we will list it together with the other two operations to emphasize that we are discussing a set together with three operations. Also, to help emphasize the distinction between lattices and lattices which are Boolean algebras, we will use the letter B as the generic symbol for the set of a Boolean algebra; that is, $[B; -, \vee, \wedge]$ will stand for a general Boolean algebra.

Example 13.3.3. Let A be any set, and let $B = \mathcal{P}(A)$. Then $[B; c, \cup, \cap]$ is a Boolean algebra. Here, c stands for the complement of an element of B with respect to A.

This is a key example for us since all finite Boolean algebras and many infinite Boolean algebras look like this example for some A. In fact, a glance at the basic Boolean algebra laws in Table 13.3.1, in comparison with the set laws of Chapter 4 and the basic laws of logic of Chapter 3, indicates that all three systems behave the same; that is, they are isomorphic.

The "pairing" of the above laws reminds us of the principle of duality, which we state for a Boolean algebra.

Definition: Principle of Duality for Boolean Algebras. *Let $[B; -, \vee, \wedge]$ be a Boolean algebra (under $\leq$), and let S be a true statement for $[B; -, \vee, \wedge]$. If S^* is obtained from S by replacing $\leq$ by $\geq$ (this is* *equivalent to turning the graph upside down), $\vee$ by $\wedge$, $\wedge$ by $\vee$, 0 by 1, and 1 by 0, then S* is also a true statement.*

TABLE 13.3.1
Basic Boolean Algebra Laws

Commutative Laws

1. $a \vee b = b \vee a$	1.' $a \wedge b = b \wedge a$

Associative Laws

2. $a \vee (b \vee c) = (a \vee b) \vee c$	2.' $a \wedge (b \wedge c) = (a \wedge b) \wedge c$

Distributive Laws

3. $a \wedge (b \vee c) = (a \wedge b) \vee (a \wedge c)$	3.' $a \vee (b \wedge c) = (a \vee b) \wedge (a \vee c)$

Identity Laws

4. $a \vee 0 = 0 \vee a = a$	4.' $a \wedge 1 = 1 \wedge a = a$

Complement Laws

5. $a \vee \bar{a} = 1$	5.' $a \wedge \bar{a} = 0$

Idempotent Laws

6. $a \vee a = a$	6.' $a \wedge a = a$

Null Laws

7. $a \vee 1 = 1$	7.' $a \wedge 0 = 0$

Absorption Laws

8. $a \vee (a \wedge b) = a$	8.' $a \wedge (\vee b) = a$

DeMorgan's Laws

9. $\overline{a \vee b} = \bar{a} \wedge \bar{b}$	9.' $\overline{a \wedge b} = \bar{a} \vee \bar{b}$

Involution Law

10. $\bar{\bar{a}} = a$

Example 13.3.4. The laws 1' through 9' are the duals of the Laws 1 through 9 respectively. Law 10 is its own dual.

We close this section with some comments on notation. The notation for operations in a Boolean algebra is derived from the algebra of logic. However, other notations are used. These are summarized in the following chart:

This Text (Mathematician's) Notation	Set Notation	Computer Designer's Notation	Read as
$\vee$	$\cup$	$+$	"join," "or," "sum,"
$\wedge$	$\cap$	$\cdot$	"meet," "and," "product,"
$-$	c	$-$	complement
$\leq$	$\subseteq$	$\leq$	

Mathematicians most frequently use the notation of the text, and, on occasion, use set notation for Boolean algebras. Thinking in terms of sets may be easier for some people. Computer designers traditionally use the "+" and " · " notation, the reason being that in the past, most computers did not have the symbols $\vee$, $\wedge$. In this latter notation, DeMorgan's Laws become:

$$(9) \quad \overline{a + b} = \bar{a} \cdot \bar{b}$$

and

$$(9') \quad \overline{a \cdot b} = \bar{a} + \bar{b}.$$

EXERCISES FOR SECTION 13.3

A Exercises

1. Determine the complement of each element $B \in M$ in Example 13.3.1. Is this lattice a Boolean algebra? Why?

2. (a) Determine the complement of each element of D_6 in $[D_6; \vee, \wedge]$.
   (b) Repeat Part a using the lattice in Example 13.2.2.
   (c) Repeat Part a using the lattice in Exercise 1 of Section 13.1.
   (d) Are the lattices in Parts a, b, and c Boolean algebras? Why?

3. Determine which of the lattices of Exercise 5 of Section 13.1 are Boolean algebras.

4. Let $A = \{a, b\}$ and $B = \mathcal{P}(A)$.
   (a) Prove that $[B; c, \cup, \cap]$ is a Boolean algebra.
   (b) Write out the operation tables for the Boolean algebra.

5. It can be shown that the following statement, S, holds for any Boolean algebra $[B; -, \vee, \wedge]$: $(a \wedge b) = a$ if $a \leq b$.
   (a) Write the dual, S^*, of the statement S.
   (b) Write the statement S and its dual, S^*, in the language of sets.
   (c) Are the statements in Part b true for all sets?
   (d) Write the statement S and its dual, S^*, in the language of logic.
   (e) Are the statements in Part d true for all propositions?

6. State the dual of:

    (a) $a \vee (b \wedge a) = a.$
    (b) $a \vee ((\overline{b} \vee a) \wedge b) = 1.$
    (c) $\overline{(a \wedge \overline{b})} \wedge b = a \vee b.$

    B Exercises

7. Formulate a definition for isomorphic Boolean algebras.

13.4 Atoms of a Boolean Algebra

In this section we will look more closely at previous claims that every finite
Boolean algebra is isomorphic to an algebra of sets. We will show that every
finite Boolean algebra has 2^n elements for some n with precisely n generators,
called *atoms*.

Consider the Boolean algebra $[B; -, \vee, \wedge]$, whose graph is:

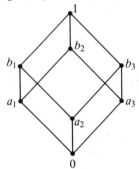

FIGURE 13.4.1
Illustration of atom concept

We note that $1 = a_1 \vee a_2 \vee a_3$, $b_1 = a_1 \vee a_2$, $b_2 = a_1 \vee a_3$, and
$b_3 = a_2 \vee a_3$; that is, each of the elements above level one can be described
completely and uniquely in terms of the elements on level one. The a_i's have
uniquely generated the non-zero elements of B much like a basis in linear
algebra generates the elements in a vector space. We also note that the a_i's
are the immediate successors of the 0-element. In any Boolean algebra, the
immediate successors of the 0-element are called *atoms*. Let A be any non-
empty set. In the Boolean algebra $[\mathscr{P}(A); c, \cup, \cap]$ (over $\subseteq$), the singleton
sets are the generators, or atoms, of the algebraic structure since each element
$\mathscr{P}(A)$ can be described completely and uniquely as the join or union of
singleton sets.

Definition: Atom. *A non-zero element a in a Boolean algebra*
$[B; -, \vee, \wedge]$ *is called an atom if for every $x \in B$, $x \wedge a = a$ or*
$x \wedge a = 0.$

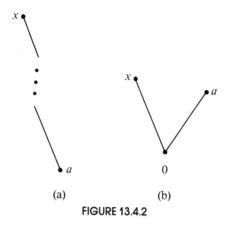

(a)   (b)

FIGURE 13.4.2

The condition that $x \wedge a = a$ tells us that x is a successor of a; that is, $a \leq x$, as depicted in Figure 13.4.2a.

The condition $x \wedge a = 0$ is true only when x and a are "not connected." This occurs when x is another atom or if x is a successor of atoms different from a, as depicted in Figure 13.4.2b.

Example 13.4.1. The set of atoms of the Boolean algebra $[D_{30}; -, \vee, \wedge]$ is $M = \{2, 3, 5\}$. To see that $a = 2$ is an atom, let x be any non-zero element of D_{30} and note that one of the two conditions $x \wedge 2 = 2$ or $x \wedge 2 = 1$ holds. Of course, to apply the definition to this Boolean algebra, we must remind ourselves that in this case the 0-element is 1, the operation $\wedge$ is *gcd*, and the poset relation $\leq$ is "divides." So if $x = 10$, we have $10 \wedge 2 = 2$ (or $2 \mid 10$), so Condition 1 holds. If $x = 15$, the first condition is not true. (Why?) However, Condition 2, $15 \wedge 2 = 1$, is true. The reader is encouraged to show that each of the elements 2, 3, and 5 satisfy the definition (see Exercise 13.4.1). Next, if we compute the join (*lcm* in this case) of all possible combinations of the atoms 2, 3, and 5, we will generate all non-zero elements of D_{30}. For example, $2 \vee 3 \vee 5 = 30$ and $2 \vee 5 = 10$. We state this concept formally in the following theorem, which we give without proof.

Theorem 13.4.1. *Let $[B; -, \vee, \wedge]$ be any finite Boolean algebra. Let $A = \{a_1, a_2, \ldots, a_n\}$ be the set of all n atoms of $[B; -, \vee, \wedge]$. Then every non-zero element in B can be expressed completely and uniquely as the join of a certain subset of A.*

We now ask ourselves if we can be more definitive about the structure of different Boolean algebras of a given order. Certainly, the Boolean algebras $[D_{30}; -, \vee, \wedge]$ and $[P(A); c, \cup, \cap]$ have the same graph (that of Figure

13.4.1), the same number of atoms, and, in all respects, look the same except for the names of the elements and the operations. In fact, when we apply corresponding operations to corresponding elements, we obtain corresponding results. We know from Chapter 11 that this means that the two structures are isomorphic as Boolean algebras. Furthermore, the graphs of these examples are exactly the same as that of Figure 13.4.1, which is an arbitrary Boolean algebra of order $8 = 2^3$.

In these examples of a Boolean algebra of order 8, we note that each had 3 atoms and $2^3 = 8$ number of elements, and all were isomorphic to $[\mathcal{P}(A);c,$ $\cup, \cap]$, where $A = \{a, b, c\}$. This leads us to the following questions:

(1) Are there any other different (non-isomorphic) Boolean algebras of order 8?
(2) What is the relationship, if any, between finite Boolean algebras and their atoms?
(3) How many different (non-isomorphic) Boolean algebras are there of order 2? Order 3? Order 4? Etc.

The answers to these questions are given in the following theorem and corollaries. We include the proofs of the corollaries since they are instructive.

Theorem 13.4.2. *Let $[B; -, \vee, \wedge]$ be any finite Boolean algebra, and let A be the set of all atoms in this Boolean algebra. Then $[B; -, \vee, \wedge]$ is isomorphic to $[\mathcal{P}(A); c, \cup, \cap]$.*

Corollary 13.4.1. *Every finite Boolean algebra $[B; -, \vee, \wedge]$ has 2^n elements for some positive integer n.*

Proof: Let A be the set of all atoms of B and let $\#A = n$. Then there are exactly 2^n elements (subsets) in $\mathcal{P}(A)$, and by Theorem 13.4.2, $[B; -, \vee, \wedge]$ is isomorphic to $[\mathcal{P}(A); c, \cup, \cap]$. #

Corollary 13.4.2. All Boolean algebras of order 2^n are isomorphic to each other. (The graph of the Boolean algebra of order 2^n is the n-cube).

Proof: By Theorem 13.4.2, every Boolean algebra of order 2^n is isomorphic to $[\mathcal{P}(A); c, \cup, \cap]$ when $\#A = n$. Hence, they are all isomorphic to each other. #

The above theorem and corollaries tell us that we can only have finite Boolean algebras of orders $2^1, 2^2, 2^3, \ldots, 2^n, \ldots$ and that all finite Boolean algebras of any given order are isomorphic. These are powerful tools in determining the structure of finite Boolean algebras. In the next section, we will try to find the easiest way of describing a Boolean algebra of any given order.

EXERCISES FOR SECTION 13.4

A Exercises

1. (a) Show that $a = 2$ is an atom of the Boolean algebra
        $[D_{30}; -, \wedge, \vee]$.
    (b) Repeat Part a for the elements 3 and 5 of D_{30}.
    (c) Verify Theorem 13.4.1 for the Boolean algebra $[D_{30}; -, \wedge, \vee]$.

2. Let $A = \{a, b, c\}$.
    (a) Rewrite the definition of atom for $[\mathscr{P}(A); c, \cup, \cap]$. What does
        $a \leq x$ mean in this example?
    (b) Find all atoms of $[\mathscr{P}(A); c, \cup, \cap]$
    (c) Verify Theorem 13.4.1 for $[\mathscr{P}(A); c, \cup, \cap]$

3. Verify Theorem 13.4.2 and its corollaries for the Boolean algebras in Exercises 1 and 2 of this section.

4. Give a description of all Boolean algebras of order 16. (Hint: Use Theorem 13.4.2.) Note that the graph of this Boolean algebra is given in Figure 9.4.5.

5. Corollary 13.4.1 states that there does not exist Boolean algebras of orders 3, 5, 6, 7, 9, etc. (orders different from 2^n). Prove that we cannot have a Boolean algebra of order 3. (Hint: Assume that $[B; -, \vee, \wedge]$ is a Boolean algebra of order 3 where $B = \{0, x, 1\}$ and show that this cannot happen by investigating the possibilities for its operation tables.)

6. (a) There are many different, yet isomorphic, Boolean algebras with two elements. Describe one such Boolean algebra that is derived from a power set, $\mathscr{P}(A)$, under $\subseteq$. Describe a second that is described from D_n, for some $n \in \mathbf{P}$, under "divides".
    (b) Since the elements of a two-element Boolean algebra must be the greatest and least elements, 1 and 0, the tables for the operations on $\{0, 1\}$ are determined by the Boolean algebra laws. Write out the operation tables for $[\{0, 1\}; -, \vee, \wedge]$.

B Exercises

7. Find a Boolean algebra with a countably infinite number of elements.

8. Prove that the direct product of two Boolean algebras is a Boolean algebra. (Hint: "Copy" the corresponding proof for groups in Section 11.6.)

13.5   Finite Boolean algebras as n-tuples of 0's and 1's

From the previous section we know that all finite Boolean algebras are of order 2^n, where n is the number of atoms in the algebra. We can therefore completely describe every finite Boolean algebra by the algebra of power sets. Is there a more convenient, or at least an alternate way, of defining finite Boolean algebras? In Chapter 11 we found that we could produce new groups by taking Cartesian products of previously known groups. We imitate this process for Boolean algebras.

The simplest non-trivial Boolean algebra is the Boolean algebra on the set $B_2 = \{0, 1\}$. The ordering on B_2 is the natural one, $0 \leq 0, 0 \leq 1, 1 \leq 1$. If we treat 0 and 1 as the truth values "false" and "true" respectively, we see that the Boolean operations $\vee$ (join) and $\wedge$ (meet) are nothing more than the logical connectives $\vee$ (or) and $\wedge$ (and). The Boolean operation, $-$, (complementation) is the logical $\sim$ (negation). In fact, this is why the symbols $-$, $\vee$, and $\wedge$ were chosen as the names of the Boolean operations. The operation tables for $[B_2; -, \vee, \wedge]$ are simply those of "or", "and," and "not," which we repeat here:

$\vee$	0	1		$\wedge$	0	1		u	$\bar{u}$
0	0	1		0	0	0		0	1
1	1	1		1	0	1		1	0

By Theorem 13.4.2 and its corollaries, all Boolean algebras of order 2 are isomorphic to this one.

We know that if we form $B_2 \times B_2 = B_2^2$ we obtain the set $\{(0, 0), (0, 1), (1, 0), (1, 1)\}$, a set of order 4. We define operations on B_2^2 the natural way, namely, componentwise, so that $(0, 1) \vee (1, 1) = (0 \vee 1, 1 \vee 1) = (1, 1)$, $(0, 1) \wedge (1, 1) = (0 \wedge 1, 1 \wedge 1) = (0, 1)$, and $\overline{(0, 1)} = (\bar{0}, \bar{1}) = (1, 0)$. We claim that B_2^2 is a Boolean algebra under the componentwise operations. Hence, $[B_2^2; -, \vee, \wedge]$ is a Boolean algebra of order 4. Since all Boolean algebras of order 4 are isomorphic to each other, we have found a simple way of describing all Boolean algebras of order 4.

It is quite clear that we can describe any Boolean algebra of order 8 by considering $B_2 \times B_2 \times B_2 = B_2^3$ and, in general, any Boolean algebra of order 2^n—that is, all finite Boolean algebras by $B_2^n = B_2 \times B_2 \times \cdots \times B_2$ (n number of times). The Hasse diagrams of the $[B_2^n; -, \vee, \wedge]$ for $n = 1$, 2, 3, 4 are given in Figure 9.4.5. Just think of the strings as n-tuples.

EXERCISES FOR SECTION 13.5

A Exercises

1. (a) Write out the operation tables for $[B_2^2; -, \vee, \wedge]$.
   (b) Draw the Hasse diagram for $[B_2^2; -, \vee, \wedge]$ and compare your results with Figure 9.4.5.
   (c) Find the atoms of this Boolean algebra.

2. (a) Write out the operation table for $[B_2^3; -, \vee, \wedge]$.
    (b) Draw the Hasse diagram for $[B_2^3; -, \vee, \wedge]$ and compare the results with Figure 9.4.5.

3. (a) List all atoms of B_2^4.
    (b) Describe the atoms of B_2^n $n \geq 1$.

    B Exercise

4. In Chapter 4, Section 4, we defined the terms *minset* and *minset normal form*. Theorem 13.4.2 tells us that we can think of any finite Boolean algebra in terms of sets. Rephrase these definitions in the language of Boolean algebra.

13.6 Boolean Expressions

In this section, we will use our background from the previous sections and set theory to develop a procedure for simplifying Boolean expressions. This procedure has considerable application to the simplification of circuits in switching theory or logical design.

Definition: Boolean Expression. *Let $[B; -, \vee, \wedge]$ be any Boolean algebra. Let $x_1, x_2, \ldots, x_k$ be variables in B; that is, variables which can assume values from B. A Boolean expression generated by $x_1, x_2, \ldots, x_k$ is any valid combination of the x_i and the elements of B with the operations of meet, join, and complementation.*

This definition, as expected, is the analog of the definition of a proposition generated by a set of propositions, presented in Section 3.2.

Each Boolean expression generated by k variables, $e(x_1, \ldots, x_k)$, defines a function $f: B^k \to B$, where $f(a_1, \ldots, a_k) = e(a_1, \ldots, a_k)$. If B is a finite Boolean algebra, then there are a finite number of functions from B^k into B. Those functions that are defined in terms of Boolean expressions are called *Boolean functions*. As we will see, there is an infinite number of Boolean expressions that define each Boolean function. Naturally, the "shortest" of these expressions will be preferred. Since electronic circuits can be described as Boolean functions with $B = B_2$, this economization is quite useful.

Example 13.6.1. Consider any Boolean algebra $[B; -, \vee, \wedge]$ of order 2. How many functions $f: B^k \to B$ are there? First, all Boolean algebras of order 2 are isomorphic to $[B_2; -, \vee, \wedge]$, so we want to determine the number of functions $f: B_2^k \to B_2$. If we consider a Boolean function of two variables, x_1 and x_2, we note that each variable has two possible values 0 and 1, so there are 2^2 ways of assigning these two values to the $k = 2$ variables. Hence, the table below has $2^2 = 4$ rows. So far we have a table such as that labeled 13.6.1.

TABLE 13.6.1
General Form of Boolean Function $f(x_1, x_2)$ of Example 13.6.2

x_1	x_2	$f(x_1, x_2)$
0	0	?
0	1	?
1	0	?
1	1	?

How many possible different function values $f(x_1, x_2)$ can there be? To list a few: $f_1(x_1, x_2) = x_1$, $f_2(x_1, x_2) = x_2$, $f_3(x_1, x_2) = x_1 \lor x_2$, $f_4(x_1, x_2) = (x_1 \land \overline{x}_2) \lor x_2$, $f_5(x_1, x_2) = x_1 \land x_2 \lor \overline{x}_2$, etc. Each of these will give a table like that of Table 13.6.1. The tables for f_1 and f_3 appear in Table 13.6.2.

Two functions are different if and only if their tables (values) are different for at least one row. Of course by using the basic laws of Boolean algebra we can see that $f_3 = f_4$. Why? So if we simply list by brute force all "combinations" of x_1 and x_2, we will obtain unnecessary duplication. However, we note that for any combination of the variables x_1 and x_2 there are only two possible values for $f(x_1, x_2)$, namely 0 or 1. Thus, we could write 2^m different functions (tables) for 2 variables where $m = 2^2 = $ number of rows. For this example, there are $2^{2^2} = 16$ possible different functions (tables).

Now let's count the number of different Boolean functions in a more general setting. We will consider two cases: first, when $B = B_2$, and second, when B is any finite Boolean algebra with 2^n elements.

Let $B = B_2$. Each function $f: B^k \to B$ is defined in terms of a table having 2^k rows. Therefore, since there are two possible images for each element of B^k, there are 2 raised to the 2^k different functions. *Every one of these functions is a Boolean function.*

Now suppose that $\#B = 2^n > 2$. A function from B^k into B can still be defined in terms of a table. There are $(\#B)^k$ rows to each table and $\#B$ possible images for each row. Therefore, there are 2^n raised to the power 2^{nk} different functions. If $n > 1$, then not every one of these functions is a Boolean function.

TABLE 13.6.2
Boolean Functions f_1 and f_2 of Example 13.6.1

x_1	x_2	$f_1(x_1, x_2) = x_1$	x_1	x_2	$f_3(x_1, x_2) = x_1 \lor x_2$
0	0	0	0	0	0
0	1	0	0	1	1
1	0	1	1	0	1
1	1	1	1	1	1

Notice that the numbers of functions that we have just derived could be obtained by simply applying Exercise 5 of Section 7.1. Since all Boolean algebras are isomorphic to a Boolean algebra of sets, the analogs of statements in sets are useful in Boolean algebras.

Definition: Minterm. *A Boolean expression generated by* $x_1, x_2, \ldots, x_k$ *which has the form*

$$\bigwedge_{i=1}^{k} y_i$$

where each y_i may be either x_i or $\overline{x}_i$, is called a minterm generated by x_1, $x_2, \ldots, x_k$.

There are 2^k different minterms generated by $x_1, \ldots, x_k$.

Definition: Minterm Normal Form. *A Boolean expression generated by* $x_1, \ldots, x_k$ *is in minterm normal form if it is the join of expressions of the form $a \wedge m$, where $a \in B$ and m is a minterm. That is, it is of the form*

$$\bigvee_{j=1}^{p} (a_j \wedge m_j),$$

where $p = 2^k$ and $m_1, m_2, \ldots, m_p$ are the minterms generated by $x_1, \ldots, x_k$.

If $B = B_2$, then each a_j in a minterm normal form is either 0 or 1. Therefore, $a_j \wedge m_j$ is either 0 or m_j.

Theorem 13.6.1. *Let $e(x_1, \ldots, x_k)$ be a Boolean expression over B. There exists a unique minterm normal form $M(x_1, \ldots, x_k)$ that is equivalent to $e(x_1, \ldots, x_k)$ in the sense that e and M define the same function from B^k into B.*

The uniqueness in this theorem does not include the possible ordering of the minterms in M (commonly referred to as "uniqueness up to the order of minterms"). The proof of this theorem would be quite lengthy but not very instructive, so we will leave it to the interested reader to attempt. The implications of the theorem are very interesting, however.

If $\#B = 2^n$, then there are 2^n raised to the 2^k different minterm normal forms. Since each different minterm normal form defines a different function, there are a like number of Boolean functions from B^k into B. If $B = B_2$, there are as many Boolean functions (2 raised to the 2^k) as there are functions. However, if $2^n > 2$, there are fewer Boolean functions than there are functions from B^k into B, since there are 2^n raised to the 2^{nk} functions from B^k into B. The significance of these results is that any desired function can be obtained using electronic circuits having 0 or 1 (off or on, positive or negative) values, but more complex, multivalued circuits would not have this flexibility.

We will close this section by examining minterm normal forms for expressions over B_2, since they are a starting point for circuit economization.

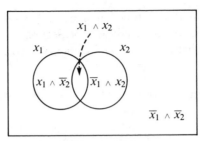

FIGURE 13.6.1
Diagram with the usual translation of notation

Example 13.6.2. Consider the Boolean expression $f(x_1, x_2) = x_1 \vee \overline{x_2}$. One method of determining the minterm normal form of f is to think in terms of sets. Consider the diagram with the usual translation of notation in Figure 13.6.1. Then $f(x_1, x_2) = (\overline{x_1} \wedge \overline{x_2}) \vee (x_1 \wedge \overline{x_2}) \vee (x_1 \wedge x_2)$.

Example 13.6.3. Consider the function $f : B_2^3 \rightarrow B_2$ defined by Table 13.6.3. The minterm normal form for f can be obtained by taking the join of minterms that correspond to rows that have an image value of 1. If $f(a_1, a_2, a_3) = 1$, then include the minterm $y_1 \wedge y_2 \wedge y_3$ where

$$y_j = \begin{cases} x_j \text{ if } a_j = 1 \\ \overline{x_j} \text{ if } a_j = 0 \end{cases}$$

Therefore,

$$f(x_1, x_2, x_3) = (\overline{x_1} \wedge \overline{x_2} \wedge \overline{x_3}) \vee (\overline{x_1} \wedge x_2 \wedge x_3) \vee (x_1 \wedge x_2 \wedge \overline{x_3}).$$

The minterm normal form is a first step in obtaining an economical way of expressing a given Boolean function. For functions of more than three variables, the above set theory approach tends to be awkward. Other procedures

TABLE 13.6.3
Boolean Function of $f(a_1, a_2, a_3)$ of Example 13.6.3

a_1	a_2	a_3	$f(a_1, a_2, a_3)$
0	0	0	1
0	0	1	0
0	1	0	0
0	1	1	1
1	0	0	0
1	0	1	0
1	1	0	1
1	0	1	0

are used to write the normal form. The most convenient is the Karnaugh map, a discussion of which can be found in any logical design/switching theory text (see, for example, Hill and Peterson).

EXERCISES FOR SECTION 13.6

A Exercises

1. (a) Write the 16 possible functions of Example 13.6.1. (Hint: Find all possible joins of minterms generated by x_1 and x_2.)
   (b) Write out the tables of several of the above Boolean functions to show that they are indeed different.
   (c) Determine the minterm normal form of

$$f_1(x_1, x_2) = x_1 \lor x_2,$$
$$f_2(x_1, x_2) = \overline{x_1} \lor \overline{x_2},$$
$$f_3(x_1, x_2) = 0, f_4(x_1, x_2) = 1.$$

2. Consider the Boolean expression $f(x_1, x_2, x_3) = (\overline{x_3} \land x_2) \lor (\overline{x_1} \land x_3)$ $\lor (x_2 \land x_3)$ on $[B_2; -, \lor, \land]$.
   (a) Simplify this expression using basic Boolean algebra laws.
   (b) Write this expression in minterm normal form.
   (c) Write out the table for the given function defined by f and compare it to the tables of the functions in Parts a and b.
   (d) How many possible different functions in three variables on $[B_2; -, \lor, \land]$ are there?

B Exercise

3. Let $[B; -, \lor, \land]$ be a Boolean algebra of order 4, and let f be a Boolean function of two variables on B.
   (a) How many elements are there in the domain of f?
   (b) How many different Boolean functions are there of two variables? Three variables?
   (c) Determine the minterm normal form of $f(x_1, x_2) = x_1 \lor x_2$.
   (d) If $B = \{0, a, b, 1\}$, define a function from B^2 into B that is not a Boolean function.

3.7 A Brief Introduction to the Application of Boolean Algebra to Switching Theory

The algebra of switching theory is Boolean algebra. The standard notation used for Boolean algebra operations in most logic design/switching theory texts is $+$ for $\lor$ and $\bullet$ for $\land$. Complementation is as in this text. Therefore, $(x_1 \land \overline{x_2}) \lor (x_1 \land x_2) \lor (\overline{x_1} \land x_2)$ becomes $x_1 \bullet \overline{x_2} + x_1 \bullet x_2 + \overline{x_1} \bullet x_2$, or

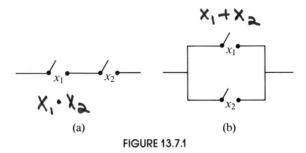

(a)   (b)

FIGURE 13.7.1

simply $x_1\overline{x_2} + x_1x_2 + \overline{x_1}x_2$. All concepts developed previously for Boolean algebras hold. The only change is purely notational. We make the change in this section solely to introduce the reader to another frequently used notation. Obviously, we could have continued the discussion with our previous notation.

The simplest switching device is the on-off switch. If the switch is closed, on, current will pass through it; if it is open, off, current will not pass through it. If we designate on by true or the logical, or Boolean, 1, and off by false, the logical, or Boolean, 0, we can describe electrical circuits containing switches by logical, or Boolean, expressions. The expression $x_1 \cdot x_2$ represents the situation in which a series of two switches appears in a circuit (see Figure 13.7.1a). In order for current to flow through the circuit, both switches must be on, i.e., have the value 1.

Similarly, a pair of parallel switches, as in Figure 13.7.1b, is described algebraically by $x_1 + x_2$. Many of the concepts in Boolean algebra can be thought of in terms of switching theory. For example, the distributive law in Boolean algebra (in $+$, $\cdot$ notation) is: $x_1 \cdot (x_2 + x_3) = x_1 \cdot x_2 + x_1 \cdot x_3$. Of course, this says the expression on the left is always equivalent to that on the right. The switching circuit analogue of the above statement is that Figure 13.7.2a is equivalent (as an electrical circuit) to Figure 13.7.2b.

The circuits in a digital computer are composed of large quantities of switches that can be represented as in Figure 13.7.2 or can be thought of as boxes or gates with two or more inputs (except for the NOT gate) and one

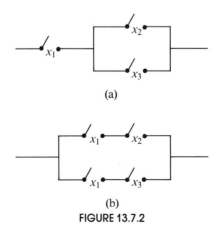

(a)

(b)

FIGURE 13.7.2

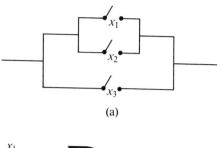

(a)

$$f(x_1, x_2, x_3) = x_1 + x_2 + x_3$$

(b)

FIGURE 13.7.3

output. These are often drawn as in Figure 13.7.3. For example, the OR gate, as the name implies, is the logical/Boolean OR function. The on-off switch function in Figure 13.7.3a in gate notation is Figure 13.7.3b.

Either diagram indicates that the circuit will conduct current if and only if $f(x_1, x_2, x_3)$ is true, or 1. We list the gate symbols which are widely used in switching theory in Figure 13.7.4 with their names. The names mean, and are read, exactly as they appear. For example, NAND means "not x_1 and x_2," or algebraically, $\overline{x_1 \wedge x_2}$, or $\overline{x_1 \cdot x_2}$.

Operation		Symbol		Logical/Boolean Function	
	read	input	output	Mathematics notation	Switch Theory notation
AND	and	x_1 x_2	$f(x_1, x_2) = x_1 x_2$	$f(x_1, x_2) = x_1 \wedge x_2$	$f(x_1, x_2) = x_1 \cdot x_2$
OR	or	x_1 x_2	$f(x_1, x_2) = x_1 + x_2$	$f(x_1, x_2) = x_1 \vee x_2$	$f(x_1, x_2) = x_1 + x_2$
NOT	not	x_1	$f(x_1) = \overline{x_1}$	$f(x_1) = \overline{x_1}$	$f(x_1) = \overline{x_1}$
NAND	not and	x_1 x_2	$f(x_1, x_2) = \overline{x_1 + x_2}$	$f(x_1, x_2) = \overline{x_1 \wedge x_2}$	$f(x_1, x_2) = \overline{x_1 \cdot x_2}$
NOR	not or	x_1 x_2	$f(x_1, x_2) = \overline{x_1 + x_2}$	$f(x_1, x_2) = \overline{x_1 \vee x_2}$	$f(x_1, x_2) = \overline{x_1 + x_2}$
Exclusive OR	Exclusive or	x_1 x_2	$f(x_1, x_2) = x_1 \oplus x_2$	$f(x_1, x_2) = x_1 \oplus x_2$	$f(x_1, x_2) = x_1 \oplus x_2$

FIGURE 13.7.4

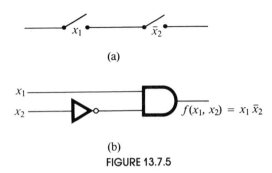

(a)

(b)

FIGURE 13.7.5

The circuit in Figure 13.7.5a can be described by gates. To do so, simply keep in mind that the Boolean function $f(x_1, x_2) = x_1 \cdot \bar{x}_2$ of this circuit contains two operations. The operation of complementation takes precedence over that of "and," so we have Figure 13.7.5b.

Example 13.7.1. The switching circuit in Figure 13.7.6a can be expressed through the logic, or gate, circuit in Figure 13.7.6b.

We leave it to the reader to analyze both figures and to convince him- or herself that they do describe the same circuit. The circuit can be described

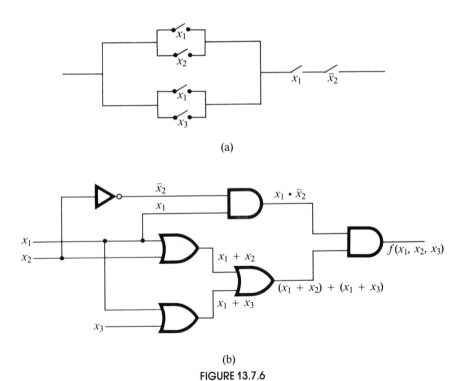

(a)

(b)

FIGURE 13.7.6

algebraically as

$$f(x_1, x_2, x_3) = ((x_1 + x_2) + (x_1 + x_3)) \cdot x_1 \cdot \overline{x_2}.$$

We can use basic Boolean algebra laws to simplify or minimize this Boolean Function (circuit):

$$\begin{aligned}
f(x_1, x_2, x_3) &= ((x_1 + x_2) + (x_1 + x_3)) \cdot x_1 \cdot \overline{x_2} \\
&= (x_1 + x_2 + x_3) \cdot x_1 \cdot \overline{x_2} \\
&= x_1 \cdot x_1 \cdot \overline{x_2} + x_2 \cdot x_1 \cdot \overline{x_2} + x_3 \cdot x_1 \cdot \overline{x_2} \\
&= x_1 \cdot \overline{x_2} + 0 \cdot x_1 + x_3 \cdot x_1 \cdot \overline{x_2} \\
&= x_1 \cdot \overline{x_2} + x_3 \cdot x_1 \cdot \overline{x_2} \\
&= x_1 \cdot (\overline{x_2} + \overline{x_2} \cdot x_3) \\
&= x_1 \cdot \overline{x_2} \cdot (1 + x_3) \\
&= x_1 \cdot \overline{x_2}.
\end{aligned}$$

The circuit for f may be described as in Figure 13.7.5. This is a less expensive circuit since it involves considerably less hardware.

The table for f is:

x_1	x_2	x_3	$f(x_1, x_2, x_3)$
0	0	0	0
0	0	1	0
0	1	0	0
0	1	1	0
1	0	0	1
1	0	1	1
1	1	0	0
1	1	1	0

The Venn diagram which represents f is the shaded portion in Figure 13.7.7. From this diagram, we can read off the minterm normal form of f:

$$f(x_1, x_2, x_3) = x_1 \cdot \overline{x_2} \cdot \overline{x_3} + x_1 \cdot \overline{x_2} \cdot x_3.$$

The circuit (gate) diagram appears in Figure 13.7.8.

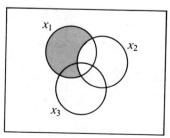

FIGURE 13.7.7

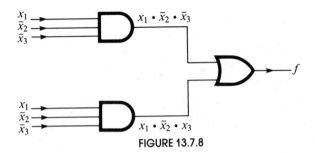

FIGURE 13.7.8

How do we interpret this? We see that $f(x_1, x_2, x_3) = 1$ when $x_1 = 1$, $x_2 = 0$, and $x_3 = 0$ or $x_3 = 1$. Current will be conducted through the circuit when switch x_1 is on, switch x_2 is off, and when switch x_3 is either off or on.

We close this section with a brief discussion of minimization, or reduction, techniques. We have discussed two in this text: algebraic (using basic Boolean rules) reduction and the minterm normal form technique. Other techniques are discussed in switching theory texts. When one reduces a given Boolean function, or circuit, it is possible to obtain a circuit which does not look simpler, but may be more cost effective, and is, therefore, simpler with respect to time. We illustrate with an example.

Example 13.7.2. The Boolean function of Figure 13.7.9a is $f(x_1, x_2, x_3, x_4) = x_1 \cdot (\overline{x_2} \cdot (\overline{x_3} \cdot x_4))$, which can also be diagrammed as in Figure 13.7.9b.

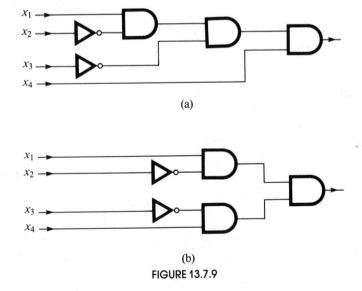

(a)

(b)

FIGURE 13.7.9

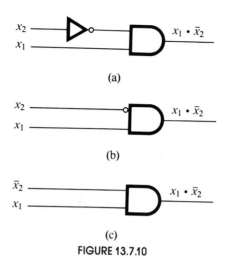

(a)

(b)

(c)

FIGURE 13.7.10

Is Circuit b simpler than Circuit a? Both circuits contain the same number of gates, so the hardware costs (costs per gate) would be the same. Hence, intuitively, we would guess that they are equivalent with respect to simplicity. However, the signals x_3 and x_4 in Circuit a pass through three levels of gating before reaching the output. All signals in Circuit b go through only two levels of gating (disregard the NOT gate when counting levels). Each level of logic (gates) adds to the time delay to the development of a signal at the output. In computers, we want the time delay to be as small as possible. Frequently, speed can be increased by decreasing the number of levels in a circuit. However, this frequently forces a larger number of gates to be used, thus increasing costs. One of the more difficult jobs of a design engineer is to balance off speed with hardware costs (number of gates).

One final remark on notation: The circuit in Figure 13.7.10a can be written as in Figure 13.7.10b, or simply as in Figure 13.7.10c.

EXERCISES FOR SECTION 13.7

A Exercises

1. (a) Write all inputs and outputs from Figure 13.7.11 and show that its Boolean function is $f(x_1, x_2, x_3) = \overline{((x_1 + x_2) \cdot x_3)} \cdot (x_1 + x_2)$.
   (b) Simplify f algebraically.
   (c) Find the minterm normal form of f.
   (d) Draw and compare the circuit (gate) diagram of Parts b and c above.
   (e) Draw the on-off switching diagram of f in Part a.
   (f) Write the table of the Boolean function f in Part a and interpret the results.

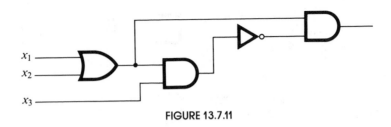

FIGURE 13.7.11

2. Given Figure 13.7.12:

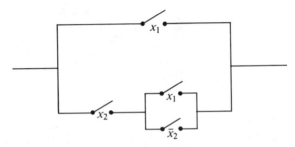

FIGURE 13.7.12

(a) Write the Boolean function which represents the given on-off circuit.

(b) Show that the Boolean function obtained in answer to Part a can be reduced to $f(x_1, x_2) = x_1$. Draw the on-off circuit diagram of this simplified representation.

(c) Draw the circuit (gate) diagram of the given on-off circuit diagram.

(d) Determine the minterm normal of the Boolean function found in answer to Part a or given in Part b; they are equivalent.

(e) Discuss the relative simplicity and advantages of the circuit gate diagrams found in answer to Parts c and d.

3. (a) Write the circuit (gate) diagram of $f(x_1, x_2, x_3) = (x_1 \cdot x_2 + x_3) \cdot (x_2 + x_3) + x_3$.

(b) Simplify the function in Part a by using basic Boolean algebra laws.

(c) Write the circuit (gate) diagram of the result obtained in Part b.

(d) Draw the on-off switch diagrams of Parts a and b.

4. Consider the Boolean function $f(x_1, x_2, x_3, x_4) = x_1 + (x_2 \cdot (\overline{x_1} + x_4) + x_3 \cdot (\overline{x_2} + \overline{x_4}))$.

(a) Simplify f algebraically.

(b) Draw the switching (on-off) circuit of f and the reduction of f obtained in Part a.

(c) Draw the circuit (gate) diagram of f and the reduction of f obtained in answer to Part a.

14

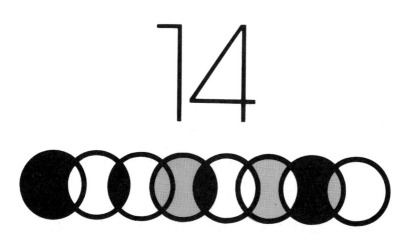

MONOIDS AND AUTOMATA

14

At first glance, the two topics that we will discuss in this chapter seem totally unrelated. The first is monoid theory, which we touched upon in Chapter 11. The second is automata theory, in which computers and other machines are described in abstract terms. After short independent discussions of these topics, we will describe how the two are related in the sense that each monoid can be viewed as a machine and each machine (as described here) has a monoid associated with it.

14.1 Monoids

Recall the definition of a monoid:

Definition: Monoid. *A monoid is a set M together with a binary operation * with the properties*

(a) *is associative: $(a*b)*c = a*(b*c)$ for all a, b, c ∈ M, and*
(b) *has an identity: there exists e ∈ M such that for all a ∈ M,*
   $a * e = e * a = a.$

Note: Since the requirements for a group contain the requirements for a monoid, every group is a monoid.

Example 14.1.1.

(a) The power set of any set together with any one of the operations intersection, union, or symmetric difference is a monoid.
(b) The set of integers, **Z** with multiplication is a monoid. With addition, **Z** is also a monoid.
(c) The set of $n \times n$ matrices over the integers, $M_n(\mathbf{Z})$, with matrix multiplication is a monoid. This follows from the fact that matrix multiplication is associative and has an identity, I_n. This is an example of a non-commutative monoid since there are matrices, A and B, for which AB and BA are different.
(d) $[\mathbf{Z}_n, \times_n]$, $n \geq 2$, is a monoid with identity 1.
(e) Let X be a non-empty set. The set of all functions from X into X, often denoted X^X, is a monoid over function composition. In Chapter 7, we saw that function composition is associative. The function $i: X \to X$ defined by $i(a) = a$ is the identity element for this system. This is another example of a non-commutative monoid, provided $\#X$ is greater than 1.

   If X is finite, $\#(X^X) = \#X^{\#X}$. For example, if $B = \{0, 1\}$, $\#(B^B) = 4$. The functions z, u, i, and t, defined by the graphs in Figure

364

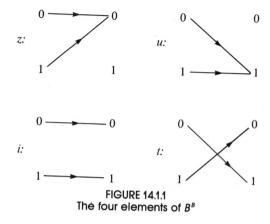

FIGURE 14.1.1
The four elements of B^B

14.1.1, are the elements of B^B. This monoid is not a group. Do you know why? One reason that B^B is non-commutative is that $tz \neq zt$, since $(tz)(0) = 1$ and $(zt)(0) = 0$.

GENERAL CONCEPTS AND PROPERTIES OF MONOIDS

Virtually all of the group concepts that were discussed in Chapter 11 are applicable to monoids. When we introduced subsystems, we saw that a submonoid of monoid M is a subset of M—that is, it itself is a monoid with the operation of M. To prove that a subset is a submonoid, you can apply the following algorithm.

Theorem/Algorithm 14.1.1. *Let $[M; *]$ be a monoid and K is a nonempty subset of M. K is a submonoid of M if and only if:*

(a) $a, b \in k$, then $a * b \in K$ (K is closed under $*$), and
(b) the identity of M belongs to K.

Often we will want to discuss the smallest submonoid that includes a certain subset S of a monoid M. This submonoid can be defined recursively by the following definition.

Definition: Submonoid Generated by a Set. *If S is a subset of monoid $[M; *]$, the submonoid generated by S, (S), is defined by:*

(a) (Basis) (i) $a \in S \Rightarrow a \in (S)$, and (ii) the identity of M belongs to (S);
(b) (Recursion), $a, b \in (S) \Rightarrow a * b \in (S)$.

Note: If $S = \{a_1, a_2, \ldots, a_n\}$, we write $(a_1, a_2, \ldots, a_n)$ in place of $(\{a_1, a_2, \ldots, a_n\})$.

Example 14.1.2.

(a) In $[\mathbf{Z}; +]$, $(2) = \{0, 2, 4, 6, 8, \ldots\}$.
(b) The power set of $\mathbf{Z}$, $\mathcal{P}(\mathbf{Z})$, over union is a monoid with identity $\emptyset$. If $S = \{\{1\}, \{2\}, \{3\}\}$, then (S) is the power set of $\{1, 2, 3\}$. If $S = \{\{n\} : n \in \mathbf{Z}\}$, then (S) is the set of finite subsets of the integers.

MONOID ISOMORPHISMS

Two monoids are *isomorphic* if and only if there exists a translation rule between them so that any true proposition in one monoid is translated to a true proposition in the other.

Example 14.1.3. $M_1 = [\mathcal{P}\{1, 2, 3\}; \cap]$ is isomorphic to $M_2 = [\mathbf{Z}_2^3; \bullet]$, where the operation in M_2 is componentwise mod 2 multiplication. A translation rule is that if $A \subseteq \{1, 2, 3\}$, then it is translated to (d_1, d_2, d_3) where $d_i = 1$ if i belongs to A; otherwise, $d_i = 0$. Two cases of how this translation rule works are:

$\{1, 2, 3\}$ is the identity for M_1, and $\{1, \;\; 2\} \cap \{2, 3\} \;\; = \;\; \{2\}$

$\updownarrow$ $\updownarrow \;\; \updownarrow \;\; \updownarrow$ $\updownarrow$

$(1, 1, 1)$ is the identity for M_2, and $(1, 1, 0) \bullet (0, 1, 1) = (0, 1, 0)$.

A more precise definition of a monoid isomorphism is identical to the definition of a group isomorphism (see Section 11.7).

EXERCISES FOR SECTION 14.1

A Exercises

1. For each of the subsets of the indicated monoid, determine whether the subset is a submonoid.
   (a) $S_1 = \{0, 2, 4, 6\}$ and $S_2 = \{1, 3, 5, 7\}$ in $[\mathbf{Z}_8; x_8]$.
   (b) $\{f \in \mathbf{N}^{\mathbf{N}} : f(n) \leq n, \forall n \in \mathbf{N}\}$ and $\{f \in \mathbf{N}^{\mathbf{N}} : f(1) = 2\}$ in $\mathbf{N}^{\mathbf{N}}$.
   (c) $\{A \subseteq \mathbf{Z} : A \text{ is finite}\}$ and $\{A \subseteq \mathbf{Z} : A^c \text{ is finite}\}$ in $[\mathcal{P}(\mathbf{Z}); \cup]$.

2. For each subset, describe the submonoid that it generates.
   (a) $\{3\}$ and $\{0\}$ in $[\mathbf{Z}_{12}; \times_{12}]$
   (b) $\{5\}$ in $[\mathbf{Z}_{25}; \times_{25}]$
   (c) the set of prime numbers and $\{2\}$ in $[\mathbf{P}; \cdot]$
   (d) $\{3, 5\}$ in $[\mathbf{N}; +]$

B Exercises

3. Definition: Stochastic Matrix. *An $n \times n$ matrix of real numbers is called stochastic if and only if each entry is non-negative and the sum of*

entries in each column is 1. Prove that the set of stochastic matrices is a monoid over matrix multiplication.

4. Prove Theorem 14.1.1.

14.2 Free Monoids and Languages

In this section, we will introduce the concept of a language. Languages are subsets of a certain type of monoid, the free monoid over an alphabet. After defining a free monoid, we will discuss languages and some of the basic problems relating to them. We will also discuss the common ways in which languages are defined.

Let A be a non-empty set, which we will call an *alphabet*. Our primary interest will be in the case where A is finite; however, A could be infinite for most of the situations that we will describe. The elements of A are called *letters* or *symbols*. Among the alphabets that we will use are $B = \{0, 1\}$, ASCII = the set of ASCII characters, and PAS = the Pascal character set (whichever one you use).

Definition: Strings over an Alphabet. *A string of length n, n $\geq$ 1, over A is a sequence of n letters from A: $a_1 a_2 \cdots a_n$. The null string, λ, is defined as the string of length zero containing no letters. The set of strings of length n over A is denoted by A^n. The set of all strings over A is denoted A^*.*

Notes:

(a) If the length of string s is n, we write $|s| = n$.
(b) The null string is not the same as the empty set, although they are similar in many ways.
(c) $A^* = A^\circ \cup A^1 \cup A^2 \cup A^3 \cup \cdots$ and if $i \neq j$, $A^i \cap A^j = \emptyset$; i.e. $\{A^\circ, A^1, A^2, A^3, \ldots\}$ is a partition of A^*.

Theorem 14.2.1. *If A is countable, then A* is countable.*

Proof:

Case 1. Given the alphabet $B = \{0, 1\}$, we can define a bijection from the positive integers into B^*. Each positive integer has a binary expansion $d_k d_{k-1} \cdots d_1 d_0$, where each d_j is 0 or 1 and $d_k = 1$. If n has such a binary expansion, then $2^k \leq n < 2^{k+1}$. We define $f : \mathbf{P} \to B^*$ by $f(n) = f(d_k d_{k-1} \cdots d_1 d_0) = d_{k-1} \cdots d_1 d_0$, where $f(1) = \lambda$. Every one of the 2^k strings of length k is the image of exactly one of the integers between 2^k and $2^{k+1} - 1$. from its definition, f is clearly a bijection; therefore, B^* is countable.

Case 2:   *A is Finite.* We will describe how this case is handled with an example first and then give the general proof. If $A = \{a, b, c, d, e\}$, then we can code the letters in A into strings from B^3. One of the coding schemes (there are many) is $a \leftrightarrow 000$, $b \leftrightarrow 001$, $c \leftrightarrow 010$, $d \leftrightarrow 011$, and $e \leftrightarrow 100$. Now every string in A^* corresponds to a different string in B^*; for example, ace would correspond with 000010100. The cardinality of A^* is equal to the cardinality of the set of strings that can be obtained from this encoding system. The possible coded strings must be countable, since they are a subset of a countable set (B^*); therefore A^* is countable.

If $\#A = m$, then the letters in A can be coded using a set of fixed-length strings from B^*. If $2^{k-1} < m \le 2^k$, then there are at least as many strings of length k in B^k as there are letters in A. Now we can associate each letter in A with an element of B^k. Then any string in A^* corresponds to a string in B^*. By the same reasoning as in the example above, A^* is countable.

Case 3:   *A is Countably Infinite.* We will leave this case as an exercise. #

FREE MONOIDS OVER AN ALPHABET

The set of strings over any alphabet is a monoid under concatenation.

Definition: Concatenation. *Let $a = a_1a_2 \cdots a_m$ and $b = b_1b_2 \cdots b_n$ be strings of length m and n, respectively. The concatenation of a with b, ab, is the string of length $m + n$: $a_1a_2 \cdots a_mb_1b_2 \cdots b_n$.*

Notes:

(a) The null string is the identity element of $[A^*$; concatenation]. Henceforth, we will denote the monoid of strings over A by A^*.

(b) Concatenation is non-commutative, provided $\#A > 1$.

(c) If $\#A_1 = \#A_2$, then the monoids A_1^* and A_2^* are isomorphic. An isomorphism can be defined using any bijection $f{:}A_1 \to A_2$. If $a = a_1a_2 \cdots a_n \in A_1^*$, $f^*(a) = f(a_1)f(a_2) \cdots f(a_n)$ defines a bijection from A_1^* into A_2^*. We will leave it to the reader to convince him or herself that for all a, b, $\in A_1^*$, $f^*(ab) = f^*(a)f^*(b)$.

LANGUAGES

The languages of the world—English, German, Russian, Chinese, etc.—are called natural languages. In order to communicate in any one of them, you must first know the letters of the alphabet and then know how to combine the letters in meaningful ways. A *formal language* is an abstraction of this situation.

Definition: Formal Language. *If A is an alphabet, a formal language over A is a subset of A*.*

Example 14.2.1.

(a) English can be thought of as a language over the set of letters A, $B, \cdots Z$ (upper and lower case) and other special symbols, such as punctuation marks and the blank. Exactly what subset of the strings over this alphabet defines the English language is difficult to pin down exactly. This is a characteristic of natural languages that we try to avoid with formal languages.

(b) The set of all ASCII stream files can be defined in terms of a language over ASCII. An ASCII stream file is a sequence of zero or more lines followed by an end-of-file symbol. A line is defined as a sequence of ASCII characters that ends with the two characters CR (carriage return) and LF (line feed). The end-of-file symbol is system-dependent; for example, CTRL/C is a common one.

(c) The set of all syntactically correct Pascal programs is a language over PAS. The set of Pascal text files is a language over PAS $\cup$ {eoln, eof}. The strings that belong to this language can be best described as the set of all strings over PAS $\cup$ {eoln} concatenated with the eof marker. Recall that in Pascal, eoln and eof are not Char type values, but only markers. Despite this fact, they can still be considered part of the alphabet that defines text files.

(d) A few languages over B are $L_1 = \{s \in B^*$; s has exactly as many 1's as it has 0's$\}$, $L_2 = \{1s0 : s \in B^*\}$ and $L_3 = (0, 01) =$ the submonoid of B^* generated by $\{0, 01\}$.

TWO FUNDAMENTAL PROBLEMS: RECOGNITION AND GENERATION

The generation and recognition problems are basic to computer programming. Given a language, L, the programmer must know how to write (or generate) a syntactically correct program that solves a problem. On the other hand, the compiler must be written to recognize whether a program contains any syntax errors.

The Recognition Problem: Design an algorithm that determines the truth of $s \in L$ in a finite number of steps for all $a \in A^*$. Any such algorithm is called a *recognition algorithm*.

Definition: Recursive Language. *A language is recursive if there exists a recognition algorithm for it.*

Example 14.2.2.

(a) The language of syntactically correct Pascal programs is recursive. Any Pascal compiler must include a syntax checker which solves the recognition problem.

(b) The three languages in Example 14.2.1 are all recursive. Recognition algorithms for L_1 and L_2 should be easy for you to imagine. The reason a recognition algorithm for L_3 might not be obvious is that L_3's definition is more cryptic. It doesn't tell us what belongs to L_3, just what can be used to create strings in L_3. This is how many languages are defined. With a second description of L_3, we can easily design a recognition algorithm. $L_3 = \{s \in B^*; s = \lambda$ or s starts with a 0 and has no consecutive 1's$\}$.

Algorithm 14.2.1:   Recognition Algorithm for L_3. Let $s = s_1 s_2 \cdots s_n \in B^*$. This algorithm determines the truth value of $s \in L_3$. The truth value is returned as the value of Word.

(1) Word := true
(2) If $n > 0$ then
    If $s_1 = 1$ then Word := false
               else for $i := 2$ to n do
                        if $s_{i-1} = 1$ and $s_i = 1$ then Word := false

The Generation Problem.   Design an algorithm that generates or produces any string in L. Here we presume that A is either finite or countably infinite; hence, A^* is countable by Theorem 14.2.1. and $L \subseteq A^*$ must be countable. Therefore, the generation of L amounts to creating a list of strings in L. The list may be either finite or infinite, and you must be able to show that every string in L appears somewhere in the list.

Theorem 14.2.2.

    (a) If A is countable, then there exists a generating algorithm for A^*.
    (b) If L is a recursive language over a countable alphabet, then there exists a generating algorithm for L.

Proof:

(a) Part a follows from the fact that A^* is countable; therefore, there exists a complete list of strings in A^*.
(b) To generate all strings of L, start with a list of all strings in A^* and an empty list, W, of strings in L. For each string s, use a recognition algorithm (one exists since L is recursive) to determine whether $s \in L$. If s is in L, add it to W; otherwise "throw it out." Then go to the next string in the list of A^*. #

Example 14.2.3.   Since all of the languages in Example 14.2.2 are recursive, they must have generating algorithms. The one given in the proof of Theorem 14.2.2 is not generally the most efficient. You could probably design more efficient generating algorithms for L_2 and L_3; however, a better generating algorithm for L_1 is not quite so obvious.

The recognition and generation problems can vary in difficulty depending on how a language is defined and what sort of algorithms we allow ourselves to use. This is not to say that the means by which a language is defined determines whether it is recursive. It just means that the truth of "L is recursive" may be more difficult to determine with one definition than with another. We will close this section with a discussion of grammars, which are standard forms of definition for a language. When we restrict ourselves to only certain types of algorithms, we can affect our ability to determine whether $s \in L$ is true. In defining a recursive language, we do not restrict ourselves in any way in regard to the type of algorithm that will be used. In Section 14.3, we will consider machines called *finite automata*, which can only perform simple algorithms.

PHASE STRUCTURE GRAMMARS AND LANGUAGES

One common way of defining a language is by means of a *phase structure grammar* (or grammar for short). The set of strings that can be produced using the grammar rules is called the *phase structure language* (of the grammar).

Example 14.2.4. We can define the set of all strings over B for which all 0's precede all 1's as follows. Define the starting symbol S and establish rules that S can be replaced with any of the following: λ, $0S$, or $S1$. These replacement rules are usually called *production (or rewriting) rules* and are usually written in the format $S \rightarrow \lambda$, $S \rightarrow 0S$, and $S \rightarrow S1$. Now define L to be the set of all strings that can be produced by starting with S and applying the production rules until S no longer appears. The strings in L are exactly the ones that are described above.

Definition: Phase Structure Grammar. *A phase structure grammar consists of four components:*

(1) A non-empty finite set of terminal characters, T. If the grammar is defining a language over A, T is a subset of $A*$.
(2) A finite set of non-terminal characters, N.
(3) A starting symbol, $S \in N$.
(4) A finite set of production rules, each of the form $X \rightarrow Y$, where X and Y are strings over $A \cup N$ such that $X \neq Y$ and X contains at least one non-terminal symbol.

If G is a phase structure grammar, $L(G)$ is the set of strings that can be obtained by starting with S and applying the production rules a finite number of times until no non-terminal characters remain. If a language can be defined by a phase structure grammar, then it is called a *phase structure language*.

Example 14.2.5. The language over B consisting of strings of alternating 0's and 1's is a phase structure language. It can be defined by the following grammar:

(1) Terminal characters: λ, 0, and 1,
(2) Non-terminal characters: S, T, and U,
(3) Starting symbol: S,
(4) Production rules: $S \to T$, $S \to U$, $S \to \lambda$, $S \to 0$, $S \to 1$, $S \to 0T$, $S \to 1U$, $T \to 10T$, $T \to 10$, $U \to 01U$, $U \to 01$.

We can verify that a string such as 10101 belongs to the language by starting with S and producing 10101 using the production rules a finite number of times: $S \to 1U \to 101U \to 10101$.

Example 14.2.6. Let G be the grammar with components:

(1) Terminal symbols = all letters of the alphabet and the digits 0 through 9,
(2) Non-terminal symbols = $\{I, X\}$,
(3) Starting symbol: I,
(4) Production rules: $I \to \alpha$, where α is any letter, $I \to \alpha X$ for any letter α, $X \to \beta X$ for any letter or digit β, and $X \to \beta$ for any letter or digit β.

There are a total of 124 production rules for this grammar. The language $L(G)$ consists of all valid Pascal identifiers.

Backus-Naur form (BNF). A popular alternate form of defining the production rules in a grammar is BNF. If the production rules $A \to B_1, A \to B_2$, ..., $A \to B_n$ are part of a grammar, they would be written in BNF as $A ::= B_1 | B_2 | \cdots | B_n$. The symbol $|$ in BNF is read as "or," while the $::=$ is read as "is defined as." Additional notations of BNF are that $\{x\}$, represents zero or more repetitions of x and $[y]$ means that y is optional.

Example 14.2.7. A BNF version of the production rules for a Pascal identifier is

$$\text{letter} ::= a | b | c | \cdots | y | z$$
$$\text{digit} ::= 0 | 1 | \cdots | 9$$
$$\text{I} ::= \text{letter} \{\text{letter} | \text{digit}\}.$$

Example 14.2.8. An arithmetic expression can be defined in BNF. For simplicity, we will consider only expressions obtained using addition and multiplication of integers. The terminal symbols are (,), $+$, $*$, $-$ and the digits 0 through 9. The non-terminal symbols are E (for expression), T (term), F (factor), and N (number). The starting symbol is E.

$$E ::= E + T \,|\, T$$
$$T ::= T * F \,|\, F$$
$$F ::= (E) \,|\, N$$
$$N ::= [-]\text{digit} \{\text{digit}\}.$$

One particularly simple type of phase structure grammar is the regular grammar.

Definition: Regular Grammar. *A regular (right-hand form) grammar is a grammar whose production rules are all of the form $A \rightarrow t$ or $A \rightarrow tB$, where A and B are non-terminal and t is terminal. A left-hand form grammar allows only $A \rightarrow t$ and $A \rightarrow Bt$. A language that has a regular phase structure language is called a regular language.*

Example 14.2.9.

(a) The set of Pascal identifiers is a regular language since the grammar that we defined the set by is a regular grammar.
(b) The language of all strings for which all 0's precede all 1's (Example 14.2.4) is regular; however, the grammar that we defined this set by is not regular. Can you define these strings with a regular grammar?
(c) The language of arithmetic expressions is not regular.

EXERCISES FOR SECTION 14.2

A Exercises

1. (a) If a computer is being designed to operate with a character set of 350 symbols, how many bits must be reserved for each character? Assume that each character will use the same number of bits.
   (b) Do the same for 3500 symbols.

2. It was pointed out in the text that the null string and the null set are different. The former is a string and the latter is a set, two different kinds of objects. Discuss how the two are similar.

3. Prove that if a language over A is recursive, then its complement is also recursive.

4. What set of strings are defined by the following grammars?
   (a) (1) Terminal symbols: λ, 0, and 1
       (2) Nonterminal symbols: S and E
       (3) Starting symbol: S
       (4) Production rules: $S \rightarrow 0S0$, $S \rightarrow 1S1$, $S \rightarrow E$, $E \rightarrow \lambda$, $E \rightarrow 0$, $E \rightarrow 1$

(b) (1) Terminal symbols: λ, a, b, and c
　　(2) Nonterminal symbols: S, T, U, and E
　　(3) Starting symbol: S
　　(4) Production rules: $S \rightarrow aS$, $S \rightarrow T$, $T \rightarrow bT$, $T \rightarrow U$, $U \rightarrow cU$, $U \rightarrow E$, $E \rightarrow \lambda$

5. Define the following languages over B with phase structure grammars. Which of these languages are regular?

　(a) the strings with an odd number of characters
　(b) the strings of length 4 or less
　(c) the strings with more zeros than ones
　(d) the strings with an even number of ones

6. Define the set of strings in B^* for which all 0's precede all 1's with a regular grammar.

7. Use BNF to define the grammars in Exercise 4.

B Exercise

8. (a) Prove that if X_1, X_2, . . . is a countable sequence of countable sets, then the union of these sets,

$$\bigcup_{i=1}^{\infty} X_i,$$

is countable.
　(b) Using the fact that the countable union of countable sets is countable, prove that if A is countable, then A^* is countable.

C Exercise

9. (a) Write a Pascal procedure to perform Algorithm 14.2.1, where S is of type

```
String = RECORD
            len : 0..100; ( * length of S * )
            bit : ARRAY[1..100] OF '0'..'1'
         END
```

　(b) Why should you use a WHILE loop instead of a FOR loop in your procedure?

14.3　Automata, Finite-State Machines

In this section, we will introduce the concept of an abstract machine. The machines that we will examine will (in theory) be capable of performing many of the tasks that are associated with digital computers. One such task is solving the recognition problem for a language. We will concentrate on one class of machines, finite-state machines (finite automata). And we will see that they are precisely the machines that are capable of recognizing strings in a regular grammar.

Given an alphabet X, we will imagine a string in X^* to be encoded on a tape which we will call an *input tape*. When we refer to a tape, we might imagine a strip of material that is divided into segments, each of which can contain either a letter or a blank.

The typical abstract machine includes an input device, the *read head*, which is capable of reading the symbol from the segment of the input tape that is currently in the read head. Some more advanced machines have a read/write head that can also write symbols onto the tape. The movement of the input tape after reading a symbol depends on the machine. With a finite-state machine, the next segment of the input tape is always moved into the read head after a symbol has been read. Most machines (including finite-state machines) also have a separate *output tape* that is written on with a *write head*. The output symbols come from an output alphabet, Y, that may or may not be equal to the input alphabet. The most significant component of an abstract machine is its *memory structure*. This structure can range from a finite number of bits of memory (as in a finite-state machine) to an infinite amount of memory that can be stored in the form of a tape that can be read from and written on (as in a Turing machine).

Definition: Finite-State Machine. *A finite-state machine is defined by a quintet (S, X, Z, w, t) where*

(1) $S = \{s_1, s_2, \ldots, s_r\}$ *is the state set, a finite set that corresponds to the set of memory configurations that the machines can have at any time.*

(2) $X = \{x_1, x_2, \ldots, x_m\}$ *is the input alphabet.*

(3) $Z = \{z_1, z_2, \ldots, z_n$ *is the output alphabet.*

(4) $w : X \times S \to Z$ *is the output function, which specifies which output symbol $w(x, s) \in Z$ is written onto the output tape when the machine is in state s and the input symbol x is read.*

(5) $t : X \times S \to S$ *is the next-state (or transition) function, which specifies which state $t(x, s) \in S$ the machine should enter when it is in state s and it reads the symbol x.*

Example 14.3.1. Many mechanical devices, such as vending machines, can be thought of as finite-state machines. For simplicity, assume that a vending machine dispenses packets of gum, spearmint (S), peppermint (P), and bubble (B), for 25¢ each. We can define the input alphabet to be {deposit 25¢, press S, press P, press B} and the state to be {Locked, Select}, where the deposit of a quarter unlocks the release mechanism of the machine and allows you to select a flavor of gum. We will leave it to the reader to imagine what the output alphabet, output function, and next-state function would be. You are also invited to let your imagination run wild and include such features as a coin-return lever and change maker.

Example 14.3.2. The following machine is called a *parity checker*. It recognizes whether or not a string in B^* contains an even number of 1's. The memory structure of this machine reflects the fact that in order to check the parity of a string, we need only keep track of whether an odd or even number of 1's has been detected.

(1) The input alphabet is $B = \{0, 1\}$.
(2) The output alphabet is also B.
(3) The state set is {even, odd}.
(4–5) The following table defines the output and next-state functions:

x	s	$z(x, s)$	$t(x, s)$
0	even	0	even
0	odd	1	odd
1	even	1	odd
1	odd	0	even

Note how the value of the most recent output at any time is an indication of the current state of the machine. Therefore, if we start in the even state and read any finite input tape, the last output corresponds to the final state of the parity checker and tells us the parity of the string on the input tape. For example, if the string 11001010 is read from left to right, the output tape, also from left to right, will be 10001100. Since the last character is a 0, we know that the input string has even parity.

An alternate method for defining a finite-state machine is with a transition diagram. A *transition diagram* is a directed graph that contains a node for each state and edges that indicate the transition and output functions. An edge (s_i, s_j) that is labeled x/z indicates that in State s_i, the input x results in an output of z and the next state is s_j. That is, $w(x, s_i) = z$ and $t(x, s_i) = s_j$. The transition diagram for the parity checker appears in Figure 14.3.1. In later examples, we will see that if different inputs, x_i and x_j, while in the same state, result in the same transitions and outputs, we label a single edge $x_i, x_j/z$ instead of drawing two edges with labels x_i/z and x_j/z.

One of the most significant features of a finite-state machine is that it retains no information about its past states that can be accessed by the machine itself. For example, after we input a tape encoded with the symbols 01101010 into the parity checker, the current state will be even, but we have no indication within the machine whether or not it has always been in even state. Note how

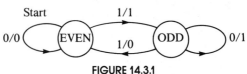

FIGURE 14.3.1
Transition diagram for a parity checker

the output tape is not considered part of the machine's memory. In this case, the output tape does contain a "history" of the parity checker's past states. We always assume that the finite-state machine has no way of recovering the output sequence for later use.

Example 14.3.5. Consider the following simplified version of the game of baseball. To be precise, this machine describes one-half inning of a simplified baseball game. Suppose that in addition to home plate, there is only one base instead of the usual three bases. Also, assume that there are only two outs per inning instead of the usual three. Our input alphabet will consist of the types of hits that the batter could have: out (O), double play (DP), single (S), and home run (HR). The input DP is meant to represent a batted ball that would result in a double play (two outs), if possible. The input DP can then occur at any time. The output alphabet is the numbers 0, 1, and 2 for the number of runs that can be scored as a result of any input. The state set contains the current situation in the inning, the number of outs, and whether a base runner is currently on the base. The list of possible states is then 00 (for 0 outs and 0 runners), 01, 10, 11, and end (when the half-inning is over). The transition diagram for this machine appears in Figure 14.3.2.

Let's concentrate on one state. If the current state is 01, 0 outs and 1 runner on base, each input results in a different combination of output and next state. If the batter hits the ball very poorly (a double play) the output is zero runs and the inning is over (the limit of two outs has been made). A simple out also results in an output of 0 runs and the next state is 11, one out and one runner on base. If the batter hits a single, one run scores (output = 1) while the state remains 01. If a home run is hit, two runs are scored (output = 2) and the next state is 00. If we had allowed three outs per inning, this graph would only be marginally more complicated. The usual game with three bases would be quite a bit more complicated, however.

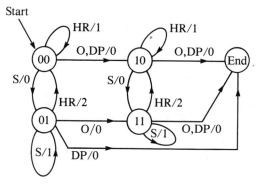

FIGURE 14.3.2
Transition diagram for a simplified game of baseball

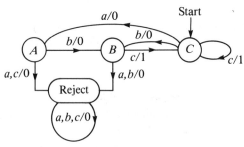

FIGURE 14.3.3

RECOGNITION IN REGULAR LANGUAGES

As we mentioned at the outset of this section, finite-state machines can recognize strings in a regular language. Consider the language L over $\{a, b, c\}$ that contains the strings of positive length in which each a is followed by b and each b is followed by c. One such string is $bccabcbc$. This language is regular. A grammar for the language would be non-terminal symbols $\{A, B, C\}$ with starting symbol C and production rules $A \to bB$, $B \to cC$, $C \to aA$, $C \to bC$, $C \to cC$, and $C \to c$. A finite-state machine (Figure 14.3.3) that recognizes this language can be constructed with one state for each non-terminal symbol and an additional state (Reject) that is entered if any invalid production takes place. At the end of an input tape that encodes a string in $\{a, b, c\}^*$, we will know when the string belongs to L based on the final output. If the final output is 1, the string belongs to L and if it is 0, the string does not belong to L. In addition, recognition can be accomplished by examining the final state of the machine. The input string belongs to the language if and only if the final state is C.

The construction of this machine is quite easy: note how each production rule translates into an edge between states other than Reject. For example, $C \to bB$ indicates that in State C, an input of b places the machine into State B. Not all sets of production rules can be as easily translated to a finite-state machine. Another set of production rules for L is $A \to aB$, $B \to bC$, $C \to cA$, $C \to cB$, $C \to cC$, and $C \to c$. Techniques for constructing finite-state machines from production rules is not our objective here. Hence we will only expect you to experiment with production rules until appropriate ones are found.

Example 14.3.4. A finite-state machine can be designed to add positive integers of any size. Given two integers in binary form, $a = a_n a_{n-1} \cdots a_1 a_0$ and $b = b_n b_{n-1} \cdots b_1 b_0$, the machine will read the input sequence, which is obtained from the digits of a and b reading from right to left,

$$a_0 b_0 (a_0 +_2 b_0) , \ldots , a_n b_n (a_n +_2 b_n),$$

followed by the special input 111. Note how all possible inputs except the last one must even parity (contain an even number of ones). The output sequence

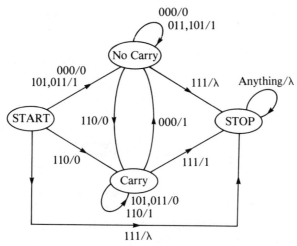

FIGURE 14.3.4
Transition diagram for a binary adder

is the sum of a and b, starting with the units digit, and comes from the set $\{0, 1, \lambda\}$. The transition diagram for this machine appears in Figure 14.3.4.

EXERCISES FOR SECTION 14.3

A Exercises

1. Draw a transition diagram for the vending machine described in Example 14.3.1.

2. Construct finite-state machines that recognize the regular languages that you identified in Section 14.2.

3. What is the input set for the machine in Example 14.3.4?

4. What input sequence would be used to compute the sum of 1101 and 0111 (binary integers)? What would the output sequence be?

B Exercise

5. *The Gray Code Decoder.* The finite-state machine defined in Figure 14.3.5 has an interesting connection with the Gray Code (Section 9.4). Given a string $x = x_1 x_2 \cdots x_n \in B^n$, we may ask where x appears in G_n. Starting in Copy state, the input string x will result in an output string $z \in B^n$, which is the binary form of the position of x in G_n. Positions are numbered from 0 to $2^n - 1$.

   (a) In what positions (0–31) do 10110, 00100, and 11111 appear in G_5?
   (b) Prove that the Gray Code Decoder always works. (Hint: Use induction on n.)

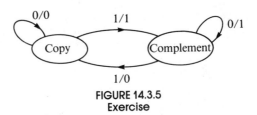

FIGURE 14.3.5
Exercise

14.4 The Monoid of a Finite-State Machine

In this section, we will see how every finite-state machine has a monoid associated with it. For any finite-state machine, the elements of its associated monoid correspond to certain input sequences. Because only a finite number of combinations of states and inputs is possible for a finite-state machine, there is only a finite number of input sequences that summarize the machine. This idea is illustrated best with a few examples.

Example 14.4.1. Consider the parity checker. The following table summarizes the effect on the parity checker of strings in B^1 and B^2. The row labeled "Even" contains the final state and final output as a result of each input string in B^1 and B^2 when the machine starts in the even state. Similarly, the row labeled "Odd" contains the same information for input sequences when the machine starts in the odd state.

Input String	0	1	00	01	10	11
Even	(Even, 0)	(Odd, 1)	(Even, 0)	(Odd, 1)	(Odd, 1)	(Even, 0)
Odd	(Odd, 1)	(Even, 1)	(Odd, 1)	(Even, 1)	(Even, 0)	(Odd, 1)
Same Effect as			0	1	1	0

Note how, as indicated in the last row, the strings in B^2 have the same effect as certain strings in B^1. For this reason, we can summarize the machine in terms of how it is affected by strings of length 1. The actual monoid that we will now describe consists of a set of functions, and the operation on the functions will be based on the concatenation operation.

Let T_0 be the final effect (state and output) on the parity checker of the input 0. Similarly, T_1 is defined as the final effect on the parity checker of the input 1. More precisely,

$$T_0 \text{ (even)} = \text{(even, 0)} \quad \text{and} \quad T_0 \text{ (odd)} = \text{(odd, 1)},$$

while

$$T_1 \text{ (even)} = \text{(odd, 1)} \quad \text{and} \quad T_1 \text{ (odd)} = \text{(even, 0)}.$$

In general, we define the operation on a set of such functions as follows: if s, t are input sequences and T_s and T_t are functions as above, then $T_s T_t = T_{st}$, that is, the result of the function that summarizes the effect on the machine by the concatenation of s with t. Since, for example, 01 has the same effect on the parity checker as 1, $T_0 T_1 = T_{01} = T_1$. We don't stop our calculation at T_{01} because we want to use the shortest string of inputs to describe the final result. A complete table for the monoid of the parity checker is

	T_0	T_1
T_0	T_0	T_1
T_1	T_1	T_0

What is the identity of this monoid? The monoid of the parity checker is isomorphic to the monoid $[\mathbf{Z}_2, +_2]$.

This operation may remind you of the composition operation on functions, but there are two principal differences. The domain of T_s is not the codomain of T_t, and the functions are read from left to right unlike in composition, where they are normally read from right to left.

You may have noticed that the output of the parity checker echoes the state of the machine and that we could have looked only at the effect on the machine as the final state. The following example has the same property, hence we will only consider the final state.

Example 14.4.2. The transition diagram for the machine that recognizes strings in B^* that have no consecutive 1's appears in Figure 14.4.1. Note how it is similar to the graph in Figure 9.1.1. Only a "reject state" has been added, for the case when an input of 1 occurs while in State a. We construct a similar table to the one in the previous example to study the effect of certain strings

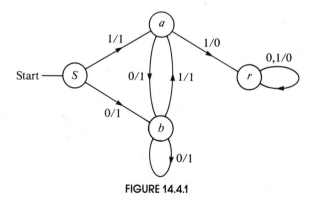

FIGURE 14.4.1

on this machine. This time, we must include strings of length 3 before we recognize that no "new effects" can be found.

Inputs	0	1	00	01	10	11	000	000	010	011	100	101	110	111
s	b	a	b	a	b	r	b	a	b	r	b	a	r	r
a	b	r	b	a	r	r	b	a	b	r	r	r	r	r
b	b	a	b	a	b	r	b	a	b	r	b	a	r	r
r	r	r	r	r	r	r	r	r	r	r	r	r	r	r
Same as	0						0	01	0	11	10	1	11	11

The following table summarizes how combinations of the strings 0, 1, 01, 10, and 11 affect this machine.

	T_0	T_1	T_{01}	T_{10}	T_{11}
T_0	T_0	T_{01}	T_{01}	T_0	T_{11}
T_1	T_{10}	T_{11}	T_1	T_{11}	T_{11}
T_{01}	T_0	T_{11}	T_{01}	T_{11}	T_{11}
T_{10}	T_{10}	T_1	T_1	T_{10}	T_{11}
T_{11}	T_{11}	T_{11}	T_{11}	T_{11}	T_{11}

All the results in this table can be obtained using the previous table. For example,

$$T_{10}T_{01} = T_{1001} = T_{100}T_1 = T_{10}T_1 = T_{101} = T_1, \text{ and}$$
$$T_{01}T_{01} = T_{0101} = T_{010}T_1 = T_0T_1 = T_{01}.$$

Note that none of the elements that we have listed in this table serves as the identity for our operation. This problem can always be remedied by including the function that corresponds to the input of the null string, T_λ. Since the null string is the identity for concatenation of strings, $T_sT_\lambda = T_\lambda T_s = T_s$ for all input strings s.

Example 14.4.3. A finite-state machine called the *unit-time delay machine* does not echo its current state, but prints its previous state. For this reason, when we find the monoid of the unit-time delay machine, we must consider both state and output. The transition diagram of this machine appears in Figure 14.4.2.

Input	0	1	00	01	10	11	100 or 000	101 or 001	110 or 010	111 or 011
0	(0, 0)	(1, 0)	(0, 0)	(1, 0)	(0, 1)	(1, 1)	(0, 0)	(1, 0)	(0, 1)	(1, 1)
1	(0, 1)	(1, 1)	(0, 0)	(1, 0)	(0, 1)	(1, 1)	(0, 0)	(1, 0)	(0, 1)	(1, 1)
Same as							00	01	10	11

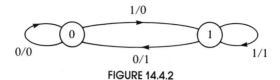

1/0

0/0   0   1   1/1

0/1

FIGURE 14.4.2

Again, since no new outcomes were obtained from strings of length 3, only strings of length 2 or less contribute to the monoid of the machine. The table for the strings of positive length shows that we must add T_λ to obtain a monoid.

	T_0	T_1	T_{00}	T_{01}	T_{10}	T_{11}
T_0	T_{00}	T_{01}	T_{00}	T_{01}	T_{10}	T_{11}
T_1	T_{10}	T_{11}	T_{00}	T_{01}	T_{10}	T_{11}
T_0	T_{00}	T_{01}	T_{00}	T_{01}	T_{10}	T_{11}
T_{01}	T_{10}	T_{11}	T_{00}	T_{01}	T_{10}	T_{11}
T_{10}	T_{00}	T_{01}	T_{00}	T_{01}	T_{10}	T_{11}
T_{11}	T_{10}	T_{11}	T_{00}	T_{01}	T_{10}	T_{11}

EXERCISES FOR SECTION 14.4

A Exercise

1. For each of the following transition diagrams, write out tables for their associated monoids. Identify the identity in terms of a string of positive length, if possible. (Hint: Where the output echos the current state, the output can be ignored.)

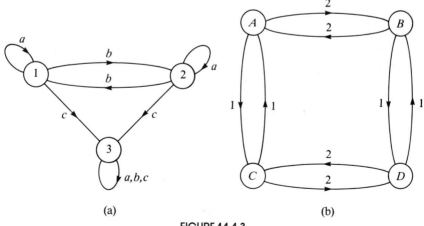

(a)   (b)

FIGURE 14.4.3

B Exercise

2. What common monoids are isomorphic to the monoids obtained in the previous exercise?

    C Exercise

3. Can two finite-state machines with non-isomorphic transition diagrams have isomorphic monoids?

14.5 The Machine of a Monoid

Any finite monoid $[M;*]$ can be represented in the form of a finite-state machine with input and state sets equal to M. The output of the machine will be ignored here, since it would echo the current state of the machine. Machines of this type are called *state machines*. It can be shown that whatever can be done with a finite-state machine can be done with a state machine; however, there is a trade-off. Usually, state machines that perform a specific function are more complex than general finite-state machines.

Definition: Machine of a Monoid.   *If $[M;*]$ is a finite monoid, then the machine of M, denoted m(M), is the state machine with state set M, input set M, and next-state function $t : M \times M \rightarrow$ defined by $t(s, x) = s * x$.*

Example 14.5.1.   We will construct the machine of the monoid $[\mathbf{Z}_2; +_2]$. As mentioned above, the state set and the input set are both $\mathbf{Z}_2$. The next state function is defined by $t(s, x) = s +_2 x$. The transition diagram for $m(\mathbf{Z}_2)$ appears in Figure 14.5.1. Note how it is identical to the transition diagram of the parity checker, which has an associated monoid that was isomorphic to $[\mathbf{Z}_2; +_2]$.

Example 14.5.2.   The transition diagram of the monoids $[\mathbf{Z}_2; \times_2]$ and $[\mathbf{Z}_3; +_3]$ appear in Figure 14.5.2.

Example 14.5.3.   Let U be the monoid that we obtained from the unit-time delay machine (Example 14.4.3). We have seen that the machine of the monoid of the parity checker is essentially the parity checker. Will we obtain a unit-time delay machine when we construct the machine of U? We can't expect to get exactly the same machine because the unit-time delay machine is not a state machine and the machine of a monoid is a state machine. However, we will see that our new machine is capable of telling us what input was received in the previous time period. The operation table for the monoid

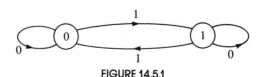

FIGURE 14.5.1

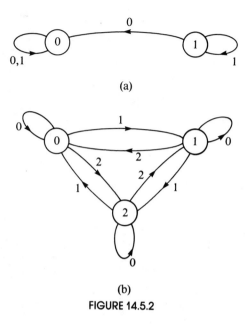

(a)

(b)

FIGURE 14.5.2

serves as a table to define the transition function for the machine. The row headings are the state values, while the column headings are the inputs. If we were to draw a transition diagram with all possible inputs, the diagram would be too difficult to read. Since U is generated by the two elements, T_0 and T_1, we will include only those inputs. Suppose that we wanted to read the transition function for the input T_{01}. Since $T_{01} = T_0 T_1$, in any state, s, $t(s, T_{01}) = t(t(s, T_0), T_1)$. The transition diagram appears in Figure 14.5.3.

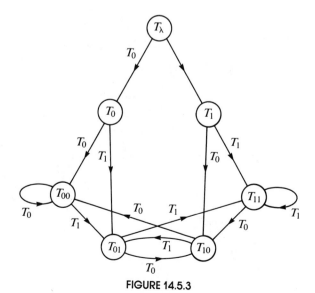

FIGURE 14.5.3

If we start reading a string of 0's and 1's while in state T_λ and are in state T_{ab} at any one time, the input from the previous time period (not the input that sent us into T_{ab}, the one before that) is a. In states T_λ T_0 and T_1, no previous input exists.

EXERCISES FOR SECTION 14.5

A Exercise

1. Draw the transition diagrams for the machines of the following monoids:
   (a) $[\mathbf{Z}_4; +_4]$
   (b) $[\mathbf{Z}_2; \times_2]^2$
   (c) the monoid obtained in Example 14.4.3

B Exercise

2. Even though a monoid may be infinite, we can visualize it as an infinite-state machine provided that it is generated by a finite number of elements. For example, the monoid $B*$ is generated by 0 and 1. A section of its transition diagram can be obtained by allowing input only from the generating set (Figure 14.5.4a). The monoid of integers under addition is generated by the set $\{-1, 1\}$. The transition diagram for this monoid can be visualized by drawing a small portion of it, as in Figure 14.5.4b.
   (a) Draw a transition diagram for $\{a, b, c\}*$.
   (b) Draw a transition diagram for $[\mathbf{Z} \times \mathbf{Z}; \text{componentwise addition}]$.

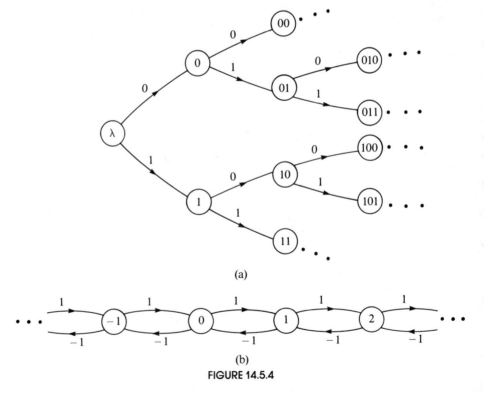

(a)

(b)

FIGURE 14.5.4

chapter

15

GROUP THEORY
AND APPLICATIONS

15 GOALS

In Chapter 11, Algebraic Systems, groups were introduced as a typical algebraic system. The associated concepts of subgroup, group isomorphism, and direct products of groups were also introduced. Groups were chosen for that chapter because they are among the simplest types of algebraic systems. Despite this simplicity, group theory abounds with interesting applications, many of which are of interest to the computer scientist. In this chapter we intend to present the remaining important concepts in elementary group theory and some of their applications.

15.1 Cyclic Groups

Groups are classified according to their size and structure. A group's structure is revealed by a study of its subgroups and other properties (e.g., whether it is abelian) that might give an overview of it. Cyclic groups have the simplest structure of all groups.

Definitions: Cyclic Group, Generator. *Group G is cyclic if there exists $a \in G$ such that the cyclic subgroup generated by a, (a), equals all of G. That is, $G = \{na \mid n \in Z\}$, in which case a is called a generator of G. The reader should note that additive notation is used for G.*

Example 15.1.1. $\mathbf{Z}_{12} = [\mathbf{Z}_{12}, +_{12}]$, where $+_{12}$ is addition modulo 12, is a cyclic group. To prove this statement, all we need to do is demonstrate that some element of $\mathbf{Z}_{12}$ is a generator. One such element is 5; that is, $(5) = \mathbf{Z}_{12}$. One more obvious generator is 1. In fact, 1 is a generator of every $[\mathbf{Z}_n; +_{12}]$. The reader is asked to prove that if an element is a generator, then its inverse is also a generator. Thus, $-5 = 7$ and $-1 = 11$ are the other generators of $\mathbf{Z}_{12}$.

Figure 15.1.1 is an example of "string art" that illustrates how 5 generates $\mathbf{Z}_{12}$. Twelve tacks are placed along a circle and numbered. A string is tied to tack 0, and is then looped around every fifth tack. As a result, the numbers of the tacks that are reached are exactly the ordered multiples of 5 modulo 12: 5, 10, 3, . . . , 7, 0. Note that if every seventh tack were used, the same artwork would be obtained. If every third tack were connected, the resulting loop would only use four tacks; thus 3 does not generate $\mathbf{Z}_{12}$.

Example 15.1.2. The group of additive integers is $[\mathbf{Z}; +]$ is cyclic:

$$\mathbf{Z} = (1) = \{n \cdot 1 \mid n \in \mathbf{Z}\}.$$

This observation does not mean that every integer is the product of an integer times 1. It means that

11   0

10   1

9   2

8   3

7   4

6   5

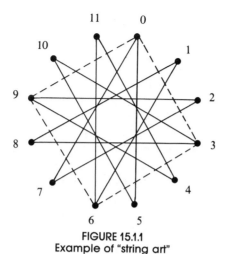

FIGURE 15.1.1
Example of "string art"

$$\mathbf{Z} = \{0\} \cup \{\underbrace{1 + \cdots + 1}_{n \text{ terms}} | n \in N\} \cup \{\underbrace{(-1) + \cdots + (-1)}_{-n \text{ terms}} | -n \in \mathbf{N}\}.$$

Theorem 15.1.1.   *If* $[G, *]$ *is cyclic, then it is abelian.*

Proof: Let a be any generator of G and let $b, c \in G$. By the definition
of the generator of a group, there exist integers m and n such that $b = ma$
and $c = na$. Thus

$$
\begin{aligned}
b * c &= (ma) * (na) \\
&= (m + n) \cdot a \quad \text{By Theorem 11.3.7(ii)} \\
&= (n + m) \cdot a \quad [\mathbf{Z}; +] \text{ is abelian} \\
&= (n\,a) * (m\,a) \\
&= c * b. \ \#
\end{aligned}
$$

One of the first steps in proving a property of cyclic groups is to use the
fact that there exists a generator. Then every element of the group can be
expressed as some multiple of the generator. Take special note of how this is
used in theorems of this section.

Up to now we have used only additive notation to discuss cyclic groups.
Theorem 15.1.1 actually justifies this practice since it is customary to use
additive notation when discussing abelian groups. Of course, some concrete
groups for which we employ multiplicative notation are cyclic. If one of its
elements, a, is a generator,

$$(a) = \{a^n | n \in \mathbf{Z}\}$$

is all of the group.

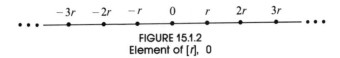

FIGURE 15.1.2
Element of [r], 0

Example 15.1.3. The group of positive integers modulo 11 with modulo 11 multiplication, $[\mathbf{Z}_{11}^*, \times_{11}]$, is cyclic. One of its generators is 6: $6^1 = 6$, $6^2 = 3$, $6^3 = 7, \ldots , 6^9 = 2$, and $6^{10} = 1$, the identity.

Example 15.1.4. The real numbers with addition, $[\mathbf{R}, +]$, is a non-cyclic group. The proof of this statement requires a bit more generality since we are saying that for all $r \in \mathbf{R}$, (r) is a proper subset of $\mathbf{R}$. If r is non-zero, the multiples of r are distributed over the real line, as in Figure 15.1.2. It is clear then that there are many real numbers, like $r/2$, that are not in (r).

The following theorem shows that a cyclic group can never be very complicated.

Theorem 15.1.2. *If G is a cyclic group, then G is either finite or countably infinite. If G is finite and $\#G=n$, it is isomorphic to $[\mathbf{Z}_n, +_n]$. If G is infinite, it is isomorphic to $[\mathbf{Z}, +]$.*

Proof:

Case 1: $\#G < \infty$. If a is a generator of G and $\#G = n$, define Ø: $\mathbf{Z}_n \to G$ by

$$\text{Ø } (k) = ka \qquad (k \in \mathbf{Z}_n)$$

Since (a) is finite, we can use the fact that the elements of (a) are the first n non-negative multiples of a. From this observation, we see that Ø is a surjection. A surjection between finite sets of the same cardinality must be a bijection. Finally, if $p, q \in \mathbf{Z}_n$

$$\text{Ø } (p) + \text{Ø } (q) = pa + qa$$
$$= (p + q)a = \begin{cases} (p + q)a & \text{if } p + q < n \\ (p + q - n)a & \text{if } p + q \geq n \end{cases}$$
$$= \text{Ø } (p +_n q).$$

Therefore Ø is an isomorphism from $\mathbf{Z}_n$ into G.

Case 2: $\#G = \infty$. We will leave this case for the reader to prove. Many abstract algebra texts supply a proof for this case. #

The proof of Theorem 15.1.3 makes use of the division property for integers that we first saw in Section 11.4: if m, n are integers, $m > 0$, there exist unique integers q (quotient) and r (remainder) such that $n = qm + r$ and $0 \leq r < m$.

Theorem 15.1.3. *Every subgroup of a cyclic group is cyclic.*

Proof: Let G be cyclic with generator a and let $H \leq G$. If $H = \{e\}$, H has e as a generator. We may now assume that $\#H \geq 2$ and $a \neq e$. Let $ma = c$ be the least positive multiple of a that belongs to H. (This is the key step. It lets us get our hands on a generator of H.) We will now show that c generates H. Suppose that $(c) \neq H$. Then there exists $b \in H$ such that $b \notin (c)$. Now, since b is in G, there exists $n \in \mathbf{Z}$ such that $b = na$. We now apply the division property

$$b = na = (qm + r)a = (qm)a + ra,$$

where $0 < r < m$. Note that $r \neq 0$, for otherwise b would clearly be in $[c]$. Remember, $na = b$ and $(qm)a = qc$ are both in H. Thus,

$$ra = na - (qm)a \in H.$$

This contradicts our choice of c, which was to be the least positive multiple of a in H because $0 < r < m$.   #

Example 15.1.5. The only proper subgroups of $\mathbf{Z}_{10}$ are $H_1 = \{0, 5\}$ and $H_2 = \{0, 2, 4, 6, 8\}$, $H_1 = (5)$, while $H_2 = (2) = (4) = (6) = (8)$. The generators of $\mathbf{Z}_{10}$ are 1, 3, 7, and 9.

Example 15.1.6. With the exception of $\{0\}$, all subgroups of $\mathbf{Z}$ are isomorphic to $\mathbf{Z}$. If $H \leq \mathbf{Z}$, then H is the cyclic subgroups generated by the least positive element of H.

We now cite a useful theorem for computing the order of cyclic subgroups of a cyclic group:

Theorem 15.1.4. *If G is a cyclic group of order n and a is a generator of G, the order of ka is n/d, where d is the greatest common divisor of n and k.*

The proof of this theorem is left to the reader.

Example 15.1.7. To compute the order of (18) in $\mathbf{Z}_{30}$, we first observe that 1 is a generator of $\mathbf{Z}_{30}$ and $18 = 18(1)$. The greatest common divisor of 18 and 30 is 6. Hence, the order of (18) is $30/6$, or 5.

APPLICATION: FAST ADDERS

At this point, we will introduce the idea of a *fast adder*, a recent application (Winograd, 1965) to an ancient theorem, the Chinese Remainder Theorem. We will present only an overview of the theory and rely primarily on examples. The interested reader can refer to Dornhoff and Hohn for details.

Out of necessity, integer addition with a computer is addition modulo n, for n some larger number. Consider the case where n is small, like 64. Then addition involves the addition of six-digit binary numbers. Consider the process of adding 31 and 1. Assume the computer's adder takes as input two bit strings $a = (a_5, a_4, \ldots, a_0)$ and $b = (b_5, b_4, \ldots, b_0)$ and outputs $s = (s_5, s_4, \ldots, s_0)$, the sum of a and b. Then, if $a = 31 = (0, 1, 1, 1, 1, 1)$ and $b = 1 = (0, 0, 0, 0, 0, 1)$, s will be $(1, 0, 0, 0, 0, 0)$, or 32. The output $s_5 = 1$ cannot be determined until all other outputs have been determined. If addition is done with a finite-state machine, as in Example 14.3.5, the time required to obtain s will be six time units, where one time unit is the time it takes to get an output from the machine. Theoretically, this time can be decreased, but the explanation would require a long digression and our relative results would not change that much. In general, we use the rule that the number of time units needed to perform addition modulo n is $\lceil \log_2 n \rceil$.

Now we will introduce a hypothetical problem that we will use to illustrate the idea of a fast adder. Suppose that we had to add 1000 numbers modulo $27720 = 8 \cdot 9 \cdot 5 \cdot 7 \cdot 11$. By the rule above, since $2^{14} < 27720 < 2^{15}$, each addition would take 15 time units. If the sum is initialized to zero, 1000 additions would be needed; thus, 15,000 time units would be needed to do the additions. We can improve this time dramatically by applying the Chinese Remainder Theorem.

The Chinese Remainder Theorem (CRT).   *Let $n_1, n_2, \ldots, n_p$ be powers of distinct primes and $n = n_1 n_2 \cdots n_p$. Define*

$$\theta: \mathbf{Z}_n \to \mathbf{Z}_{n_1} \times \cdots \times \mathbf{Z}_{n_p}$$

by $\theta(k) = (k_1, \ldots, k_p)$, where for $1 \le i \le p$,

$$(a) \quad 0 \le k_i < n_i \text{ and}$$
$$(b) \quad k \equiv k_i \ (mod \ n_i).$$

θ is an isomorphism from $\mathbf{Z}_n$ into $\mathbf{Z}_{n_1} \times \cdots \times \mathbf{Z}_{n_p}$.

This theorem can be stated in several different forms, and its proof can be found in many abstract algebra tests, including Fisher and Dornhoff and Hohn.

Example 15.1.8.   As we saw in Chapter 11, $\mathbf{Z}_6$ is isomorphic to $\mathbf{Z}_2 \times \mathbf{Z}_3$. This is the smallest case to which the CRT can be applied. An isomorphism between $\mathbf{Z}_6$ and $\mathbf{Z}_2 \times \mathbf{Z}_3$ is

$$
\begin{array}{ll}
\theta(0) = (0, 0) & \theta(3) = (1, 0) \\
\theta(1) = (1, 1) & \theta(4) = (0, 1) \\
\theta(2) = (0, 2) & \theta(5) = (1, 2)
\end{array}
$$

Now consider our problem of adding in $\mathbf{Z}_{27720}$. Let $G = \mathbf{Z}_8 \times \mathbf{Z}_9 \times \mathbf{Z}_5 \times \mathbf{Z}_7 \times \mathbf{Z}_{11}$. The CRT gives us an isomorphism between $\mathbf{Z}_{27720}$ and G.

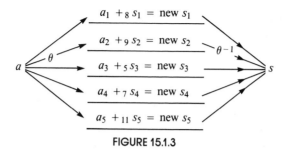

$$a_1 +_8 s_1 = \text{new } s_1$$

$$a_2 +_9 s_2 = \text{new } s_2$$

$$a_3 +_5 s_3 = \text{new } s_3$$

$$a_4 +_7 s_4 = \text{new } s_4$$

$$a_5 +_{11} s_5 = \text{new } s_5$$

FIGURE 15.1.3

The basic idea behind the fast adder, illustrated in Figure 15.1.3, is to make use of this isomorphism.

After each of the s_i's is initialized to zero, the summands are decomposed, one at a time, into a quintuple $(a_1, \ldots, a_5)$ in G. Addition in G can be done in parallel so that each new subtotal in the form of the quintuple $(s_1, \ldots, s_5)$ takes only as long to compute as it takes to add in $\mathbf{Z}_{11}$. By the time rule that we have established, the addition of 1000 numbers can be done in 1000 $\lceil \log_2 11 \rceil = 4000$ time units, or in less than one-third the time.

Two more factors must still be considered, however. How easy is it to determine $\theta(a)$ and $\theta^{-1}(s_1, s_2, \ldots, s_5)$? We must compute $\theta(a)$ one thousand times, and, if it requires a sizable amount of time, there may not be any advantage to the fast adder. The computation of an inverse is not as time-critical since it must be done only once, after the final sums are determined in G.

The determination of $\theta(a)$ is not a major problem. If the values of $\theta(1)$, $\theta(10)$, $\theta(100)$, $\theta(1000)$, and $\theta(10000)$ are stored, $a = d_4 d_3 d_2 d_1 d_0$ ten and input is done decimally, then

$$\theta(a) = d_0 \theta(1) + d_1 \theta(10) + d_2 \theta(100) + d_3 (1000) + d_4 \theta(10000)$$

by the fact that θ is an isomorphism. The components of $\theta(a)$ can be computed economically using this formula so as not to slow down the actual adding process. For different forms of input, similar schemes can be devised.

The computation of $\theta^{-1}(s_1, s_2, \ldots, s_5)$ is simplified by the fact that θ^{-1} is also an isomorphism. The final sum is $s_1 \theta^{-1}(1, 0, 0, 0, 0) + s_2 \theta^{-1}(0, 1, 0, 0, 0) + \cdots + s_5 \theta^{-1}(0, 0, 0, 0, 1)$. The arithmetic in this expression is in $\mathbf{Z}_{27720}$ and is more time consuming. However, as was noted above, it need only be done once. This is why the fast adder is only practical in situations where many additions must be performed to get a single sum.

To illustrate the potential of fast adders, consider the problem of addition modulo

$$n = 2^5 \cdot 3^3 \cdot 5^2 \cdot 7^2 \cdot 11 \cdot 13 \cdot 17 \cdot 19 \cdot 23 \cdot 29 \cdot 31 \cdot 37 \cdot 41 \cdot 43 \cdot 47,$$

$n > 3 \times 10^{21}$. Each addition using the usual modulo n addition with full adders would take 72 time units. By decomposing each summand into 15-tuples according to the CRT, the time is reduced to $\lceil \log_2 49 \rceil = 6$ time units per addition.

EXERCISES FOR SECTION 15.1

A Exercises

1. What generators besides 1 does $[\mathbf{Z}, +]$ have?

2. Without doing any multiplications, determine how many generators $[\mathbf{Z}_{11}; +_{11}]$ (Example 15.1.3) has.

3. Prove that if $\#G > 2$ and G is cyclic, G has at least two generators.

4. If you wanted to list the generators of $\mathbf{Z}_n$, you would only have to test the first $n/2$ positive integers. Why?

5. Which of the following groups are cyclic? Explain.
   (a) $[\mathbf{Q}, +]$
   (b) $[\mathbf{R}^+, \cdot]$
   (c) $[6\mathbf{Z}, +]$ where $6\mathbf{Z} = \{6n \mid n \in \mathbf{Z}\}$
   (d) $\mathbf{Z} \times \mathbf{Z}$
   (e) $\mathbf{Z}_2 \times \mathbf{Z}_3 \times \mathbf{Z}_{25}$

6. For each group and element, determine the order of the cyclic subgroup generated by the element:
   (a) $\mathbf{Z}_{25}$, 15
   (b) $\mathbf{Z}_4 \times \mathbf{Z}_9$, (2, 6) (apply Exercise 8)
   (c) $\mathbf{Z}_{64}$, 2

B Exercises

7. How can Theorem 15.1.4 be applied to list the generators of $\mathbf{Z}_n$? What are the generators of $\mathbf{Z}_{25}$? Of $\mathbf{Z}_{256}$?

8. Prove that if the greatest common divisor of n and m is 1, then (1, 1) is a generator of $\mathbf{Z}_n \times \mathbf{Z}_m$, and, hence, $\mathbf{Z}_n \times \mathbf{Z}_m$ is isomorphic to $\mathbf{Z}_{nm}$.

9. (a) Illustrate how the fast adder can be used to add the numbers 21, 5, 7, and 15 using the isomorphism between $\mathbf{Z}_{77}$ and $\mathbf{Z}_7 \times \mathbf{Z}_{11}$.
   (b) If the same isomorphism is used to add the numbers 25, 26, and 40, what would the result be, why would it be incorrect, and how would the answer differ from the answer in Part a?

C Exercise

10. Write a Pascal program to list the generators of $\mathbf{Z}_n$. Make use of the fact that you need only test the first $n/2$ positive integers.

15.2 Cosets and Factor Groups

Consider the group $[\mathbf{Z}_{12}; +_{12}]$. As we saw in the previous section, we can picture its cyclic properties with the string art of Figure 15.1.1. Here we will be interested in the non-generators, like 3. The solid lines in Figure 15.2.1

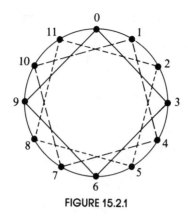

FIGURE 15.2.1

show that only one-third of the tacks have been reached by starting at zero and jumping to every third tack. The numbers of these tacks correspond to $(3) = \{0, 3, 6, 9\}$. What happens if you start at one of the unused tacks and again jump to every third tack? The two broken paths on Figure 15.2.1 show that identical squares are produced. The tacks are thus partitioned into very similar subsets. The subsets of $\mathbf{Z}_{12}$ that they correspond to are $\{0, 3, 6, 9\}$, $\{1, 4, 7, 10\}$, and $\{2, 5, 8, 11\}$. These subsets are called *cosets*. In particular, they are called *cosets of the subgroup* $\{0, 3, 6, 9\}$. We will see that under certain conditions, cosets of a subgroup can form a group of their own. Before pursuing this example any further we will examine the general situation.

Definition: Coset. *If $[G, *]$ is a group, $H \leq G$ and $a \in G$, the left coset $a * H$ of H is $\{a * h \mid h \in H\}$. The right coset $H * a$ of H is $\{h * a \mid h \in H\}$.*

Notes:

(a) H itself is a left and right coset since $e * H = H * e = H$.
(b) If G is abelian, $a * H = H * a$ and the left-right distinction for cosets can be dropped. We will use left coset notation in that situation.

Definition: Representative. *Any element of a coset is called a representative of that coset.*

One might wonder whether a is in any way a special representative of $a * H$ since it seems to define the coset. It is not, as we shall see.

Theorem 15.2.1. *If $b \in a * H$, then $a * H = b * H$, and if $b \in H * a$, then $H * a = H * b$.*

Remark: A Duality Principle. A duality principle can be formulated concerning cosets because left and right cosets are defined in such similar ways. Any theorem about left and right cosets will yield a second theorem when "left" and "right" are exchanged for "right" and "left".

Proof: In light of the remark above, we need only prove the first part of this theorem. Suppose that $x \in a * H$. By the definition of $a * H$, there exists h_1 and h_2 in H such that $b = a * h_1$ and $x = a * h_2$. Given these two equations, we need only find a way of expressing x as "b times an element of H." Then we will have proven that $a * H \subseteq b * H$.

$$b = a * h_1 \Rightarrow a = b * h_1^{-1}$$

and

$$x = a * h_2 = (b * h_1^{-1}) * h_2 = b * (h_1^{-1} * h_2).$$

Since $h_1, h_2 \in H, h_1^{-1} * h_2 \in H$, we are done with this part of the proof. In order to show that $b * H \subseteq a * H$, one can follow essentially the same steps, which we will let the reader fill in. #

Example 15.2.1. In Figure 15.2.1, you can start at either 1 or 7 and obtain the same path by taking jumps of three tacks in each step. Thus,

$$1 +_{12} \{0, 3, 6, 9\} = 7 +_{12} \{0, 3, 6, 9\} = \{1, 4, 7, 10\}.$$

The set of left (or right) cosets of a subgroup partition a group in a special way:

Theorem 15.2.2. *If $[G, *]$ in a group and $H \leq G$, the set of left cosets of H is a partition of G. In addition, all of the left cosets of H have the same cardinality. The same is true for right cosets.*

Proof: That every element of G belongs to a left coset is clear because $a \in a * H$ for all $a \in G$. If $a * H$ and $b * H$ are left cosets, they are either equal or disjoint. Suppose that $(a * H) \cap (b * H)$ is non-empty and c belongs to the intersection. Then

$$c \in a * H \Rightarrow a * H = c * H$$

and

$$c \in b * H \Rightarrow b * H = c * H.$$

Hence

$$a * H = b * H.$$

We complete the proof by showing that each left coset has the same cardinality as H. To do this, we simply observe that if $a \in G, r : H \rightarrow a * H$

defined by $r(h) = a * h$ is a bijection. We will leave the proof of this statement to the reader. #

The function r has a nice interpretation in terms of our opening example. If $a \in Z_{12}$, the graph of $\{0, 3, 6, 9\}$ is rotated $30a°$ to coincide with one of the three cosets of $\{0, 3, 6, 9\}$.

Counting Formula. *If $\#G < \infty$, $H < G$, the number of left cosets* $= \dfrac{\#G}{\#H}$. For this reason we use G/H to denote the set of left cosets of H.

Example 15.2.2. The set of multiples of four, $4\mathbf{Z}$, is a subgroup of $[\mathbf{Z}, +]$. Four distinct cosets of $4\mathbf{Z}$ partition the integers. They are $4\mathbf{Z}$, $1 + 4\mathbf{Z}$, $2 + 4\mathbf{Z}$, and $3 + 4\mathbf{Z}$, where, for example, $1 + 4\mathbf{Z} = \{1 + 4k \,|\, k \in \mathbf{Z}\}$. $4\mathbf{Z}$ can also be written $0 + 4\mathbf{Z}$.

Distinguished Representatives. Although we have seen that any representative can describe a coset, it is often convenient to select a *distinguished representative* from each coset. The advantage to doing this is that there is a unique name for each coset in terms of its distinguished representative. In numeric examples such as the one above, the *distinguished representative* is the smallest non-negative representative. Remember, this is purely a convenience and there is absolutely nothing wrong in writing $-203 + 4\mathbf{Z}$, $5 + 4\mathbf{Z}$ or $621 + 4\mathbf{Z}$ in place of $1 + 4\mathbf{Z}$.

Before completing the main thrust of this section, we will make note of a significant implication of Theorem 15.2.2. Since a finite group is divided into cosets of a common size by any subgroup, we can conclude:

Lagrange's Theorem. *The order of a subgroup of a finite group must divide the order of the group.*

The proof is left to the reader.

One immediate implication of Lagrange's Theorem is that if p is prime, $\mathbf{Z}_p$ has no proper subgroups.

We will now describe the operation on cosets which will, under certain circumstances, result in a group. For most of this section, we will assume that G is an Abelian group. This is one condition that guarantees that the set of left cosets will form a group.

Definition: Operation on Cosets. *Let C and D be left cosets of H, a subgroup of G. Then $C \otimes D = (c * d) * H$, where c and d are representatives of C and D, respectively.* The operation $\otimes$ is called the *operation induced on left cosets by* $*$.

In Theorem 15.2.3, later in this section, we prove that if G is an abelian group, $\otimes$ is indeed an operation. In practice, if the group G is an additive group, the symbol $\otimes$ is replaced by $+$, as in the following example.

TABLE 15.2.1
Coset Operation—Table of Example 15.2.3

+	$\bar{0}$	$\bar{1}$	$\bar{2}$	$\bar{3}$
$\bar{0}$	$\bar{0}$	$\bar{1}$	$\bar{2}$	$\bar{3}$
$\bar{1}$	$\bar{1}$	$\bar{2}$	$\bar{3}$	$\bar{0}$
$\bar{2}$	$\bar{2}$	$\bar{3}$	$\bar{0}$	$\bar{1}$
$\bar{3}$	$\bar{3}$	$\bar{0}$	$\bar{1}$	$\bar{2}$

Example 15.2.3. In regard to the cosets described in Example 15.2.2, if we rename $0 + 4\mathbf{Z}$, $1 + 4\mathbf{Z}$, $2 + 4\mathbf{Z}$, and $3 + 4\mathbf{Z}$ with the symbols $\bar{0}$, $\bar{1}$, $\bar{2}$, and $\bar{3}$, $\bar{1} + \bar{3} = \bar{0}$ because $9 \in \bar{1}$, $7 \in \bar{3}$, and $9 + 7 \in 16 + 4\mathbf{Z} = \bar{0}$. Our choice of the representatives $\bar{1}$ and $\bar{3}$ was completely arbitrary. Since $C \otimes D$ (or $\bar{1} + \bar{3}$ in this case) can be computed in many ways, it is necessary to show that the choice of representatives does not affect the result. When $C \otimes D$ is always independent of our choice of representatives, we say that "$\otimes$ is well defined." $+$ is a well-defined operation on the left cosets of $4\mathbf{Z}$ and is summarized in Table 15.2.1. Do you notice anything familiar?

Example 15.2.4. Consider the real numbers. $[\mathbf{R}; +]$, and its subgroup of integers, $\mathbf{Z}$. Every element of $\mathbf{R}/\mathbf{Z}$ has the same cardinality as $\mathbf{Z}$. Let s, $t \in \mathbf{R}$. $s \in t + \mathbf{Z}$ if s can be written $t + n$ for some $n \in \mathbf{Z}$:

$$s = t + n \Rightarrow s - t = n.$$

Hence s and t belong to the same coset if they differ by an integer. (See Exercise 6 for a generalization of this fact.)

Now consider the coset $0.25 + \mathbf{Z}$. Real numbers that differ by an integer from 0.25 are $1.25, 2.25, 3.25, \ldots$ and $-0.75, -1.75$, and $-2.75. \ldots$
If any real number is selected, there exists a representative of its coset that is greater than or equal to 0 and less than 1. We will call that representative the distinguished representative of the coset. For example, 43.125 belongs to the coset represented by 0.125; $-6.382 + \mathbf{Z}$ has 0.618 as its distinguished representative. The operation on $\mathbf{R}/\mathbf{Z}$ is commonly called *addition modulo 1*. A few calculations in $\mathbf{R}/\mathbf{Z}$ are

$$(0.1 + \mathbf{Z}) + (0.48 + \mathbf{Z}) = 0.58 + \mathbf{Z} \quad \text{As expected}$$
$$(0.7 + \mathbf{Z}) + (0.31 + \mathbf{Z}) = 0.01 + \mathbf{Z}$$
$$-(0.41 + \mathbf{Z}) = -0.41 + \mathbf{Z} = 0.59 + \mathbf{Z}.$$

In General, $-(a + \mathbf{Z}) = (1 - a) + \mathbf{Z}$.

Example 15.2.5. Consider $F = (\mathbf{Z}_4 \times \mathbf{Z}_2)/(0, 1)$. Since $\mathbf{Z}_4 \times \mathbf{Z}_2$ is of order 8 and $(0, 1)$ is $\{(0, 0), (0, 1)\}$, each element of F is a coset containing two ordered pairs. We will leave it to the reader to verify that the four distinct cosets are

$$(0, 0) + ((0, 1)),$$
$$(1, 0) + ((0, 1)),$$
$$(2, 0) + ((0, 1)), \text{ and}$$
$$(3, 0) + ((0, 1)).$$

The reader can also verify that F is isomorphic to $\mathbf{Z}_4$, since F is cyclic. An educated guess should give you a generator.

Example 15.2.6. Consider the group $\mathbf{Z}_2^4 = \mathbf{Z}_2 \times \mathbf{Z}_2 \times \mathbf{Z}_2 \times \mathbf{Z}_2$. Let $H = ((1, 0, 1, 0))$. $\#H = 2$; hence, $\mathbf{Z}_2^4/H$ has 8 elements. A typical element

$$C = (0, 1, 1, 1) + H = \{0, 1, 1, 1), (1, 1, 0, 1)\}.$$

Since $2(0, 1, 1, 1) = (0, 0, 0, 0)$, $2C = H$, the identity of $\mathbf{Z}_2^4/H$. The orders of all non-identity elements of $\mathbf{Z}_2^4/H$ are all 2, and it can be shown that the factor group is isomorphic to $\mathbf{Z}_2^3$.

Theorem 15.2.3. *If G is an abelian group, and $H \leq G$, the operation induced on cosets of H by the operation of G is well defined.*

Proof: Suppose that a, b, and a', b' are two choices for representatives of cosets C and D. That is to say that a, $a' \in C$, b, $b' \in D$. We will show that $a * b$ and $a' * b'$ are representatives of the same coset. Theorem 15.2.1 implies that $C = a * H$ and $D = b * H$, thus

$$a' \in a * H \text{ and } b' \in b * H.$$

Then there exists h_1, $h_2 \in H$ such that

$$a' = a * h_1 \text{ and } b' = b * h_2$$

and

$$a' * b' = (a * h_1) * (b * h_2)$$
$$= (a * b) * (h_1 * h_2)$$

by various group properties.
This last expression for $a' * b'$ implies that

$$a' * b' \in (a * b) * H \text{ since } h_1 * h_2 \in H. \; \#$$

Exercise. In the proof of Theorem 15.2.3, where does the assumption that G is abelian come into play?

Theorem 15.2.4. *Let G be a group and $H \leq G$. If the operation induced on left cosets of H by the operation of G is well defined, then the set of left cosets form a group under that operation.*

Proof: Let C_1, C_2, and C_3 be the left cosets with representatives r_1, r_2, and r_3, respectively. The value of $C_1 \otimes (C_2 \otimes C_3)$ and $(C_1 \otimes C_2) \otimes C_3$ are determined by $r_1 * (r_2 * r_2)$ and $(r_1 * r_2) * r_3$. By the associativity of $*$ in G, the two cosets' products must be equal. Therefore, the induced operation is associative. As for the identity and inverse properties, there is no surprise. The identity coset is H, or $e * H$, the coset that contains G's identity. If C is a coset with representative a—i.e., $C = a * H$—then C^{-1} is $a^{-1} * H$.

$$(a * H) \otimes (a^{-1} * H) = (a * a^{-1}) * H = e * H$$
$$= \text{identity coset.}$$

Definition: Factor Group.   *Let G be a group and $H \leq G$. If the set of left cosets of H form a group, then that group is called the factor group of G modulo H. It is denoted G/H.*

Note: If G is abelian, then every subgroup of G yields a factor group. We will delay further consideration of the non-abelian case to Section 15.4.

Remark on Notation: It is customary to use the same symbol for the operation of G/H as for the operation on G.

EXERCISES FOR SECTION 15.2

A Exercises

1.   Consider $\mathbf{Z}_{10}$ and the subsets of $\mathbf{Z}_{10}$, $\{0, 1, 2, 3, 4\}$ and $\{5, 6, 7, 8, 9\}$. Why is the operation induced on these subsets by modulo ten addition not well defined?

2.   For each group and subgroup, what is G/H isomorphic to?
     (a) $G = \mathbf{Z}_4 \times \mathbf{Z}_2$ $H = ((2, 0))$
           Compare to Example 15.2.5
     (b) $G = [\mathbf{C}, +]$ $H = \mathbf{R}$
     (c) $G = \mathbf{Z}_{20}$ $H = (8)$

3.   Can you think of a group G, with a subgroup H such that $\#H = 6$ and $\#(G/H) = 6$? Is your answer unique?

4.   Real addition modulo r $(r > 0)$ can be described as the operation induced on cosets of (r) by ordinary addition. Describe a system of distinguished representatives for the elements of $\mathbf{R}/(r)$.

5. Consider the trigonometric function sine. Given that $\sin(x + 2\pi k) = \sin x$ for all $x \in R$ and $k \in Z$, show how the distinguished representatives of $R/(2\pi)$ can be useful in developing an algorithm for calculating the sine of a number.

B Exercise

6. Prove that if G is a group, $H \leq G$ and $a, b \in G$, $a \in b * H$ if and only if $b^{-1} * a \in H$.

C Exercise

7. Write a Pascal function

```
MODDAD (X, Y: REAL) : REAL
```

to do real addition modulo R, where R is a global real variable and MODADD returns the distinguished representative of the coset containing $X + Y$.

5.3 Permutation Groups

At the risk of boggling the reader's mind, we will now examine groups whose elements are functions. Recall that a *permutation on a set A* is a bijection from A into A. Suppose that $A = \{1, 2, 3\}$. There are $3! = 6$ different permutations on A. They are listed in Table 15.3.1. The matrix form for describing a function on a finite set is to list the domains across the top row and the image of each element directly below it. For example $r_1(1) = 2$.

The operation that will give $\{i, r_1, r_2, f_1, f_2, f_3\}$ a group structure is function composition. Consider the "product" $r_1 f_3$:

$$(r_1 f_3)(1) = r_1(f_3(1)) = r_1(2) = 3$$
$$(r_1 f_3)(2) = r_1(f_3(2)) = r_1(1) = 2$$
$$(r_1 f_3)(3) = r_1(f_3(3)) = r_1(3) = 1$$

TABLE 15.3.1
Elements of S_3

$$i = \begin{pmatrix} 1 & 2 & 3 \\ 1 & 2 & 3 \end{pmatrix} \qquad f_1 = \begin{pmatrix} 1 & 2 & 3 \\ 1 & 3 & 2 \end{pmatrix}$$

$$r_1 = \begin{pmatrix} 1 & 2 & 3 \\ 2 & 3 & 1 \end{pmatrix} \qquad f_2 = \begin{pmatrix} 1 & 2 & 3 \\ 3 & 2 & 1 \end{pmatrix}$$

$$r_2 = \begin{pmatrix} 1 & 2 & 3 \\ 3 & 1 & 2 \end{pmatrix} \qquad f_3 = \begin{pmatrix} 1 & 2 & 3 \\ 2 & 1 & 3 \end{pmatrix}$$

TABLE 15.3.2
Table for S_3

	i	r_1	r_2	f_1	f_2	f_3
i	i	r_1	r_2	f_1	f_2	f_3
r_1	r_1	r_2	i	f_3	f_1	f_2
r_2	r_2	i	r_1	f_2	f_3	f_1
f_1	f_1	f_2	f_3	i	r_1	r_2
f_2	f_2	f_3	f_1	r_2	i	r_1
f_3	f_3	f_1	f_2	r_1	r_2	i

The images of 1, 2, and 3 under $r_1 f_3$ and f_2 are identical. Thus, by the definition of equality for functions, we can say $r_1 f_3 = f_2$. The complete table for the operation of function composition is given in Table 15.3.2. We don't even need the table to verify that we have a group:

(a) Function composition is associative (see Chapter 7).
(b) The identity for the group is i. If g is any one of the permutations on A and $x \in A$,

$$gi(x) = g(i(x)) = g(x)$$

and

$$ig(x) = i(g(x)) = g(x).$$

(c) If a permutation is displayed in matrix form, its inverse can be obtained by exchanging the two rows and rearranging the columns so that the top row is in order. The first step is actually sufficient to obtain the inverse, but the sorting of the top row makes it easier to recognize the inverse.

Example 15.3.1.   If

$$f = \begin{pmatrix} 1 & 2 & 3 & 4 & 5 \\ 5 & 3 & 2 & 1 & 4 \end{pmatrix},$$

$$f^{-1} = \begin{pmatrix} 5 & 3 & 2 & 1 & 4 \\ 1 & 2 & 3 & 4 & 5 \end{pmatrix} = \begin{pmatrix} 1 & 2 & 3 & 4 & 5 \\ 4 & 3 & 2 & 5 & 1 \end{pmatrix}.$$

Note from Table 15.3.2 that this group is non-abelian. Remember, non-abelian is the negation of abelian. The existence of two elements that don't commute is sufficient to make a group non-abelian. In this group, r_1 and f_3 is one such pair: $r_1 f_3 \neq f_3 r_1$.

Definition: Symmetric Group.   *Let A be a non-empty set. The set of all permutations on A with the operation of function composition is called the symmetric group on A, denoted S_A.*

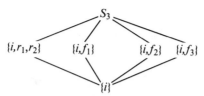

FIGURE 15.3.1
Lattice diagram for subgroups of S_3

Our main interest will be in the case where A is finite. The size of A is more significant than the elements, and we will denote by S_k the symmetric group on any set of cardinality k, $k \geq 1$.

Example 15.3.2. Our opening example was the group S_3. It is the smallest non-abelian group. For that reason, all of its proper subgroups are abelian: in fact, they are all cyclic. Figure 15.3.1 shows the Hasse diagram for the subgroups of S_3.

Example 15.3.3. The only abelian symmetric groups are S_1 and S_2, with 1 and 2 elements, respectively. The elements of S_2 are

$$i = \begin{pmatrix} 1 & 2 \\ 1 & 2 \end{pmatrix} \quad \text{and } a = \begin{pmatrix} 1 & 2 \\ 2 & 1 \end{pmatrix}.$$

S_2 is isomorphic to $\mathbf{Z}_2$.

Theorem 15.3.1. *For $k \geq 1$, $\#S_k = k!$. For $k \geq 3$, S_k is non-abelian.*

Proof: The first part of the theorem follows from the extended rule of products (Chapter 2). We leave the proof of the second part to the reader after the following hint: consider f and g in S_n, where $f(1) = 2$, $f(2) = 3$, $f(3) = 1$, and $f(j) = j$ for $3 \leq j \leq n$, and g is defined similarly. #

Example 15.3.4. Dihedral Groups. The dihedral groups are a family of subgroups of the symmetric groups. If $k \geq 3$, the dihedral group, D_k, is a subgroup of S_k. The permutations in D_k can be best described in terms of a regular n-gon ($n = 3$: equilateral triangle, $n = 4$: square, $n = 5$: a regular pentagon, . . .). Here we will only pursue the case of D_4. If a square is fixed in space, there are many motions of the square that will, at the end of the motion, not change the apparent position of the square. The actual changes in position can be seen if the corners of the square are labeled. In Figure 15.3.2, the initial labeling scheme is shown, along with the four axes of symmetry of the square.

It might be worthwhile making a square like this with a sheet of paper. Be careful to label the back so that the numbers match up. Two motions of the square will be considered equivalent if the square is in the same position after

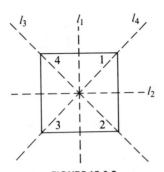

FIGURE 15.3.2
Axes of symmetry of the square

performing either motion. There are eight distinct motions. The first four are
$0°$, $90°$, $180°$, and $270°$ clockwise rotations of the square, and the other four
are the $180°$ flips along the axes l_1, l_2, l_3, and l_4. We will call the rotations i,
r_1, r_2, and r_3, respectively, and the flips f_1, f_2, f_3, and f_4, respectively. Figure
15.3.3 illustrates l_1 and f_1 and lists the permutations that they correspond to.
One application of D_4 is in the design of a letter-facing machine.

Imagine letters entering a conveyer belt to be postmarked. They are placed
on the conveyer belt at random so that two sides are parallel to the belt.
Suppose that a postmarker can recognize a stamp in the top right corner of the

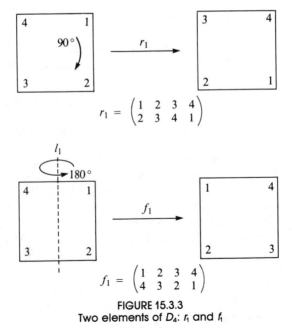

FIGURE 15.3.3
Two elements of D_4: r_1 and f_1

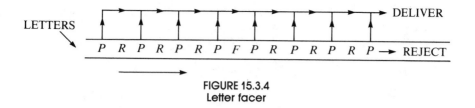

FIGURE 15.3.4
Letter facer

envelope, on the side facing up. In Figure 15.3.4, a sequence of machines is shown that will recognize a stamp on any letter, no matter what position the letter starts in. The letter P stands for a postmarker. The letters R and F stand for rotating and flipping machines that perform the motions of r_1 and f_1.

The arrows pointing up indicate that if a letter is postmarked, it is taken off the conveyer belt for delivery. If a letter reaches the end, it must not have a stamp. Letter-facing machines like this have been designed (see Gallian's paper). One economic consideration is that R-machines tend to cost more than F-machines. R-machines also tend to damage more letters. Taking these facts into consideration, the reader is invited to design a better letter-facing machine. Assume that R-machines cost \$800 and F-machines cost \$500. Be sure that all corners of incoming letters will be examined as they go down the conveyer belt.

Cyclic Notation.   A second way of describing a permutation is by means of cycles, which we will introduce first with an example. Consider $f \in S_8$:

$$f = \begin{pmatrix} 1 & 2 & 3 & 4 & 5 & 6 & 7 & 8 \\ 8 & 2 & 7 & 6 & 5 & 4 & 1 & 3 \end{pmatrix}.$$

Consider the images of 1 when f is applied repeatedly. The images $f(1), f^2(1),$ $f^3(1), \ldots$ are 8, 3, 7, 1, 8, 3, 7, $\ldots$ If $j \geq 1$ and $f^j(1)$ is any element on Figure 15.3.5, then $f^{j+1}(1) = f(f^j(1))$ is the next element in the indicated direction. Of course the images of 8, 3, and 7 could be described in exactly the same way.

The numbers 1, 8, 3, and 7 make up a cycle of length 4. Figure 15.3.5 illustrates how the cycle can be represented. Representation a is more descriptive, but we will use b because it is more universally recognized. For Repre-

$$
\begin{array}{ccc}
& 8 & \\
1 & & 3 \\
& 7 &
\end{array}
\qquad \text{or} \qquad
\begin{array}{l}
(8, 3, 7, 1) \\
(3, 7, 1, 8) \\
(7, 1, 8, 3) \\
(1, 8, 3, 7)
\end{array}
$$

(a)   (b)

FIGURE 15.3.5
Cycle of length 4

sentation b, the last number in the list has as its image the first number. The other cycles of f are (2), $(4, 6)$, and (5). We can express f as a product of disjoint cycles

$$f = (1, 8, 3, 7)(2)(4, 6)(5)$$

or

$$f = (1, 8, 3, 7)(4, 6),$$

where the absence of 2 and 5 implies that $f(2) = 2$ and $f(5) = 5$.

Disjoint Cycles.   We say that two cycles are *disjoint* if no number appears in both cycles, as is the case in our expressions for f above. Disjoint cycles can be written in any order. Thus, we could also say that

$$f = (4, 6)(1, 8, 3, 7).$$

We will now consider the composition of permutations written in cyclic form, again by an example. Suppose that $f = (1, 8, 3, 7)(4, 6)$ and $g = (1, 5, 6)(8, 3, 7, 4)$ are elements of S_8. To calculate fg, we start with simple concatenation:

(P) $fg = (1, 8, 3, 7)(4, 6)(1, 5, 6)(8, 3, 7, 4).$

Although this is a valid expression for fg, our goal is to express fg as a product of disjoint cycles as f and g were individually written. We will start with the cycle that contains 1. The four cycles in (P) are read from right to left. The first cycle does not contain 1; thus we move on to the second. The image of 1 under that cycle is 5. Now we move on to the next cycle in (P), looking for 5, which doesn't appear. The fourth cycle does not contain a 5 either; so $fg(1) = 5$. At this point, we would have written

$$fg = (1, 5$$

on paper. We repeat the steps to determine $(fg)(5)$. This time the second cycle of (P) moves 5 to 6 and then the third cycle moves 6 to 4. Therefore, $(fg)(5) = 4$. We continue until the cycle $(1, 5, 4, 3)$ is completed by determining that $(fg)(3) = 1$. The process is repeated starting with any number that does not appear in the cycle(s) that have already obtained. The final result for our example is

$$fg = (1, 5, 4, 3)(6, 8, 7).$$

Since $f(2) = 2$ and $g(2) = 2$, we need not include the cycle (2).

Example 15.3.5.

(a) $(1, 2, 3, 4)(1, 2, 3, 4) = (1, 3)(2, 4)$.
(b) $(1, 4)(1, 3)(1, 2) = (1, 2, 3, 4)$.

Note that the cyclic notation does not indicate the set which is being permuted. The examples above could be in S_5, where the image of 5 is 5. This ambiguity is usually overcome by making the context clear at the start of a discussion.

Definition: Transposition. *A transposition is a cycle of length 2.*

Example 15.3.6. $f = (1, 4)$ and $g = (4, 5)$ are transpositions in S_5. $fg = (4, 5, 1)$ and $gf = (1, 5, 4)$ are not transpositions; thus, the set of transpositions is not closed under composition. Since f^2 and g^2 are both equal to the identity permutation, f and g are their own inverses. In fact, every transposition is its own inverse.

Theorem 15.3.2. *Every cycle of length greater than 2 can be expressed as a product of transpositions.*

Instead of a formal proof, we will indicate how the product of transpositions can be obtained. The key fact that is needed is that if $(a_1, a_2, \ldots , a_k)$ is a cycle of length k, it is equal to the product

$$(a_1, a_k) \ldots (a_1, a_3)(a_1, a_2).$$

Example 11.3.5b illustrates this fact. Of course, a product of cycles can be written as a product of transpositions just as easily by applying the rule above to each cycle. For example,

$$(1, 3, 5, 7)(2, 4, 6) = (1, 7)(1, 5)(1, 3)(2, 6)(2, 4).$$

Unlike the situation with disjoint cycles, we are not free to change the order of the transpositions.

The proofs of the following two theorems appear in many abstract algebra texts.

Theorem 15.3.3. *Every permutation on a finite set can be expressed as the product of an even number of transpositions or an odd number of transpositions, but not both.*

Theorem 15.3.3 suggests that S_n can be partitioned into its "even" and "odd" elements.

Example 15.3.7. The even permutations of S_3 are i, r_1, and r_2, $\{i, r_1, r_2\} \leq S_3$. In general:

Theorem 15.3.4. *Let $n \geq 2$. The set of even permutations in S_n is a proper subgroup of S_n called the alternating group on $\{1, 2, \ldots n\}$, denoted A_n. The order of A_n is $n!/2$.*

Proof: In this proof, the letters s and t stand for transpositions and p, q are even non-negative integers.

If $f, g \in A_n$, we can write fg as $s_1 s_2 \ldots s_p t_1 t_2 \ldots t_q$. Since $p + q$ is even, $fg \in A_n$.

By Theorem 11.5.2, we have proven that $A_n \leq S_n$. To prove the final assertion, let B_n be the set of odd permutations and let $t = (1, 2)$. Define $\theta :$ $A_n \to B_n$ by $\theta(f) = ft$. Suppose that $\theta(f) = \theta(g)$. Then $ft = gt$ and by the cancellation law, $f = g$. Hence, θ is an injection. If $h \in B_n$, h is the image of an element of A_n; therefore, θ is also a surjection. Specifically, h is the image of ht.

$$\theta(ht) = (h\ t)\ t \quad \textit{Why?}$$
$$= h(t\ t) \quad \textit{Why?}$$
$$= hi \quad \textit{Why?}$$
$$= h \quad \textit{Why?}$$

Since θ is a bijection, $\#A_n = \#B_n = \dfrac{n!}{2}$. #

Example 15.3.8. Consider the sliding-tile puzzles pictured in Figure 15.3.6. Each numbered square is a tile and the dark square is a gap. Any tile that is adjacent to the gap can slide into the gap. In most versions of this puzzle, the tiles are locked into a frame so that they can be moved only in the manner described above. The object of the puzzle is to arrange the tiles as they appear in Configuration a. Configurations b and c are typical starting points. We propose to show why the puzzle can be solved starting with b, but not with c.

We will associate any configuration of the puzzle with an element of S_{16}. Imagine that a tile numbered 16 fills in the gap. If f is any configuration of the puzzle, i is Configuration a, and $1 \leq k \leq 16$,

$$f(k) = \text{the number that appears in the position of } k \text{ in } i.$$

If we call Configurations b and c by the names f_1 and f_2 respectively,

$$f_1 = (1, 5, 3, 7)(2, 6, 4, 8)(9, 10)(11, 14, 13, 12)(15)(16)$$

and

$$f_2 = (1, 5, 3, 7, 15)(2, 6, 4, 8)(9, 10)(11, 14, 13, 12)(16).$$

1	2	3	4
5	6	7	8
9	10	11	12
13	14	15	

(a)

5	6	7	8
3	4	1	2
10	9	14	11
12	13	15	

(b)

5	6	7	8
3	4	15	2
10	9	14	11
12	13	1	

(c)

FIGURE 15.3.6

How can we interpret the movement of one tile as a permutation? Consider what happens when the 12 tile of i slides into the gap. The result is a configuration that we would interpret as $(12,16)$, a single transposition. Now if we slide the 8 tile into the old 12 position, the result is

1	2	3	4
5	6	7	
9	10	11	8
13	14	15	12

FIGURE 15.3.7

or $(8, 16, 12)$. Hence, by "exchanging" the tiles 8 and 16, we have obtained $(8, 16)(12, 16) = (8, 16, 12)$.

Every time you slide a tile into the gap, the new permutation is a transposition composed with the old permutation. Now observe that to start with i and terminate after a finite number of moves with the gap in its original position, you must make an even number of moves. Thus, any permutation that leaves 16 fixed, such as f_1 or f_2, cannot be solved if it is odd. Note that f_2 is an odd permutation; thus, Puzzle c can't be solved. The proof that all even permutations, such as f_1, can be solved is left to the interested reader to pursue.

EXERCISES FOR SECTION 15.3

A Exercises

1. Given

$$f = \begin{pmatrix} 1 & 2 & 3 & 4 \\ 2 & 1 & 4 & 3 \end{pmatrix}, \ g = \begin{pmatrix} 1 & 2 & 3 & 4 \\ 2 & 3 & 4 & 1 \end{pmatrix}, \text{ and } h = \begin{pmatrix} 1 & 2 & 3 & 4 \\ 3 & 2 & 4 & 1 \end{pmatrix},$$

compute

(a) fg (e) h^{-1}
(b) gh (f) $h^{-1}gh$
(c) $(fg)h$ (g) f^{-1}
(d) $f(gh)$

2. Write f, g, and h from Exercise 1 as products of disjoint cycles and determine whether each is odd or even.

3. Do the left cosets of $A_3 = \{i, r_1, r_2\}$ form a group under the induced operation on left cosets of A_3? What about the left cosets of (f_1)?

4. The dihedral group, D_3 is equal to S_3. Can you give a geometric explanation why?

5. Complete the list of elements of D_4 and write out a table for the group's operation.

6. List the subgroups of D_4 in a lattice diagram. Are they all cyclic? What are the subgroups of D_4 isomorphic to?

7. Design a better letter-facing machine (see Example 15.3.4).

8. How can you verify that a letter-facing machine does indeed check every corner of a letter? Can it be done on paper without actually sending letters through it?

9. Complete the proof of Theorem 15.3.1.

10. How many elements are there in D_5? Describe them geometrically.

11. Prove by induction that if $r \geq 1$ and each t_i is a transposition, then

$$(t_1 \, t_2 \cdots t_r)^{-1} = t_r t_{r-1} \cdots t_2 t_1 \, .$$

12. (a) Prove that the tile puzzles corresponding to $A_{16} \cap \{f \in S_{16} \mid f(16) = 16\}$ are solvable.
    (b) If $f(16)\lambda \neq 16$, how can you determine whether f's puzzle is solvable?

13. How many left cosets does $A_n (n \geq a)$ have?

14. Prove that S_3 is isomorphic to R_3, the group of 3 by 3 rook matrices (see Section 11.2 exercises).

15.4 Normal Subgroups and Group Homomorphisms

Our goal in this section is to answer an open question and introduce a related concept. The question is: When are left cosets of a subgroup a group under the induced operation? This question is open for non-abelian groups. Now that we have some examples to work with, we can try a few experiments.

NORMAL SUBGROUPS

Example 15.4.1. $A_3 = \{i, r_1, r_2\}$ is a subgroup of S_3, and its left cosets are A_3 itself and $B_3 = \{f_1, f_2, f_3\}$. Whether $\{A_3, B_3\}$ is a group boils down to determining whether the induced operation is well defined. At this point we will ask for reader participation.

(a) Copy Table 15.3.2, the table for S_3.
(b) Shade in all occurrences of the elements of B_3 : f_1, f_2 and f_1. We will call these elements the gray elements and the elements of A_3 the white ones.

Now consider the process of computing A_3B_3. The coset "product" is obtained by selecting one white element and one gray element. Note that white "times"gray is always gray. Thus, A_3B_3 is well defined. Similarly, the other three possible products are well defined. The table for the factor group S_3/A_3 is

	A_3	B_3
A_3	A_3	B_3
B_3	B_3	A_3

S_3/A_3 is just $\mathbf{Z}_2$ in disguise. Note that A_3 and B_3 are also the right cosets of A_3.

Example 15.4.2. Now let's try the left cosets of (f_1) in S_3. There are three of them. Will we get a complicated version of $\mathbf{Z}_3$? The left cosets are

$$C_0 = i(f_1) = \{i, f_1\}$$
$$C_1 = r_1(f_1) = \{r_1, f_3\}$$
$$C_2 = r_2(f_1) = \{r_2, f_2\}.$$

The reader might be expecting something to go wrong eventually, and here it is. To determine C_1C_2 we can choose from four pairs of representatives:

$$r_1 \in C_1, r_2 \in C_2 \qquad\qquad r_1r_2 = i \in C_0$$
$$r_1 \in C_1, f_2 \in C_2 \qquad\qquad r_1f_2 = f_1 \in C_0$$
$$f_3 \in C_1, r_2 \in C_2 \qquad\qquad f_3r_2 = f_2 \in C_2$$
$$f_3 \in C_1, f_2 \in C_2 \qquad\qquad f_3f_2 = r_2 \in C_2.$$

This time, we don't get the same coset for each pair of representatives. Therefore, the induced operation is not well defined and no factor group is obtained.

Commentary: This last development changes our course of action. If we had gotten a factor group from $\{C_0, C_1, C_2\}$, we might have hoped to prove that every collection of left cosets forms a group. Now our question is: How can we determine whether we will get a factor group? Of course, this question is equivalent to: When is the induced operation well defined? There was only one step in the proof of Theorem 15.2.3, where we used the fact that G was abelian. We repeat the equations here:

$$a' * b' = (a * h_1) * (b * h_2)$$
$$= (a * b) * (h_1 * h_2),$$

since G is abelian.

The last step was made possible by the fact that $h_1 * b = b * h_1$. As the proof continued, we used the fact that $h_1 * h_2$ was in H and so $a' * b'$ is

$(a * b) * h$ for some h in H. All that we really needed in the "abelian step" was that

$$h_1 * b = b * (\text{something in } H) = b * h_3 .$$

Then, since H is closed under G's operation, $h_3 * h_2$ is an element of H. The consequence of this observation is included in the following theorem, the proof of which can be found in any abstract algebra text.

Theorem 15.4.1. *If $H \leq G$, then the operation induced on left cosets of H by the operation of G is well defined if any one of the following conditions is true:*

  (a) G is abelian.
  (b) If $h \in H$, $a \in G$, then there exists $h' \in H$ such that $h * a = a * h'$.
  (c) If $h \in H$, $a \in G$, then $a^{-1} * h * a \in H$.
  (d) Every left coset of H is equal to a right coset of H.
  (e) $\#H = \#G/2$.

Example 15.4.3.   The right cosets of $(f_1) \leq S_3$ are $\{i, f_1\}$, $\{r_1, f_2\}$, and $\{r_2, f_3\}$. These are not the same as the left cosets of f_1. In addition, $f_2^{-1} f_1 f_2 = f_2 f_1 f_2 = f_3 \notin (f_1)$.

Definition: Normal Subgroup.   *If G is a group, $H \leq G$, then H is called a normal subgroup of G, denoted $H \triangleleft G$, if it satisfies Condition d of Theorem 15.4.1.*

   It can be shown that Conditions b and c of Theorem 15.4.1 are equivalent to Condition d.

Example 15.4.4.   By condition e of Theorem 15.4.1, A_n is a normal subgroup of S_n and S_n/A_n is isomorphic to Z_2.

Example 15.4.5. A_5, a group in its own right with 60 elements, has many proper subgroups, but none are normal.

Example 15.4.6.   Let G be the set of two by two invertible matrices.

$$\begin{bmatrix} a & b \\ c & d \end{bmatrix}$$

belongs to G if $ad-bc \neq 0$. G is a group with matrix multiplication (introduced in Chapter 11).

$$H_1 = \left\{ \begin{bmatrix} a & 0 \\ 0 & a \end{bmatrix} : a \neq 0 \right\}$$

is a normal subgroup of G.

$$H_2 = \left\{ \begin{bmatrix} a & 0 \\ 0 & b \end{bmatrix} : ab \neq 0 \right\}$$

is also a subgroup, but is not normal.

Example 15.4.7 The improper subgroups $\{e\}$ and G of any group G are normal subgroups. $G / \{e\}$ is isomorphic to G.

HOMOMORPHISMS

Think of the word *isomorphism*. Chances are, one of the first images that comes to mind is an equation like

$$H : \theta(x) \# \theta(y) = \theta(x * y).$$

An isomorphism must be a bijection, but H is the algebraic feature of an isomorphism. Here we will examine functions that satisfy equations of this type.

Many homomorphisms are useful since they point out similarities between the two groups (or, on the universal level, two algebraic systems) involved.

Consider the groups $[\mathbf{R}^3; +]$ and $[\mathbf{R}^2; +]$. Every time you use a camera, you are trying to transfer the essence of something three-dimensional onto a photograph—that is, something two-dimensional.

If you show a friend a photo you have taken, that person can appreciate much of what you saw, even though a "dimension" is lacking. The "picture-taking" map is a function $f : \mathbf{R}^3 \to \mathbf{R}^2$ defined by $f(x_1, x_2, x_3) = (x_1, x_2)$. This function is not a bijection, but it does satisfy the equation $f(x + y) = f(x) + f(y)$ for $x = (x_1, x_2, x_3)$ and $y = (y_1, y_2, y_3)$. Such a function is called a *homomorphism*, and when homomorphism exists between two groups, the groups are called *homomorphic*—that is, similar. A question that arises with groups, or other algebraic structures, which we claim are homomorphic, or similar, is: How similar are they? When we say that two groups are isomorphic—that is, identical—the map which we use to prove this is unimportant. However, when we say that two groups are homomorphic, the map used gives us a measure of the group's similarities (or dissimilarities). For example, the maps:

$f_1 : \mathbf{R}^3 \to \mathbf{R}^3$ defined by $f_1(x_1, x_2, x_3) = (x_1, x_2, x_3)$,

$f_2 : \mathbf{R}^3 \to \mathbf{R}^3$ defined by $f_2(x_1, x_2, x_3) = (x_1, x_2, 0)$, and

$f_3 : \mathbf{R}^3 \to \mathbf{R}^3$ defined by $f_3(x_1, x_2, x_3) = (0, 0, 0)$

are all homomorphisms. Think of them all as "picture-taking" maps, or cameras. The first camera gives us a three-dimensional picture, the ideal, actually an isomorphism. The second gives us the usual two-dimensional picture, certainly something quite worthwhile. The third collapses the whole scene onto a point, a "black dot," which gives no idea of the original structure. Hence, the knowledge that two groups are homomorphic really gives little information about the similarities in the structures of the two groups. The homomorphism itself illustrates how closely the two structures resemble each other. For this reason, the term *homomorphic* is rarely used (unlike *isomorphic*), and the functions, the *homomorphisms*, are studied.

Definition: Homomorphism. *Let $[G; *]$ and $[G'; \#]$ be groups. $\theta : G \to G'$ is a homomorphism if $\theta(x * y) = \theta(x) \# \theta(y)$ for all $x, y \in G$.*

Example 15.4.8. Define $\alpha : Z_6 \to Z_3$ by $\alpha(n) = n(1)$, where $n \in Z_6$ and $n(1)$ is the sum of n ones in Z_3. Therefore, $\alpha(0) = 0$, $\alpha(1) = 1$, $\alpha(2) = 2$, $\alpha(3) = 1 + 1 + 1 = 0$, $\alpha(4) = 1$, and $\alpha(5) = 2$. If $n, m \in Z_6$,

$$\alpha(n +_6 m) = (n + m)(1)$$
$$= n(1) +_3 m(1)$$
$$= \alpha(n) +_3 \alpha(m).$$

Remark on Notation: For the remainder of this section we will leave out the operation symbols when discussing homomorphisms in general.

Theorem 15.4.2. *A few properties of homomorphisms are that if $\theta : G \to G'$ is a homomorphism, then:*

(a) $\theta(e) = \theta(\text{identity of } G) = \text{identity of } G' = e'$.
(b) $\theta(a^{-1}) = \theta(a)^{-1}$ for all $a \in G$.
(c) If $H \leq G$, then $\theta(H) = \{\theta(h) \mid h \in H\} \leq G'$.

Since a homomorphism need not be an injection, some other elements of G may have also e' as their image. Property c is true for the case of $H = G$. That is, $\theta(G) \leq G'$. This is of interest when θ is not a surjection.

Proof:

(a) Let a be any element of G. Then $\theta(a) \in G'$.

$$\theta(a) \, e' = \theta(a) \qquad \text{Definition of } e'$$
$$= \theta(a \, e) \qquad \text{Definition of } e$$
$$= \theta(a) \, \theta(e) \qquad \text{Definition of a homomorphism}$$
$$\text{by cancellation, } e' = \theta(e)$$

(b) Again, let $a \in G$.

$$e' = \theta(e) = \theta(a \, a^{-1}) = \theta(a) \, \theta(a^{-1})$$

Hence,

$$\theta(a)^{-1} = \theta(a^{-1}).$$

(c) Let $b_1, b_2 \in \theta(H)$. Then there exists $a_1, a_2 \in H$ such that $\theta(a_1) = b_1$, $\theta(a_2) = b_2$. Recall that a compact necessary and sufficient condition for $H \leq G$ is that $xy^{-1} \in H$ for all $x, y \in H$. Now we apply the same fact in G':

$$b_1 b_2^{-1} = \theta(a_1)\, \theta(a_2^{-1})$$
$$= \theta(a_1)\, \theta(a_2^{-1}) \text{ by Property b of this theorem}$$
$$= \theta(a_1\, a_2^{-1}) \in \theta(H),$$

since $a_1\, a_2^{-1} \in H$, and we can conclude that $\theta(H) \leq G$. #

Example 15.4.9. Define $\pi: \mathbf{Z} \to \mathbf{Z}/4\mathbf{Z}$ by $\pi(n) = n + 4\mathbf{Z}$. π is a homomorphism. The image of the subgroup $4\mathbf{Z}$ is the single coset $0 + 4\mathbf{Z}$, the identity of the factor group. Homomorphisms of this type are called *natural homomorphisms*. The following theorems will verify that π is a homomorphism and also show the connection between homomorphisms and normal subgroups. The reader can find more detail and proofs in most abstract algebra texts.

Theorem 15.4.3. *If $H \lhd G$, then the function*

$$\pi : G \to G/H$$

defined by $\pi(a) = aH$ is a homomorphism, called the natural homomorphism.

Remark: Every normal subgroup gives us a homomorphism.

Theorem 15.4.4. *Let $\theta : G \to G'$ be a homomorphism from G into G. Then the set of all elements of G whose image is the identity of G' is a normal subgroup of G.*

Definition: Kernel. *The subgroup of Theorem 15.4.4 is called the kernel of θ, denoted ker θ.*

Remark: Every homomorphism gives us a normal subgroup.

Theorem 15.4.5. *If $\theta : G \to G'$ is a homomorphism, then $\theta(G)$ is isomorphic to $G/\ker \theta$.*

This last theorem is often called the *Fundamental Theorem of Group Homomorphisms*.

Example 15.4.10. Define $\theta : \mathbf{Z} \to \mathbf{Z}_{10}$ by $\theta(n) = $ the remainder from dividing n by 10. The three previous theorems imply the following:

(15.4.3) $\pi : \mathbf{Z} \to \mathbf{Z}/10\mathbf{Z}$ defined by $\pi(n) = n + 10\mathbf{Z}$ is a homomorphism.

(15.4.4) $\{n \mid \theta(n) = 0\} = \{10n \mid n \in \mathbf{Z}\} = 10\mathbf{Z} \lhd \mathbf{Z}$.

(15.4.5) $\mathbf{Z}/10\mathbf{Z}$ is isomorphic to $\mathbf{Z}_{10}$.

Example 15.4.11. Let G be the same group of two by two invertible matrices as in Example 15.4.6. Define $\emptyset: G \to G$ by $\emptyset(A) = A/(det A)^2$. We will let the reader verify that $\emptyset$ is a homomorphism. The theorems above imply:

$$(15.4.4) \ ker \ \emptyset = \{A \mid \emptyset(A) = I\} = \left\{ \begin{bmatrix} a & o \\ o & a \end{bmatrix} \middle| a \neq 0 \right\} \lhd G.$$

This verifies our statement in Example 15.4.6. As in that example, let ker $\emptyset = H_1$.

(15.4.5) G/H_1 is isomorphic to $\left\{ \begin{bmatrix} a & b \\ c & b \end{bmatrix} \middle| ad - bc = 1 \right\}$.

(15.4.3) $\pi : G \to G/H_1$ defined, naturally, by $\pi(A) = AH_1$ is a homomorphism.

Remark: For the remainder of this section, we will be examining certain kinds of homomorphisms that will play a part in our major application to homomorphism, coding theory.

Example 15.4.12.   Consider $\emptyset : \mathbf{Z}_2^2 \to \mathbf{Z}_2^3$ defined by $\emptyset(a, b) = (a, b, a +_2 b)$. If $(a_1, b_1), (a_2, b_2) \in \mathbf{Z}_2^2$,

$$\begin{aligned}
\emptyset((a_1, b_1) + (a_2, b_2)) &= \emptyset(a_1 +_2 a_2, b_1 +_2 b_2) \\
&= (a_1 +_2 a_2, b_1 +_2 b_2, a_1 +_2 a_2 +_2 b_1 +_2 b_2) \\
&= (a_1, b_1, a_1 +_2 b_1) + (a_2, b_2, a_2 +_2 b_2) \\
&= \emptyset(a_1, b_1) + \emptyset(a_2, b_2).
\end{aligned}$$

Since $\emptyset(a, b) = (0, 0, 0)$ implies that $a = 0$ and $b = 0$, the kernel of $\emptyset$ is $\{(0, 0)\}$. By previous theorems, $\emptyset(\mathbf{Z}_2^2) = \{(0, 0, 0), (1, 0, 1), (0, 1, 1), (1, 1, 0)\}$ is isomorphic to $\mathbf{Z}_2^2$.

We can generalize the previous example as follows: If $n, m \geq 1$ and A an m by n matrix of 0's and 1's (elements of $\mathbf{Z}_2$), then $\emptyset : \mathbf{Z}_2^m \to \mathbf{Z}_2^n$ defined by

$$\emptyset(a_1, a_2, \ldots , a_m) = (a_1, \ldots , a_m)A$$

is a homomorphism. This is true because matrix multiplication is distributive over addition. The only new idea here is that computation is done in $\mathbf{Z}_2$ where $1 +_2 1 = 0$. If $a = (a_1, \ldots , a_m)$ and $b = (b_1, \ldots , b_m)$,

$$(a + b)A = aA + bA.$$

Therefore, $\emptyset(a + b) = \emptyset(a) + \emptyset(b)$.

EXERCISES FOR SECTION 15.4

A Exercises

1. Show that D_4 has one normal subgroup, but that (f_1) is not normal.

2. Which of the following functions are homomorphisms?
   (a) $\theta_1 : \mathbf{R}^* \to \mathbf{R}^+$ defined by $\theta_1(a) = |a|$.
   (b) $\theta_2 : \mathbf{Z}_5 \to \mathbf{Z}_2$ defined by $\theta_2(n) = 0$ if n is even and $\theta_2(n) = 1$ if n is odd.
   (c) Same as b but $\theta_3 : \mathbf{Z}_8 \to \mathbf{Z}_2$.
   (d) $\theta_4 : S_4 \to S_4$ defined by $\theta_4(f) = f^2$.

3. Prove that the function ϕ in Example 15.4.11 is a homomorphism.

4. Define $\alpha : \mathbf{Z}_2^3 \to \mathbf{Z}_2^4$ by
   $$\alpha(a_1, a_2, a_3) = (a_1, a_2, a_3, a_1 +_2 a_2 +_2 a_3)$$
   and
   $\beta : \mathbf{Z}_2^4 \to \mathbf{Z}_2$ by
   $$\beta(b_1, b_2, b_3, b_4) = b_1 +_2 b_2 +_2 b_3 +_2 b_4$$
   Describe the function $\beta\alpha$. Is it a homomorphism?

5. How many homomorphisms are there from
   (a) $\mathbf{Z}_{12}$ into $\mathbf{Z}_{12}$?
   (b) $\mathbf{Z}_{12}$ into $\mathbf{Z}_4$?
   (c) $\mathbf{Z}_9$ into $\mathbf{Z}_8$?

6. Express $\emptyset$ in Example 15.4.12 in matrix form.

B Exercises

7. Prove that if G is an abelian group, then $q(x) = x^2$ defines a homomorphism from G into G. Is q ever an isomorphism?

8. Prove that if $\theta : G \to G'$ is a homomorphism, and $H \lhd G$, then $\theta(H) \lhd \theta(G)$. Is it also true that $\theta(H) \lhd G'$?

9. Prove that if $\theta : G \to G'$ is a homomorphism, and $H' \leq \theta(G)$, then $\theta^{-1}(H') \leq G$.

15.5 Coding Theory—Group Codes

In this section, we will introduce the basic ideas involved in coding theory and consider solutions of a coding problem by means of group codes.

A Transmission Problem. Imagine a situation in which information is being transmitted between two points. The information takes the form of high and low pulses (for example, radio waves or electric currents) which we will label 1 and 0.

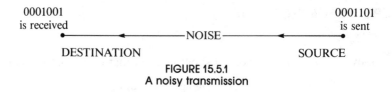

FIGURE 15.5.1
A noisy transmission

Blocks. As these pulses are sent and received, they are grouped together in blocks of fixed length. Fixed-length blocks are not always the optimal method of representing information, but they will make our discussion easier. Huffman codes are an example of a variable-length block system (see, for example, Aho, Hopcroft, and Ullman). The length determines how much information can be contained in one block. If the length is r, there are 2^r different values that a block can have. If the information being sent takes the form of text, each block might be a character. In that case, the length of a block may be seven, so that $2^7 = 128$ block values can represent letters (both upper and lower case), digits, punctuation, etc. Figure 15.5.1 illustrates the problem that can be encountered if information is transmitted between two points.

Noise. Noise is a fact of life for anyone who tries to transmit information. Fortunately, in most situations, we could expect a high percentage of the pulses that are sent to be received properly. However, when large numbers of pulses are transmitted, there are usually some errors due to noise. For the remainder of the discussion, we will make assumptions about the nature of the noise and the message that we want to send. Henceforth, we will refer to the pulses as *bits*.

BINARY SYMMETRIC CHANNEL

We will assume that our information is being sent along a *binary symmetric channel*. By this we mean that any single bit that is transmitted will be received improperly with a certain fixed probability, p. The value of p is usually quite small. We will assume that $p = 0.001$, which, in the real world, would be considered large. Since $1 - p = 0.999$, we can expect 99.9% of all bits to be properly received.

Suppose that our message consists of 3,000 bits of information, to be sent in blocks of three bits each. Two factors will be considered in evaluating a method of transmission. The first is the probability that the message is received with no errors. The second is the number of bits that will be transmitted in order to send the message. This quantity is called the rate of transmission:

$$\text{Rate} = \frac{\text{Message Length}}{\text{Bits Transmitted}}.$$

As you might expect, as we devise methods to improve the probability of success, the rate will decrease.

Case 1: Raw information. Suppose that we ignore the noise and transmit the message as is. The probability of success is

$$(0.999)^{3000} = 0.0497.$$

The rate of $3000/3000 = 1$ certainly doesn't offset this poor probability.

THE CODING PROCESS

Our strategy will be to send an encoded message across the binary symmetric channel. The encoding will be done in such a way that small errors can be identified and corrected. This idea is illustrated in Figure 15.5.2.

In our examples, the functions that will correspond to our encoding and decoding devices will all be homomorphisms between Cartesian products of $\mathbf{Z}_2$.

Case 2: An Error-Detecting Code. Suppose that each block of three bits $a = (a_1, a_2, a_3)$ is encoded according to the function

$$e : \mathbf{Z}_2^3 \to \mathbf{Z}_2^4,$$

where

$$e(a) = (a_1, a_2, a_3, a_1 +_2 a_2 +_2 a_3).$$

When the encoded block is received, the first three bits are probably part of the message (it is correct approximately 99.7% of the time), but the added bit that is sent will make it possible to detect single errors in the block.

Note that when $e(a)$ is transmitted, the sum of its components is

$$a_1 +_2 a_2 +_2 a_3 +_2(a_1 +_2 a_2 +_2 a_3) = 0$$

If any single bit is garbled by noise, the sum of the bits will be 1. The last bit of $e(a)$ is called the *parity bit*. A parity error occurs if the sum of the received bits is 1. Since more than one error is unlikely when p is small, a high percentage of all errors can be detected.

At the receiving end, the decoding function acts on the four-bit block $b = (b_1, b_2, b_3, b_4)$ according to

$$d(b) = (b_1, b_2, b_3, b_1 +_2 b_2 +_2 b_3 +_2 b_4).$$

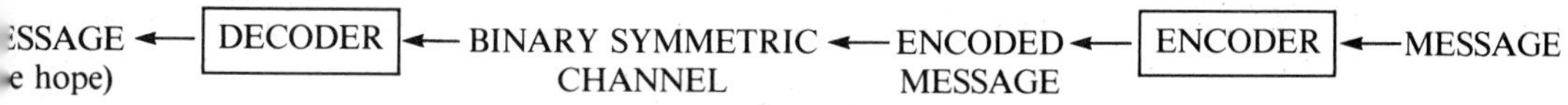

ESSAGE ◄— | DECODER | ◄— BINARY SYMMETRIC ◄— ENCODED ◄— | ENCODER | ◄— MESSAGE
e hope)   CHANNEL   MESSAGE

FIGURE 15.5.2
The coding process

The fourth bit is called the *parity-check bit*. If no parity error occurs, the first three bits are recorded as part of the message. If a parity error occurs, we will assume that a retransmission of that block can be requested. This request can take the form of automatically having the parity-check bit of $d(b)$ sent back to the source. If 1 is received, the previous block is retransmitted; if 0 is received, the next block is sent. This assumption of two-way communication is a major one, but it is necessary to make this coding system useful. It is reasonable to expect that the probability of a transmission error in the opposite direction is also 0.001.

Without going into the details, we will report that the probability of success is approximately 0.990 and the rate is approximately $3/5$. The rate includes the transmission of the parity-check bit to the source.

Case 3:   An Error-Correcting Code. For our final case, we will consider a coding process that can correct errors at the receiving end so that only one-way communication is needed. Before we begin, recall that every element of $\mathbf{Z}_2^n$, $n \geq 1$, is its own inverse; that is, $-b = b$. Therefore, $a - b = a + b$.

The three-bit message blocks are difficult to transmit because they are so similar to one another. If a and b are in $\mathbf{Z}_2^3$, their difference, $a +_2 b$, can be thought of as a measure of how close they are. If a and b differ in only one bit position, one error can change one into the other. The encoding that we will introduce takes a block $a = (a_1, a_2, a_3)$ and produces a block of length 6 called the *code word of a*. The code words are selected so that they are farther from one another than the messages are. In fact, each code word will differ from each other code word by at least three bits. As a result, any single error will not push a code word close enough to another code word to cause confusion. Now for the details. Let

$$G = \begin{bmatrix} 1 & 0 & 0 & 1 & 1 & 0 \\ 0 & 1 & 0 & 1 & 0 & 1 \\ 0 & 0 & 1 & 0 & 1 & 1 \end{bmatrix}$$

and

$$a = (a_1, a_2, a_3).$$

Define $e : \mathbf{Z}_2^3 \to \mathbf{Z}_2^6$ by

$$\begin{aligned} e(a) &= a\, G \\ &= (a_1, a_2, a_3, a_4, a_5, a_6), \end{aligned}$$

where

$$\begin{aligned} a_4 &= a_1 +_2 a_2, \\ a_5 &= a_1 +_2 a_3, \text{ and} \\ a_6 &= a_2 +_2 a_3. \end{aligned}$$

Note that e is a homomorphism. If a and b are distinct elements of $\mathbf{Z}_2^3$, then $c = a + b$ has at least one coordinate equal to 1. Now consider the difference between $e(a)$ and $e(b)$:

$$\begin{aligned} e(a) + e(b) &= e(a + b) \\ &= e(c) \\ &= (c_1, c_2, c_3, c_4, c_5, c_6). \end{aligned}$$

Whether c has 1, 2, or 3 ones, $e(c)$ must have at least three ones; therefore $e(a)$ and $e(b)$ differ in at least three bits.

Now consider the problem of decoding the code words. Imagine that a code word, $e(a)$, is transmitted, and $b = (b_1, b_2, b_3, b_4, b_5, b_6)$ is received. At the receiving end, we know the formula for $e(a)$, and if no error has occurred in transmission,

$$\begin{aligned} b_1 &= a_1 \\ b_2 &= a_2 \\ b_3 &= a_3 && b_1 +_2 b_2 +_2 b_4 = 0 \\ b_4 &= a_1 +_2 a_2 && \Rightarrow && b_1 +_2 b_3 +_2 b_5 = 0 \text{ and} \\ b_5 &= a_1 +_2 a_3 && b_2 +_2 b_3 +_2 b_6 = 0 \\ b_6 &= a_2 +_2 a_3 \end{aligned}$$

The three equations on the right are called *parity-check equations*. If any of them is not true, an error has occurred. This error checking can be described in matrix form. Let

$$P = \begin{bmatrix} 1 & 1 & 0 \\ 1 & 0 & 1 \\ 0 & 1 & 1 \\ 1 & 0 & 0 \\ 0 & 1 & 0 \\ 0 & 0 & 1 \end{bmatrix}$$

P is called the *parity-check matrix*. Now define $p : \mathbf{Z}_2^6 \to \mathbf{Z}_2^3$ by $p(b) = bP$. $p(b)$ is called the *syndrome* of the received block. For example,

$$p(0, 1, 0, 1, 0, 1) = (0, 0, 0) \text{ and } p(1, 1, 1, 1, 0, 0) = (1, 0, 0).$$

Note that p is also a homomorphism. If the syndrome of a block is $(0, 0, 0)$, we can be almost certain that the message block is (b_1, b_2, b_3).

Next we turn to the method of correcting errors. Despite the fact that there are only eight code words, one for each three-bit block value, the set of possible received blocks is $\mathbf{Z}_2^6$, with 64 elements. Suppose that b is not a code word, but that it differs from a code word by exactly one bit. In other words, it is the result of a single error in transmission. Suppose that w is the code word that b is close to and that they differ in the first bit. Then

$$b + w = (1, 0, 0, 0, 0, 0)$$

and

$$p(b) = p(b) + p(w) \qquad \text{Since } p(w) = (0, 0, 0)$$
$$ = p(b + w) \qquad\quad \text{Since } p \text{ is a homomorphism}$$
$$ = p(1, 0, 0, 0, 0, 0)$$
$$ = (1, 1, 0).$$

Note that we haven't specified b or w, only that they differ in the first bit. Therefore, if b is received and $p(b) = (1, 1, 0)$, the transmitted code word was probably $b + (1, 0, 0, 0, 0, 0)$ and the message block was $(b_1 +_2 1, b_2, b_3)$. The same analysis can be done if b and w differ in any of the other five bits.

This process can be described in terms of cosets. Let W be the set of code words; that is, $W = e(\mathbf{Z}_2^3)$. W is a subgroup of $\mathbf{Z}_2^6$. Consider the factor group $\mathbf{Z}_2^6/W$:

$$\#(\mathbf{Z}_2^6/W) = (\#\mathbf{Z}_2^6) / (\#W) = 64/8 = 8.$$

Suppose that b_1 and b_2 are representatives of the same coset. Then $b_1 = b_2 + w$ for some w in W. Therefore,

$$p(b_1) = p(b_1) + p(w) \quad \text{Since } p(w) = (0, 0, 0)$$
$$ = p(b_1 + w)$$
$$ = p(b_2)$$

b_1 and b_2 have the same syndrome.

Finally, suppose that d_1 and d_2 are distinct and both have only a single coordinate equal to 1. Then $d_1 + d_2$ has exactly two ones. Note that the identity of $\mathbf{Z}_2^6$, $(0, 0, 0, 0, 0, 0)$, must be in W. Since $d_2 + d_2$ differs from the identity by two bits, $d_1 + d_2 \notin W$. Hence d_1 and d_2 belong to distinct cosets. The reasoning above serves as a proof of the following theorem.

Theorem 15.5.1. *There is a system of distinguished representatives of $\mathbf{Z}_2^6/W$ such that each of the six-bit blocks having a single one is a distinguished representative of its own coset.*

Now we can describe the error-correcting process. First match each of the blocks with a single one with its syndrome. In addition, match the identity of W with the syndrome $(0, 0, 0)$ (see Table 15.5.1). Since there are eight cosets of W, select any representative of the eighth coset to be distinguished. This is the coset with syndrome $(1, 1, 1)$.

When block b is received, you need only:

(1) Compute the syndrome, $p(b)$, and
(2) Add to b the error correction that matches $p(b)$.

We will conclude this case by computing the probability of success for our hypothetical situation. It is

TABLE 15.5.1
Error-Correction Table

Syndrome			Error Correction					
0	0	0	0	0	0	0	0	0
1	1	0	1	0	0	0	0	0
1	0	1	0	1	0	0	0	0
1	1	1	0	0	1	0	0	0
1	0	0	0	0	0	1	0	0
0	1	0	0	0	0	0	1	0
0	0	1	0	0	0	0	0	1
1	1	1	1	0	0	0	0	1

$$((0.999)^6 + 6(0.999)^5(0.001))^{1000} \approx 0.985.$$

The rate for this method is $1/2$.

EXERCISES FOR SECTION 15.5

A Exercises

1. If the error-detecting code is being used, how would you act on the following received blocks?
   (a) $(1, 0, 1, 1)$
   (b) $(1, 1, 1, 1)$
   (c) $(0, 0, 0, 0)$

2. Express the encoding and decoding functions for the error-detecting code using matrices.

3. If the error-correcting code is being used, how would you decode the following blocks? Expect a problem with one of these. Why?
   (a) $(1, 0, 0, 0, 1, 1)$
   (b) $(1, 0, 1, 0, 1, 1)$
   (c) $(0, 1, 1, 1, 1, 0)$
   (d) $(0, 0, 0, 1, 1, 0)$

4. Describe how the triple-repetition code with encoding function $e(a_1, a_2, a_3) = (a_1, a_1, a_1, a_2, a_2, a_2, a_3, a_3, a_3)$ can allow us to correct a single error. What is the probability of success for the $p = 0.001$, 3000-bit situation? What is the rate?

B Exercise

5. Describe an error-detecting code for message blocks of nine bits.

chapter

16

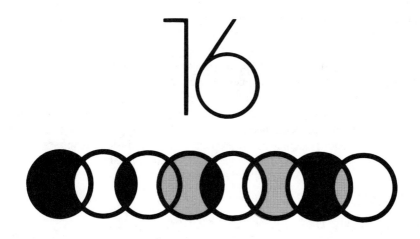

AN INTRODUCTION TO
RINGS AND FIELDS

16

Goals

In our early elementary-school days we began the study of mathematics by learning addition and multiplication on the set of positive integers. We then extended this to operations on the set of all integers. Subtraction and division are defined in terms of addition and multiplication. Later we investigated the set of real numbers under the operations of addition and multiplication. Hence, it is quite natural to investigate those structures on which we can define these two fundamental operations, or operations similar to them. The structures similar to the set of integers are called *rings*, and those similar to the set of real numbers are called *fields*.

In coding theory, unstructured coding is at best awkward. Therefore, highly structured codes are needed. The theory of finite fields is essential in the development of such structured codes. We will discuss basic facts about finite fields and introduce the reader to polynomial algebra.

16.1 Rings—Basic Definitions and Concepts

As mentioned in our goals, we would like to investigate algebraic systems whose structure imitates that of the integers.

Definition: Ring. *A ring is a set R together with two binary operations, denoted by the symbols + and · , such that the following axioms are satisfied:*

(1) $[R; +]$ *is an abelian group.*
(2) *Multiplication is associative on R.*
(3) *Multiplication is distribution over addition; that is, for all a, b, c $\in$ R, the left distributive law, a(b + c) = ab + ac, and the right distributive law, (b + c)a = ba + ca, hold.*

Comments:

(1) A ring is designated as $[R; +, \cdot]$ or as just plain R if the operations are understood.
(2) The symbols + and · stand for arbitrary operations, not just "regular" addition and multiplication. These symbols are referred to by the usual names. For simplicity, we will write ab instead of $a \cdot b$ if it is not ambiguous.
(3) Since part of the definition of a ring is that $[R; +]$ is an abelian group, additive notation is used for groups, as opposed to the multiplicative notation we used in Chapter 11. In particular, the group identity is designated by 0 rather than by e and is customarily called the "zero" of the ring. The group inverse is written in additive notation: $-a$ rather than a^{-1}.

(4) Most importantly, one must continually keep in mind that all rings must be abelian groups under $+$. If a set is a ring, then all theorems and facts about groups proven and discussed in Chapter 11 are true for that set under addition.

We now look for some examples of rings. Certainly all the additive abelian groups of Chapter 11 are likely candidates for rings.

Example 16.1.1. $[\mathbf{Z}; +, \cdot]$ is a ring, where $+$ and $\cdot$ stand for regular addition and multiplication on $\mathbf{Z}$. From Chapter 11, we already know that $[\mathbf{Z}; +]$ is an abelian group, so we need only check Parts 2 and 3 of the definition of a ring. From elementary algebra, we know that the associative law under multiplication and the distributive laws are true for $\mathbf{Z}$. This is our main example of an infinite ring.

Example 16.1.2. $[\mathbf{Z}_n; +_n, \times_n]$ is a ring. The properties of modular arithmetic on $\mathbf{Z}_n$ were described in Section 11.4, and they give us the information we need to convince ourselves that $[\mathbf{Z}_n; +_n, \times_n]$ is a ring. This example is our main example of finite rings of different orders.

Definition: Commutative Ring. *A ring in which the commutative law holds under the operation of multiplication is called a commutative ring.*

It is common practice to use the word *abelian* when referring to the commutative law under addition and the word *commutative* when referring to the commutative law under the operation of multiplication.

Definition: Unity. *A ring $[R; +, \cdot]$ which has a multiplicative identity is called a ring with unity. The multiplicative identity itself is called the unity of the ring. More formally, if there exists an element, designated by 1, such that for all $x \in R$, $x1 = 1x = x$, then R is called a ring with unity.*

Example 16.1.3. The rings in Examples 16.1.1 and 16.1.2 are commutative rings with unity, the unity in both cases being the number 1. The ring $[M_{m\times m}(\mathbf{R}); +, \cdot]$ is a noncommutative ring with unity, the unity being the identity matrix I.

DIRECT PRODUCTS OF RINGS

Let $R_1, R_2, \ldots, R_n$ be rings under the operations of $+_1, +_2, \ldots, +_n$ and $\cdot_1, \cdot_2, \ldots, \cdot_n$ respectively. Let

$$P = \overset{n}{\underset{i=1}{\times}} R_i$$

and

$$a = (a_1, a_2, \ldots, a_n), b = (b_1, b_2, \ldots, b_n) \in P.$$

From Chapter 11 we know that P is an abelian group under the operation of componentwise addition:

$$a + b = (a_1 +_1 b_1, a_2 +_2 b_2, \ldots, a_n +_n b_n).$$

We also define multiplication on P componentwise:

$$ab = (a_1 \cdot_1 b_1, a_2 \cdot_2 b_2, \ldots, a_n \cdot_n b_n).$$

To show that P is a ring under the above operations, we need only show that the (multiplicative) associative law and the distributive laws hold. This is indeed the case, and we leave it as an exercise. If each of the rings R_i is commutative, then P is commutative, and if each of the R_i contains unity, then P is a ring with unity which is the n-tuple consisting of the unities of each of the R_i's.

Example 16.1.4. Since $[\mathbf{Z}_4; +_4, \times_4]$ and $[\mathbf{Z}_3, +_3, \times_3]$ are rings, then $\mathbf{Z}_4 \times \mathbf{Z}_3$ is a ring, where for example,

$$(2, 1) + (2, 2) = (2 +_4 2, 1 +_3 2) = (0, 0)$$

and

$$(3, 2)(2, 2) = (3 \times_4 2, 2 \times_3 2) = (2, 1).$$

To determine the unity, if it exists, in the ring $\mathbf{Z}_4 \times \mathbf{Z}_3$, we look for the element (m, n) such that for all elements $(x, y) \in \mathbf{Z}_4 \times \mathbf{Z}_3$, $(x, y) = (x, y)$ $(m, n) = (m, n)(x, y)$, or, equivalently, $(x \times_4 m, y \times_3 n) = (m \times_4 x, n \times_3 y) = (x, y)$. So we want m such that $x \times_4 m = m \times_4 x = x$ in the ring $\mathbf{Z}$. The only element m in $\mathbf{Z}_4$ that satisfies this equation is the element $m = 1$. Similarly, we obtain a value of 1 for n. So the unity of $\mathbf{Z}_4 \times \mathbf{Z}_3$, which is unique by Exercise 4 of this section, is $(1, 1)$. We leave to the reader to verify that this ring is commutative.

Hence, products of rings are analogous to products of groups or products of Boolean algebras.

We now consider the extremely important concept of multiplicative inverses. Certainly many basic equations in elementary algebra (e.g., $2x = 3$) cannot be solved without this concept. We introduce the main idea here and develop it more completely in the next section.

Example 16.1.5. The equation $2x = 3$ has a solution in the ring $[\mathbf{R}; +, \cdot]$ but does not have a solution in $[\mathbf{Z}; +, \cdot]$. Since, to solve this equation, we must multiply both sides of the equation $2x = 3$ by the multiplicative inverse of 2. This number, 2^{-1}, exists in $\mathbf{R}$ but does not exist in $\mathbf{Z}$. We formalize this important idea in a definition which by now should be quite familiar to you.

Definition: Multiplicative Inverse. *Let $[R; +, \cdot]$ be a ring with unity, designated by the symbol 1. If $u \in R$ and there exists an element designated by u^{-1} in R such that $uu^{-1} = u^{-1}u = 1$, then u is said to have a multiplicative* *inverse, namely* u^{-1}. *It is common to call a ring element that possesses a multiplicative inverse a unit of the ring. The set of all units of a ring R is denoted by U(R).*

Example 16.1.6. In the rings $[\mathbf{R}; +, \cdot]$ and $[\mathbf{Q}; +, \cdot]$ every non-zero element has a multiplicative inverse. The only elements in $\mathbf{Z}$ that have multiplicative inverses are -1 and 1. That is, $U(\mathbf{R}) = \mathbf{R}^*$, $U(\mathbf{Q}) = \mathbf{Q}^*$, and $U(\mathbf{Z}) = \{-1, 1\}$.

Example 16.1.7. Let us find the multiplicative inverses, when they exist, of each element of the ring $[\mathbf{Z}_6; +_6, \times_6]$. If $u = 3$, we want an element u^{-1} such that $3 \times_6 u^{-1} = 1$. We do not have to check whether $u^{-1} \times_6 3 = 1$ since $\mathbf{Z}_6$ is commutative. If we try each of the six elements 0, 1, 2, 3, 4, 5 of $\mathbf{Z}_6$, we find that none of them satisfies the above equation, so 3 does not have a multiplicative inverse in $\mathbf{Z}_6$. However, if $u = 5$, since $5 \times_6 u^{-1}$ does equal 1 for $u^{-1} = 5$, 5 does have a multiplicative inverse in $\mathbf{Z}_6$, namely itself. The following table gives all multiplicative inverses in $\mathbf{Z}_6$.

u	u^{-1}
0	does not exist
1	1
2	does not exist
3	does not exist
4	does not exist
5	5

Isomorphism is a universal concept that is important in every algebraic structure. Two rings are isomorphic as rings if and only if they have the same cardinality and if they behave exactly the same under corresponding operations. They are essentially the same ring. For this to be true, they must behave the same as groups (under $+$) and they must behave the same under the operation of multiplication.

Definition: Ring Isomorphism. *Let $[R; +, \cdot]$ and $[R'; +', \cdot']$ be rings. The ring R is isomorphic to the ring R' if and only if there exists a map, $f : R \rightarrow R'$, called a ring isomorphism, such that*

(1) f is one-to-one and onto,
(2) $f(a + b) = f(a) +' f(b)$ for all a, b $\in$ R, and
(3) $f(a \cdot b) = f(a) \cdot' f(b)$ for all a, b $\in$ R.

Conditions 1 and 2 tell us that f is a group isomorphism.

Therefore, to show that two rings are isomorphic, we must produce a map, called an *isomorphism*, which satisfies the above definition. Sometimes it is quite difficult to find a map which works. This does not necessarily mean that no such isomorphism exists, but simply that we cannot find it.

This leads us to the problem of how to show that two rings are not isomorphic. This is a universal concept. It is true for any algebraic structure and was discussed in Chapter 11. To show that two rings are not isomorphic, we must demonstrate that they behave differently under one of the operations. We illustrate through several examples.

Example 16.1.8. Consider the rings $[\mathbf{Z}; +, \cdot]$ and $[2\mathbf{Z}; +. \cdot]$. In Chapter 11 we showed that as groups, the two sets $\mathbf{Z}$ and $2\mathbf{Z}$ with addition were identical, were isomorphic. The group isomorphism that proved this was the map $f : \mathbf{Z} \to 2\mathbf{Z}$, defined by $f(n) = 2n$. Is f a ring isomorphism? We need only check whether $f(mn) = f(m) f(n)$ for all m and n in $\mathbf{Z}$. Let $m, n \in \mathbf{Z}$. Then $f(mn) = 2mn$ and $f(m) f(n) = (2m) (2n) = 4mn$, so this map f is not a ring isomorphism. This does not necessarily mean that the two rings $\mathbf{Z}$ and $2\mathbf{Z}$ are not isomorphic, but simply that the above map does not illustrate that they are. We could proceed and try to determine another function f to see whether it is a ring isomorphism, or we could try to show that $\mathbf{Z}$ and $2\mathbf{Z}$ are not isomorphic as rings. To do the latter, we must find something different about the ring structure of $\mathbf{Z}$ and $2\mathbf{Z}$.

We already know that they behave identically under addition, so if they are different as rings, it must have something to do with how they behave under the operation of multiplication. Let's begin to develop a checklist of how the two rings could differ:

(1) Do they have the same cardinality? Yes.
(2) Are they both commutative? Yes.
(3) Are they both rings with unity? No.

$\mathbf{Z}$ is a ring with unity, namely the number 1; $2\mathbf{Z}$ is not. Hence, they are not isomorphic as rings.

Example 16.1.9. Next consider whether $[2\mathbf{Z}; +, \cdot]$ and $[3\mathbf{Z}; +, \cdot]$ are isomorphic. Because of the previous example, we guess not. Checklist Items 1 through 3 above do not help us. Why? We add another checklist item:

(4) Find an equation that makes sense in both rings, which is solvable in one and not the other.

The equation $x + x = x \cdot x$ makes sense in both rings. However, this equation has a solution $x = 2$ in the ring $2\mathbf{Z}$ but does not have a solution in the ring $3\mathbf{Z}$. Thus we have an equation solvable in one ring which cannot be solved in the other, so they cannot be the same ring up to isomorphism.

Another universal concept that applies to the theory of rings is that of a subsystem. A *subring* of a ring $[R; +, \cdot]$ is any subset S of R which is a ring under the operations of R. First, for S be a subring of the ring R, S must be a subgroup of the group $[R; +]$. Also, S must be closed under $\cdot$, satisfy the associative law (under $\cdot$), and satisfy the distributive laws. But since R is a ring, the associative and distributive laws are true for every element in R, and, in particular, for all elements in S, since $S \subseteq R$. We have just proven the following theorem:

Theorem 16.1.1. *A subset S of a ring $[R; +, \cdot]$ is a subring of R if and only if:*

(1) $[S; +]$ *is a subgroup of the group $[R; +]$, which by Theorem 11.5.1, means we must show:*

(a) *If $a, b \in S$, then $a + b \in S$.*
(b) *$0 \in S$.*
(c) *If $a \in S$, then $-a \in S$.*

(2) *S is closed under multiplication: if $a, b \in S$, then $a \cdot b \in S$.*

Example 16.1.10. $2\mathbf{Z}$ is a subring of the ring $[\mathbf{Z}; +, \cdot]$ since $[2\mathbf{Z}; +]$ is a subgroup of the group $[\mathbf{Z}; +]$ and since $2m, 2n \in 2\mathbf{Z} \Rightarrow (2m)(2n) = 2(2mn) \in 2\mathbf{Z}$.

Several of the basic facts that we are familiar with are true for any ring. The following theorem lists a few of the elementary properties of rings.

Theorem 16.1.2. *Let $[R; +, \cdot]$ be a ring, $a, b \in R$. Then:*

(1) $a \cdot 0 = 0 \cdot a = 0$.
(2) $a \cdot (-b) = (-a) \cdot b = -(a \cdot b)$.
(3) $(-a) \cdot (-b) = a \cdot b$.

Proof of Part 1:

$$a \cdot 0 = a \cdot (0 + 0)$$
$$= a \cdot 0 + a \cdot 0 \text{ (left distributive law)}.$$

Hence if we add $-(a \cdot 0)$ to both sides of the above (or by the cancellation law for the group $[R; +]$), we obtain $a \cdot 0 = 0$. Similarly $0 \cdot a = 0$.

Proof of Part 2: Before we begin the proof of Part 2, recall that the inverse of each element of the group $[R; +]$ is unique. Hence the inverse of the element $a \cdot b$ is unique and it is designated by $-(a \cdot b)$.

Therefore, to prove that $a \cdot (-b) = -(a \cdot b)$, we need only show

$$a \cdot (-b) + a \cdot b = 0$$
$$a \cdot (-b) + a \cdot b = a \cdot (-b + b) \qquad \text{Why?}$$
$$= a \cdot 0 \qquad \text{Why?}$$
$$= 0 \qquad \text{Why?}$$

Similarly, it can be shown that $(-a) \cdot b = -(a \cdot b)$. This completes the proof of Part 2. We leave the proof of Part 3 to the reader (see Exercise 8 of this section). #

Example 16.1.11. Compute $2 \cdot (-2)$ in the ring $(\mathbf{Z}_6; +_6, \times_6]$. $2 \times_6 (-2) = -(2 \times_6 2) = -(4) = 2$, since the additive inverse of 4 (mod 6) is 2. Of course, we could have done the problem directly as $2 \times_6 (-2) = 2 \times_6 4 = 2$.

As the example above illustrates, Theorem 16.1.2 is a modest beginning in the study of which algebraic manipulations are possible in the solution of problems in rings. A fact in elementary algebra that is used frequently in problem solving is the cancellation law. We know that the cancellation laws are true under addition for any ring (Theorem 11.3.5).

Are the cancellation laws true under multiplications? More specifically, let $[R; +, \cdot]$ be a ring and let $a, b, c \in R$ with $a \neq 0$. When can we cancel the a's in the equation $a \cdot b = a \cdot c$? We can certainly do so if a^{-1} exists, but this need not be the case. The answer to this question is found with the following definition and Theorem 16.1.3.

Definition: Divisors of Zero. *Let $[R; +, \cdot]$ be a ring. If a and b are two non-zero elements of R such that $a \cdot b = 0$, then a and b are called divisors of zero.*

Example 16.1.12a. In the ring $[\mathbf{Z}_8; +_8, \times_8]$, the numbers 4 and 2 are divisors of zero since $4 \times_8 2 = 0$. The number 6 is also a divisor of zero since $6 \times_8 6 = 0$.

Example 16.1.12b. In the ring $[M_{2 \times 2}(\mathbf{R}); +, \cdot]$, the matrices

$$A = \begin{bmatrix} 0 & 0 \\ 0 & 1 \end{bmatrix} \text{ and } B = \begin{bmatrix} 0 & 1 \\ 0 & 0 \end{bmatrix}$$

are divisors of zero since $AB = 0$.

Example 16.1.13. $[\mathbf{Z}; +, \cdot]$ has no divisors of zero.

Theorem 16.1.3. *The (multiplicative) cancellation law holds in a ring $[R; +, \cdot]$ if and only if R has no divisors of zero.*

We prove the theorem using the left cancellation law, namely that if $a \neq 0$ and $a \cdot b = a \cdot c$, then $b = c$ for all $a, b, c \in R$. The proof is similar using the right cancellation law.

Proof: ($\Rightarrow$) Assume the left cancellation law holds in R and assume that a and b are two elements in R such that $a \cdot b = 0$. We must show that either $a = 0$ or $b = 0$. To do this, assume that $a \neq 0$ and show that b must be 0.

$a \cdot b = 0 \Rightarrow a \cdot b = a \cdot 0$ (by Theorem 16.1.1) $\Rightarrow b = 0$ by the cancel-lation law.

($\Leftarrow$) Conversely, assume that R has no divisors of 0 and prove that the cancellation law must hold. To do this, assume that $a, b, c \in R$, $a \neq 0$, such that $a \cdot b = a \cdot c$ and show that $b = c$.

$$
\begin{aligned}
a \cdot b = a \cdot c &\Rightarrow a \cdot b - a \cdot c = 0 & \text{Why?} \\
&\Rightarrow a \cdot (b - c) = 0 & \text{Why?} \\
&\Rightarrow b - c = 0 & \text{Why?} \\
&\Rightarrow b = c & \text{Why? \#}
\end{aligned}
$$

Hence, the only time that the cancellation laws hold in a ring is when there are no divisors of zero. The commutative rings with unity in which the above is true are given a special name.

Definition: Integral Domain. *A commutative ring with unity containing no divisors of zero is called an integral domain. Integral domains are usually designated generically by the letter* D.

We state the following two useful facts without proof.

Theorem 16.1.4. *The element m in the ring $[\mathbf{Z}_n, +_n, \times_n]$ is a divisor of zero if and only if m is not relatively prime to n (i.e., $gcd(m, n) \neq 1$).*

Corollary: *If p is a prime, then $\mathbf{Z}_p$ has no divisors of zero.*

Example 16.1.14. $[\mathbf{Z}; +, \cdot]$, $[\mathbf{Z}_p; +_p, \times_p]$, $[\mathbf{Q}; +, \cdot]$, $[\mathbf{R}; +, \cdot]$, and $[\mathbf{C}; +, \cdot]$ are all integral domains. The key example of an infinite integral domain is $[\mathbf{Z}; +, \cdot]$. In fact, it is from $\mathbf{Z}$ that the term *integral domain* is derived. The main example of a finite integral domain is $[\mathbf{Z}_p; +_p, \times_p]$, when p is prime.

We close this section with the verification of an observation that was made in Chapter 11, namely that the product of two algebraic systems may not be an algebraic system of the same type.

Example 16.1.15. Both $[\mathbf{Z}_2; +_2, \times_2]$ and $[\mathbf{Z}_3; +_3, \times_2]$ are integral do-mains. Consider the product $\mathbf{Z}_2 \times \mathbf{Z}_3$. The set $\mathbf{Z}_2 \times \mathbf{Z}_3$ is a commutative ring with unity (see Exercise 13). However, $(1, 0) \cdot (0, 2) = (0, 0)$, so $\mathbf{Z}_2 \times \mathbf{Z}_3$ has divisors of zero and is therefore not an integral domain.

EXERCISES FOR SECTION 16.1

A Exercises

1. Review the definition of rings to show that the following are rings. The operations involved are the usual operations defined on the sets.

  (a) $[\mathbf{Z}; +, \cdot]$
  (b) $[\mathbf{Q}; +, \cdot]$
  (c) $[\mathbf{C}; +, \cdot]$
  (d) $[M_{2\times2}(\mathbf{R}); +, \cdot]$
  (e) $[M_{n\times n}(\mathbf{R}); +, \cdot]$
  (f) $[\mathbf{Z}_2; +_2, \times_2]$
  (g) $[\mathbf{Z}_6; +_6, \times_6]$
  (h) $[\mathbf{Z}_8; +_8, \times_8]$
  (i) $[\mathbf{Z}_5; +_5, \times_5]$
  (j) $[\mathbf{Z} \times \mathbf{Z}; +, \cdot]$
  (k) $[\mathbf{Z}_2^3; +, \cdot]$

2. Which of the above rings are commutative? Are they rings with unity? Determine the unity of the above rings.

3. Determine the units (those elements which have multiplicative inverses) for each of the rings in Exercise 1.

4. Prove: If R is a ring with unity, then this unity is unique.

5. Determine all multiplicative inverses of the units in each of the rings with unity in Exercise 1.

6. Show that the following rings are not isomorphic:
    (a) $[\mathbf{Z}; +, \cdot]$ and $[M_{2\times2}(\mathbf{R}); +, \cdot]$
    (b) $[3\mathbf{Z}; +, \cdot]$ and $[4\mathbf{Z}; +, \cdot]$
    (c) $(\mathbf{R}; +, \cdot]$ and $[\mathbf{Q}; +, \cdot]$
    (d) $[\mathbf{Z}_2 \times \mathbf{Z}_2; +, \cdot]$, and $[\mathbf{Z}_4; +, \cdot]$

7. (a) Show that $3\mathbf{Z}$ is a subring of the ring $[\mathbf{Z}; +, \cdot]$.
    (b) Find all subrings of $\mathbf{Z}_8$.
    (c) Find all subrings of $\mathbf{Z}_2 \times \mathbf{Z}_2$.

8. Prove Part 3 of Theorem 16.1.2.

9. Perform the indicated operations (a) as they appear and (b) using Theorem 16.1.2 for the ring $[\mathbf{Z}_8; +_8, \times_8]$:
    (i) $2 \times_8 (-4)$ (iii) $(-3) \times_8 5$
    (ii) $(-2) \times_8 (-4)$ (iv) $(-3) \times_8 5 + (-3) \times_8 (-5)$

10. Verify the validity of Theorem 16.1.2 by finding examples of elements a, b, and $c\,(a \neq 0)$ in the following rings, where $a \cdot b = a \cdot c$ and yet $b \neq c$:
    (a) $\mathbf{Z}_8$
    (b) $\mathbf{Z}_{12}$
    (c) $M_{2\times2}(\mathbf{R})$

11. (a) Determine all solutions of the equation $x^2 - 5x + 6 = 0$ in $\mathbf{Z}_{12}$. Why are there more than two solutions? (That is, find all elements of $\mathbf{Z}_{12}$ which satisfy this equation.)

(b) Find all solutions of the equation in Part a in $\mathbf{Z}$. Can there be any more than two solutions to this equation in $\mathbf{Z}$?

12. Prove the corollary to Theorem 16.1.4.

13. (a) Prove that the ring $\mathbf{Z}_2 \times \mathbf{Z}_3$ is commutative and has unity.
    (b) Determine all divisors of zero for the ring $\mathbf{Z}_2 \times \mathbf{Z}_3$.
    (c) Give another example illustrating the fact that the product of two integral domains may not be an integral domain. Is there an example where the product is an integral domain?

B Exercises

14. Prove the multiplicative, associative, and distributive laws for the ring
$$\underset{i=1}{\overset{n}{\times}} R_i .$$
If each R_1 is commutative, is
$$\underset{i=1}{\overset{n}{\times}} R_i$$
commutative? Proof?

15. Ring isomorphism is an equivalence relation. Explain the meaning of this statement.

16. Solve the equation $x^2 + 4x + 4 = 0$:
    (a) in $\mathbf{Z}_{12}$
    (b) in $\mathbf{Z}$
    (c) in $M_{2\times2}(\mathbf{R})$
    (d) in $\mathbf{Z}_3$

17. Solve the equation $ax - ac = -ax$ for x, for each of the rings in Exercise 16, if possible.

18. (a) For any ring $[R; + , \cdot]$, expand $(a + b)(c + d)$ for $a, b, c, d \in R$.
    (b) If each R; is commutative, prove that $(a + b)^2 = a^2 + 2ab + b^2$ for all $a, b \in R$.

19. (a) Let R be a commutative ring with unity. Prove by induction that for $n \geq 1$,
$$(a + b)^n = \sum_{k=0}^{n} \binom{n}{k} a^k b^{n-k}.$$
    (b) Simplify $(a + b)^5$ in $\mathbf{Z}_5$.

16.2 Fields

Although the algebraic structures of rings and integral domains are widely used and play an important part in the applications of mathematics, we still cannot solve the simple equation $ax = b, a \neq 0$ in all rings or in all integral domains. Yet this is one of the first equations we learn to solve in elementary algebra and its solvability is basic to enumerable questions. Certainly, if we wish to solve a wide range of problems in a system we need at least all of the laws true for rings and the cancellation laws together with the ability to solve the equation $ax = b$, $a \neq 0$. We summarize the above in a definition and list several theorems without proof which will place this concept in the context of the previous section.

Definition: Field. *A field is a commutative ring with unity such that each non-zero element has a multiplicative inverse. A field is frequently designated generically by the letter F.*

Example 16.2.1. $[\mathbf{Q}; +, \cdot]$, $(\mathbf{R}; +, \cdot]$, $[\mathbf{C}; +, \cdot]$, and $[\mathbf{Z}_p; +_p, \times_p]$ (p a prime) are all fields.

   Reminder: Since every field is a ring, all facts and concepts that are true for rings are true for any field.

Theorem 16.2.1.   *Every field is an integral domain.* (Of course the converse is not true: consider $[\mathbf{Z}; +, \cdot]$).

Theorem 16.2.2.   *Every finite integral domain is a field.*

Theorem 16.2.3.   *If p is a prime, then $\mathbf{Z}_p$ is a field.* (This is immediate from Theorem 16.2.2.)

   One fact that Theorem 16.2.1 reminds us of is that the cancellation laws must be true for any field. Theorem 16.2.3 gives us a large number of finite fields, but we must be cautious. This theorem does not tell us that all finite fields are of the form $\mathbf{Z}_p$, p a prime. To see this, let's try to construct a field of order 4.

Example 16.2.2.   First the field must contain the additive and multiplicative identities, 0 and 1, so, without loss of generality, we can assume that the field we are looking for is of the form $F = \{0, 1, a, b\}$. Since there are only two non-isomorphic groups of order 4, we have only two choices for the group table for $[F; +]$. Let's try the following table:

+	0	1	a	b
0	0	1	a	b
1	1	0	b	a
a	a	b	0	1
b	b	a	1	0

Next, by Theorem 16.1.2, Part 1, and since 1 is the unity of F, the table for the operation must look like:

·	0	1	a	b
0	0	0	0	0
1	0	1	a	b
a	0	a	—	—
b	0	b	—	—

Hence, to complete the table, we have only four elements to find, and, since F must be commutative, this reduces our options to three. Next, each non-zero element of F must have a unique multiplicative inverse. The inverse of a must be either a itself or b. If $a^{-1} = a$, then $b^{-1} = b$. (Why?) But $a^{-1} = a \Rightarrow a \cdot a = 1$. But if $a \cdot a = 1$, then $a \cdot b$ is equal to a or b. In either case, by the cancellation law, we obtain $a = 1$ or $b = 1$. So $a^{-1} = b$ and hence $b^{-1} = a$. To determine the final two products of the table, simply note that by closure, $a \cdot a \neq a$ (also $b \cdot b$) must be elements of F different from 0 and 1. (Why?) $a \cdot a \neq a$ since, if this were true, by the cancellation law, we have $a = 1$ so $a \cdot a = b$ and $b \cdot b = a$. Hence, our multiplication table for F is:

·	0	1	a	b
0	0	0	0	0
1	0	1	a	b
a	0	a	b	1
b	0	b	1	a

The table listing the multiplicative inverse of each non-zero element is:

u	u^{-1}
1	1
a	b
b	a

We leave it to the reader to convince him- or herself, if it is not already clear, that $[F; +, \cdot]$, as described above, is a field. Hence, we have produced a field of order 4 and 4 is not a prime.

However, even though not all finite fields are isomorphic to $\mathbf{Z}_p$, for some prime p it can be shown that every field F must have either:

(1) a subfield isomorphic to $\mathbf{Z}_p$ for some prime p, or
(2) a subfield isomorphic to $\mathbf{Q}$.

In particular, if F is a finite field, a subfield of F must exist that is isomorphic to $\mathbf{Z}_p$. One can think of all fields as being constructed from either $\mathbf{Z}_p$ or $\mathbf{Q}$.

Example 16.2.3. $[R; +, \cdot]$ is a field, and it contains a subfield isomorphic to $[Q; +, \cdot]$, namely Q itself.

Example 16.2.4. Example 16.2.2 should have a subfield isomorphic to Z_p for some prime p. From the tables, we note that the subset $\{0, 1\}$ of $\{0, 1, a, b\}$ under the given operations of F behaves exactly like $[Z_2; +_2, \times_2]$. (Note we are not using Boolean arithmetic in Z_2 but arithmetic modulo 2.) Hence, the field in Example 16.2.2 has a subfield isomorphic to $[Z_2; +_2, \times_2]$. Does it have a subfield isomorphic to a larger field, say Z_3? We claim not and leave this investigation to the reader (see Exercise 3 of this section).

We close this section with a brief discussion of isomorphic fields. Again, since a field is a ring, the definition of isomorphism of fields is the same as that of rings. It can be shown that if f is a field isomorphism, then $f(a^{-1}) = (f(a))^{-1}$; that is, inverses are mapped onto inverses under any field isomorphism. A major question to try to solve is: How many different non-isomorphic finite fields (or rings, for that matter) are there of any given order? If p is a prime, it seems clear from our discussions that all fields of order p are isomorphic to Z_p. But how many non-isomorphic fields are there, if any, of order 4, 6, 8, etc.? The answer is given in the following theorem, whose proof is beyond the scope of this text.

Theorem 16.2.4.

(1) *Any finite field F has order p^n for a prime p and a positive integer n.*
(2) *For any prime p and any positive integer n there is a field of order p^n.*
(3) *Any two fields of order p^n are isomorphic. This field of order p^n is frequently referred to as the Galois field of order p^n and it is designated by $GF(p^n)$. Evariste Galois (1811–32) was a pioneer in the field of abstract algegra.*

This theorem tells us there is a field of order $2^2 = 4$, and there is only one such field up to isomorphism. That is, all such fields of order 4 are isomorphic to that given in Example 16.2.2.

EXERCISES FOR SECTION 16.2

A Exercises

1. Write out the addition, multiplication, and "inverse" tables for each of the following fields:
   (a) $[Z_2; +_2, \times_2]$
   (b) $[Z_3; +_3, \times_3]$
   (c) $[Z_5, +_5, \times_5]$

2. Show that the set of units of the fields in Exercise 1 form a group under the operation of the multiplication of the given field. Recall that a unit is an element which has a multiplicative inverse.

3. (a) Determine whether the field in Example 16.2.2 contains a subfield isomorphic to $[\mathbf{Z}_3; +_3, \times_3]$ by investigating the field table of F.

4. Write out the operation tables for $[\mathbf{Z}_2^2; +_2, \times_2]$. Is $\mathbf{Z}_2^2$ a ring? An integral domain? A field? Explain.

5. Determine all values x from the given field which satisfy the given equation:
   (a) $x + 1 = -1$ over $\mathbf{Z}_2$, $\mathbf{Z}_3$ and $\mathbf{Z}_5$
   (b) $2x + 1 = 2$ over $\mathbf{Z}_3$ and $\mathbf{Z}_5$
   (c) $3x + 1 = 2$ over $\mathbf{Z}_5$

6. (a) Prove that if $\mathbf{Z}_p$ and $\mathbf{Z}_q$ are fields, then $\mathbf{Z}_p \times \mathbf{Z}_q$ is never a field.
   (b) Can $\mathbf{Z}_p^n$ be a field for any prime p and any positive integer $n \geq 2$?

7. The following are equations over $\mathbf{Z}_2$. Their coefficients come solely from $\mathbf{Z}_2$. Determine all solutions over $\mathbf{Z}_2$; that is, find all numbers in $\mathbf{Z}_2$ that satisfy the equations:
   (a) $x^2 - 1 = 0$
   (b) $x^2 + 1 = 0$
   (c) $x^3 + x^2 + x + 1 = 0$
   (d) $x^3 + x + 1 = 0$

8. Determine the number of different fields, if any, of all orders 1, 2, 3, . . . , 15. Wherever possible, describe these fields via a known field.

9. (a) Let $\mathbf{Q}(\sqrt{2}) = \{a + b\sqrt{2} \mid a, b \in \mathbf{Q}\}$. Prove that $[\mathbf{Q}(\sqrt{2}); +, \cdot]$ is a field where $+, \cdot$ stand for "regular" addition and multiplication.
   (b) Show that $\mathbf{Q}$ is a subfield of $\mathbf{Q}(\sqrt{2})$. For this reason, $\mathbf{Q}(\sqrt{2})$ is called an *extension field of* $\mathbf{Q}$.
   (c) Show that all the roots of the equation $x^2 - 2 = 0$ lie in the extension field $\mathbf{Q}(\sqrt{2})$.
   (d) Do the roots of the equation $x^2 - 3 = 0$ lie in this field? Explain.

16.3 Polynomial Rings

In the previous sections we mentioned the solutions of equations over different rings and fields. To solve the equation $x^2 - 2 = 0$ over the field of the real numbers means to find all solutions of this equation which are in this particular field $\mathbf{R}$. This statement can be replaced as follows: Determine all $x \in \mathbf{R}$ such that the polynomial $f(x) = x^2 - 2$ is equal to zero, or, equivalently, find all zeros of the polynomial $f(x) = x^2 - 2$ in $\mathbf{R}$. The process of solving equations is frequently expressed via polynomials. For this reason, this section will concentrate on polynomials. We will develop concepts using the general setting of polynomials over rings since results proven over rings are true for fields (and integral domains). The reader should keep in mind that in most cases we are just formalizing concepts which he or she learned in high-school algebra over the field of reals.

Definition: Polynomial over R. Let $[R; +, \cdot]$ be a ring. A polynomial, $f(x)$, over R is an expression of the form

$$f(x) = \sum_{i=0}^{n} a_i x^i = a_0 + a_i x + a_2 x^2 + \cdots + a_n x^n, \ n \geq 0,$$

where $a_0, a_1, a_2, \ldots, a_n \in R$ and $a_m = 0$ when $m > n$. If $a_n \neq 0$, then the degree of $f(x)$ is n. If $f(x) = 0$, then the degree of $f(x)$ is $-\infty$. If the degree of $f(x)$ is n, we write deg $(f(x)) = n$.

Comments:

(1) The above definition is that of a polynomial in the single variable, or "indeterminate" x. One can also define polynomials in more than one indeterminate.

(2) The set of all polynomials in the indeterminate x with coefficients in R is denoted by $R[x]$.

(3) Note that $R \subseteq R[x]$. The elements of R are called constant polynomials.

(4) R is called the *ground ring* for $R[x]$.

Example 16.3.1a. $f(x) = 3$, $g(x) = 2 - 4x + 7x^2$, and $h(x) = 2 + x^4$ are all polynomials in $\mathbf{Z}[x]$. Their degrees are 0, 2, and 4, respectively.
  Addition and multiplication of polynomials are performed as in high school algebra. However, we must do our computations in the ring over which we are considering the polynomials.

Example 16.3.1b.   In $\mathbf{Z}_2[x]$, if $f(x) = x + 1$ and $g(x) = x + 1$, then $f(x) + g(x) = (x + 1) + (x + 1) = (1 +_2 1)x + (1 +_2 1) = 0x + 0 = 0$ and $f(x)g(x) = (x + 1) \cdot (x + 1) = x^2 + (1 +_2 1)x + 1 = x^2 + 1$.

  However, for the same polynomials as above, $f(x)$ and $g(x)$ in $\mathbf{Z}[x]$, we have $f(x) + g(x) = (x + 1) + (x + 1) = (1 + 1)x + (1 + 1) = 2x + 2$ and $f(x)g(x) = x^2 + (1 + 1)x + 1 = x^2 + 2x + 1$.
  The important fact to keep in mind is that addition and multiplication in $R[x]$ depend solely on addition and multiplication in R. The x's merely serve the purpose of "place holders." All computations are done over the given ring.
  We summarize in the following theorem:

Theorem 16.3.1.   Let $[R; +, \cdot]$ be a ring. Then:

(1) $R[x]$ is a ring under the usual operations of polynomial addition and multiplication, which depend on (are induced by) the operations in R.

(2) If R is a commutative ring, then $R[x]$ is a commutative ring.

(3) If R is a ring with unity, 1, then $R[x]$ is a ring with unity (the unity in $R[x]$ is $1 + 0x + 0x^2 + \cdots$).

(4) If R is an integral domain, then $R[x]$ is an integral domain.

(5) If F is a field, then $F[x]$ is not a field. However, $F[x]$ is an integral domain.

Proof of Part 5. $F[x]$ is not a field since for $x \in F[x]$, $x^{-1} = 1/x \notin F[x]$. Hence not all non-zero elements in $F[x]$ have multiplicative inverses in $F[x]$. Every field F is an integral domain. By Part 4, $F[x]$ is an integral domain. #

The proofs for Parts 1 through 4 are not difficult but rather long, so we omit them. For those inclined to prove them, we include the formal definitions of addition and multiplication in $R[x]$.

Definition: Addition in $R[x]$. *Let $f(x) = a_0 + a_1x + a_2x^2 + \cdots + a_mx^m$ and $g(x) = b_0 + b_1x + b_2x^2 + \cdots + a_nx^n$ be elements in $R[x]$ so that $a_i \in R$ and $b_i \in R$ for all i. Without loss of generality, assume that the $deg(f(x)) > deg(g(x))$; that is, $m > n$. Then $f(x) + g(x) = c_0 + c_1x + c_2x^2 + \cdots + c_mx^m$, where $c_i = a_i + b_i$ for $i = 0, 1, 2, \ldots, m$.*

Definition: Multiplication in $R[x]$. *Let $f(x)$ and $g(x)$ be as above. Then*

$$f(x) \cdot g(x) = d_0 + d_1x = d_2x^2 + \cdots + d_px^p$$

where

$$d_s = \sum_{i=0}^{s} a_ib_{s-i},$$

and $p = m + n$.

Example 16.3.2. Let $f(x) = 2 + x^2$ and $g(x) = -1 + 4x + 3x^2$. Compute $f(x) \bullet g(x)$ in $\mathbf{Z}[x]$. Of course this product can be obtained by the usual methods of high-school algebra. We will, for illustrative purposes, use the above definition. Using the notation of the above definition, $a_0 = 2$, $a_1 = 0$, $a_2 = 1$, and $b_0 = -1$, $b_1 = 4$, and $b_2 = 3$. We want to compute the coefficients d_0, d_1, d_2, d_3, and d_4. We will compute d_3, the coefficients of the x^3 term of the product, and leave the remainder to the reader (see Exercise 2 of this section):

$$d_3 = \sum_{i=0}^{3} a_ib_{3-i} = a_0b_3 + a_1b_2 + a_2b_1 + a_3b_0$$
$$= (2) \bullet (0) + (0) \bullet (3) + (1) \bullet (4) + (0) \bullet (-1) = 4.$$

From high-school algebra we all learned the standard procedure for dividing a polynomial $f(x)$ by a second polynomial $g(x)$. This process of polynomial long division is referred to as *the division property for polynomials*. Under this scheme we continue to divide until the result is a quotient $q(x)$ and a remainder $r(x)$ whose degree is strictly less than that of the divisor $g(x)$. This property is valid over any field.

Example 16.3.3. Let $f(x) = 1 + x + x^3$ and $g(x) = 1 + x$ be two polynomials in $\mathbf{Z}_2[x]$. Let us divide $f(x)$ by $g(x)$. Keep in mind that we are in $\mathbf{Z}_2[x]$ and that, in particular, the additive inverse of 1, that is, $-1 = 1$ in $\mathbf{Z}_2$.

$$
\begin{array}{r}
x^2 + x \\
x + 1 \overline{\smash{)}\, x^3 + 0x^2 + x + 1} \\
\underline{x^3 + x^2 } \\
+ \ x^2 + x \\
\underline{+ \ x^2 + x } \\
0x + 1
\end{array}
$$

so

$$
\frac{x^3 + x + 1}{x + 1} = x^2 + x + \frac{1}{x + 1},
$$

or equivalently,

$$
x^3 + x + 1 = (x + 1)(x^2 + 1) + 1.
$$

That is, $f(x) = g(x) \cdot q(x) + r(x)$ where $\deg(r(x)) = 0$ is strictly less than the $\deg(g(x)) = 1$.

Example 16.3.4. Let $f(x) = 1 + x^4$ and $g(x) = 1 + x$ be polynomials in $\mathbf{Z}_2[x]$. Let us divide $f(x)$ by $g(x)$:

$$
\begin{array}{r}
x^3 + x^2 + x + 1 \\
x + 1 \overline{\smash{)}\, x^4 + 0x^3 + 0x^2 + 0x + 1} \\
\underline{x^4 + x^3 } \\
+ \ x^3 \\
\underline{+ \ x^3 - x^2 } \\
x^2 \\
\underline{x^2 + x } \\
+ \ x + 1 \\
\underline{+ \ x + 1} \\
0
\end{array}
$$

Thus $(x^4 + 1) = (x + 1)(x^3 + x^2 + x + 1)$.

Since we have 0 as a remainder, $x + 1$ must be a factor of $x^4 + 1$, as in high-school algebra. Also, since $x + 1$ is a factor of $x^4 + 1$, $x = -1$ (which is 1 in $\mathbf{Z}_2$) is a zero of the polynomial $f(x) = x^4 + 1$ (that is, a root of the equation $x^4 + 1 = 0$). Of course we could have determined that $x = 1$ is a zero of $f(x)$ simply by computing $f(1) = 1^4 + 1 = 1 + 1 = 0$.

Before we summarize the main results obtained in the previous examples, we should probably consider what could have happened if we had performed divisions of polynomials in the ring $\mathbf{Z}[x]$ rather than over the field $\mathbf{Z}_2$. For example, $f(x) = x^2 - 1$ and $g(x) = 2x - 2$ are both elements of the ring $\mathbf{Z}[x]$, yet the quotient

$$\frac{x^2 - 1}{2x - 2} = \frac{1}{2}x + \frac{1}{2}$$

is not a polynomial over $\mathbf{Z}$ but a polynomial over the field $\mathbf{Q}$. For this reason it would be wise to describe all results over the field F rather than over the ring R.

Theorem 16.3.2: Division Property for $F[x]$. *Let $[F; +, \cdot]$ be a field and let $f(x)$ and $g(x)$ be two elements of $F[x]$ with $g(x) \neq 0$. Then there exist unique polynomials $q(x)$ and $r(x)$ in $F[x]$ such that $f(x) = g(x) \, q(x) + r(x)$ and where $deg(r(x)) < deg(g(x))$.*

Theorem 16.3.2 can be proven by induction on $deg(f(x))$.

Theorem 16.3.3. *Let $[F; +, \cdot]$ be a field. An element $a \in F$ is a zero of $f(x) \in F[x]$ if and only if $x - a$ is a factor of $f(x)$ in $F[x]$.*

Proof: ($\Rightarrow$) Assume that $a \in F$ is a zero of $f(x) \in F[x]$. We wish to show that $x - a$ is a factor of $f(x)$. To do so, apply the division property to $f(x)$ and $g(x) = x - a$. Hence, there exist unique polynomials $q(x)$ and $r(x)$ from $F[x]$ such that $f(x) = (x - a) \cdot q(x) + r(x)$ and the $deg(r(x)) < deg(x - a) = 1$, so $r(x) = c \in F$, that is, a constant. Also a is a zero of $f(x)$ means $f(a) = 0$. So $f(x) = (x - a) \, q(x) + c$ becomes $0 = f(a) = (a - a) \, q(a) + c$. Hence $c = 0$, so $f(x) = (x - a) \cdot q(x)$, and $x - a$ is a factor of $f(x)$. The reader should note that a critical point of the proof of this half of the theorem was the part of the division property which states that $deg(r(x)) < deg(g(x))$.
   ($\Leftarrow$) See Exercise 6. #

Theorem 16.3.4. *A non-zero polynomial $f(x) \in F[x]$ of degree n can have at most n zeros.*

Proof: Let $a \in F$ be a zero of $f(x)$. Then $f(x) = (x - a) \cdot q_1(x)$ by Theorem 16.3.3. If $b \in F$ is a zero of $q_1(x)$, then again by Theorem 16.3.3, $f(x) = (x - a)(x - b) \, q_2(x)$. Continue this process, which must terminate in at most n steps. #
   From Theorem 16.3.3 we can obtain yet another insight into the problems associated with solving polynomial equations; that is, finding the zeros of a polynomial. The theorem states that an element $a \in F$ is a zero of $f(x) \in F[x]$ if and only if $x - a$ is a factor of $f(x)$. The initial important idea here is that the zero a is from the ground field F. Second, a is a zero only if $(x - a)$ is a factor of $f(x)$ in $F[x]$—that is, only when $f(x)$ can be factored (or reduced) to the product of $(x - a)$ times some other polynomial in $F[x]$.

Example 16.3.5. Consider the polynomial $f(x) = x^2 - 2$ taken as being in $\mathbf{Q}[x]$. From high-school algebra we know that $x^2 - 2$ has two zeros (or roots), namely $\pm\sqrt{2}$, and $x^2 - 2$ can be factored (reduced) as $(x - \sqrt{2})$ $(x + \sqrt{2})$. However, we are working with $f(x) = x^2 - 2$ in $\mathbf{Q}[x]$. Since $\sqrt{2} \notin \mathbf{Q}$, and $(x - \sqrt{2})$ and $(x + \sqrt{2})$ are not elements in $\mathbf{Q}[x]$, therefore $x^2 - 2$ does not have a zero in $\mathbf{Q}$ since it cannot be factored over $\mathbf{Q}$. When this happens, we say that the polynomial is irreducible over $\mathbf{Q}$.

The problem of factoring polynomials is tied hand-in-hand with that of the reducibility of polynomials. We give a precise definition of this concept.

Definition: Irreducible over F. *Let $[F; +, \bullet]$ be a field and let $f(x) \in F[x]$ be a non-constant polynomial. $f(x)$ is called irreducible over F if and only if $f(x)$ cannot be expressed as a product of two (or more) polynomials, both from $F[x]$ and both of degree lower than that of $f(x)$.*

Example 16.3.6. The polynomial $f(x) = x^4 + 1$ of Example 16.3.4 is reducible over $\mathbf{Z}_2$ since $x^4 + 1 = (x + 1)(x^3 - x^2 + x - 1)$.

Example 16.3.7. Is the polynomial $f(x) = x^3 + x + 1$ of Example 16.3.3 reducible over $\mathbf{Z}_2$? From Example 16.3.3 we know that $x + 1$ is not a factor of $x^3 + x + 1$, and from high-school algebra we realize that a cubic (also second-degree) polynomial is reducible if and only if it has a linear (first-degree) factor. (Why?) Does $f(x) = x^3 + x + 1$ have any other linear factors? Theorem 16.3.1 gives us a quick way of determing this since $x - a$ is a factor of $x^3 + x + 1$ over $\mathbf{Z}_2$ if and only if $a \in \mathbf{Z}_2$ is a zero of $x^3 + x + 1$. So $x^3 + x + 1$ is reducible over $\mathbf{Z}_2$ if and only if it has a zero in $\mathbf{Z}_2$. Since $\mathbf{Z}_2$ has only two elements, $-$ and 1, this is easy enough to check. $f(0) = 0^3 + 0 + 1 = 1$ and $f(1) = 1^3 + 1 + 1 = 1$, so neither 0 nor 1 is a zero of $f(x)$ over $\mathbf{Z}_2$. Hence, $x^3 + x + 1$ is irreducible over $\mathbf{Z}_2$.

From high-school algebra we know that $x^3 + x + 1$ has three zeros from some field. Can we find this field? To be more precise, can we construct (find) the field which contains $\mathbf{Z}_2$ and all zeros of $x^3 + 1 + 1$? We will consider this task in the next section.

We close this section with a final analogy. Prime numbers play an important role in mathematics. The concept of irreducible polynomials (over a field) is analogous to that of a prime number. Just think of the definition of a prime number. A useful fact concerning primes is: If p is a prime and if $p \mid ab$, then $p \mid a$ or $p \mid b$. We leave it to the reader to think about the veracity of the following: If $p(x)$ is an irreducible polynomial (over a field F) and $p(x) \mid a(x) b(x)$ (both polynomials in $F[x]$), then $p(x) \mid a(x)$ or $p(x) \mid q(x)$.

EXERCISES FOR SECTION 16.3

A Exercises

1. Let $f(x) = 1 + x$ and $g(x) = 1 + x + x^2$. Compute the following sums and products in the indicated rings by using (a) high-school algebra and (b) the definitions given in this section
    (i) $f(x) + g(x)$ and $f(x) \cdot g(x)$ in $\mathbf{Z}[x]$
    (ii) $f(x) + g(x)$ and $f(x) \cdot g(x)$ in $\mathbf{Z}_2[x]$
    (iii) $(f(x) \cdot g(x)) f(x)$ in $\mathbf{Z}[x]$
    (iv) $(f(x) \cdot g(x)) \cdot f(x)$ in $\mathbf{Z}_2[x]$
    (v) $f(x) \cdot f(x) + f(x) \cdot g(x)$ in $\mathbf{Z}_2[x]$

2. Complete Example 16.3.2.

3. Prove that:
    (a) The ring $\mathbf{R}$ is a subring of the ring $\mathbf{R}[x]$.
    (b) The ring $\mathbf{Z}[x]$ is a subring of the $\mathbf{Q}[x]$.
    (c) The ring $\mathbf{Q}[x]$ is a subring of the ring $\mathbf{R}[x]$.

4. (a) Find all zeros of $x^4 + 1$ in $\mathbf{Z}_3$.
    (b) Find all zeros of $x^5 + 1$ in $\mathbf{Z}_5$.

5. Determine which of the following are reducible over $\mathbf{Z}_2$. Explain.
    (a) $f(x) = x^3 + 1$.
    (b) $g(x) = x^3 + x^2 + x$.
    (c) $h(x) = x^3 + x + 1$.
    (d) $k(x) = x^4 + 1$. (Be careful.)

6. Prove the second half of Theorem 16.3.3.

7. Give an example of the contention made in the last paragraph of this section.

8. Determine all zeros of $x^4 + 3x^3 + 2x + 4$ in $\mathbf{Z}_5[x]$.

9. Show that $x^2 - 3$ is irreducible over $\mathbf{Q}$ but reducible over the field of real numbers.

B Exercise

10. The definition of a vector space given in Chapter 13 holds over any field F, not just over the field of real numbers, where the elements of F are called scalars.
    (a) Show that $F[x]$ is a vector space over F.
    (b) Find a basis for $F[x]$ over F.
    (c) What is the dimension of $F[x]$ over F?

11. Prove Theorem 16.3.2.

12. (a) Show that the field $\mathbf{R}$ of real numbers is a vector space over $\mathbf{R}$. Find a basis for this vector space. What is dim $\mathbf{R}$ over $\mathbf{R}$?

(b) Repeat Part a for an arbitrary field F.

(c) Show that **R** is a vector space over **Q**.

16.4 Field Extensions

From high-school algebra we realize that to solve a polynomial equation means to find its roots (or, equivalently, to find the zeros of the polynomials). From Example 16.3.5 of the previous section we know that the zeros may not lie in the given ground field. Hence, to solve a polynomial really involves two steps: first, find the zeros, and second, find the field in which the zeros lie. For economy's sake we would like this field to be the smallest field which contains all the zeros of the given polynomial. To illustrate this concept, let us reconsider Example 16.3.5.

Example 16.4.1. Let $f(x) = x^2 - 2 \in$ **Q**$[x]$. It is important to remember that we are considering $x^2 - 2$ over **Q**, no other field. We would like to find all zeros of $f(x)$ and the smallest field, call it S, which contains them. The zeros are $x = \pm \sqrt{2}$, neither of which is an element of **Q**. The set S we are looking for must:

(1) contain **Q** as a subfield,

(2) contain all zeros of $f(x) = x^2 - 2$, and

(3) be a field.

Certainly $\sqrt{2}$ must be an element of S, and, if S is to be a field, the sum, product, difference, and quotient of elements in S must be in S. So $\sqrt{2}$, $(\sqrt{2})^2$, $(\sqrt{2})^3$, $\sqrt{2} + \sqrt{2}$, $\sqrt{2} - \sqrt{2}$, and $\sqrt{2} \div \sqrt{2}$ must all be elements of S. Further, since S contains **Q** as a subset, any element of **Q** combined with $\sqrt{2}$ under any field operation must be an element of S. Hence, every element of the form $a + b\sqrt{2}$, where a and b can be any elements in **Q**, is an element of S. We leave to the reader to show that S is a field (see Exercise 1 of this section). We note that the second zero of $x^2 - 2$, namely $-\sqrt{2}$, is an element of S. To see this, simply take $a = 0$ and $b = -1$. The field S is frequently designated as **Q**$(\sqrt{2})$, and it is referred to as an *extension field of* **Q**. Note that the polynomial $x^2 - 2 = (x + \sqrt{2})(x - \sqrt{2})$ factors into linear factors or splits in **Q**$(\sqrt{2})[x]$; that is, all coefficients of both factors are elements of the field **Q**$(\sqrt{2})$.

Example 16.4.2. Consider the polynomial $g(x) = x^2 + x + 1 \in$ **Z**$_2[x]$. Let's repeat the previous example for $g(x)$ over **Z**$_2$. First, $g(0) = 1$ and $g(1) = 1$, so none of the elements of **Z**$_2$ are zeros of $g(x)$. Hence, the zeros of $g(x)$ must lie in an extension field of **Z**$_2$. By Theorem 16.3.3, $g(x) = x^2 + x + 1$ can have at most two zeros. Let α be a zero of $g(x)$. Then the extension field S of **Z**$_2$ must contain $\alpha \cdot \alpha = \alpha^2$, α^3, $\alpha + \alpha$, $\alpha - \alpha$, etc. But, since $g(\alpha) = 0$, we have $\alpha^2 + \alpha + 1 = 0$, or, equivalently,

$\alpha^2 = -\alpha - 1 = \alpha + 1$ (remember, we are working in $\mathbf{Z}_2$). Note the recurrence relation.

So far the extension field S of $\mathbf{Z}_2$ is the set $\{0, 1, \alpha, \alpha + 1\}$. For S to be a field, all possible sums, products, differences, and quotients of elements in S must be in S. Let's try a few: $\alpha + \alpha = \alpha(1 +_2 1) = \cdot 0 = 0$, which is in S: $-\alpha = \alpha$, which is in S; $\alpha + \alpha + \alpha = \alpha$, which is in S. In fact, $n\alpha$ is in S for all possible positive integers n. Next $\alpha^3 = \alpha^2 \cdot \alpha = (\alpha + 1)\alpha = \alpha^2 + \alpha = (\alpha + 1) + \alpha = 2\alpha + 1 = 0 + 1 = 1$, which is in S. It is not difficult to see that α^n is in S for all positive n. Does S contain all zeros of $x^2 + x + 1$? Remember, $g(x)$ can have at most two distinct zeros and we called one of them α, so if there is a second, it must be $\alpha + 1$. To see if $\alpha + 1$ is indeed a zero of $g(x)$, simply compute $f(\alpha + 1) = (\alpha + 1)^2 + (\alpha + 1) + 1 = \alpha^2 + 1 + \alpha + 1 + 1 = \alpha^2 + \alpha + 1 = \alpha + 1 + \alpha + 1 = 0$, so $\alpha + 1$ is also a zero of $x^2 + x + 1$. Hence, $S = \{0, 1, \alpha, \alpha + 1\}$ is the smallest field which contains $\mathbf{Z}_2 = \{0, 1\}$ as a subfield and all zeros of $x^2 + x + 1$. This extension field is designated by $\mathbf{Z}_2(\alpha)$. Note that $x^2 + x + 1$ splits in $\mathbf{Z}_2(\alpha)$; that is, it factors into linear factors in $\mathbf{Z}_2(\alpha)$. We also observe that $\mathbf{Z}_2(\alpha)$ is a field containing exactly four elements. By Theorem 16.2.4, we expected that $\mathbf{Z}_2(\alpha)$ would be of order p^n for some prime p and positive integer n. Also recall that all fields of order p^n are isomorphic. Hence, we have described all fields of order $2^2 = 4$ by finding the extension field of a polynomial which is irreducible over $\mathbf{Z}_2$.

The reader might feel somewhat uncomfortable with the results obtained in Example 16.4.2. In particular, what is α? Can we describe it through a known quantity? All we know about α is that it is a zero of $g(x)$ and that $\alpha^2 = \alpha + 1$. We could also say that $\alpha(\alpha + 1) = 1$, but we really expected more. However, should we expect more? In Example 16.4.1, α, or $\sqrt{2}$, if you prefer, is a number we are more comfortable with, but all we really know about it is that $\sqrt{2}$ is such that $\alpha^2 = 2$. Similarly, the zero, α, that the reader will obtain in Exercise 2 of this section is $\alpha = i$. Here again, this α is simply a symbol, and all we know about it is that $\alpha^2 = -1$. Hence, the result obtained in Example 16.4.2 is not really that strange.

The reader should be aware that we have just scratched the surface in the development of topics in polynomial rings. One area of significant applications is in coding theory. For a discussion of algebraic coding theory, the reader is referred to Hamming, Guiasu, or Dornhoff and Hohn.

EXERCISES FOR SECTION 16.4

A Exercises

1. (a) Use the definition of a field to show that $\mathbf{Q}(\sqrt{2})$ is a field.
    (b) Use the definition of vector space to show that $\mathbf{Q}(\sqrt{2})$ is a vector space over $\mathbf{Q}$.
    (c) Prove that $\{1, \sqrt{2}\}$ is a basis for the vector space $\mathbf{Q}(\sqrt{2})$ over $\mathbf{Q}$, and, therefore, the dimension of vector space $\mathbf{Q}(\sqrt{2})$ over $\mathbf{Q}$ is 2.

2. (a) Determine the splitting field of $f(x) = x^2 + 1$ over $\mathbf{R}$. This means consider the polynomial $f(x) = x^2 + 1 \in \mathbf{R}[x]$ and find the smallest field which contains $\mathbf{R}$ and all the zeros of $x^2 + 1$. Designate this field by $\mathbf{R}(i)$.
   (b) $\mathbf{R}(i)$ is more commonly referred to be a different name. What is it?
   (c) Show that $\{1, i\}$ is a basis for the vector space $\mathbf{R}(i)$ over $\mathbf{R}$. What is the dimension of this vector space (over $\mathbf{R}$)?

3. Determine the splitting field of $x^4 - 5x^2 + 6$ over $\mathbf{Q}$.

4. (a) Factor $x^2 + x + 1$ into linear factors in $\mathbf{Z}_2(\alpha)$.
   (b) Write out the field tables for the field $\mathbf{Z}_2(\alpha)$ and compare the results to the tables of Example 16.2.2.
   (c) Cite a theorem and use it to show why the results of Part b were to be expected.

5. (a) Show that $x^3 + x + 1$ is irreducible over $\mathbf{Z}_2$.
   (b) Determine the splitting field of $x^3 + x + 1$ over $\mathbf{Z}_2$.
   (c) Use Theorem 16.2.4 to illustrate that you have described all fields of order 2^3.

6. (a) List all polynomials of degree 1, 2, 3, and 4 over $\mathbf{Z}_2 = GF(2)$.
   (b) Use your results in Part a and list all irreducible polynomials of degree 1, 2, 3, and 4.
   (c) Determine the splitting fields of each of the polynomials in Part b.
   (d) What is the order of each of the splitting fields obtained in Part c? Explain your results using Theorem 16.2.4.

16.5 Power Series

In Section 16.3 we found that a typical element in the ring $R[x]$ is an expression of the form

$$f(x) = \sum_{i=0}^{n} a_i x^i = a_0 + a_i x + a_2 x^2 + \cdots + a_n x^n,$$

called a polynomial of degree n. In Section 8.5 we defined a generating function of a sequence s with terms $s_o, s_1, \ldots$ as the infinite sum

$$G(s, z) = \sum_{i=0}^{\infty} s_i z^i = s_0 + s_1 z^1 + s_2 z^2 + \cdots$$

The main difference between these two expressions, disregarding notation of course, is that the latter is an infinite expression and the former is a finite expression. In this section we will compare the algebra of infinite expressions like $G(s, z)$, called *power series*, to that of polynomials.

Definition: Power Series. *Let $[R; +, \cdot]$ be a ring. A power series $f(x)$ over R is an expression of the form*

$$f(x) = \sum_{i=0}^{\infty} a_i x^i$$

where $a_0, a_1, a_2, \ldots \in R$. The set of all expressions of the above form is designated by $R[[x]]$.

Our first observation in our comparison of $R[x]$ and $R[[x]]$ is that $R[x]$ is a subset, in fact a subring, of $R[[x]]$, since the polynomial $a_0 + a_1 x + a_2 x^2 + \cdots + a_n x^n$ of degree n, which is of course an element of $R[x]$, can be thought of as an infinite expression where $a_i = 0$ for $i > n$. So every element in $R[x]$ is an element of $R[[x]]$. In fact, the observant reader will note that an element of $R[x]$ can be defined as an expression of the form

$$f(x) = \sum_{i=0}^{\infty} a_i x^i$$

where all but a finite number of the a_i equal 0.

$R[[x]]$ is given a ring structure by defining addition and multiplication in $R[[x]]$ as we did in $R[x]$, with the modification that, since we are dealing with infinite expressions, the sums and products will remain infinite expressions which we can determine term by term, as was done in Section 16.3.

Example 16.5.1: Example 8.5.3 Revisited. Let

$$f(x) = \sum_{i=0}^{\infty} i x^i = 0 + 1x^1 + 2x^2 + 3x^3 + \cdots$$

and

$$g(x) = \sum_{i=0}^{\infty} 2^i x^i = 2^0 + 2^1 x + 2^2 x^2 + 2^3 + x^3 + \cdots$$

be elements in $\mathbf{Z}[[x]]$. Let us compute $f(x) + g(x)$ and $f(x) g(x)$.

$$f(x) + g(x) = (0 + 1) + (1 + 2^1)x + (2 + 2^2)x^2 +$$

$$(3 + 2^3) + x^3 + \cdots + = \sum_{i=0}^{\infty} (i + 2^i)x^i.$$

Of course it would have been easier to work directly as follows:

$$f(x) + g(x) = \sum_{i=0}^{\infty} i x^i + \sum_{i=0}^{\infty} 2^i x^i = \sum_{i=0}^{\infty} (i + 2^i)x^i .$$

$$f(x) \cdot g(x) = d_0 + d_1 x^1 + d_2 x^2 + \cdots$$

where

$$d_s = \sum_{i=0}^{s} a_i b_{s-i}$$

and where the a's and b's are the coefficients of $f(x)$ and $g(x)$ respectively. So

$$d_0 = \sum_{i=0}^{0} a_i b_{0-i} = a_0 b_0 = 0 \cdot 1 = 0$$

$$d_1 = \sum_{i=0}^{1} a_i b_{1-i} = a_0 b_1 + a_1 b_0 = 0 \cdot 2 + 1 \cdot 1 = 1$$

$$d_2 = \sum_{i=0}^{2} a_i b_{2-i} = a_0 b_2 + a_1 b_1 + a_2 b_0$$

$$= 0 \cdot 4 + 1 \cdot 2 + 2 \cdot 1 = 4.$$

Hence, $f(x) \cdot g(x) = 0 + 1\,x + 4\,x^2 + \cdots$

It can be shown (see Exercise 1 of this section) that $f(x) \cdot g(x)$ can also be expressed as:

$$f(x)\,g(x) = \sum_{s=0}^{\infty} \left(\sum_{i=0}^{s} a_i b_{s-i} \right) x^s$$

$$= \sum_{s=0}^{\infty} \left(\sum_{i=0}^{s} i \cdot 2^{s-i} \right) x^s$$

We have shown that addition and multiplication in $R[[x]]$ is virtually identical to that in $R[x]$. Similarly, Theorem 16.3.1 holds for $R[[x]]$.

We are most interested in the situation when the set of coefficients forms a field. Theorem 16.3.1 for power series indicates that when F is a field, $F[[x]]$ is an integral domain. The reason that $F[[x]]$ is not a field is the same as that for $F[x]$, namely that x does not have multiplicative inverse in $F[[x]]$.

The difference between $F[x]$ and $F[[x]]$ becomes even more apparent when one studies which elements are units (have multiplicative inverses) in each. First we prove that the only units in $F[x]$ are the non-zero constants—that is, the non-zero elements of F.

Theorem 16.5.1. *Let $[F; +, \cdot]$ be a field. $f(x)$ is a unit in $F[x]$ if and only if $f(x)$ is a non-zero element of F.*

Proof: ($\Rightarrow$) Let $f(x)$ be a unit in $F[x]$. Then $f(x)$ has a multiplicative inverse, call it $g(x)$, such that $f(x) \cdot g(x) = 1$. Hence, the *deg* $(f(x) \cdot g(x)) = deg\,(1) = 0$. But *deg* $(f(x) \cdot g(x)) = deg\,f(x) + deg\,g(x)$. So *deg* $f(x) + deg\,g(x) = 0$, and since the degree of a polynomial is always non-negative, this can only happen when the *deg* $f(x) = deg\,g(x) = 0$. Hence, $f(x)$ is a constant, an element of F, which is a unit if and only if it is non-zero.

($\Leftarrow$) If $f(x)$ is a non-zero element of F, then it is a unit since F is a field. #

Before we proceed to categorize the units in $F[[x]]$, we remind the reader that two power series $a_0 + a_1 x + a_2 x^2 + \cdots$ and $b_0 + b_1 x + b_2 x^2 + \cdots$ are equal if and only if corresponding coefficients are equal.

Theorem 16.5.2. *Let $[F; +, \cdot]$ be a field. Then $f(x) = a_0 + a_1 x + a_2 x^2 + \cdots = \sum_{i=0}^{\infty} a_i x^i$ is a unit in $F[[x]]$ if and only if $a_0 \neq 0$.*

Proof: ($\Rightarrow$) If

$$f(x) = \sum_{i=0}^{\infty} a_i x^i$$

is a unit in $F[[x]]$, then there exists a power series

$$g(x) = \sum_{i=0}^{\infty} b_i x^i \in F[[x]]$$

such that $f(x) \cdot g(x) = 1$. Since corresponding coefficients are equal, we must have $a_0 b_0 = 1$. Hence, $a_0 \neq 0$. Why?

($\Leftarrow$) Assume that $a_0 \neq 0$. To prove that $f(x)$ is a unit in $F[[x]]$ we need to find

$$g(x) = \sum_{i=0}^{\infty} b_i x^i \text{ in } F[[x]]$$

such that $f(x) \cdot g(x) = 1$. If we use the formula for the coefficients $d_0, d_1, d_2, \ldots$ of the product $f(x) \cdot g(x)$ and equate coefficients, we will obtain

$$d_0 = a_0 b_0 = 1$$
$$d_1 = a_0 b_1 + a_1 b_0 = 0$$
$$d_2 = a_0 b_2 + a_1 b_1 + a_2 b_0 = 0$$

$$\cdot$$
$$\cdot$$
$$\cdot$$

$$d_s = a_0 b_s + a_1 b_{s-1} + \cdots + a_s b_0 = 0$$

$$\cdot$$
$$\cdot$$
$$\cdot$$

Therefore, the existence of $g(x)$ is equivalent to the existence of a solution $b_0, b_1, b_2, \ldots$ to the above system of equations. Next since $a_0 \neq 0$, we can solve Equation 1 for b_0. (Note that if $a_0 = 0$, we cannot proceed further.) Then we can continue to Equation 2 and solve for b_1, then b_2, b_3, etc. Hence,

$$b_0 = a_0^{-1}$$
$$b_1 = -a_0^{-1} (a_1 b_0)$$
$$b_2 = -a_0^{-1} (a_1 b_1 + a_2 b_0)$$

$$\cdot$$
$$\cdot$$
$$\cdot$$

$$b_s = -a_0^{-1} (a_1 b_{s-1} + a_2 b_{s-2} + \cdots + a_s b_0).$$

Therefore, the power series

$$g(x) = \sum_{i=0}^{\infty} b_i x^i$$

is an expression whose coefficients lie in F and which satisfies the statement $f(x) \cdot g(x) = 1$. Hence, $g(x)$ is the multiplicative inverse of $f(x)$.

Example 16.5.2.   Let

$$f(x) = 1 + 2x + 3x^2 + \cdots$$

$$= \sum_{i=0}^{\infty} (1 + i)x^i$$

be an element of $Q[[x]]$. Then, by Theorem 16.5.2, since $a_0 = 1 \neq 0$, the inverse of $f(x)$, call it $g(x)$, exists. To compute $g(x)$, we can either redo the procedure outlined in the above theorem or use the formula for the b_i's. Using the formulas for the b_i's we obtain:

$$b_0 = 1$$
$$b_1 = -1(2 \cdot 1) = -2$$
$$b_2 = -1(2 \cdot (-2) + 3 \cdot 1) = 1$$
$$b_3 = -1 (2 \cdot 1 + 3 \cdot (-1) + 4 \cdot 1) = 0$$
$$\cdot$$
$$\cdot$$
$$\cdot$$

Hence, $g(x) = 1 - 2x + 1x^2$ is the multiplicative inverse of $f(x)$. We leave to the reader the verification of this by the product formula—that is, to show that $f(x) \cdot g(x) = 1$.

If we compare Theorems 16.5.1 and 16.5.2, we note that while the only elements in $F[x]$ which are units are the non-zero constants of F, the units in $F[[x]]$ are every single expression in x where $a_0 \neq 0$. So certainly $F[[x]]$ contains a wider variety of units than $F[x]$. Yet $F[[x]]$ is not a field, since $x \in F[[x]]$ is not a unit by Theorem 16.5.2. So concerning the algebraic stucture of $F[[x]]$, we know that it is an integral domain which contains $F[x]$. If we allow our power series to take on negative exponents—that is, consider expressions of the form

$$f(x) = \sum_{i=-\infty}^{\infty} a_i x^i,$$

where all but a finite number of terms with a negative index equal zero—then these expressions are called *extended power series*. The set of all such expressions is a field, call it E. This set does contain, for example, the inverse of x namely $1/x$. It can be shown that each non-zero element of E is a unit.

EXERCISES FOR SECTION 16.5

A Exercises

1. Let

$$f(x) = \sum_{i=0}^{\infty} a_i x^i \text{ and } g(x) = \sum_{i=0}^{\infty} b_i x^i$$

be elements of $R[[x]]$. Let

$$f(x)\, g(x) = \sum_{s=0}^{\infty} d_s x^s.$$

(a) Apply the distributive law repeatedly to $(a_0 + a_1 x + a_2 x^2 + \cdots)$ $\cdot (b_0 + b_1 x + b_2 x^2 + \cdots)$ to obtain the formula

$$d_s = \sum_{i=0}^{s} a_i b_{s-i}$$

for the coefficients of $f(x) \cdot g(x)$. Hence, you have shown that

$$f(x) \cdot g(x) = \sum_{s=0}^{\infty} \left(\sum_{i=0}^{s} a_i b_{s-i} \right) x^s.$$

(b) Apply the above formula to the product in Example 16.5.1 and show that the result is the same as that obtained in this example.

2. Restate Theorem 16.3.1 for $R[[x]]$

3. Show that if F is a field, $F[[x]]$ is not a field.

4. (a) Prove that for any integral domain D, the following can be proven:

$$f(x) = \sum_{i=0}^{\infty} a_i x^i$$

is a unit in $D[[x]]$ if and only if a_o is a unit in D. (Hint: imitate the proof of Theorem 16.5.2.)

(b) Compare the statement in Part a to that in Theorem 16.5.2.

(c) Give an example of the statement in Part a in $Z[[x]]$.

5. Use the formula for the product to verify that the expression $g(x)$ of Example 16.5.2 is indeed the inverse of $f(x)$.

6. (a) Determine the inverse of $f(x) = 1 + x + x^2 + \cdots$ in $Q[[x]]$. Verify.

(b) Repeat Part a with $f(x)$ taken in $Z_2[[x]]$.

(c) Use the method outline in Chapter 8 to show that the power series $f(x) = 1 + x + x^2 + \cdots \in Q[[x]]$ is the rational generating function $1/(1 - x)$. What is the inverse of this function? Compare your results in those in Part a.

7. (a) Determine the inverse of

$$f(x) = \sum_{i=0}^{\infty} 2^i x^i \text{ in } Q[[x]].$$

(b) Use the procedures in Chapter 8 to find a rational generating function for $f(x)$ in Part a. Find the multiplicative inverse of this function.

8. Let $a(x) = 1 + 3x + 9x^2 + 27x^3 + \cdots$ and $b(x) = 1 + x + x^2 + x^3 + \cdots$ both in $\mathbf{R}[[x]]$.

(a) What are the first four terms (counting the constant term as the 0th term) of $a(x) + b(x)$?

(b) Find a closed form expression for $a(x)$.

(c) What are the first four terms of $a(x) b(x)$?

B Exercise

9. Write as an extended power series:
(a) $(x^4 - 1x^5)^{-1}$
(b) $(x^2 - 2x^3 + x^4)^{-1}$

Appendix

●━━━━━━━━━━━━━━━━━━━━━━━━━●

DETERMINANTS

GOALS

In Chapter 5 we defined the determinant of a 2×2 matrix for the sole purpose of providing some hands-on experience in the computation of inverses of 2×2 matrices. In this appendix we will define the determinant of an $n \times n$ matrix, summarize the main properties of a determinant, and give an alternate procedure for finding the inverse of an $n \times n$ matrix. We will not offer an in-depth mathematical treatment of determinants.

A.1 Definition of a Determinant

Associated with every $n \times n$ matrix is a number called the *determinant of the matrix*. This number gives a great deal of information about the matrix, not the least of which is whether the matrix is invertible. We first recall the definition of the determinant of a 2×2 matrix and give that of a 1×1 matrix. If A is a matrix, then the determinant of A is designated by $|A|$ or $det(A)$.

Definitions: Determinant of a 1×1 and a 2×2 Matrix.

(1) *If $A = [A_{11}]$, then $det(A) = A_{11}$.*
(2) *If*

$$A = \begin{bmatrix} A_{11} & A_{12} \\ A_{21} & A_{22} \end{bmatrix},$$

then

$$det(A) = \begin{vmatrix} A_{11} & A_{12} \\ A_{21} & A_{22} \end{vmatrix} = A_{11}A_{22} - A_{12}A_{21}.$$

Example A.1.1. If $A = [-5]$, then $det(A) = -5$. If

$$A = \begin{bmatrix} 1 & 2 \\ -1 & 2 \end{bmatrix},$$

then $det(A) = 2 - (-2) = 4$.

We now proceed to define the determinant of an $n \times n$ matrix, $n > 2$. This definition requires two preliminary definitions—those of minor and cofactor.

Definition: Minor. *Let $A = [A_{ij}]$ be an $n \times n$ matrix. The determinant of the $(n - 1) \times (n - 1)$ matrix formed by omitting the ith row and jth column of a is the minor of the element A_{ij}, and it is designated by the number M_{ij}.*

Example A.1.2.   Let

$$A = \begin{bmatrix} 1 & 2 & -1 & 3 \\ 2 & 0 & 1 & 4 \\ 0 & 3 & -1 & 5 \\ 3 & 2 & 1 & 1 \end{bmatrix}.$$

Then

$$M_{33} = \begin{vmatrix} 1 & 2 & 3 \\ 2 & 0 & 4 \\ 3 & 2 & 1 \end{vmatrix}.$$

Definition: Cofactor. *The cofactor C_{ij} of the ith row, jth column element of the $n \times n$ matrix A is the number $C_{ij} = (-1)^{i+j} M_{ij}$.*

The signs $(-1)^{i+j}$ can easily be remembered by noting that they form the following checkerboard pattern:

$$\begin{bmatrix} + & - & + & - & \cdot & \cdot & \cdot \\ - & + & - & + & \cdot & \cdot & \cdot \\ + & - & + & - & \cdot & \cdot & \cdot \\ - & + & - & + & \cdot & \cdot & \cdot \\ \cdot & & & & & & \\ \cdot & & & \cdot & & \cdot & \cdot \end{bmatrix}$$

Example A.1.3.   Let A be as in Example A.1.2. Then:

$$C_{33} = (-1)^{3+3} \begin{vmatrix} 1 & 2 & 3 \\ 2 & 0 & 4 \\ 3 & 2 & 1 \end{vmatrix} = \begin{vmatrix} 1 & 2 & 3 \\ 2 & 0 & 4 \\ 3 & 2 & 1 \end{vmatrix}$$

and

$$C_{41} = (-1)^{4+1} \begin{vmatrix} 2 & -1 & 3 \\ 0 & 1 & 4 \\ 3 & -1 & 5 \end{vmatrix} = - \begin{vmatrix} 2 & -1 & 3 \\ 0 & 1 & 4 \\ 3 & -1 & 5 \end{vmatrix}.$$

Definition: Determinant. *Let A be an $n \times n$ matrix, $n \geq 2$. The determinant of A is defined by*

$$det(A) = \sum_{j=1}^{n} A_{1j} C_{1j}.$$

Example A.1.4. Let

$$A = \begin{bmatrix} A_{11} & A_{12} & A_{13} \\ A_{21} & A_{22} & A_{23} \\ A_{31} & A_{32} & A_{33} \end{bmatrix}.$$

Then

$$det(A) = \begin{vmatrix} A_{11} & A_{12} & A_{13} \\ A_{21} & A_{22} & A_{23} \\ A_{31} & A_{32} & A_{33} \end{vmatrix}$$

$$= A_{11} C_{11} + A_{12} C_{12} + A_{13} C_{13}$$

$$= A_{11} (-1)^{1+1} M_{11} + A_{12} (-1)^{1+2} M_{12} + A_{13} (-1)^{1+3} M_{13}$$

$$= A_{11} \begin{vmatrix} A_{22} & A_{23} \\ A_{32} & A_{33} \end{vmatrix} - A_{12} \begin{vmatrix} A_{21} & A_{23} \\ A_{31} & A_{33} \end{vmatrix} + A_{13} \begin{vmatrix} A_{21} & A_{22} \\ A_{31} & A_{32} \end{vmatrix}.$$

Example A.1.5. Let

$$A = \begin{vmatrix} 1 & 2 & -3 \\ 2 & 0 & 4 \\ 3 & 2 & 1 \end{vmatrix}.$$

Then

$$det(A) = 1 \begin{vmatrix} 0 & 4 \\ 2 & 1 \end{vmatrix} - 2 \begin{vmatrix} 2 & 4 \\ 3 & 1 \end{vmatrix} + (-3) \begin{vmatrix} 2 & 0 \\ 3 & 2 \end{vmatrix}$$

$$= 1(0 - 8) - 2(2 - 12) - 3(4 - 0)$$

$$= 0.$$

Our definition of a determinant involves what is called *expansion* along the first row of the matrix A. A very useful tool is that the determinant of a matrix can be found by expanding along any row or any column of A.

Example A.1.6. Let A be as in Example A.1.5. If we expand down the second column of A, we obtain:

$$det(A) = -2 \begin{vmatrix} 2 & 4 \\ 3 & 1 \end{vmatrix} + 0 \begin{vmatrix} 1 & -3 \\ 3 & 1 \end{vmatrix} - 2 \begin{vmatrix} 1 & -3 \\ 2 & 4 \end{vmatrix}$$

$$= -2(-10) + 0 - 2(10)$$

$$= 0.$$

If we compare Examples A.1.5 and A.1.6, we note that since A contained a 0 in Column 2, expanding down Column 2 was more efficient than expanding along Row 1. This is important, since if A is large, the evaluation of $det(A)$ can be quite time consuming, as the following example illustrates.

Example A.1.7. The 5×5 matrix

$$A = \begin{vmatrix} 1 & 2 & -1 & 3 & 1 \\ -1 & 5 & 2 & 1 & 1 \\ 2 & 1 & -1 & 3 & 2 \\ -2 & 1 & 1 & 1 & 1 \\ 3 & 6 & 5 & 2 & 3 \end{vmatrix}$$

looks quite innocent. If we expand along Row 1, we have:

$$|A| = 1M_{11} - 2M_{12} - 1M_{13} - 3M_{14} + 1M_{15},$$

where each M_{ij} is the determinant of a 4×4 matrix, whose evaluation involves a four-termed expression, each term of which requires the evaluation of the determinant of a 3×3 matrix, the computation of which is like those of the previous examples. This is certainly a time-consuming project, even for this relatively small matrix. In fact there are $5! = 120$ multiplications alone needed to evaluate the determinant of a 5×5 matrix. Fortunately, certain properties of the determinant allow us to simplify the process of the evaluation of $det(A)$ considerably. We list some of the main properties of the determinant in the following theorem:

Theorem A.1.1. *Let A be an $n \times n$ matrix.*

(i)   *det(A) can be found by expanding along any row or any column.*

(ii)   *If a row (or column) of A consists entirely of zeros, then det(A) = 0.*

(iii)   *If two rows (or columns) of A are interchanged, the det(A) changes sign.*

(iv)   *If a matrix A has two equal rows (or columns) then det(A) = 0.*

(v)   *If any row (or column) of A is a scalar multiple of any other row (or column) of A, then det(A) = 0.*

(vi)   *The value of a determinant is unchanged if a multiple of one row (or column) of A is added to another row (or column) of A.*

(vii)   *If one row (or column) of a matrix A is multiplied by a constant c, then the value of det(A) is multiplied by c.*

(viii)   *det(A • B) = (det A) (det B).*

(ix)   *det(I) = 1 where I is the $n \times n$ identity matrix.*

(x)   *(det A^{-1}) = (det A)$^{-1}$, if A^{-1} exists.*

Obviously, the evaluation of the determinant of a matrix is relatively easy if that matrix contains a row (or column) with a large number of zeros.

Property *vi* gives us the mechanism to effect this for any matrix and, hence, Property *vi* is probably the most widely used property of determinants. We illustrate with an example.

Example A.1.7. Determine the $det(A)$ if

$$A = \begin{bmatrix} 1 & 1 & 2 \\ 2 & 1 & 4 \\ 3 & 5 & 1 \end{bmatrix}.$$

By repeated applications of Property *vi*, we have:

$$det(A) = \begin{vmatrix} 1 & 1 & 2 \\ 2 & 1 & 4 \\ 3 & 5 & 1 \end{vmatrix}$$

$$= \begin{vmatrix} 1 & 1 & 2 \\ 0 & -1 & 0 \\ 0 & 2 & 5 \end{vmatrix} \quad (R_2 := -2\,R_1 + R_2)$$

$$= \begin{vmatrix} 1 & 1 & 2 \\ 0 & -1 & 0 \\ 0 & 2 & -5 \end{vmatrix} \quad (R_3 := -3\,R_1 + R_3)$$

Hence, if we expand the last form of $|A|$ down Column 1, we obtain

$$det(A) = 1 \begin{vmatrix} -1 & 0 \\ 2 & -5 \end{vmatrix} = 5.$$

Example A.1.8. Let

$$A = \begin{bmatrix} 1 & 2 & 0 & -1 \\ 3 & 2 & 5 & 2 \\ 1 & -1 & 1 & 5 \\ 2 & 3 & -3 & 2 \end{bmatrix}$$

In this case, Row 1 (also Column 3) contains a zero, so it appears that we should expand along Row 1 (or down Column 3). Let's choose expansion along Row 1. We first try to obtain as many zeros as possible in this row using Property *vi*. We use the convention that C_i stands for the ith column of the matrix.

$$det(A) = \begin{vmatrix} 1 & 2 & 0 & -1 \\ 3 & 2 & 5 & 2 \\ 1 & -1 & 1 & 5 \\ 2 & 3 & -3 & 2 \end{vmatrix}$$

$$= \begin{vmatrix} 1 & 0 & 0 & 0 \\ 3 & -4 & 5 & 5 \\ 1 & -3 & 1 & 6 \\ 2 & -1 & -3 & 4 \end{vmatrix} \quad \begin{matrix} C_2 := -2C_1 + C_2 \\ C_4 := C_1 + C_2 \end{matrix}$$

So far we have

$$det(A) = 1 \begin{vmatrix} -4 & 5 & 5 \\ -3 & 1 & 6 \\ -1 & -3 & 4 \end{vmatrix} = 1\,M_{11}.$$

We will simplify M_{11} through Property vi using the element 1 in the Row 2, Column 2 of M_{11} as a "pivot" to obtain as many zeros as possible in Row 2. Of course there are a variety of other approaches which are as simple.

$$M_{11} = \begin{vmatrix} -4 & 5 & 5 \\ -3 & 1 & 6 \\ -1 & 3 & 4 \end{vmatrix}$$

$$= \begin{vmatrix} 11 & 5 & -25 \\ 0 & 1 & 0 \\ -10 & 3 & 22 \end{vmatrix} \qquad \begin{matrix} C_1 := 3C_2 + C_1 \\ C_3 := -6C_2 + C_3 \end{matrix}$$

$$= 1 \begin{vmatrix} 11 & -25 \\ -10 & 22 \end{vmatrix}$$

$$= 1(-8) = -8.$$

Hence, $det(A) = 1\,M_{11} = 1(-8) = -8.$

EXERCISES FOR SECTION A.1

A Exercises

1. Use the definition of the determinant of an $n \times n$ matrix to derive the formula for a 2×2 determinant.

2. Is Theorem A.1.1, Part ii, an if and only if statement?

3. Use Property vii to prove that if A is an $n \times n$ matrix and c is a constant, $det(cA) = c^n\,det(A)$.

4. (a) Let

$$A = \begin{bmatrix} A_{11} & 0 & 0 \\ 0 & A_{22} & 0 \\ 0 & 0 & A_{33} \end{bmatrix}.$$

   Compute $det(A)$.
   (b) Let A be an arbitrary $n \times n$ diagonal matrix. What is $det(A)$?

5. (a) Consider

$$\begin{bmatrix} A_{11} & A_{12} & A_{13} \\ 0 & A_{22} & A_{23} \\ 0 & 0 & A_{33} \end{bmatrix}.$$

What is its determinant? A matrix of this form is called an (*upper*) *triangular matrix*.

(b)   What is the value of the determinant of any upper triangular matrix?

6.   Evaluate $det(A)$ for each of the following matrices:

(a) $\begin{bmatrix} 2 & -3 \\ 1 & 1 \end{bmatrix}$ (b) $[-1/2]$

(c) $\begin{bmatrix} 2 & 0 & 1 \\ 4 & 0 & 1 \\ 1 & 1 & 1 \end{bmatrix}$

(d) $\begin{bmatrix} 1 & 3 & -2 \\ 1 & 2 & 1 \\ 4 & 1 & -1 \end{bmatrix}$

(e) $\begin{bmatrix} 2 & 3 & 1 & -1 \\ 1 & -5 & -3 & 2 \\ 1 & 1 & 0 & 1 \\ 2 & 5 & 1 & 2 \end{bmatrix}$

(f) $\begin{bmatrix} 1 & 7 & 3 & -4 \\ 2 & 1 & 5 & 2 \\ 3 & 21 & 9 & -12 \\ 4 & 5 & 0 & 0 \end{bmatrix}$

(g) $\begin{bmatrix} 1 & 0 & 0 \\ 0 & \cos\theta & -\sin\theta \\ 0 & \sin\theta & \cos\theta \end{bmatrix}$

(h) $\begin{bmatrix} 1 & 3 & 0 \\ -2 & -1 & 4 \\ 0 & 3 & 2 \end{bmatrix}$

(i) $\begin{bmatrix} 1 & -3 & 2 \\ 0 & -4 & 3 \\ 2 & 1 & 1 \end{bmatrix}$

(j) $\begin{bmatrix} 1 & -2 & 1 \\ -3 & 4 & 2 \\ 2 & -1 & -4 \end{bmatrix}$

(k) $\begin{bmatrix} 1 & -2 & 1 & 0 \\ -3 & 4 & 2 & -2 \\ 2 & -1 & -4 & 1 \\ 0 & -4 & 3 & 1 \end{bmatrix}$

A.2   Cramer's Rule and the Cofactor Method for Inverses

Consider the system of equations given in Example 12.1.1:

$$x_1 + 2x_2 + x_3 = 1$$
$$2x_1 + x_2 + x_3 = 4$$
$$2x_1 + 2x_2 + x_3 = 3$$

If we let

$$A = \begin{bmatrix} 4 & 2 & 1 \\ 2 & 1 & 1 \\ 2 & 2 & 1 \end{bmatrix}, X = \begin{bmatrix} x_1 \\ x_2 \\ x_3 \end{bmatrix} \text{ and } B = \begin{bmatrix} 1 \\ 4 \\ 3 \end{bmatrix},$$

then this system can be expressed as $AX = B$. We will outline a technique called *Cramer's Rule* for solving this system. In general, let $AX = B$ be a system of n linear equations, n unknowns. Note that this forces A to be a square matrix of order n.

Theorem A.2.1.:   Cramer's Rule. *Let $AX = B$ be a system of n linear equations, n unknowns, with $det(A) \neq 0$. Let A(i) denote the matrix obtained by replacing the ith column of A by B. Then the unique solution to the given system is:*

$$x_1 = \frac{det\, A(1)}{det(A)}, \; x_2 = \frac{det\, A(2)}{det(A)}, \; \cdots \; , \; x_n = \frac{det\, A(n)}{det(A)}$$

Example A.2.1.   For the above system,

$$A(1) = \begin{bmatrix} 1 & 2 & 1 \\ 4 & 1 & 1 \\ 3 & 2 & 1 \end{bmatrix},$$

$$A(2) = \begin{bmatrix} 4 & 1 & 1 \\ 2 & 4 & 1 \\ 2 & 3 & 1 \end{bmatrix},$$

$$A(3) = \begin{bmatrix} 4 & 2 & 1 \\ 2 & 1 & 4 \\ 2 & 2 & 3 \end{bmatrix}.$$

So that

$$x_1 = \frac{det\,A(1)}{det(A)} = \frac{\begin{vmatrix} 1 & 2 & 1 \\ 4 & 1 & 1 \\ 3 & 3 & 3 \end{vmatrix}}{\begin{vmatrix} 4 & 2 & 1 \\ 2 & 1 & 1 \\ 2 & 2 & 1 \end{vmatrix}} = (2)/(-2) - 1$$

$$x_2 = \frac{det\,A(2)}{det(A)} = \frac{\begin{vmatrix} 4 & 1 & 1 \\ 2 & 4 & 1 \\ 2 & 3 & 1 \end{vmatrix}}{\begin{vmatrix} 4 & 2 & 1 \\ 2 & 1 & 1 \\ 2 & 2 & 1 \end{vmatrix}} = \frac{2}{-2} = -1$$

$$x_3 = \frac{det\,A(3)}{det(A)} = \frac{\begin{vmatrix} 4 & 2 & 1 \\ 2 & 1 & 4 \\ 2 & 2 & 3 \end{vmatrix}}{\begin{vmatrix} 4 & 2 & 1 \\ 2 & 1 & 1 \\ 2 & 2 & 1 \end{vmatrix}} = \frac{-14}{-2} = 7$$

An alternate method of determining the inverse of a matrix is called the *adjoint method*. This method is impractical for matrices of order greater than 3. We mention it here solely for the edification of the reader. One feature worth noting is that this method expresses A^{-1} in terms of the factor $(det\,A)$. This fact shows us that A is invertible if and only if $det(A) \neq 0$, a statement made in Chapter 5.

Definition: Transpose. *Let A be an $m \times n$ matrix. The transpose of A, denoted A^T, is an $n \times m$ matrix with $(A^T)_{ij} = A_{ji}$.*

In simple terms, the rows of A are the columns of A^T, and vice versa.

Definition. Adjoint. *The adjoint of an $n \times n$ matrix A, designated by adj(A), is the transpose of the matrix of cofactors of the elements of A.*

Theorem A.2.2. *Let A be an $n \times n$ matrix with $det(A) \neq 0$. Then A is invertible and $A^{-1} = \dfrac{1}{det(A)}$ adj(A) where adj $A = C^T$, C being the matrix of cofactors of A.*

Example A.2.2. Let

$$A = \begin{bmatrix} 1 & 1 & 2 \\ 2 & 1 & 4 \\ 3 & 5 & 1 \end{bmatrix},$$

the matrix of Example 12.2.1. The matrix C of cofactors of A are computed as in Example A.1.2 as follows:

$$C_{11} = (-1)^{1+1} M_{11} = 1 \begin{vmatrix} 1 & 4 \\ 5 & 1 \end{vmatrix} = -19$$

$$C_{12} = (-1)^{1+2} M_{12} = -1 \begin{vmatrix} 2 & 4 \\ 3 & 1 \end{vmatrix} = 10$$

$$C_{13} = (-1)^{1+3} M_{13} = 1 \begin{vmatrix} 2 & 1 \\ 3 & 5 \end{vmatrix} = 7$$

$$C_{21} = (-1)^{2+1} M_{21} = -1 \begin{vmatrix} 1 & 2 \\ 5 & 1 \end{vmatrix} = 9$$

$$C_{22} = (-1)^{2+2} M_{22} = 1 \begin{vmatrix} 1 & 2 \\ 3 & 1 \end{vmatrix} = -5$$

$$C_{23} = (-1)^{2+3} M_{23} = -1 \begin{vmatrix} 1 & 1 \\ 3 & 5 \end{vmatrix} = -2$$

$$C_{31} = (-1)^{3+1} M_{31} = 1 \begin{vmatrix} 1 & 2 \\ 1 & 4 \end{vmatrix} = 2$$

$$C_{32} = (-1)^{3+2} M_{32} = -1 \begin{vmatrix} 1 & 2 \\ 2 & 4 \end{vmatrix} = 0$$

$$C_{33} = (-1)^{3+3} M_{33} = 1 \begin{vmatrix} 1 & 1 \\ 2 & 1 \end{vmatrix} = -1$$

Therefore

$$C = \begin{bmatrix} -19 & 10 & 7 \\ 9 & -5 & -2 \\ 2 & 0 & -1 \end{bmatrix}$$

and

$$adj\, A = C^T = \begin{bmatrix} -19 & 9 & 2 \\ 10 & -5 & 0 \\ 7 & -2 & -1 \end{bmatrix}$$

so that

$$A^{-1} = \frac{1}{det(A)} \, adj \, a$$

$$= \frac{1}{5} \begin{bmatrix} -19 & 9 & 2 \\ 10 & -5 & 0 \\ 7 & -2 & -1 \end{bmatrix}.$$

EXERCISES FOR SECTION A.2

A Exercises

1. A requirement for Cramer's Rule is that $det(A) \neq 0$ for the matrix of coefficients A. What can you say about a system of n equations and n unknowns if $det(A) = 0$?

2. Solve each of the systems of equations given in Exercise 1 of Section 12.1 using Cramer's Rule.

3. Use the cofactor method to find the inverse, if it exists, of each of the matrices in Exercise 2 of Section 12.2.

B Exercise

4. Prove that the equation of a straight line in the xy plane through the points (x_1, y_1) and (x_2, y_2) can be expressed as:

$$\begin{vmatrix} x & y & 1 \\ x_1 & y_1 & 1 \\ x_2 & y_x & 1 \end{vmatrix} = 0.$$

Solutions and Hints to Selected Exercises

Chapter 1

2. (a) $\{2, 1\}$ (b) ϕ (c) $\{i, -i\}$
3. (a) $\{2k + 1 : k \in Z, 2 \le k \le 39\}$ (b) $\{x \in Q : -1 < x < 1\}$
   (c) $\{2n : n \in Z\}$
4. (a) True   (b) False   (c) True   (d) True   (e) False
   (f) True   (g) False   (h) False
5. (a) $\{2, 3\}$ (b) $\{0, 2, 3\}$ (c) $\{0, 2, 3\}$ (d) $\{0, 1, 2, 3, 5, 9\}$
   (e) ϕ (f) $\{0\}$ (g) $\{(1, 8),(5,8),(9,8)\}$ (h) ϕ
   (i) $\{1, 4, 5, 6, 7, 8, 9\}$ (j) $\{0, 2, 3, 4, 6, 8\}$ (k) $\{(0, 2),(0, 3),$
   $(2, 2),(2, 3),(3, 2),(3, 3)\}$ (l) $\{(2, 2),(2, 3),(3, 2),(3, 3)\}$
   (m) $\{(2, 2, 2),(2, 2, 3),(2, 3, 2),(2, 3, 3),(3, 2, 2),(3, 2, 3),(3, 3, 2),$
   $(3, 3, 3)\}$
6. (a) False   (b) False   (c) True   (d) True   (e) True
   (f) True   (g) False   (h) False
7. (a) $\{3, 4, 5, 6, 7, 8, 9\}$ (b) $\{3, 6, 9\}$ (c) $\{0, 1, 2\}$
8. (a) $\{+00, +01, +10, +11, -00, -01, -10, -11\}$ (b) 16 and 512
10. (a) 7000   (b) 15700   (c) 15000   (d) 650   (e) 300
    (f and g) Cannot be determined with the information given.
11. When $A = B$ or when either A or B is empty.
13. If S is of Type SET OF 0..11, FOR I:= 0 TO 11 DO IF (I IN S) THEN WRITE(I)

Chapter 2

1. $9 \times 10 \times 10 \times 26 \times 26 \times 26 = 15,818,400$
2. $C(10, 3)C(25, 4) = 1,518,000$
3. (a) $2^8 = 256$ (b) $C(8, 3) = 56$
   (c) $C(8, 8) + C(8, 6) + C(8, 4) + C(8, 2) + C(8, 0) = 128$
4. (a) $3P(3; 2) = 18$ (b) $2P(3; 2) = 12$
5. (a) $7!$ (b) $C(6, 2)4! = 6!/2 = 360$ (c) $C(4, 2) = 6$
7. $C(10, 7) + C(10, 8) + C(10, 9) + C(10, 10)$
8. $2^3 - 1 = 7$ and 4
9. $C(4, 2) \times C(2, 1) \times C(3, 2) \times C(7, 1) = 252$
10. (a) 12   (b) The set $A \times B$ would represent the possible outcomes produced by flipping a coin and then rolling a die.
11. (a) $C(3, 2) = 3$ (b) $3! = 6$
13. $998^3 = (10000 - 2)^3 = 1 \times 10000^3 - 3 \times 10000^2 \times 2 + 3 \times 10000 \times 2^2 - 1 \times 2^3 = 994,011,992$
14. $P(1000, 3)$

15. $C(385, 5) \times C(380, 5)$

16. (a) $C(8, 2) \times 6! = 20160$ (b) $20160 - 7! = 15120$

17. 3^{10}

18. $3 \times 2 \times 3 = 18$

19. $2^5 - C(5, 4) - C(5, 5) = 27$

20. 5^6

21. $(3)(2)(2) = 12$

23. (a) $C(10, 2) = 45$ (b) $C(10, 3) = 120$

24. $2^{n-1} - 1$ and $2^n - 2$

25. $C(n, 2) \times C(n - 2, 2) \times \cdots \times C(4, 2) \times C(2, 2)$. Simplify this expression.

26. (a)

```
PROGRAM MAXFACTORIAL (OUTPUT)
VAR
  F,N: INTEGER;
BEGIN
  WRITELN('MAXINT = ',MAXINT);
  N := 1;
  F := 1;
  WHILE F< = (MAXINT DIV(N + 1) DO
  BEGIN
    N := N + 1;
    F := F*N
  END;
WRITELN(N,'! = ',F)
END.
```

On our computer, the results obtained were

```
MAXINT = 21847976710655
16! = 2092278988800
```

Sections 3.1 and 3.2

1. (a)   1110   (b)   0010   (c)   00000001   (d)   00010111
    (e)   1110   (f)   0111111111111111

2. (a) $\sim (p \wedge r) \vee s$ (b) $(p \vee q) \wedge (r \vee q)$

Section 3.3

1. $a \Leftrightarrow e, d \Leftrightarrow f, g \Leftrightarrow h$

2. a, c, e

3. b

4. 2^4 (the number of possible truth tables)

5. $\sim (\sim p \wedge \sim q)$

6. (a)   0001101   (b) $p \wedge (\sim q \vee r)$ (c) $p \wedge r$ (d) p (The solutions to the last three parts are not unique.)

8.
```
FUNCTION CONDITIONAL (P,Q : BOOLEAN) : BOOLEAN;
BEGIN
    CONDITIONAL :=(NOT P) OR Q
END.
```

Section 3.5

1. (a)

p	q	$(p \lor q) \land \sim q$	$((p \lor q) \land \sim q) \to p$
0	0	0	1
0	1	0	1
1	0	0	1
1	1	0	1

(b)

p	q	$(p \to q) \land \sim q$	$\sim p$	$((p \to q) \land (\sim q)) \to \sim p$
0	0	1	1	1
0	1	0	1	1
1	0	0	0	1
1	1	0	0	1

2. (a) Direct proof:
    (1) $d \to (a \lor c)$
    (2) d
    (3) $a \lor c$
    (4) $a \to b$
    (5) $\sim a \lor b$
    (6) $c \to b$
    (7) $\sim c \lor b$
    (8) $(\sim a \lor b) \land (\sim c \lor b)$
    (9) $(\sim a \land \sim c) \lor b$
    (10) $\sim(a \lor c) \lor b$
    (11) $(a \lor c)\sim b$
    (12) b #
    Indirect proof:
    (1) b Negated conclusion
    (2) $a \to b$ Premise
    (3) $\sim a$ Contrapositive (1), (2)
    (4) $c \to b$ Premise
    (5) $\sim c$ Contrapositive (1), (4)
    (6) $(\sim a \land \sim c)$ Conjuctive (3), (5)
    (7) $\sim(a \lor c)$ DeMorgan's law (6)
    (8) $d \to (a \lor c)$ Premise
    (9) $\sim d$ Contrapositive (7), (8)
    (10) d Premise
    (11) 0   (9), (10) #
    (b) Direct proof:
    (1) $(p \to q) \land (r \to s)$
    (2) $p \to q$
    (3) $(q \to t) \land (s \to u)$

(4) $q \rightarrow t$
(5) $p \rightarrow t$
(6) $r \rightarrow s$
(7) $s \rightarrow u$
(8) $r \rightarrow u$
(9) $p \rightarrow r$
(10) $p \rightarrow u$
(11) $p \rightarrow (t \wedge u)$ Use $(x \rightarrow y) \wedge (x \rightarrow z) \Leftrightarrow x \rightarrow (y \wedge z)$.
(12) $\sim(t \wedge u) \rightarrow \sim p$
(13) $\sim(t \wedge u)$
(14) $\sim p$ #

Indirect proof:
(1) p
(2) $p \rightarrow q$
(3) q
(4) $q \rightarrow t$
(5) t
(6) $\sim(t \wedge u)$
(7) $\sim t \vee \sim u$
(8) $\sim u$
(9) $s \rightarrow u$
(10) $\sim s$
(11) $r \rightarrow s$
(12) $\sim r$
(13) $p \rightarrow r$
(14) r
(15) 0 #

(c) Direct proof:
(1) $\sim s \vee p$ Premise
(2) s Added premise (conditional conclusion)
(3) $\sim(\sim s)$ Involution (2)
(4) p Disjunctive simplification (1), (3)
(5) $p \rightarrow (q \rightarrow r)$ Premise
(6) $q \rightarrow r$ Detachment (4), (5)
(7) q Premise
(8) r Detachment (6), (7) #

Indirect proof:
(1) $\sim(s \rightarrow r)$ Negated conclusion
(2) $\sim(\sim s \vee r)$ Conditional equivalence (1)
(3) $s \wedge \sim r$ DeMorgan (2)
(4) s Conjuctive simplification (3)
(5) $\sim s \vee p$ Premise
(6) $s \rightarrow p$ Conditional equivalence (5)
(7) p Detachment (4), (6)
(8) $p \rightarrow (q \rightarrow r)$ Premise
(9) $q \rightarrow r$ Detachment (7), (8)
(10) q Premise
(11) r Detachment (9), (10)
(12) $\sim r$ Conjuctive simplification (3)
(13) 0 Conjuction (11), (12) #

(d)   Direct proof:
- (1) $p \rightarrow q$
- (2) $q \rightarrow r$
- (3) $p \rightarrow r$
- (4) $p \vee r$
- (5) $\sim p \vee r$
- (6) $(p \vee r) \wedge (\sim p \vee r)$
- (7) $(p \wedge \sim p) \vee r$
- (8) $0 \vee r$
- (9) r #

Indirect proof:
- (1) $\sim r$ Negated conclusion
- (2) $p \vee r$ Premise
- (3) p (1), (2)
- (4) $p \rightarrow q$ Premise
- (5) q Detachment (3), (4)
- (6) $q \rightarrow r$ Premise
- (7) r Detachment (5), (6)
- (8) 0 (1), (2) #

3.   (a) not valid   (b) valid
4. $p_1 \rightarrow p_k$ and $p_k \rightarrow p_{k+1}$ implies $p_1 \rightarrow p_{k+1}$. It takes two steps to get to $p_1 \rightarrow p_{k+1}$ from $p_1 \rightarrow p_k$. This means it takes $2(100-1)$ steps to get to $p_1 \rightarrow p_{100}$ (subtract 1 because) $p_1 \rightarrow p_2$ is stated as a premise). A final step is needed to apply detachment to imply p_{100}.

Section 3.6

1.   (a) $\{\{1\}, \{3\}, \{1, 3\}, \phi\}$
  (b) $\{\{3\}, \{3, 4\}, \{3, 2\}, \{2, 3, 4\}\}$
  (c) $\{\{1\}, \{1, 2\}, \{1, 3\}, \{1, 4\}, \{1, 2, 3\}, \{1, 2, 4\}, \{1, 3, 4\}, \{1, 2, 3, 4\}\}$
  (d) $\{\{2\}, \{3\}, \{4\}, \{2, 3\}, \{2, 4\}, \{3, 4\}\}$
  (e) $\{A \subseteq U : \#A = 2\}$
2.   (a) $T_p = \{2, 3, 5, 7, 11, 13, 17, 19, 23, 29, 31\}$.
    $T_q = \{1, 3, 9, 27, \ldots\} = \{3^k : k \text{ in } \mathbf{P}\}$.
    $T_r = \{1, 3, 9, 27\}$.
  (b) r implies q.
3.   There are $2^3 = 8$ subsets of U, allowing for the possibility of 2^8 non-equivalent propositions over U.
4.   (a) $\{9, 10, 11, \ldots\}$ (b) ϕ (c) $\{1, 2, 3, 6, 9, 18\}$
  (d) $\{1, 2, 3, 4, 5, 6, 7, 8, 9, 18\}$
5. s is odd and $(s-1)(s-3)(s-5)(s-7) = 0$.
6.   (a) $T_p = \{1, 4, 9, 16, 25, 36, 49, 64, 81\}$.
    $T_q = \{1, 2, 4, 8, 16, 32, \ldots\}$.
  (b) $\{1, 4, 16, 64\}$
7. b and c

Section 3.7

2.   Let $p(n)$ be the proposition $1(1!) + \cdots + n(n!) = (n+1)! - 1$. Assume that for some $n \geq 1$, $p(n)$ is true. Then

$1(1!) + \cdots + n(n!) + (n + 1)((n + 1)!) = ((n + 1)! - 1) + (n + 1)(n + 1)!$
$$= ((n + 1)! + (n + 1)(n + 1)!) - 1$$
$$= (n + 1)!(1 + n + 1) - 1$$
$$= (n + 2)! - 1.$$

Also, $p(1)$ is clearly true since $1(1!) = (1 + 1)! - 1$.

4. Let S_n be the set of strings of zeros and ones of length n (we assume that $\#S_n = 2^n$ is known), E_n = the even strings, and E_n^c the odd strings. The problem is to prove that for $n \geq 1$, $\#E_n = 2^{n-1}$. Clearly $\#E_1 = 1$, and, if, for some $n \geq 1$, $\#E_n = 2^{n-1}$, it follows that $\#E_{n+1} = 2^n$ by the following reasoning:

$$E_{n+1} = \{1s : s \text{ in } E_n^c\} \cup \{0s : s \text{ in } E_n\}$$

Therefore, $\#E_{n+1} = \#E_n^c + \#E_n = \#S_n = 2^n$. #

6. Assume that for $n(n \geq 1)$ persons, $(n - 1)n/2$ handshakes take place. If one more person enters the room, he or she will shake hands with n people.

$$(n - 1)n/2 + n = (n^2 - n + 2n)/2 = n(n + 1)/2$$
$$= ((n + 1) - 1)(n + 1)/2.$$

Also, for $n = 1$, there are no handshakes: $(1 - 1)(1)/2 = 0$. #

9. Hint: $8^{n+1} = (5 + 3)8^n$.

Section 3.8

1. (a) There is a book with a cover that is not blue.   (b) Every mathematics book that is published in the United States has a blue cover.   (c) There exists a mathematics book with a cover that is not blue.   (d) There exists a book that appears in the bibliography of every mathematics book.
    (e) $(\forall x)(B(x) \rightarrow M(x))$ (f) $(\exists x)(M(x) \land \sim U(x))$
    (g) $(\exists x)((\forall y)(\sim R(x, y))$
2. $\sim(\exists x)_Q(x^2 = 3)$
3. $(\forall a)_R((\forall b)_R((\exists x)_R(a + x = b)))$
4. (a) Every subset of U has a cardinality different from its complement. (True)   (b) There is a pair of disjoint subsets of U both having cardinality 5. (False)   (c) $A - B = B^c - A^c$ is a tautology. (True)
6. The equation $4u^2 - 9 = 0$ has a solution in the integers. (False)
7. $(\forall a)_Q(\forall b)_Q(a + b$ is a rational number.)

Section 4.1

2. (a) Assume that $x \in A$ (condition of the conditional conclusion $A \subseteq C$). Since $A \subseteq B$, $x \in B$ by the definition of $\subseteq$. $B \subseteq C$ and $x \in B$ implies that $x \in C$. Therefore, if $x \in A$, then $x \in C$. #
    (b) (Proof that $A - B \subseteq A \cap B^c$) Let x be in $A - B$. Therefore, x is in A, but it is not in B; i.e., $x \in A$ and $x \in B^c \Rightarrow x \in A \cap B^c$. #
    (c) ($\Rightarrow$) Assume that $A \subseteq B$ and $A \subseteq C$. Let $x \in A$. By the two premises, $x \in B$ and $x \in C$. Therefore, by the definition of intersection, $x \in B \cap C$. #
    (d) ($\Rightarrow$) (Indirect) Assume that $A \subseteq B$ and B^c is not a subset of A^c. Therefore,

there exists $x \in B^c$ that does not belong to A^c. $x \notin A^c \Rightarrow x \in A$. Therefore, $x \in A$ and $x \in B$, a contradiction to the assumption that $A \subseteq B$. #

3. (a) If $A = \mathbf{Z}$ and $B = \phi$, $A - B = \mathbf{Z}$, while $B - A = \phi$.
    (b) If $A = \{U\}$ and $B = \{1\}$, $(0, 1) \in A \times B$, but $(0, 1) \notin B \times A$.
    (c) Let $A = \phi$, $B = \{0\}$, and $C = \{1\}$.

Section 4.2

2. (a) $A \cup (B - A) = A \cup (B \cap A^c)$ by Exercise 2b of Section 4.1
    $= (A \cup B) \cap (A \cup A^c)$ by the distributive law
    $= (A \cup B) \cap U$ by the null law
    $= (A \cup B)$ by the identity law #

    (c) Select any element, x, in $A \cap C$, which exists since $A \cap C$ is not empty.

    x in $A \cap C \Rightarrow x$ in A and x in C
    $\Rightarrow x$ in B and x in C
    $\Rightarrow x$ in $B \cap C$
    $\Rightarrow B \cap C \neq \phi$ #

    (e) $A - (B \cup C) = A \cap (B \cup C)^c$
    $= A \cap (B^c \cap C^c)$
    $= (A \cap B^c) \cap (A \cap C^c)$
    $= (A - B) \cap (A - C)$ #

Section 4.3

2. a, c, and d
3. No. By this definition it is possible that an element of A_i might not belong to A.
5. Hint: Partition the set of fractions into blocks, where each block contains fractions that are numerically equivalent. Describe how you would determine whether two fractions belong to the same block. Redefine the rational numbers to be this partition. Each rational number is a set of fractions.
6. $\{A_1, A_2, \ldots, A_n\}$ is a partition of A. Hence by the definition of a partition, the sets of A_i are non-empty and disjoint and

$$A = \bigcup_{i=1}^{n} A_i.$$

By the distributive laws, $B \cap A = B \cap (\bigcup_{i=1}^{n} A_i) = \bigcup_{i=1}^{n} (B \cap A_i)$, and it follows that the proposed partition contains all elements of $A \cap B$. Next it must be shown that the blocks are disjoint. If $i \neq j$, $(A_i \cap B) \cap (A_j \cap B) = (A_i \cap A_j) \cap (B \cap B) = (A_i \cap A_j) \cap B = \phi \cap B = \phi$ since A_i and A_j are disjoint.

Section 4.4

1. (a) $\{1\}$, $\{2, 3, 4, 5\}$, $\{6\}$, $\{7, 8\}$, $\{9, 10\}$ (b) 2^5, as compared with 2^{10}. $\{1, 2\}$ is one of the 992 sets that can't be generated.
2. $\{\lambda, 1\}$, $\{0\}$, $\{10, 11\}$, $\{00, 01\}$

4. (c) $B_1^c = (B_1^c \cap B_2 \cap B_3) \cup (B_1^c \cap B_2 \cap B_3^c) \cup (B_1^c \cap B_2^c \cap B_3)$
$\cup (B_1^c \cap B_2^c \cap B_3^c)$, $B_1 \cap B_2 = (B_1 \cap B_2 \cap B_3) \cup (B_1 \cap B_2 \cap B_3^c)$
6. 2^n

Section 5.3

1. (a) $AB = \begin{bmatrix} -3 & 6 \\ 9 & -13 \end{bmatrix}$ $BA = \begin{bmatrix} 2 & 3 \\ -7 & -18 \end{bmatrix}$ (b) $\begin{bmatrix} 1 & 0 \\ 5 & -2 \end{bmatrix}$

   (c) $\begin{bmatrix} 3 & 0 \\ 15 & -6 \end{bmatrix}$ (d) $\begin{bmatrix} 18 & -15 & 15 \\ -39 & 35 & -35 \end{bmatrix}$ (e) $\begin{bmatrix} -12 & 7 & -7 \\ 21 & -6 & 6 \end{bmatrix}$

   (f) $B + 0 = B$ (g) $\begin{bmatrix} 0 & 0 \\ 0 & 0 \end{bmatrix}$ (h) $\begin{bmatrix} 0 & 0 \\ 0 & 0 \end{bmatrix}$ (i) $\begin{bmatrix} 5 & -5 \\ 10 & 15 \end{bmatrix}$

2. $\begin{bmatrix} 1/2 & 0 \\ 0 & 1/3 \end{bmatrix}$

4. $A^3 = \begin{bmatrix} 1 & 0 & 0 \\ 0 & 8 & 0 \\ 0 & 0 & 27 \end{bmatrix}$ $A^{15} = \begin{bmatrix} 1 & 0 & 0 \\ 0 & 32768 & 0 \\ 0 & 0 & 14348907 \end{bmatrix}$

6. Hint: To save time and avoid errors, use the answer to Part a in doing Parts c and e.

Section 5.4

1. (a) $\begin{bmatrix} -1/5 & 3/5 \\ 2/5 & -1/5 \end{bmatrix}$ (b) $\begin{bmatrix} 1 & 3 \\ 0 & 1 \end{bmatrix}$ (c) No inverse exists.

   (d) $A^{-1} = A$ (e) $\begin{bmatrix} 1/3 & 0 & 0 \\ 0 & 2 & 0 \\ 0 & 0 & -1/5 \end{bmatrix}$

5. (b) $1 = det(I) = det(AA^{-1}) = det(A)det(A^{-1})$. Now solve for $det(A^{-1})$.
8. (a) The basis step is given as a premise. Assume that for some m greater than or equal to 1, $A^m = BD^mB^{-1}$. Then

$$A^{m+1} = (BDB^{-1})(BD^mB^{-1})$$
$$= BD(B^{-1}B)D^mB^{-1}$$
$$= BDID^mB^{-1}$$
$$= BDD^mB^{-1}$$
$$= BD^{m+1}B^{-1} \ \#$$

Section 5.6

3. (a) $A^2 = A$ and $det(A) \neq 0 \Rightarrow A^{-1}$ exists, and if you multiply the equation $A^2 = A$ on both sides by A^{-1}, you obtain $A = I$.
   (b) Counterexample: $A = \begin{bmatrix} 1 & 0 \\ 0 & -1 \end{bmatrix}$

Section 6.1

1. (a) (2, 4), (2, 8) (b) (2, 3)(2, 4)(5, 8) (c) (1, 1)(2, 4)
2. (a) rs = {(1, 3), (1, 5), (3, 5)} (b) sr = {(1, 3)(1, 5)(3, 5)}

3. (a) $r = \{(1, 2)(2, 3)(3, 4)(4, 5)\}$ (b) $r^2 = \{(1, 3), (2, 4), (3, 5)\} = \{(x, y) : y = x + 2, x, y \in A\}$ (c) $r^3 = \{(1, 4), (2, 5)\} = \{(x, y) : y = y + 3, x, y \in A\}$

4. $st = \{(1, 1)\}$, $ts = \mathbf{Z} \times \mathbf{Z}$

5. (a) 27   (b) 3^n

6. Let $(a, c) \in r_1 r_3$. Then $(\exists b)_A$ such that $(a, b) \in r_1$ and $(b, c) \in r_3$. But $r_1 \subseteq r_2$, so $(a, b) \in r_2$. Therefore $(a, c) \in r_2 r_3$. #

Section 6.2

1.

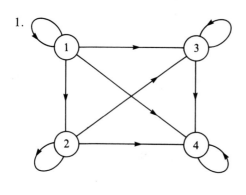

2.

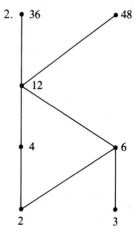

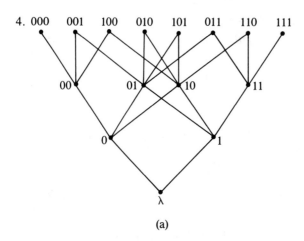

4.

(a)

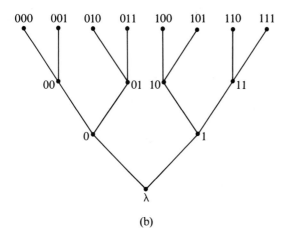

(b)

5. A Hasse diagram cannot be used because not every set is related to itself. Also, $\{a\}$ and $\{b\}$ are related in both directions.

Section 6.3

1.

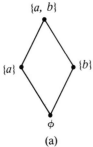

(a)

6

2 • • 3

1

(b)

(c) The graphs are the same if we disregard the names of the vertices.

3. (a) (i) reflexive
            not symmetric
            not antisymmetric
            transitive

        (ii) reflexive
             not symmetric
             antisymmetric
             transitive

        (iii) not reflexive
              symmetric
              not antisym-
              metric

        (iv) not reflexive
             symmetric
             antisymmetric
             transitive

        (v) reflexive
            symmetric
            not antisym-
            metric
            transitive

        (vi) reflexive
             not symmetric
             antisymmetric
             transitive

        (vii) not reflexive
              not symmetric
              not antisymmetric
              not transitive

(b) Graphs ii and vi show partial ordering relations. Graph v is an equivalence relation.

4. (a) It is an equivalence relation but not a partial ordering relation since r is not antisymmetric.

(b) Not transitive; therefore it is neither.

5. (b) $c(0) = \{0\}$, $c(1) = \{1, 2, 3\} = c(2) = c(3)$

(c) $c(0) \cup c(1) = A$ and $c(0) \cap c(1) = \phi$

(d) Let A be any set and let r be an equivalence relation on A. Let a be any element of A. $a \in c(a)$ since r is reflexive, so each element of A is in some equivalence class. Therefore, the union of all equivalence classes gives A. Next we show that any two equivalence classes are either identical or disjoint and we are done. Let $c(a)$ and $c(b)$ be two equivalence classes and assume that $c(a) \cap c(b) \neq \phi$. We want to show that $c(a) = c(b)$. To show that $c(a) \subseteq c(b)$, let $x \in c(a)$. $x \in c(a) \Rightarrow arx$. Also, there exists an element, y, of A that is in the intersection of $c(a)$ and $c(b)$ by our assumption. Therefore,

$$ary \text{ and } bry \Rightarrow ary \text{ and } yrb \ (r \text{ is symmetric})$$
$$\Rightarrow arb \text{ (transitive)}.$$

Next,

$$arx \text{ and } arb \Rightarrow xra \text{ and } arb$$
$$\Rightarrow xrb$$
$$\Rightarrow brx$$
$$\Rightarrow x \in c(b). \text{ Similarly, } c(b) \subseteq c(a). \ \#$$

7. (a) No   (b) Yes   (c) No

9. No, r is not reflexive or transitive.

Section 6.4

1. (a)

$$\begin{array}{c} \\ 1 \\ 2 \\ 3 \\ 4 \end{array} \begin{array}{ccc} 4 & 5 & 6 \\ \left[\begin{array}{ccc} 0 & 0 & 0 \\ 1 & 0 & 0 \\ 0 & 1 & 0 \\ 0 & 0 & 1 \end{array}\right] \end{array} \quad \text{and} \quad \begin{array}{c} \\ 4 \\ 5 \\ 6 \end{array} \begin{array}{ccc} 6 & 7 & 8 \\ \left[\begin{array}{ccc} 0 & 0 & 0 \\ 1 & 0 & 0 \\ 0 & 1 & 0 \end{array}\right] \end{array}$$

(b)

$$\begin{array}{c} \\ 1 \\ 2 \\ 3 \\ 4 \end{array} \begin{array}{ccc} 6 & 7 & 8 \\ \left[\begin{array}{ccc} 0 & 0 & 0 \\ 0 & 0 & 0 \\ 1 & 0 & 0 \\ 0 & 1 & 0 \end{array}\right] \end{array}$$

3. $R : xry$ if and only if $|x - y| = 1$.

$S : xsy$ if and only if x is less than y.

5. The diagonal entries of the matrix for such a relation must be 1. When the three entries above the diagonal are determined, the entries below are also determined. Therefore, the answer is 2^3.

7. (b)

$$PQ = \begin{matrix} & \begin{matrix} 1 & 2 & 3 & 4 \end{matrix} \\ \begin{matrix} 1 \\ 2 \\ 3 \\ 4 \end{matrix} & \begin{bmatrix} 0 & 1 & 0 & 1 \\ 1 & 0 & 1 & 0 \\ 0 & 1 & 0 & 1 \\ 1 & 0 & 1 & 1 \end{bmatrix} \end{matrix}$$

$$P^2 = \begin{matrix} & \begin{matrix} 1 & 2 & 3 & 4 \end{matrix} \\ \begin{matrix} 1 \\ 2 \\ 3 \\ 4 \end{matrix} & \begin{bmatrix} 1 & 0 & 1 & 0 \\ 0 & 1 & 0 & 1 \\ 1 & 0 & 1 & 0 \\ 0 & 1 & 0 & 1 \end{bmatrix} \end{matrix} = Q^2$$

8. (a) If (a, b) belongs to r^2, then there exists c in A such that arc and crb. By the transitive property, we conclude that arb; i.e. $(a, b) \in r$. # (b) The relation "less than" on the integers is one example.

Section 6.5

4. (a) $r^+ = \{(1,2), (1, 5), (1, 3), (1, 4), (5, 3), (5, 4), (5, 2), (5, 5), (3, 4), (3, 2), (3, 5), (3, 3), (4, 5), (4, 3), (4, 4), (4, 2)\}$.

5. (a) **Definition:** Reflexive Closure. *Let r be a relation on A. A reflexive closure of r is the smallest reflexive relation that contains r.*

Theorem: *The reflexive closure of r is the union of r with $\{(x, x) : x \in A\}$.*

6. (a) $aS^+ b$ if and only if a is less than b. (b) $aR^+ b$ if and only if $a - b$ is even.

Section 7.1

1. (a) Yes (b) Yes (c) No (d) No (e) Yes
2. (a) $C_S = \{(a, 1), (b, 1), (c, 0)\}$
   (b) $C_S = \{a, 1), (b, 0), (c, 1), (d, 0), (e, 1)\}$.
   (c) $C_\phi = \{(a, 0), (b, 0), (c, 0)\}$ and $C_A = \{(a, 1), (b, 1), (c, 1)\}$.
4. (a) $\{4x + 1 : x \in \mathbf{Z}\}$ (b) $\mathbf{N} = \{0, 1, 2, 3, \ldots\}$ (c) $\mathbf{Z}$
5. $\#B^{\#A}$
6. (b) $\{0\}, \{1, -1\}, \{2, -2\}, \{3, -3\}, \ldots$

Section 7.2

1. The only one-to-one function and the only onto function is f.
3. (a) onto but not one to one $(f_1(0) = f_1(1))$ (b) one to one and onto
   (c) one to one but not onto (d) onto but not one to one (e) one to one but not onto (f) one to one but not onto
4. (b) f is not a surjection since ϕ is not in the range of f.
7. (a) $f(n) = n$, for example. (b) $f(n) = 1$, for example.
   (c) None exists. (d) None exists.
10. (a) bijection (b) neither (c) bijection
12. (a) Use $s : \mathbf{N} \rightarrow \mathbf{P}$ defined by $s(x) = x + 1$.
    (b) Use the function $f : \mathbf{N} \rightarrow \mathbf{Z}$ defined by $f(x) = x/2$ if x is even and $f(x) = -(x + 1)/2$ if x is odd.

(c) Method 1: List the string of length zero, then all strings of length one, then all strings of length two, . . . Clearly, every string will appear in this list. Method 2: Define $h : \mathbf{P} \rightarrow$ strings as follows: If x is a positive integer with binary representation $b_n b_{n-1} \cdots b_1 b_0$, with $b_n = 1$, let $h(x)$ equal $b_{n-1} \cdots b_1 b_0$.

Section 7.3

2. $f^2(1) = 1, f^2(2) = 2, f^2(3) = 3$ $(f^2 = i_A)$
    $f^3(1) = 2, f^3(2) = 1, f^3(3) = 3$ $(f^3 = f)$
    $f^4 = f^2$ and $f^{-1} = f$

4. (a) $usd(n) = (n-1)^2 + 1 = n^2 - 2n + 2$ (b) $sud(n) = n^2$
    (c) $dsu(n) = (n+1)^2 - 1 = n^2 + 2n$

6. (a) $g^{-1}(4) = \{2, -2\}, g^{-1}(0) = \{0\}, g^{-1}(-1) = \phi$
    (b) $\{x \in \mathbf{R} : 0 < x \leq 1\}$

8. (a) f and g (b) f and h (c) $fg(k) = 2k + 1, \ gf(k) = 2k + 2,$
    $gh(k) = $ the least even integer greater than or equal to k, $hg(k) = [k]$,
    $h^2(k) = [k/4]$

9. **Theorem:** *If $f : A \rightarrow B$ and f has an inverse, then that inverse is unique.*

    Proof: Suppose that g and h are both inverses of f. Let b be an element of B.

    $$g(b) = (g(i_B))(b)$$
    $$= (g(fh))(b)$$
    $$= ((gf)h)(b)$$
    $$= (gf)(h(b))$$
    $$= (i_A)(h(b))$$
    $$= h(b) \Rightarrow g = h \ \#$$

Section 8.1

4. (a) Execution 1: BSRCH(1, 9): MID = 5
        Execution 2: BSRCH(1, 4): MID = 2
        Execution 3: BSRCH(3, 4): MID = 3 (Desired record found at this point.)
    (b) Execution 1: BSRCH(1, 9): MID = 5
        Execution 2: BSRCH(1, 4): MID = 2
        Execution 3: BSRCH(3, 4): MID = 3
        Execution 4: BSRCH(4, 4): MID = 4
        Execution 5: BSRCH(4, 3) (First parameter greater than second parameter $\Rightarrow$ unsuccessful search.)

5. The basis is not reached in a finite number of steps if you try to compute $f(x)$ for x not zero.

Section 8.2

2. (a) $Q(k) - Q(k-1) = 2,$
    (b) $A(k) - 2A(k-1) + A(k-2) = 2$

3. Imagine drawing line k in one of the infinite regions that it passes through. That

infinite region is divided into two infinite regions by line k. As line k is drawn through every one of the $k - 1$ previous lines, you enter another region that line k divides. Therefore, the number of regions is increased by k.

5. For n greater than zero, $M(n) = M(n - 1) + 1$, and $M(0) = 0$.

Section 8.3

1. $S(k) = 2 + 9^k$
2. $S(k) = (2/3) 3^k + (1/3)6^k$
3. $S(k) = 6(1/4)^k$
4. $S(k) = (2 + k)10^k$
5. $S(k) = k^2 - 10k + 25$
6. $S(k) = 11(-2)^k + 4 \cdot 3^k + 5$
7. $S(k) = (3 + k)5^k$
8. $S(k) = -2 \cdot 3^k + 2 \cdot 2^k + 1 = -2 \cdot 3^k + 2^{k+1} + 1$
9. $S(k) = 12 + 3k + (k^2 + 7k + 22)2^{k-1}$
10. $S(k) = a(1 - r^k)/(1 - r)$
11. $P(k) = 4(-3)^k + 2^k - 5^{k+1}$
12. $P(k) = k^2/2 + k/2 + 1$
14. (a) $S(n) = n(n + 1)/2$ (b) $Q(n) = n(2n + 1)(n + 1)/6$
    (c) $P(n) = 2 - (1/2)^n$ (d) $T(n) = [n(n + 1)/2]^2$
15. (a) $D(n) = 2D(n - 1) + 1$ for n greater than or equal to 2 and $D(1) = 0$.   (b) $D(n) = 2^{n-1} - 1$

Section 8.4

1. (a) $S(n) = 1/n!$ (c) $U(k) = 1/k$
2. (a) $T(n) = 3 (\lfloor \log_2 n \rfloor + 1)$ (c) $V(n) = \lfloor \log_8 n \rfloor + 1$

Section 8.5

1. (a) $3/(1 - 2z) + 2/(1 + 2z), 3 \cdot 2^k + 2(-2)^k$
   (b) $-1/(1 - 5z) + 7/(1 - 6z), 7 \cdot 6^k - 5^k$
   (c) $10/(1 - z) + 12/(2 - z), 10 + 6(1/2)^k$
3. $G(F; z) = 1/(1 - z - z^2)$
4. Coefficients of z^0 through z^5 in $(1 + 5z)(2 + 4z)(3 + 3z)(4 + 2z) (5 + z)$
5. Coefficient of z^{40} in $(z + z^4 + z^9 + z^{16} + z^{25} + z^{36})^{10}$

Section 9.1

2.

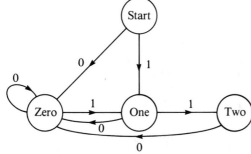

4.

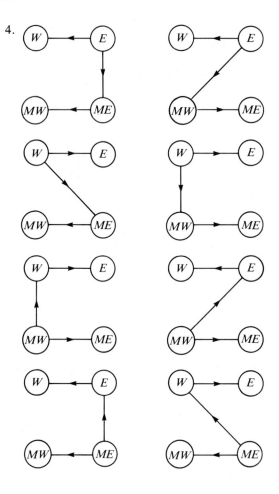

5.   No, the maximum number of edges would be $(7)(8)/2 = 28$.
6. a and f, b and i, c and g, d and e, h and i
7.   (a) $(n - 1)n/2$ (b) $n - 1$

Section 9.2

1.   (a)   The appropriate data structure would probably be an edge list since the graph is sparse.

```
PTR = ↑Route;
Route =     RECORD
                destination: city;
                carrier: airline;
                miles: Integer;
                next: ptr
            END
        Airport: ARRAY[Index] OF city;
```

2. Edge = ARRAY[1 , ,Max, 1 , ,Max] OF (none, red,
    blue, yellow, green)

Section 9.3

1.

k	1	2	3	4	5	6	
V[k].found	T	T	T	F	F	T	
V[k].from	2	5	6	*	*	5	
Depth Set	2	1	2	*	*	1	(* = undefined)

3. If the number of vertices is n, there can be $(n - 1)(n - 2)/2$ vertices with one vertex not connected to any others. One more edge and connectivity is assured.

4.

k	1	2	3	4	5	6	7	8	9
V[k].name	a	b	c	d	e	f	g	h	i
V[k].found	T	T	T	T	T	T	T	T	T
V[k].from	4	1	2	3	4	2	3	4	7
Depth set	4	1	2	3	4	2	3	4	4

(Elements in depth sets were put into alphabetical order.)

Section 9.4

2. Only c
5. Any bridge between two land masses will be sufficient.
7. Theorem: Let $G = (V, E)$ be a directed graph. G has a Eulerian circuit if (a) G is connected and (b) $indeg(v) = outdeg(v)$ for all v in V. There exists a Eulerian path from v_1 to v_2 if (a) G is connected and (b) $indeg(v_1) = outdeg(v_1) + 1$, $indeg(v_2) = outdeg(v_2) - 1$, and for all other vertices in V the indegree and outdegree are equal.
9. A round-robin tournament graph is rarely Eulerian. It will be Eulerian if it has an odd number of vertices and each vertex (team) wins exactly as many times as it loses. Every round-robin tournament graph has a Hamiltonian circuit. This can be proven by induction on the number of vertices.

Section 9.5

1. The circuit would be Boston, Providence, Hartford, Concord, Montpelier, Augusta, Boston. It *does* matter where you start. If you start in Concord, for example, your mileage will be higher.
2. $n = 3$: No, the closest neighbour circuit is always optimal. $n = 4$: No, $C_{cn}/C_{opt} \leq 1.25$.
3. (a)   Optimal cost = $2\sqrt{2}$.
        Phase 1 cost = $2.4\sqrt{2}$.
        Phase 2 cost = $2.6\sqrt{2}$.
    (b)   Optimal cost = 2.60.
        Phase 1 cost = 3.00.
        Phase 2 cost = $2\sqrt{2}$.
5. (a)

i	1	2	3	4	5
$g(e_i)$	20	30	10	30	20
$h(e_i)$	21	29	9	30	20

    g and h are both maximal.

(b)   If a is any real number between 0 and 1, let f_a be defined by $f_a(e) = af_2(e) + (1 - a)g(e)$. The set of all such functions is the set of maximal flows. For example, $f_{0.2} = h$.

6. (a)   Value of maximal flow = 31.
   (b)   Value of maximal flow = 14.
   (c)   Value of maximal flow = 14.
       One way of obtaining this flow is:

Step	Flow Augmenting Path	Flow Added
1	Source, A, Sink	2
2	Source, C, B, Sink	3
3	Source, E, D, Sink	4
4	Source, A, B, Sink	1
5	Source, C, D, Sink	2
6	Source, A, B, C, D, Sink	2

7. To locate the closest neighbor among a list of k other points on the unit square requires a time proportional to k. Therefore the time required for the closest neighbor algorithm with n points is proportional to $(n - 1) + (n + 2) + \cdots + 2 + 1$, which is proportional to n^2. Since the strip algorithm takes a time proportional to $n(\log n)$, it is much faster for large values of n.

Section 9.6

3. Theorem 9.6.2 can be applied to infer that if $n \geq 5$, then a K_n is non-planar. A K_4 is the largest complete planar graph.
4. No
5. (a) 3   (b) 3   (c) 3   (d) 3   (e) 2   (f) 4
8. n
10. Since $\#E + E^c = n(n - 1)/2$, either E or E^c has $n(n - 1)/4$ elements. Assume that it is E that is largest. Since $n(n - 1)/4$ is greater than $3n - 6$ for n greater than or equal to 11, G would be non-planar. Of course, if E' is larger, then G' would be non-planar by the same reasoning.
12. Draw a graph with one vertex for each edge. If two edges in the original graph meet at the same vertex, then draw an edge connecting the corresponding vertices in the new graph.

Section 10.1

1. The numbers of trees should be: (a) 1, (b) 3, and (c) 8. The trees that connect v_c are

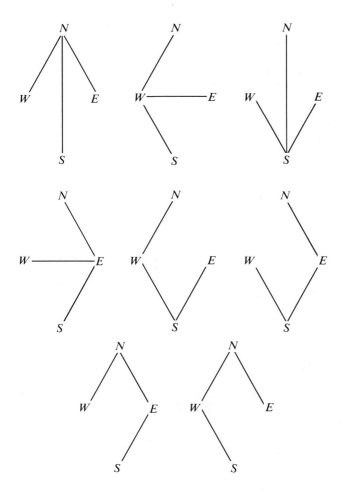

2. Suppose that $T = (V, E)$ is a non-planar tree. As we draw T in a plane, there must be an edge $e = \{v_1, v_2\}$ which cannot be drawn. Therefore, either v_1 or v_2 must be inside a finite region, which is bounded by a circuit, contradicting the definition of a tree. #

3. Hint: Use induction on $\#E$.

Section 10.2

1. It might not be most economical with respect to Objective 1. You should be able to find an example to illustrate this claim. The new system can always be

made most economical with respect to Objective 2 if the old system was designed with that objective in mind.

3. The edge $\{1, 2\}$ is not a minimal bridge between $L = \{1, 4\}$ and $R = \{2, 3\}$, but it is part of the minimal spanning tree for this graph:

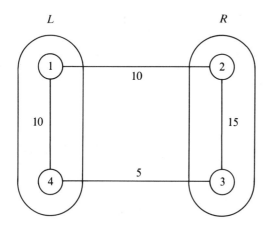

4. (a) Edges in one solution are: $\{0, 1\}$, $\{0, 2\}$, $\{0, 4\}$, $\{1, 2\}$, and $\{4, 5\}$. This solution is not unique.

5. (a) Edges in one solution are: $\{8, 7\}$, $\{8, 9\}$, $\{8, 13\}$, $\{7, 6\}$, $\{9, 4\}$, $\{13, 12\}$, $\{13, 14\}$, $\{6, 11\}$, $\{6, 1\}$, $\{1, 2\}$, $\{4, 3\}$, $\{4, 5\}$, $\{14, 15\}$, and $\{5, 10\}$. This solution is not unique.

Section 10.3

1. Locate any simple path of length d and locate the vertex in position $\lfloor d/2 \rfloor$ on the path. The tree rooted at that vertex will have minimum depth.

2.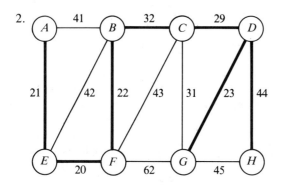

Section 10.4

1.

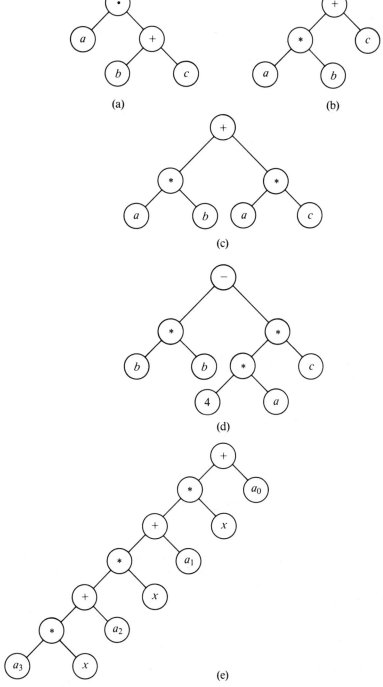

(a)

(b)

(c)

(d)

(e)

3. ($*$ = multiplication)

	Preorder	Inorder	Postorder
(a)	$* a + bc$	$a * b + c$	$abc + *$
(b)	$+ * abc$	$a * b + c$	$ab * c+$
(c)	$+ * ab * ac$	$a * b + a * c$	$ab * ac * +$

5.

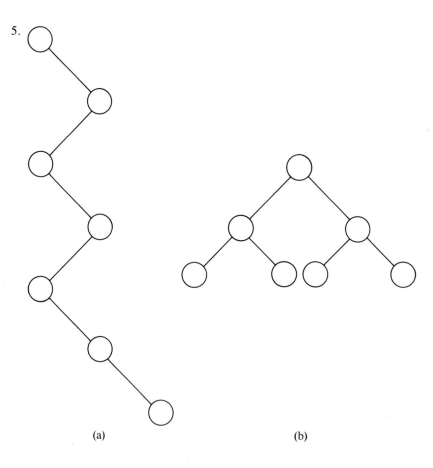

(a)   (b)

7. Hint: Use induction on the number of internal vertices. In the induction step, apply the induction hypothesis to the left and right subtrees.

Section 11.1

1. (a) Commutative, associative (Zero is the identity for addition, but it is not a positive integer.)   (b) Commutative, associative, identity(1)
(c) Commutative,   associative,   identity(1),   indempotent   (d) Commutative, associative, idempotent   (e) None.   Note:

$$2@(3@3) = 512$$
$$(2@3)@2 = 64$$
and while $a@1 = a$, $1@a = 1$.

2. Multiplication over addition
    @ over M
    @ over M
    Addition over M
    Multiplication over M
    Addition over M
    Multiplication over M
3. a, b in $A \cap B \Rightarrow a$, b in A (definition of intersection)
    $\qquad\qquad\qquad \Rightarrow a * b$ in A (closure of A)
    Similarly, a, b in $A \cap B \Rightarrow a * b$ in B. Therefore, $a * b$ in $A \cap B$.
       The set of positive integers is closed under addition and so is the set of
    negative integers, but $1 + -1 = 0$. Therefore, their union, the non-zero
    integers, is not closed under addition.

Section 11.2

2. Every group is a monoid.
3. b, d, e, and f
5. (a) $\begin{bmatrix} 1 & 0 \\ 0 & 1 \end{bmatrix}$, $\begin{bmatrix} 0 & 1 \\ 1 & 0 \end{bmatrix}$, abelian
    (b) Non-abelian
    (c) $4! = 24$, $n!$
7. e is the identity $a * b = c$, $a * c = b$, $b * c = a$, and $[V; *]$ is abelian. (This
    group is commonly called the Klein-4 group.)

Section 11.3

1. (a) f is injective: $f(x) = f(y) \Rightarrow a * x = a * y$
        $\qquad\qquad\qquad\qquad\qquad \Rightarrow x = y$ (left cancellation).
        f is surjective: $f(x) = b$ has the solution $a^{-1} * b$.
    (b) Functions of the form $f(x) = a + x$, where a is any integer are bijections
        on $\mathbf{Z}$.
2. Theorem 8.3.2 (Rephrased): *If G is a group with identity e and $a \in G$, then
    there is exactly one element of G, call it b, such that $a * b = b * a = e$.*

   Proof: That one inverse of a exists is true since G is a group. If c is a second
inverse of a, with $c \neq b$, we will reach a contradiction; i.e., that $c = b$. The
contradiction will complete the proof: $b = b * e = b * (a * c) = (b * a) * c =
e * c = c$. #
6. (a) $[\mathbf{R}^+; \cdot]$ or $[\mathbf{R}^*; \cdot]$, Theorem 11.3.6   (b) $[\mathbf{Z}; +]$, Theorem 11.3.3
    (c) $[\mathbf{M}_{3 \times 3}(\mathbf{R})$; matrix addition$]$, Theorem 11.3.5   (d) $[\mathscr{P}(\mathbf{N})$; sym-
    metric difference$]$, Theorem 11.3.1

Section 11.4

1. (a) 2   (b) 5   (c) 0   (d) 0   (e) 2   (f) 2   (g) 1
   (h) 3
2. (a) 6, 4, 1   (b) 34, 25, 10
3. (a) 1   (b) 1   (c) $m(4) = r(4)$, where $m = 11q + r$, $0 \le r < 11$
4. Hint: Prove by induction on m that you can divide any positive integer into m. That is, let $p(m)$ be "For all n greater than zero, there exist unique integers q and r such that. . . ." In the induction step, divide n into $m - n$.

Section 11.5

1. a and c
2. (a) $20\mathbf{Z}$ (b) $12\mathbf{Z}$ (c) 0
3. $\{I, R_1, R_2\}$, $\{I, F_1\}$, $\{I, F_2\}$, and $\{I, F_3\}$ are all the proper subgroups of R_3.
5. (a) (1) = (5) = $\mathbf{Z}_6$
        (2) = (4) = $\{2, 4, 0\}$
        (3) = $\{3, 0\}$
        (0) = $\{0\}$

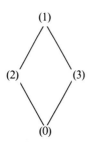

Section 11.6

1. The only two proper subgroups are $\{0, 0), (1, 0)\}$ and $\{(0, 0), (0, 1), (0, 2)\}$.
3. (a) (i) $a + b = (1, 0)$ or $(0, 1)$.
        (ii) $a + b = (1, 1)$.
   (b) (i) $a + b = (1, 0, 0), (0, 1, 0),$ or $(0, 0, 1)$.
        (ii) $a + b = (1, 1, 1)$.
   (c) (i) $a + b$ has exactly one 1.
        (ii) $a + b = $ has all 1's.

5. (a) No, 0 is not an element of $\mathbf{Z} \times \mathbf{Z}$.   (b) Yes   (c) No, $(0, 0)$ is not an element of this set.   (d) No, the set is not closed: $(1, 1) + (2, 4) = (3, 5)$ and $(3, 5)$ is not in the set.   (e) Yes

Section 11.7

1. (a) Yes, $f(n, x) = (x, n)$ for all $(n, x) \in \mathbf{Z} \times \mathbf{R}$.   (b) No, $\mathbf{Z}_2 \times \mathbf{Z}$ has a finite proper subgroup while $\mathbf{Z} \times \mathbf{Z}$ does not.   (c) No

   (d) Yes   (e) No   (f) Yes, $f(a_1, a_2, a_3, a_4) = \begin{bmatrix} a_1 & a_2 \\ a_3 & a_4 \end{bmatrix}$.

   (g) Yes, $f(a_1, a_2) = (a_1, 10^{a_2})$.   (h) Yes   (i) Yes, $f(k) = k(1, 1)$.

4. (a) $f(k) = i^k$ is an isomorphism from Z_4 into G.
5. Z_8, $Z_2 \times Z_4$, and Z_2^3. One other is the fourth dihedral group, introduced in Section 15.3
8. (a) $x \rightarrow (sgn\,(x), log|x|)$ is an isomorphism, where $sgn\,(x) = 0$ if x is positive and $sgn\,(x) = 1$ if x is negative.

Section 11.8

3. Associativity. We would expect that it is always true, since the operation in a group is always associative. However, floating point arithmetic is *not* associative.

Section 12.1

1. (a) $x_1 = 4/3$, $x_2 = 1/3$ (b) $x_1 = -5$, $x_2 = 14/5$, $x_3 = 8/5$
    (c) $x_1 = -3.8$, $x_2 = 2$, $x_3 = 1.4$ (d) $x_1 = 4$, $x_2 = 5$, $x_3 = -28$
4. (a) No solution
    (b) $\{(-3 - 0.5x_3, 11 - 4x_3, x_3) : x_3 \in \mathbf{R}\}$
    (c) $\{(6.25 - 2.5x_3, -0.75 + 0.5x_3, x_3) : x_3 \in \mathbf{R}\}$
    (d) $\{(7 - 2x_4, -4 + 1.4x_4, 2 - 0.2x_4, x_4) : x_4 \in \mathbf{R}\}$
    (e) $\{(0.5 - 1.5x_3, 2.5 + 0.5x_3 - x_4, x_3, x_4) : x_3, x_4 \in \mathbf{R}\}$

Section 12.2

2. (a) $\begin{bmatrix} 3/5 & -2/5 \\ 1/5 & 1/5 \end{bmatrix}$

    (b) $\begin{bmatrix} -8 & -4 & 1 \\ 7 & 3 & -1 \\ -5 & -2 & 1 \end{bmatrix}$

    (c) $\begin{bmatrix} 1/4 & -5/4 & 3/4 \\ 1/2 & -3/2 & 1/2 \\ -2 & 9 & -4 \end{bmatrix}$

    (d) $(1/15) \begin{bmatrix} 8 & 15 & -49 & -9 \\ 4 & 0 & -17 & 3 \\ -1 & 0 & 8 & 3 \\ 1 & 0 & 7 & -3 \end{bmatrix}$

    (e) A^{-1} does not exist.
    (f) A^{-1} does not exist.
3. (b) $(D^{-1})_{ii} = (D_{ii})^{-1}$ and $(D^{-1})_{ij} = 0$ for $i \neq j$.

Section 12.3

3. (b) Yes
6. (a) $y = 5x_1 + 6x_2$ (b) $y = 1x_1 + 0x_2$ (c) $y = 5x_1 + (-6)x_2$
7. If the matrices are named B, A_1, A_2, A_3, and A_4, then
    $B = (8/3)A_1 + (5/3)A_2 + (-9)A_3 + (23/3)A_4$.
8. $(3)1 + (0)x + (-4)x^2 + (1)x^3$
9. (a) If $x_1 = (1, 0)$, $x_2 = (0, 1)$, and $y = (b_1, b_2)$, then $y = (b_1)x_1 + (b_2)x_2$.
    If $x_1 = (2, 1)$, $x_2 = (1, 1)$, and $y = (b_1, b_2)$, then $y = (b_1 - b_2)x_1 + (2b_2 - b_1)x_2$.

If $x_1 = (3, 2)$, $x_2 = (2, 1)$, and $y = (b_1, b_2)$, then $y = (2b_2 - b_1)x_1 + (2b_2 - 3b_1)x_2$.

(b) If $y = (b_1, b_2)$ is any vector in $\mathbf{R}^2$, then $y = (4b_2 - 3b_1)x_1 + (b_2 - b_1)x_2 + (0)x_3$.

(c) One solution is to add any vector(s) to x_1, x_2, and x_3 of Part b.

(d) $2, n$

(e) If the matrices are A_1, A_2, A_3, and A_4, then

$$\begin{bmatrix} x & y \\ z & w \end{bmatrix} = xA_1 + yA_2 + zA_3 + wA_4.$$

(f) $a_0 + a_1x + a_2x^2 + a_3x^3 = a_0(1) + a_1(x) + a_2(x^2) + a_3(x^3)$.

13. (d) They are isomorphic. Once you have completed Part a of this exercise, the following translation rules will give you the answer to Parts b and c.

$$(a, b, c, d) \leftrightarrow \begin{bmatrix} a & b \\ c & d \end{bmatrix} \leftrightarrow a + bx + cx^2 + dx^3$$

Section 12.4

3. (c) You should obtain $\begin{bmatrix} 4 & 0 \\ 0 & 1 \end{bmatrix}$.

4. (a) If $P = \begin{bmatrix} 2 & 1 \\ 3 & -1 \end{bmatrix}$, then $P^{-1}AP = \begin{bmatrix} 4 & 0 \\ 0 & -1 \end{bmatrix}$.

(b) If $P = \begin{bmatrix} 1 & 1 \\ 7 & 1 \end{bmatrix}$, then $P^{-1}AP = \begin{bmatrix} 5 & 0 \\ 0 & -1 \end{bmatrix}$.

(c) If $P = \begin{bmatrix} 1 & 0 \\ 0 & 1 \end{bmatrix}$, then $P^{-1}AP = \begin{bmatrix} 3 & 0 \\ 0 & 4 \end{bmatrix}$.

(d) Hint: The two eigenvalues, $x = (1 + \sqrt{5})/2$ and $y = (1 - \sqrt{5})/2$, satisfy the equations $x + y = 1$ and $xy = -1$.

If $P = \begin{bmatrix} y & x \\ -1 & -1 \end{bmatrix}$, then $P^{-1}AP = \begin{bmatrix} x & 0 \\ 0 & y \end{bmatrix}$.

(e) If $P = \begin{bmatrix} 1 & 1 \\ 4 & -2 \end{bmatrix}$, then $P^{-1}AP = \begin{bmatrix} 0 & 0 \\ 0 & 4 \end{bmatrix}$.

(f) This matrix has one eigenvalue, 1, and the dimension of the eigenspace of 1 is one; therefore, the matrix is not diagonalizable.

6. (a) If $P = \begin{bmatrix} 1 & -1 & 1 \\ -1 & 4 & 2 \\ -1 & 1 & 1 \end{bmatrix}$, then $P^{-1}AP = \begin{bmatrix} -2 & 0 & 0 \\ 0 & 1 & 0 \\ 0 & 0 & 3 \end{bmatrix}$.

(b) Not diagonalizable. Five is a double root of the characteristic equation, but has an eigenspace with dimension only 1.

(c) If $P = \begin{bmatrix} 1 & -1 & -2 \\ -1 & 1 & 0 \\ 1 & 0 & 1 \end{bmatrix}$, then $P^{-1}AP = \begin{bmatrix} 4 & 0 & 0 \\ 0 & -2 & 0 \\ 0 & 0 & -2 \end{bmatrix}$.

(d) If $P = \begin{bmatrix} -1 & 1 & 1 \\ 0 & -2 & 1 \\ 1 & 1 & 1 \end{bmatrix}$, then $P^{-1}AP = \begin{bmatrix} 1 & 0 & 0 \\ 0 & -1 & 0 \\ 0 & 0 & 2 \end{bmatrix}$.

Section 12.5

2. Hint to part a: $X_k = A^{k-1}X_0$, where for k greater than zero,

$$X_k = \begin{bmatrix} T_k \\ T_{k-1} \end{bmatrix}, \; X_0 = \begin{bmatrix} T_1 \\ T_0 \end{bmatrix}, \text{ and } A = \begin{bmatrix} 7 & -10 \\ 1 & 0 \end{bmatrix}.$$

3. Hint: $Y_k = A^{k-1}Y_0$, where for k greater than zero,

$$Y_k = \begin{bmatrix} S_k \\ 1 \end{bmatrix}, \; Y_0 = \begin{bmatrix} S_1 \\ 1 \end{bmatrix}, \text{ and } A = \begin{bmatrix} 5 & 4 \\ 0 & 1 \end{bmatrix}.$$

4. Paths of length 6 are given by

$$A^6 = P \begin{bmatrix} 1 & 0 & 0 \\ 0 & 2^6 & 0 \\ 0 & 0 & (-1)^6 \end{bmatrix} P^{-1},$$

where P and P^{-1} are as stated in Example 12.5.2. $(A^6)_{13} = 21$, so there are 21 paths of length 6 from node 1 to node 3. $(A^6)_{22} = 22$, so there are 22 walks of length 6 from node 2 to node 2.

Section 13.1

1. (a) 1,5   (b) 5   (c) 30   (d) 30   (e) See Figure 13.4.1 with
    $0 = 1, a_1 = 2, a_2 = 3, a_3 = 5, b_1 = 6, b_2 = 10, b_3 = 15,$ and $1 = 30.$
2. $D_8 = \{1, 2, 4, 8\}, D_{50} = \{1, 2, 5, 10, 25, 50\}, D_{1001} = \{1, 7, 11, 13, 77, 91, 143, 1001\}$
5. Solution of Hasse diagram b

    (a)

lub	a_1	a_2	a_3	a_4	a_5
a_1	a_1	a_2	a_3	a_4	a_5
a_2	a_2	a_2	a_4	a_4	a_5
a_3	a_3	a_4	a_3	a_4	a_5
a_4	a_4	a_4	a_4	a_4	a_5
a_1	a_5	a_5	a_5	a_5	a_5

glb	a_1	a_2	a_3	a_4	a_5
a_1	a_1	a_1	a_1	a_1	a_1
a_2	a_1	a_2	a_1	a_2	a_2
a_3	a_1	a_1	a_3	a_3	a_3
a_4	a_1	a_2	a_3	a_4	a_4
a_1	a_1	a_2	a_4	a_4	a_5

(b) a_1 is the least element and a_5 is the greatest element.

Partial solution for Hasse diagram f:

(a) $lub(a_2, a_3)$ and $lub(a_4, a_5)$ do not exist.

(b) No greatest element exists, but a_1 is the least element.

6. If 0 and $0'$ are distinct least elements, then

$$\left.\begin{array}{l} 0 \leq 0' \text{ since 0 is least} \\ 0' \leq 0 \text{ since } 0' \text{ is least} \end{array}\right\} \Rightarrow 0 = 0' \text{ by antisymmetry, a contradiction.} \quad \#$$

Section 13.2

2. All are ordering diagrams of lattices *except* e, f, and g. The lattices of figures c and d are nondistributive.
3. See Table 13.3.1 for the statements of these laws. Most of the proofs follow from the definition of *gcd* and *lcm*.

Section 13.3

1.

B	Complement of B
ϕ	A
$\{a\}$	$\{b, c\}$
$\{b\}$	$\{a, c\}$
$\{c\}$	$\{a, b\}$
$\{a, b\}$	$\{c\}$
$\{a, c\}$	$\{b\}$
$\{b, c\}$	$\{a\}$
A	ϕ

This lattice is a Boolean algebra since it is a distributive complemented lattice.

2. (a) $\bar{1} = 6, \bar{6} = 1, \bar{3} = 2, \bar{2} = 3.$
3. a and g.
4. (a) The least element is ϕ, the greatest element is A, the distributive law holds for all sets, and the complement of each set C in B is $A-C$.
   (b) The operation tables are simply the tables for intersection, union, and set complementation.
5. (a) $S^* : (a \vee b) = a$ if $a \geq b$.
   (b) $S : A \cap B = A$ if $A \subseteq B$
       $S^* : A \cup B = A$ if $A \supseteq B$
   (c) Yes
   (d) $S : (p \wedge q) \Leftrightarrow p$ if $p \Rightarrow q$
       $S^* : (p \vee q) \Leftrightarrow p$ if $p \Leftarrow q$
   (e) Yes

7. **Definition:** Boolean Algebra Isomorphism. $[B; \wedge, \vee, -]$ *is isomorphic to* $[B'; \wedge, \vee, \sim]$ *if and only if there exists a translating function* $T : B \to B'$ *such that:*

   (a) *T is a bijection.*
   (b) *$T(a \wedge b) = T(a) \wedge T(b)$ for all a, b in B.*
   (c) *$T(a \vee b) = T(a) \vee T(b)$ for all a, b in B.*
   (d) *$T(\bar{a}) = \widetilde{T(a)}$ for all a in B.*

Section 13.4

1. (a) Hint: Mimic Example 13.4.1.
2. (a) Let M be an element of $P(A)$, $M \neq \phi$. Then M is called an atom if for every $X \subseteq \mathcal{P}(A)$, $M \subseteq X$ or $X \cap M = \phi$.
   (b) The subsets with cardinality one.
   (c) Every non-empty subset of $P(A)$ can be expressed completely and uniquely as the union of some of the atoms in Part b. If B is a subset of A, then

$$\bigcup_{a \text{ in } B} \{a\}$$

4. Hint: First decide how many atoms there are.
6. (a) Use $A = \{a\}$ and D_p, where p is any prime number, as examples.
   (b) These are just the tables for the logical operations conjunction, disjunction, and negation. (see section 13.5).

Section 13.5

1. (a)

$\vee$	(0,0)	(0,1)	(1,0)	(1,1)
(0,0)	(0,0)	(0,1)	(1,0)	(1,1)
(0,1)	(0,1)	(0,1)	(1,1)	(1,1)
(1,1)	(1,0)	(1,1)	(1,0)	(1,1)
(1,1)	(1,1)	(1,1)	(1,1)	(1,1)

$\wedge$	(0,0)	(0,1)	(1,0)	(1,1)
(0,0)	(0,0)	(0,0)	(0,0)	(0,0)
(0,1)	(0,0)	(0,1)	(0,0)	(0,1)
(1,0)	(0,0)	(0,0)	(1,0)	(1,0)
(1,1)	(0,0)	(0,1)	(1,0)	(1,1)

u	$\bar{u}$
(0,0)	(1,1)
(0,1)	(1,0)
(1,0)	(0,1)
(1,1)	(0,0)

(b) The graphs are isomorphic.   (c) (0,1) and (1,0)
3. (a) (1,0,0,0), (0,1,0,0), (0,0,1,0), and (0,0,0,1)   (b) The n-tuples with exactly one 1

Section 13.6

1. (a) $M_1(x_1, x_2) = 0$
   $M_2(x_1, x_2) = \bar{x}_1 \wedge \bar{x}_2$
   $M_3(x_1, x_2) = \bar{x}_1 \wedge x_2$
   $M_4(x_1, x_2) = x_1 \wedge \bar{x}_2$
   $M_5(x_1, x_2) = x_1 \wedge x_2$
   $M_6(x_1, x_2) = (\bar{x}_1 \wedge \bar{x}_2) \vee (\bar{x}_1 \wedge x_2) = \bar{x}_1$
   $M_7(x_1, x_2) = (\bar{x}_1 \wedge \bar{x}_2) \vee (x_1 \wedge \bar{x}_2) = \bar{x}_2$

$$M_8(x_1, x_2) = (\bar{x}_1 \wedge \bar{x}_2) \vee (x_1 \wedge x_2)$$
$$M_9(x_1, x_2) = (\bar{x}_1 \wedge x_2) \vee (x_1 \wedge \bar{x}_2)$$
$$M_{10}(x_1, x_2) = (\bar{x}_1 \wedge x_2) \vee (x_1 \wedge x_2) = x_2$$
$$M_{11}(x_1, x_2) = (x_1 \wedge \bar{x}_2) \vee (x_1 \wedge x_2) = x_1$$
$$M_{12}(x_1, x_2) = (\bar{x}_1 \wedge \bar{x}_2) \vee (\bar{x}_1 \wedge x_2) \vee (x_1 \wedge \bar{x}_2)$$
$$= \bar{x}_1 \vee \bar{x}_2$$
$$M_{13}(x_1, x_2) = (\bar{x}_1 \wedge \bar{x}_2) \vee (\bar{x}_1 \wedge x_2) \vee (x_1 \wedge x_2)$$
$$= \bar{x}_1 \vee x_2$$
$$M_{14}(x_1, x_2) = (\bar{x}_1 \wedge \bar{x}_2) \vee (x_1 \wedge \bar{x}_2) \vee (x_1 \wedge x_2)$$
$$= x_1 \vee \bar{x}_2$$
$$M_{15}(x_1, x_2) = (\bar{x}_1 \wedge x_2) \vee (x_1 \wedge \bar{x}_2) \vee (x_1 \wedge x_2)$$
$$= x_1 \vee x_2$$
$$M_{16}(x_1, x_2) = (\bar{x}_1 \wedge \bar{x}_2) \vee (\bar{x}_1 \wedge x_2) \vee (x_1 \wedge \bar{x}_2) \vee (x_1 \wedge x_2)$$
$$= 1$$

(b) The tables for the functions in Part a are:

x_1	x_2	M_1	M_2	M_3	M_4	M_5	M_6	M_7
0	0	0	1	0	0	0	1	1
0	1	0	0	0	0	0	0	0
1	0	0	0	0	1	0	0	1
1	1	0	0	0	0	1	0	0

M_8	M_9	M_{10}	M_{11}	M_{12}	M_{13}	M_{14}	M_{15}	M_{16}
1	0	0	0	1	1	1	0	1
0	1	1	0	1	1	0	1	1
0	1	0	1	1	0	1	1	1
1	0	1	1	0	1	1	1	1

(c) $f_1(x_1, x_2) = M_{15}(x_1, x_2)$
   $f_2(x_1, x_2) = M_{12}(x_1, x_2)$
   $f_3(x_1, x_2) = M_1(x_1, x_2)$
   $f_4(x_1, x_2) = M_{16}(x_1, x_2)$

2. (a) $f(x_1, x_2, x_3) = (\bar{x}_3 \wedge x_2) \vee (\bar{x}_1 \wedge x_3) \vee (x_2 \wedge x_3)$
         $= (\bar{x}_3 \wedge x_2) \vee (x_3 \wedge x_2) \vee (\bar{x}_1 \wedge x_3)$ (commutative law)
         $= [(\bar{x}_3 \vee x_3) \wedge x_2] \vee (\bar{x}_1 \wedge x_3)$ (distributive law)
         $= (1 \wedge x_2) \vee (\bar{x}_1 \wedge x_3)$ (complement law)
         $= x_2 \vee (\bar{x}_1 \wedge x_3)$ (identity law)

   (b) $f(x_1, x_2, x_3) = (x_1 \wedge x_2 \wedge x_3) \vee (x_1 \wedge x_2 \wedge \bar{x}_3) \vee (\bar{x}_1 \wedge x_2 \wedge x_3) \vee$
        $(\bar{x}_1 \wedge x_2 \wedge \bar{x}_3) \vee (\bar{x}_1 \wedge \bar{x}_2 \wedge x_3)$

   (c) The final values in the tables are all the same, verifying that the functions are all the same.

   (d) $256 = 2^m$ where $m = 2^3$

3. (a) The number of elements in the domain of f is $16 = 4^2 = (\#B)^2$.

   (b) With two variables, there are $4^4 = 256$ different Boolean functions. With three variables, there are $4^8 = 65536$ different Boolean functions.

   (c) $f(x_1, x_2) = (1 \wedge \bar{x}_1 \wedge \bar{x}_2) \vee (1 \wedge x_1 \wedge \bar{x}_2) \vee (1 \wedge \bar{x}_1 \wedge x_2) \vee$
        $(0 \wedge \bar{x}_1 \wedge \bar{x}_2)$

(d) $f:B^2 \to B$, defined by $f(0,0) = 0$, $f(0,1) = 1$, $f(1,0) = a$, $f(0,1) = a$, and $f(0,a) = b$, with the images of all other pairs in B^2 defined arbitrarily, is one such function.

Section 13.7

1. (a)

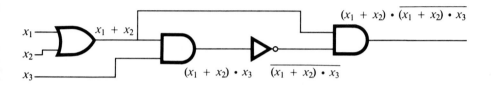

(b) $f(x_1, x_2, x_3) = \overline{((x_1 + x_2) \cdot x_3)} \cdot (x_1 + x_2)$
$$= ((x_1 + x_2) + \bar{x}_3)(x_1 + x_2)$$
$$= ((x_1 + x_2)(x_1 + x_2)) + \bar{x}_3(x_1 + x_2)$$
$$= 0 + \bar{x}_3(x_1 + x_2)$$
$$= \bar{x}_3(x_1 + x_2)$$

(c) The Venn diagram for the function is:

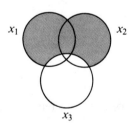

We can read off the minterm normal form from this diagram: $f(x_1, x_2, x_3) = x_1 \cdot \bar{x}_2 \cdot \bar{x}_3 + x_1 \cdot x_2 \cdot \bar{x}_3 + \bar{x}_1 \cdot x_2 \cdot \bar{x}_3$.

(d)

Simplified form:

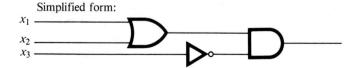

Minterm form:

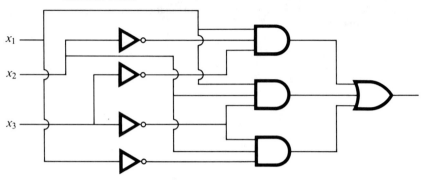

(e)

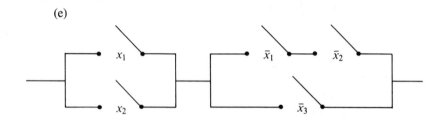

(f)

x_1	x_2	x_3	x_1+x_2	$\overline{(x_1+x_2)\cdot x_3}$	f
0	0	0	0	1	0
0	0	1	0	1	0
0	1	0	1	1	1
0	1	1	1	0	0
1	0	0	1	1	1
1	0	1	1	0	0
1	1	0	1	1	1
1	1	1	1	0	0

Current will flow only when one of the switches x_1 or x_2 is on and x_3 is off.

2.   (a) $f(x_1, x_2) = x_1 + (x_2 \cdot (x_1 + \bar{x}_2))$.

     (b) $f(x_1, x_2) = x_1 + (x_2 \cdot x_1 + x_2 \cdot \bar{x}_2)$
$$= x_1 + (x_2 \cdot x_1 + 0)$$
$$= x_1 + x_2 \cdot x_1$$
$$= x_1$$

(c)

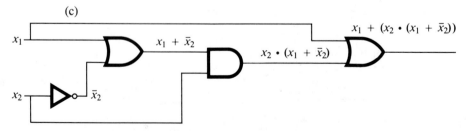

(d) $f(x_1, x_2) = x_1 \cdot \bar{x}_2 + x_1 \cdot x_2$

(e)

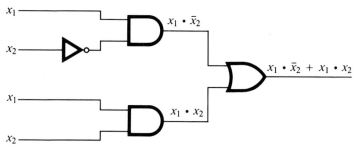

The graph in Part c has three levels of gates and that in Part e has two levels. Hence the "time cost" of the figure in Part e is slightly less. The hardware costs are the same for both circuits.

Section 14.1

1. (a) S_1 is not a submonoid since the identity of $[\mathbf{Z}_8; \times_8]$, 1, is not in S_1. S_2 is a submonoid since 1 is in S_2 and S_2 is closed under multiplication; i.e., for all $a, b \in S_2$, $a \times_8 b$ is in S_2.

    (b) The identity of $\mathbf{N}^\mathbf{N}$ is the function $i:\mathbf{N} \to \mathbf{N}$ defined by $i(a) = a$ for all a in N. If $a \in \mathbf{N}$, $i(a) = a \le a$, thus the identity of $\mathbf{N}^\mathbf{N}$ is in S_1. However, the image of 1 under any function in S_2 is 2 and thus the identity of $\mathbf{N}^\mathbf{N}$ is not in S_2, so S_2 is not a submonoid. The composition of any two functions in S_1, f and g, will be a function in S_1:

    $$fg(n) = f(g(n)) \le g(n) \text{ since } f \text{ is in } S_1$$
    $$\le n \text{ since } g \text{ is in } S_1. \ \#$$

    Thus the two conditions of a submonoid are satisfied; S_1 is a submonoid of $\mathbf{N}^\mathbf{N}$.

    (c) The first set is a submonoid, but the second is not since the null set has a non-finite complement.

2. (a) $(3) = \{1, 3, 9\}$ and $(0) = \{0, 1\}$

    (b) $(5) = \{0, 1, 5\}$

    (c) (The set of prime numbers) $= \mathbf{P}$ and $(2) = \{1, 2, 4, 8, 16, 32, \ldots\}$

    (d) $(3, 5) = \{0, 3, 5, 6, 8, 9, 10, 11, 12, 13, \ldots\} = \mathbf{N} - \{1, 2, 4, 7\}$

3. The set of $n \times n$ real matrices is a monoid under matrix multiplication. This follows from the laws of matrix algebra in Chapter 5. To prove that the set of stochastic matrices is a monoid over matrix multiplication, we need only show that the identity matrix is stochastic (this is obvious) and that the set of stochastic matrices is closed under matrix multiplication. Let A and B be $n \times n$ stochastic matrices.

$$(AB)_{ij} = \sum_{k=1}^{n} a_{ik} b_{kj}.$$

The sum of the jth column of AB is:

$$\sum_{k=1}^{n} a_{1k} b_{kj} + \sum_{k=1}^{n} a_{2k} b_{kj} + \cdots + \sum_{k=1}^{n} a_{nk} b_{kj}$$

$$= \sum_{k=1}^{n} (a_{1k}b_{kj} + a_{2k}b_{kj} + \cdots + a_{nk}b_{kj})$$

$$= \sum_{k=1}^{n} b_{kj}(a_{1k} + a_{2k} + \cdots + a_{nk}).$$

The terms in parentheses are just the sum of the elements in the columns of A, which is 1 since A is stochastic. Therefore, the sum of elements in the jth column of AB is equal to the sum $\sum_{k=1}^{n} b_{kj}$, which equals 1, since B is stochastic.

Section 14.2

1. (a) For a character set of 350 symbols, the number of bits needed for each character is the smallest n such that 2^n is greater than or equal to 350. $2^9 = 512 \geq 350 > 2^8$; therefore 9 bits are needed.
   (b) $2^{12} = 4096 \geq 3500 > 2^{11}$; therefore 12 bits are needed.
3. If s is in A^* and L is recursive, we can answer the question "Is s in L^C?" by negating the answer to "Is s in L?"
4. (a) This grammar defines the set of all strings over B for which each string is a pallindrome (same string if read forward or backward).
   (b) This grammar defines the set of all strings over a, b, c with any a's preceeding b's and c's and any b's preceding c's; i.e. the language $\{xyz : x \in \{a\}^*, y \in \{b\}^*, z \in \{c\}^*\}$.
5. (a) Terminal symbols: The null string, 0, and 1
   Nonterminal symbols: S, E
   Starting symbol: S
   Production rules: $S \rightarrow 00S$, $S \rightarrow 01S$, $S \rightarrow 10S$, $S \rightarrow 11S$, $S \rightarrow E$, $E \rightarrow 0$, $E \rightarrow 1$
   This is a regular grammar.
   (b) Terminal symbols: The null string, 0, and 1
   Nonterminal symbols: S, A, B, C
   Starting symbol: S
   Production rules: $S \rightarrow 0A$, $S \rightarrow 1A$, $S \rightarrow \lambda$, $A \rightarrow 0B$, $A \rightarrow 1B$, $A \rightarrow \lambda$, $B \rightarrow 0C$, $B \rightarrow 1C$, $B \rightarrow \lambda$, $C \rightarrow 0$, $C \rightarrow 1$, $C \rightarrow \lambda$
   This is a regular grammar.
6. Terminal symbols: The null string, 0, and 1
   Nonterminal symbols: S, E
   Starting symbol: S
   Production rules: $S \rightarrow 0S$, $S \rightarrow E$, $E \rightarrow 1E$, $E \rightarrow \lambda$
7. (a) $S ::= \quad 0S0 \mid 1S1 \mid E$
   $E ::= \quad \lambda \mid 0 \mid 1$
   (b) $S ::= \quad aS \mid T$
   $T ::= \quad bT \mid U$
   $U ::= \quad cU \mid E$
   $E ::= \quad \lambda$
8. (a) List the elements of each set X_i in a sequence $x_{i1}, x_{i2}, x_{i3}, \ldots$

x_{11}	x_{12}	x_{13}	x_{14} $\cdots$
x_{21}	x_{22}	x_{23}	x_{24} $\cdots$
x_{31}	x_{32}	x_{33}	x_{34} $\cdots$
x_{41}	x_{42}	x_{43}	x_{44} $\cdots$
$\cdots$	$\cdots$	$\cdots$	$\cdots$

Now draw arrows as shown and list the element in the union in the order established by this pattern: $x_{11}, x_{21}, x_{12}, x_{31}, x_{22}, x_{13}, x_{41}, x_{32}, x_{23}, x_{14}, \ldots$

9.
```
PROCEDURE RECOGNIZE(S: STRING; VAR WORD: BOOLEAN)
VAR  I:  INTEGER;
BEGIN
    WORD := TRUE;
    IF (S.LEN) > 0  THEN
                        IF S.BIT[1] = 1  THEN WORD := FALSE
            ELSE
                BEGIN (*ELSE*)
                    I := 3
                    WHILE I <= (S.LEN) DO
                        BEGIN
                            IF (S.BIT[I-1] = 1 ) AND (S.BIT[I] = 1 )
                                THEN
                                    BEGIN
                                        WORD := FALSE
                                        I := (S.LEN) + 1
                                    END
                                ELSE I := I + 1
                        END
                END
END;
```

Section 14.3

1. Press S, P, B/Nothing

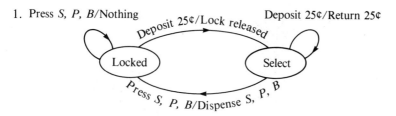

x	s	$Z(x, s)$	$t(x, s)$
Deposit 25¢	Locked	Nothing	Select
Deposit 25¢	Select	Return 25¢	Select
Press S	Locked	Nothing	Locked
Press S	Select	Dispense S	Locked
Press P	Locked	Nothing	Locked
Press P	Select	Dispense P	Locked
Press B	Locked	Nothing	Locked
Press B	Select	Dispense B	Locked

2. If the final output from these machines is 1, then the string is accepted; otherwise, the string is rejected.

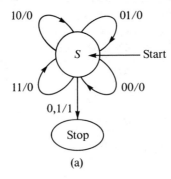

(a)

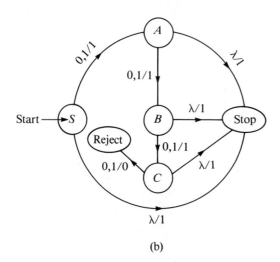

(b)

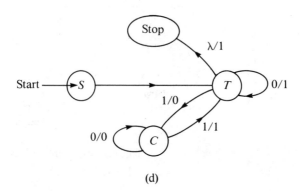

(d)

Section 14.4

1. (a)

Input String	a	b	c	aa	ab	ac
1	$(a, 1)$	$(a, 2)$	$(c, 3)$	$(a, 1)$	$(a, 2)$	$(c, 3)$
2	$(a, 2)$	$(a, 1)$	$(c, 3)$	$(a, 2)$	$(a, 1)$	$(c, 3)$
3	$(c, 3)$	$(c, 3)$	$(c, 3)$	$(c, 3)$	$(c, 3)$	$(c, 3)$

Input String	ba	bb	bc	ca	cb	cc
1	$(a, 2)$	$(a, 1)$	$(c, 3)$	$(c, 3)$	$(c, 3)$	$(c, 3)$
2	$(a, 1)$	$(a, 2)$	$(c, 3)$	$(c, 3)$	$(c, 3)$	$(c, 3)$
3	$(c, 3)$	$(c, 3)$	$(c, 3)$	$(c, 3)$	$(c, 3)$	$(c, 3)$

We can see that $T_a T_a = T_{aa} = T_a$, $T_a T_b = T_{ab} = T_b$, etc. Therefore, we have the following monoid:

	T_a	T_b	T_c
T_a	T_a	T_b	T_c
T_b	T_b	T_a	T_c
T_c	T_c	T_c	T_c

T_a is the identity.

(b)

Input String	1	2	11	12	21	22
A	C	B	A	D	D	A
B	D	A	B	C	C	B
C	A	D	C	B	B	C
D	B	C	D	A	A	D

Input String	111	112	121	122	211	212	221	222
A	C	B	B	C	B	C	C	B
B	D	A	A	D	A	D	D	A
C	B	C	C	B	C	B	B	C
D	B	C	C	B	C	B	B	C

We have the following monoid:

	T_1	T_2	T_{11}	T_{12}
T_1	T_{11}	T_{12}	T_1	T_2
T_2	T_{12}	T_{11}	T_2	T_1
T_{11}	T_1	T_2	T_{11}	T_{12}
T_{12}	T_2	T_1	T_{12}	T_{11}

T_{11} is the identity of this monoid.

2. (a) $[\mathbf{Z}_3; \times_3]$ (b) $[\mathbf{Z}_4; +_4]$

Section 14.5

1.

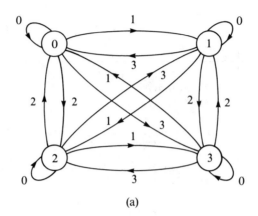

(a)

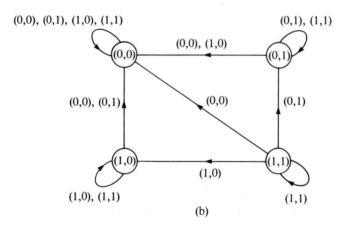

(b)

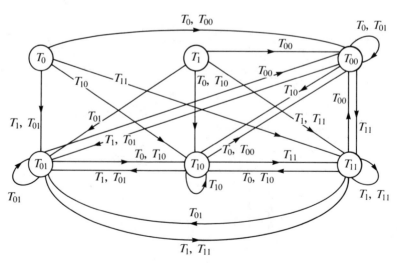

(c)

Section 15.1

1. -1
3. If $\#G = m$ is greater than 2, and $G = (a)$, then $a, a^2, a^3, \ldots, a^{m-1}, a^m = e$ are the distinct elements of G. Furthermore, $a^{-1} = a^{m-1} \neq a$. If k is between 1 and n, a^{-1} generates a^k:

$$(a^{-1})^{m-k} = (a^{m-1})^{m-k} = a^{m^2-m-mk+k} = (a^m)^{m-1-k} * a^k = e * a^k = a^k.$$

Similarly, if G is infinite and $G = (a)$, then a^{-1} generates G.
5. (a) No. Assume that $g \in \mathbf{Q}$ generates $\mathbf{Q}$. The $(q) = \{nq : n \in \mathbf{Z}\}$. But this gives us at most integer multiples of q, not every element in $\mathbf{Q}$.   (b) No. Similar reasoning to Part a.   (c) Yes. 6 is a generator of $6\mathbf{Z}$.   (d) No.   (e) Yes. $(1, 1, 1)$ is a generator of the group.
8. Consider the elements in the group generated by $(1, 1)$. They will have a 0 as a first coordinate every integer multiple of n and will have a zero as a second coordinate every integer multiple of m. Since $gcd\,(n, m) = 1$, 0 will appear simultaneously in both coordinates for the first time when $(1, 1)$ is added to itself nm times. Hence $(1, 1)$ generates nm different elements of $\mathbf{Z}_n \times \mathbf{Z}_m$; i.e., all of them. By Theorem 15.1.2, $\mathbf{Z}_n \times \mathbf{Z}_m$ is isomorphic to $\mathbf{Z}_{nm}$.

Section 15.2

2. (a) $H = \{(0,0), (2,0)\}$, $(1,0) + H$, $(0,1) + H$, and $(1,1) + H$ are the four distinct cosets in G/H. None of these cosets generates G/H; therefore G/H is not cyclic. Hence G/H must be isomorphic to $\mathbf{Z}_2 \times \mathbf{Z}_2$.
    (b) $[\mathbf{R}; +]$
    (c) $(8) = \{0, 4, 8, 12, 16\}$; therefore $\#(\mathbf{Z}_{20}/(8)) = 4$. The four cosets are $\bar{0}, \bar{1}, \bar{2}$, and $\bar{3}$. $\bar{1}$ generates all four cosets; therefore the factor group is isomorphic to $[\mathbf{Z}_4; +_4]$.
3. One solution, which is not unique, is $G = \mathbf{Z}_{36}$ and $H = (6)$.
6. $a \in bH \Rightarrow a = b * h$, for some $h \in H$
    $\quad\quad\quad \Rightarrow b^{-1} * a = h$, for some $h \in H$
    $\quad\quad\quad \Rightarrow b^{-1} * a \in H$

Section 15.3

1. (a) $\begin{pmatrix} 1 & 2 & 3 & 4 \\ 1 & 4 & 3 & 2 \end{pmatrix}$ (b) $\begin{pmatrix} 1 & 2 & 3 & 4 \\ 4 & 3 & 1 & 2 \end{pmatrix}$ (c) $\begin{pmatrix} 1 & 2 & 3 & 4 \\ 3 & 4 & 2 & 1 \end{pmatrix}$

    (d) $\begin{pmatrix} 1 & 2 & 3 & 4 \\ 3 & 4 & 2 & 1 \end{pmatrix}$ (e) $\begin{pmatrix} 1 & 2 & 3 & 4 \\ 4 & 2 & 1 & 3 \end{pmatrix}$ (f) $\begin{pmatrix} 1 & 2 & 3 & 4 \\ 3 & 1 & 4 & 2 \end{pmatrix}$

    (g) $\begin{pmatrix} 1 & 2 & 3 & 4 \\ 2 & 1 & 4 & 3 \end{pmatrix}$

3. Yes and no respectively
5. (a) Let $i = \begin{pmatrix} 1 & 2 & 3 & 4 \\ 1 & 2 & 3 & 4 \end{pmatrix}$ $r_1 = \begin{pmatrix} 1 & 2 & 3 & 4 \\ 2 & 3 & 4 & 1 \end{pmatrix}$

    $r_2 = \begin{pmatrix} 1 & 2 & 3 & 4 \\ 3 & 4 & 1 & 2 \end{pmatrix}$ $r_3 = \begin{pmatrix} 1 & 2 & 3 & 4 \\ 4 & 1 & 2 & 3 \end{pmatrix}$

$$f_1 = \begin{pmatrix} 1 & 2 & 3 & 4 \\ 4 & 3 & 2 & 1 \end{pmatrix} \qquad f_2 = \begin{pmatrix} 1 & 2 & 3 & 4 \\ 2 & 1 & 4 & 3 \end{pmatrix}$$

$$f_3 = \begin{pmatrix} 1 & 2 & 3 & 4 \\ 3 & 2 & 1 & 4 \end{pmatrix} \qquad f_4 = \begin{pmatrix} 1 & 2 & 3 & 4 \\ 1 & 4 & 3 & 2 \end{pmatrix}$$

	i	r_1	r_2	r_3	f_1	f_2	f_3	f_4
i	i	r_1	r_2	r_3	f_1	f_2	f_3	f_4
r_1	r_1	r_2	r_3	i	f_4	f_3	f_1	f_2
r_2	r_2	r_3	i	r_1	f_2	f_1	f_4	f_3
r_3	r_3	i	r_1	r_2	f_3	f_4	f_2	f_1
f_1	f_1	f_3	f_2	f_4	i	r_2	r_1	r_3
f_2	f_2	f_4	f_1	f_3	r_2	i	r_3	r_1
f_3	f_3	f_2	f_4	f_1	r_3	r_1	i	r_2
f_4	f_4	f_1	f_3	f_2	r_1	r_3	r_2	i

10. There are ten elements in D_5. Five of them correspond to the rotations of the regular pentagon (0, 72, 144, 216, and 288 degrees). The other five correspond to the five axes of symmetry of the regular pentagon.

Section 15.4

2. (a) Yes   (b) No, since $\theta_2(2 +_5 4) = \theta_2(1) = 1$, but $\theta_2(2) +_2 \theta_2(4) = 0 +_2 0 = 0$.   (c) Yes   (d) No
4. $(\beta\alpha)(a_1, a_2\, a_3) = 0$. $\beta\alpha$ is a homomorphism.
6. $\phi(a, b) = (a, b) \begin{bmatrix} 1 & 0 & 1 \\ 0 & 1 & 1 \end{bmatrix}$
8. Let $\theta(h) \in \theta(H)$ and $\theta(g) \in \theta(G)$. $\theta(g)^{-1}\, \theta(h)\, \theta(g) = \theta(g^{-1}hg)$, which is an element of $\theta(H)$ since $g^{-1}hg \in H$, because H is a normal subgroup of G. $\theta(H)$ is not always a normal subgroup of G'. For example, Let $G = Z_2$ and $G' = S_4$ and define θ by $\theta(0) = i$, and $\theta(1) = $ any transposition in S_4.

Section 15.5

1. (a) Error detected, since an odd number of 1's was received; ask for re-transmission.
   (b) No error detected; accept this block.
   (c) No error detected; accept this block.
3. (a) Syndrome = (1, 0, 1). Corrected message = (1, 1, 0).
   (b) Syndrome = (1, 1, 0). Corrected message = (0, 0, 1).
   (c) Syndrome = (0, 0, 0). No error; message = (0, 1, 1).
   (d) Syndrome = (1, 1, 0). Corrected message = (1, 0, 0).
   (e) Syndrome = (1, 1, 1). This syndrome occurs only if two bits have been switched. No reliable correction is possible.

Section 16.1

2. All but rings d and e are commutative. All of the rings have a unity element. The number 1 is the unity for all of the rings except d, e, j, and k. The unity

for $M_{n \times n}$ (**R**) is I_n. The unity for **Z** $\times$ **Z** is (1, 1). The unity for $\mathbf{Z}_2^3$ is (1, 1, 1).

3. (a) $\{1, -1\}$ (b) **Q***   (c) **C***
   (d) $\{A : A_{11}A_{22} - A_{12}A_{22} \neq 0\}$
   (e) $\{A : det(A) \neq 0\}$ (f) $\{1\}$ (g) $\{1, 5\}$
   (h) $\{1, 3, 5, 7\}$ (i) $\mathbf{Z}_5^*$ (j) $\{(a, b) : |a| = 1, |b| = 1\}$
   (k) $\{(1, 1, 1)\}$.

4. Hint: The set of units of a ring is a group under multiplication. Apply a theorem from a group theory.

6. One word hints: (a) Commutative   (b) Equation   (c) Countable
   (d) Units.

9. (a) 0   (b) 0   (c) 1   (d) 0

11. (a) 2, 3, 6, and 11   (b) The left-hand side of the equation factors into the product $(x - 2)(x - 3)$. Since **Z** is an integral domain, $x = 2$ and $x = 3$ are the only possible solutions.

16. (a) 4 and 10   (b) -2 (c) Any 2×2 matrix satisfying $(X + 21)^2 = \mathbf{O}_n$ (d) 1

Section 16.2

4.

+	(0, 0)	(0, 1)	(1, 0)	(1, 1)
(0, 0)	(0, 0)	(0, 1)	(1, 0)	(1, 1)
(0, 1)	(0, 1)	(0, 0)	(1, 1)	(1, 0)
(1, 0)	(1, 0)	(1, 1)	(0, 0)	(0, 1)
(1, 1)	(1, 1)	(1, 0)	(0, 1)	(0, 0)

·	(0, 0)	(0, 1)	(1, 0)	(1, 1)
(0, 0)	(0, 0)	(0, 0)	(0, 0)	(0, 0)
(0, 1)	(0, 0)	(0, 1)	(0, 0)	(0, 1)
(1, 0)	(0, 0)	(0, 0)	(1, 0)	(1, 0)
(1, 1)	(0, 0)	(0, 1)	(1, 0)	(1, 1)

$\mathbf{Z}_2^2$ is a ring, but not an integral domain since $(1, 0)(0, 1) = (0, 0)$. Since every field is an integral domain, $\mathbf{Z}_2^2$ couldn't be a field.

5. (a) 0(over $\mathbf{Z}_2$), 1(over $\mathbf{Z}_3$), 3(over $\mathbf{Z}_5$)   (b) 2(over $\mathbf{Z}_3$), 3(over $\mathbf{Z}_5$)
   (c) 2

7. (a) 1   (b) none   (c) 1   (d) none

Section 16.3

4. (a) None   (b) 4

5. (a) Reducible, $(x + 1)(x^2 + x + 1)$
   (b) Reducible, $x(x^2 + x + 1)$ (c) Irreducible
   (d) Reducible, $(x + 1)^4$

10. (b) $1, x\,x^2, \ldots$ (c) $F[x]$ is an infinite dimensional vector space.

11. (a) A basis is $\{1\}$. $dim_{\mathbf{R}}\mathbf{R} = 1$ (b) A basis is the set containing the unity element of F. $dim_F F = 1$.

Section 16.4

2. (a) $R(i) = \{a + bi \mid a, b \in \mathbf{R}\}$ (b) The field of complex numbers, $\mathbf{C}$
3. $(\mathbf{Q}(\sqrt{2}))(\sqrt{3}) = \{a_0 + a_1\sqrt{3} : a_0, a_1 \in \mathbf{Q}(\sqrt{2})\}$
    $= \{b_0 + b_1\sqrt{2} + b_2\sqrt{3} + b_3\sqrt{6} : \text{each } b_i \in \mathbf{Q}\}$
5. (a) $f(x) = x^3 + x + 1$ is reducible if and only if it has a factor of the form $x - a$. By Theorem 16.3.3, $x - a$ is a factor if and only if a is a zero. Neither 0 nor 1 is a zero of $f(x)$ over $\mathbf{Z}_2$.
    (b) Since $f(x)$ is irreducible over $\mathbf{Z}_2$, all zeros of $f(x)$ must lie in an extension field of $\mathbf{Z}_2$. Let c bet a zero of $f(x)$. $\mathbf{Z}_2(c)$ can be described several different ways. One method is to note that since $c \in \mathbf{Z}_2(c)$, $c^n \in \mathbf{Z}_2(c)$ for all n, so $\mathbf{Z}_2$ includes $0, 1, c, c^2, c^3, \ldots$ But $c^3 = c + 1$ since $f(c) = 0$. Further, $c^4 = c^2 + c$, $c^5 = c^2 + c + 1$, $c^6 = c^2 + 1$, and $c^7 = 1$. Further powers of c repeat preceding elements. $\mathbf{Z}_2(c) = \{0, 1, c, c^2, c + 1, c^2 + 1, c^2 + c + 1, c^2 + c\}$. The three zeros of $f(x)$ are c, c^2, and $c^2 + c$. The reader is encouraged to write out the linear factors of $f(x)$ and verify that the product of these factors is $f(x)$.
    (c) Cite Theorem 16.2.4, Part 3.

Section 16.5

2. Simply replace $R[x]$ by $R[[x]]$. All of Theorem 16.3.1 is true for $R[[x]]$.
3. Theorem 16.5.2 proves that not all non-zero elements in $F[[x]]$ are units.
6. (a) Imitate Example 15.5.2 to obtain $g(x) = 1 - x = (f)x))^{-1}$.
    (b) $f(x)^{-1} = 1 + x$ (c) Partial answer: Since by Chapter 8, $f(x) = G(S;x)$, where $S(k) = 1$ for all k, the inverse of $f(x)$ is $1 - x$.

Section A.1

2. No
4. (a) $A_{11}A_{22}A_{33}$ (b) The product of its diagonal entries
5. Same answer as Exercise 4
6. (a) 5   (b) $-1/2$ (c) 2   (d) 26
    (e) -7 (f) 0   (g) 1   (g) -2
    (i) -9 (j) -3 (k) -27

Section A.2

1. The system has either no solutions or an infinite number of solutions.
4. Expansion of the determinant along any row or column shows that the equation is a linear equation describing a line L. If you substitute (x_1, y_1) for (x, y) the determinant equals zero (by Théorem A 1.1, Part v). Therefore, (x_1, y_1) lies on L. Similarly, (x_2, y_2) is on L.

Table of Symbols

Symbol	Read As	Section Number	Some Alternative Notation
P	the set of positive integers	1.1	$\mathbf{Z}^{+}$
N	the set of natural numbers	1.1	
Z	the set of integers	1.1	
Q	the set of rational numbers	1.1	
R	the set of real numbers	1.1	
C	the set of complex numbers	1.1	
$\in$	"is an element of"	1.1	IN (Pascal)
$\#A$	cardinality of set A	1.1	$\lvert A \rvert$
$/$	"such that" or "where"	1.1	: or ;
$\subseteq$	"is a subset of"	1.2	$\leq$ (Pascal)
$\varnothing$	the null set, the empty set	1.2	{} or [] (Pascal)
$\cap$	intersection	1.3	* (Pascal)
$\cup$	union	1.3	+ (Pascal)
U	universe; the universal set	1.3	
$\notin$	"is not an element of"	1.3	
A^c	the complement of set A	1.3	$U - A$ or A'
$A \times B$	the Cartesian product of sets A and B	1.3	
A^2	Cartesian product of set A with itself	1.3	$A \times A$
A^n	the set of all n-tuples of elements from A	1.3	$A \times A \times \cdots \times A$ (n times)
$\bigcup\limits_{i=1}^{n} A_i$	$A_1 \cup A_2 \cup \cdots \cup A_n$	1.3	
$\bigcap\limits_{i=1}^{n} A_i$	$A_1 \cap A_2 \cap \cdots \cap A_n$	1.3	
$n!$	n-factorial $= (n)(n - 1) \cdots (2)(1)$	2.1	
$P(n; k)$	the number of permutations of k objects from an n element set	2.1	

Symbol	Read As	Section Number	Some Alternative Notation
$\binom{n}{k}$	the number of k element subsets of an n element set	2.1	$C(n; k)$
$\mathcal{P}(A)$	the power set of set A	2.1	2^A
$:=$	"becomes" assignment symbol in Pascal and algorithms	2.1	
$\wedge$	conjunction operation ("and")	3.1	
$\vee$	disjunction operation ("or")	3.1	
$\sim$	negation operation ("not")	3.1	
$\rightarrow$	Conditional operation ("if . . . then")	3.1	
$\leftrightarrow$	biconditional operation ("if and only if")	3.1	
1	tautology	3.3	
0	contradiction	3.3	
$\Leftrightarrow$	"is equivalent to"	3.3	
$\Rightarrow$	"implies"	3.3	
$\#$	"end of proof"	3.5	■, Q.E.D.
T_p	the truth set of p	3.6	
$\exists$	there exists, for some	3.8	
$\forall$	for all, for every	3.8	
A_{ij}	the entry in the ith row and the jth column of matrix A	5.1	
$[A_{ij}]$	Assuming $1 \le i \le m$ and $1 \le j \le n$, this is a matrix with m rows and n columns	5.1	
$M_{m \times n}(\mathbf{R})$	the set of all $m \times n$ matrices whose entries are real numbers	5.1	
I	the identity matrix	5.4	I_n (n by n)
A^{-1}	the inverse of A; A inverse	5.4	
$det(A)$	The determinant of A	5.4	$|A|$
$r \circ s$	the composition of relations r and s (r first)	6.1	
r^+	the transitive closure of relation r	6.5	
$f: A \rightarrow B$	a function f from the set A into the set B	7.1	

Symbol	Read As	Section Number	Some Alternative Notation
$f(a)$	the image of a; the image of a under f	7.1	
B^A	the set of all functions from A into B	7.1	
$\lceil x \rceil$	the ceiling of x, the smallest integer $\geq x$	7.2	
$g \circ f$	the composition of function f with function g $(g \circ f)(x) = g(f(x))$	7.3	gf
$f^{n+1}(a)$	$f(f^n(a))$	7.3	
i	the identity function on A	7.3	
f^{-1}	f inverse	7.3	
$f^{-1}(b)$	the inverse image of b	7.3	
$\lfloor x \rfloor$	the floor of x, the largest integer $\leq x$	8.3	
$\prod_{k=1}^{n} f(k)$	$f(0)f(1) \cdots f(n)$	8.4	
$G(S, z)$	generating function of sequence S	8.5	
$\downarrow$	push operation on sequences	8.5	
$\uparrow$	pop operation on sequences	8.5	
K_n	complete undirected graph with n vertices	9.1	
(v_1, v_2)	an edge in a directed graph	9.1	
$\{v_1, v_2\}$	an edge in an undirected graph	9.1	
$deg(v)$	the degree of vertex v	9.4	
$outdeg(v)$	the outdegree of vertex v	9.4	
$indeg(v)$	the indegree of vertex v	9.4	
G_n	the Gray Code for the n-cube	9.4	
$*$	a binary operation (generic)	11.1	
$a * b$	the image of the pair (a, b)	11.1	$* ab, ab *$
$[V; *_1, \ldots, *_n]$	an algebraic system where set V is the domain and $*_1, \ldots, *_n$ are operations	11.1	
λ	the null (or empty) string	11.1	

Symbol	Read As	Section Number	Some Alternative Notation
B^*	the set of all finite strings of 0's and 1's including λ	11.1	
$\oplus$	symmetric difference	11.2	
$a=b\,(mod\ n)$	"a congruent to b modulo n"	11.3	$a\equiv_n b$
$+_n$	addition modulo n	11.3	$\oplus$
$\times_n$	multiplication modulo n	11.3	$\otimes$
$\mathbf{Z}_n$	the set $\{0, 1, 2, \ldots, n-1\}$	11.3	
$W \leq V$	"W is a subsystem of V"	11.5	
(a)	cyclic subgroup generated by a	11.5	
R_i	row i of a matrix	12.1	
$\rightarrow$	row equivalent	12.1	
$\vec{x}$	The vector "x"	12.3	$\mathbf{x}$
dim	dimension	12.4	
glb	greatest lower bound	13.1	
lub	least upper bound	13.1	
D_n	the set of positive integers that are divisors of n	13.1	
$a \vee b$	"a join b", least upper bound of a and b	13.2	$a + b$
$a \wedge b$	"a meet b", greatest lower bound of a and b	13.2	$a \times b$
$[L; \vee \wedge]$	A lattice L (generic)	13.2	
$lcm\,(a, b)$	The least common multiple of a and b	13.2	
$gcd\,(a, b)$	The greatest common divisor of a and b	13.2	
$\bar{a}$	the complement of a	13.3	
A^n	the set of strings of length n over alphabet A	14.2	
A^*	The set of all strings over A	14.2	
$|s| = n$	"the length of string s is n"	14.2	
S_A	the symmetric group on A	15.3	
S_k	the symmetric group on any set of cardinality k	15.3	
$Ker\,f$	the kernel of f	15.4	

Symbol	Read As	Section Number	Some Alternative Notation
$[F; +, \cdot]$	a field under the operations $+$ and $\cdot$	16.1	
$R[x]$	polynomials over ring R	16.3	
$R[[x]]$	power series over ring R	16.5	

Bibliography

• ▬▬▬▬▬▬▬▬▬▬▬▬▬▬▬▬▬▬▬▬▬▬▬ •

References are listed in alphabetical order by author. The Code preceding each listing indicates its general subject area.

A = Algebra, abstract and linear
B = Boolean algebra and logic design
C = Combinatorics
CS = Computer science and programming
D = Discrete mathematics
G = Graph theory
L = Logic
M = Miscellaneous

[CS] Aho, Alfred V., and Jeffrey D. Ullman. *Principles of Compiler Design*. Reading, MA: Addison-Wesley, 1977.

[G] Appel, K., and W. Haken. "Every Planar Map Is 4-colorable." *Bull. Am. Math. Soc.* 82 (1976): 711–12.

[D] Arbib, M. A., A. J. Kfoury, and R. N. Moll. *A Basis for Theoretical Computer Science*. New York: Springer-Verlag, 1981.

[L] Austin, A. Keith. "An Elementary Approach to NP-Completeness." *American Math. Monthly* 90 (1983): 398–99.

[G] Beardwood, J., J. H. Halton, and J. M. Hammersley. "The Shortest Path Through Many Points." *Proc. Cambridge Phil. Soc.* 55 (1959): 299–327.

[CS] Ben-Ari, M. *Principles of Concurrent Programming*. Englewood Cliffs, NJ: Prentice-Hall, 1982.

[G] Berge, C. *The Theory of Graphs and Its Applications*. New York: Wiley, 1962.

[C] Bogart, K. P. *Introductory Combinatorics*. Marshfield, MA: Pitman, 1983.

[A] Bronson, Richard. *Matrix Methods*. New York: Academic Press, 1969.

[G] Busacker, Robert G., and Thomas L. Saaty. *Finite Graphs and Networks*. New York: McGraw-Hill, 1965.

[A] Connell, Ian. *Modern Algebra, A Constructive Introduction*. New York: North-Holland, 1982.

[CS] Denning, Peter J., Jack B. Dennis, and Joseph L. Qualitz. *Machines, Languages, and Computation*. Englewood Cliffs, NJ: Prentice-Hall, 1978.

[A] Dornhoff, L. L., and F. E. Hohn. *Applied Modern Algebra*. New York: Macmillan, 1978.

[G] Even, S. *Graph Algorithms*. Potomac, MD: Computer Science Press, 1979.

[A] Fisher, J. L. *Application-Oriented Algebra*. New York: Harper & Row, 1977.

[G] Ford, L. R., Jr., and D. R. Fulkerson. *Flows in Networks,* Princeton, NJ: Princeton Univesity Press, 1962.

[A] Fraleigh, John B. *A First Course in Abstract Algebra.* 3rd ed. Reading, MA: Addison-Wesley, 1982.

[D] Gersting, J. L. *Mathematical Structures for Computer Science.* San Francisco: W. H. Freeman, 1982.

[CS] Guiasu, S. *Information Theory with Applications.* New York: McGraw-Hill, 1977.

[CS] Hamming, R. W. *Coding and Information Theory.* Englewood Cliffs, NJ: Prentice-Hall, 1980.

[B] Hill, F. J., and G. R. Peterson. *Switching Theory and Logical Design.* 2nd ed. New York: Wiley, 1974.

[L] Hofstadter, D. R. *Godel, Escher, Bach: An Eternal Golden Braid.* New York: Basic Books, 1979.

[B] Hohn, F. E. *Applied Boolean Algebra.* 2nd ed. New York: Macmillan, 1966.

[CS] Hopcroft, J. E., and J. D. Ullman. *Formal Languages and Their Relation to Automata.* Reading, MA: Addison-Wesley, 1969.

[C] Hu, T. C. *Combinatorial Algorithms.* Reading, MA: Addison-Wesley, 1982.

[CS] Kinder, D. "Transparent Multiprocessing Boosts uC Throughput." *Electronics Design* 30 (1982): 159–67.

[CS] Knuth, D. E. *The Art of Computer Programming.* Vol. 1, *Fundamental Algorithms.* 2nd ed. Reading, MA: Addison-Wesley, 1973.

[CS] ———. *The Art of Computer Programming.* Vol. 3, *Sorting and Searching.* Reading, MA: Addison-Wesley, 1973.

[CS] ———. *The Art of Computer Programming.* Vol. 2, *Seminumerical Algorithms.* 2nd ed. Reading, MA: Addison-Wesley, 1981.

[A] Kulisch, U. W., and Miranker, W. L. *Computer Arithmetic in Theory and Practice.* New York: Academic Press, 1981.

[D] Lipschutz, S. *Discrete Mathematics.* New York: Schaum, 1976.

[D] ———. *Essential Computer Mathematics.* New York: Schaum, 1982.

[A] Lipson, J. D. *Elements of Algebra and Algebraic Computing.* Reading, MA: Addison-Wesley, 1981.

[D] Liu, C. L. *Elements of Discrete Mathematics.* New York: McGraw-Hill, 1977.

[M] Mathematical Association of America, Committee on the Undergraduate Program in Mathematics, Recommendations for a General Mathematical Sciences Program. Washington, D.C., 1981.

[CS] Miller, A. R. *Pascal Programs for Scientists and Engineers.* Berkeley, CA: Sybex, 1981.

[G] Ore, O. *Graphs and Their Uses.* New York: Random House, 1963.

[G] Parry, R. T., and H. Pferrer. "The Infamous Traveling-Salesman Problem: A Practical Approach." *Byte* 6 (July 1981): 252–90.

[D] Prather, R. E. *Discrete Mathematical Structures for Computer Science.* Boston: Houghton Mifflin, 1976.

[CS] Pyle, I. C. *The Ada Programming Language.* Englewood Cliffs, NJ: Prentice-Hall, 1981.

[L]   Quine, W. V. *The Ways of Paradox and Other Essays*. New York: Random House, 1966.

[D]   Sahni, S. *Concepts in Discrete Mathematics*. Frindley, MN: Camelot, 1981.

[L]   Solow, Daniel. *How to Read and Do Proofs*. New York: Wiley, 1982.

[CS] Standish, T. A. *Data Structure Techniques*. Reading, MA: Addison-Wesley, 1980.

[L]   Stoll, Robert R. *Sets, Logic and Axiomatic Theories*. San Francisco: W. H. Freeman, 1961.

[G]   Supowit, K. J., E. M. Reingold, and D. A. Plaisted. "The Traveling Salesman Problem and Minimum Matching in the Unit Square." *SIAM J. Computing* 12 (1983): 144–56.

[D]   Tremblay, J. P., and R. Manohar. *Discrete Mathematical Structures with Applications to Computer Science*. New York: McGraw-Hill, 1975.

[CS] Wand, Mitchell. *Induction, Recursion, and Programming*. New York: North-Holland, 1980.

[G]   Warshall, S. "A Theorem on Boolean Matrices." *Journal of the Association of Computing Machinery*, 1962, 11–12.

[A]   Williams, Gareth. *Computational Linear Algebra with Models*. 2nd ed. Boston: Allyn and Bacon, 1978.

[A]   Winograd, S. "On the Time Required to Perform Addition." *J. Assoc. Comp. Mach.* 12 (1965): 277–85.

Index

606049